Phytochemicals of Nutraceutical Importance

Phytochemicals of Nutraceutical Importance

Edited by

Dhan Prakash

Amity Institute for Herbal Research & Studies, Amity University Uttar Pradesh, India

and

Girish Sharma

Amity Center for Cancer Epidemiology & Cancer Research and Amity Institute of Biotechnology, Amity University Uttar Pradesh, India

www.cabi.org

CABI is a trading name of CAB International

CABI
Nosworthy Way
Wallingford
Oxfordshire OX10 8DE
UK

Tel: +44 (0)1491 832111
Fax: +44 (0)1491 833508
E-mail: info@cabi.org
Website: www.cabi.org

CABI
38 Chauncey Street
Suite 1002
Boston, MA 02111
USA

Tel: +1 800 552 3083 (toll free)
Tel: +1 (0)617 395 4051
E-mail: cabi-nao@cabi.org

A catalogue record for this book is available from the British Library, London, UK.

Library of Congress Cataloging-in-Publication Data

Phytochemicals of nutraceutical importance / edited by Dhan Prakash, Amity Institute for Herbal Research & Studies, Amity University-Uttar Pradesh, Girish Sharma, Amity Institute of Biotechnology, Amity University-Uttar Pradesh.
pages cm
Includes bibliographical references and index.
ISBN 978-1-78064-363-2 (hbk)
1. Functional foods. 2. Phytochemicals. 3. Nutrition. I. Prakash, Dhan, editor of compilation. II. Sharma, Girish, editor of compilation.

QP144.F85P4835 2013
613.2--dc23

2013021166

ISBN-13: 978 1 780643632

Commissioning editor: Sreepat Jain
Editorial assistant: Emma McCann
Production editor: Shankari Wilford

Typeset by SPi, Pondicherry, India
Printed and bound in the UK by CPI Group (UK) Ltd, Croyden, CR0 4YY

Contents

Contributors

Todd R. Callaway, United States Department of Agriculture, Agricultural Research Service, College Station, Texas 77845, USA.

Daniel Chan, University of Colorado Denver, Division of Medical Oncology, MS-8117, 12801 East 17th Avenue, Aurora, Colorado 80045, USA. E-mail: Dan.Chan@ucdenver.edu

Amar P. Garg, Department of Microbiology, CCS University, Meerut-250004, India.

Glenda Gobe, Centre for Kidney Disease Research, University of Queensland School of Medicine, Princess Alexandra Hospital, Woolloongabba, Brisbane, Australia 4102. E-mail: g.gobe@uq.edu.au

Armando Enrique González-Stuart, Coordinator, Center for Interdisciplinary Health Research and Evaluation, College of Health Sciences, University of Texas at El Paso, 500 W University Avenue, El Paso, Texas 79968, USA. E-mail: asgonzalez1@utep.edu

Charu Gupta, Amity Institute for Herbal Research & Studies, Amity University-Uttar Pradesh, Sector-125, Noida-201313, India. E-mail: charumicro@gmail.com

Sneh Gupta, Department of Zoology, R.G.P.G. College, Meerut-250001, India.

Raosaheb K. Kale, School of Life Sciences, Central University of Gujarat, Gandhinagar, India. School of Life Sciences, Jawaharlal Nehru University, New Delhi, India.

Rebecca Knackstedt, Department of Cellular and Molecular Pharmacology and Experimental Therapeutics, Hollings Cancer Center, Medical University of South Carolina, Charleston, South Carolina 29425, USA. E-mail: rew27@musc.edu

Neeraj Kumar, Assistant Professor, Shri Ram Murti Smarak College of Engineering and Technology (Pharmacy), Nainital Road, Bareilly-243202, U.P., India. E-mail: neerajsitm@yahoo.com

Kamal Kishore Maheshwari, Associate Professor, Department of Pharmacy, M.J.P. Rohilkhand University, Bareilly-243006, U.P., India.

Lilia Masson, Profesor Emérito de la Universidad de Chile, Santiago, Chile. Profesor Visitante Extranjero, Fundación CAPES, Universidad Federal de Rio de Janeiro, Instituto de Nutrición Josué de Castro, Rio de Janeiro, Brasil. Av. Carlos Chagas Filho 373, Prédio do CCS Bloco J/2° andar, Cidade Universitaria, CEP 21941-902, Rio de Janeiro, Brasil. E-mail: masson_lilia@yahoo.es

Vondina Moseley, Department of Cellular and Molecular Pharmacology and Experimental Therapeutics, Hollings Cancer Center, Medical University of South Carolina, Charleston, South Carolina 29425, USA.

Mónica Azucena Nazareno, CITSE-CONICET, Universidad Nacional de Santiago del Estero, Av. Belgrano (S) 1912, CP4200, Santiago del Estero, Argentina. E-mail: manazar2004@yahoo.com; nazareno@unse.edu.ar

Dhan Prakash, Amity Institute for Herbal Research and Studies, Amity University-Uttar Pradesh, Sector-125, Noida-201313, India. E-mail: dprakash_in@yahoo.com

Marcos H. Rostagno, United States Department of Agriculture, Agricultural Research Service, West Lafayette, Indiana 47907, USA.

Girish Sharma, Amity Center for Cancer Epidemiology & Cancer Research and Amity Institute of Biotechnology, Amity University Uttar Pradesh, Sector-125, Noida-201313, India. E-mail: sharmagi03@gmail.com

Sapna Sharma, Division of Nephrology, Department of Medicine, University of Chicago Medical Center, 5841 S. Maryland Avenue, Chicago, Illinois 60637, USA.

Narendra Singh, School of Life Sciences, Central University of Gujarat, Gandhinagar, India.

Pankaj Singh, Department of Biochemistry, Dr RML Avadh University, Faizabad-224 001, India.

R.L. Singh, Department of Biochemistry, Dr RML Avadh University, Faizabad-224 001, India. E-mail: drrlsingh@rediffmail.com

Rana P. Singh, School of Life Sciences, Central University of Gujarat, Gandhinagar-382030, India. School of Life Sciences, Jawaharlal Nehru University, New Delhi, India. E-mail: rana_singh@mail.jnu.ac.in; ranaps@hotmail.com

Dhanir Tailor, School of Life Sciences, Central University of Gujarat, Gandhinagar, India.

S.P. Vishwakarma, Nutraceutical Laboratory, Department of Biochemistry, Dr RML Avadh University, Faizabad-224 001, India.

Michael J. Wargovich, Department of Cellular and Molecular Pharmacology and Experimental Therapeutics, Hollings Cancer Center, Medical University of South Carolina, Charleston, South Carolina 29425 USA.

Ken Wojcikowski, Southern Cross University, Lismore, New South Wales, Australia.

Preface

The word nutraceuticals is derived from the nutrition and pharmaceuticals that provide health and medical benefits, including the prevention and treatment of disease. A potential nutraceutical is one that holds a promise of a particular health or medical benefit; such a potential nutraceutical only becomes established after there are sufficient clinical data to demonstrate such a benefit. Therefore, a nutraceutical is exhibited to have a physiological benefit or provide protection against chronic disease. Such products may range from isolated nutrients, dietary supplements and specific diets to genetically engineered foods, herbal products and processed foods. Their bioactive ingredients, the phytochemicals, sustain or promote health and occur at the crossroads of the food and pharmaceutical industries. They play a crucial role in maintaining optimal immune response, such that deficient or excessive intakes can have negative impacts on health. The growing awareness of nutraceutical benefits and shift of healthcare economics in favour of nutraceuticals brought nutraceuticals into the spotlight of government health policies in various countries. Epidemiological and animal studies suggest that the regular consumption of fruits, vegetables and whole grains reduces the risk of chronic diseases.

The present book describes evidences for protective and health-beneficial effects of phytochemicals of nutraceutical importance and is divided into six parts. Part I provides an introduction and overview of phytochemicals of nutraceutical importance. These are non-nutritive plant chemicals, bioactive constituents that sustain or promote health. They may range from isolated nutrients, dietary supplements and specific diets to genetically engineered designer foods, herbal products, processed foods and beverages. The phytochemicals, either alone or in combination, have significant therapeutic potential in curing various ailments. They play positive pharmacological effects in human health as antioxidants, antibacterial, antifungal, anti-inflammatory, anti-allergic, antispasmodic, anti-aging, antidiabetes, chemopreventive, hepatoprotective, neuroprotective, hypolipidaemic, hypotensive, diuretic, CNS stimulant, immuno-modulator, carminative, analgesic, induce apoptosis and protect from osteoporosis, DNA damage, cancer and heart diseases.

In Part II, Phytochemicals in Disease and Prevention Therapy, Chapter 2 deals with progressive chronic kidney disease (CKD), which is debilitating, generally irreversible, and is associated with considerable morbidity and mortality, especially when it progresses to end stage kidney disease (ESKD) where patients require dialysis or transplant to survive. Although conventional therapies, such as angiotensin converting enzyme inhibitors (ACEi) and angiotensin receptor blockers (ARB), do have some beneficial outcomes in blocking progression of fibrosis, they are by no means perfect therapies because, even with these drugs, CKD progression is often insidious and persistent. Phytochemicals, and other complementary therapies, may provide a beneficial adjunct to these conventional drugs. Chapter 3 deals with natural products in the prevention of cancer, investigating clues in traditional diets for potential modern-day cures. As the process of acculturation occurs globally, traditional diets are being replaced with foods typically associated with Western cultures. Traditional diets have disease-fighting compounds that need to be introduced into diets in order to restore their disease preventing abilities. Chapter 4 describes resveratrol as a chemo-preventative agent with diverse applications. It is an antioxidant synthesized by wine grapes as a natural defence against both fungal infections and UV light. Preclinical and clinical trials have established the therapeutic effects of resveratrol, including the treatment of various cancers, lipid disorders, anti-inflammatory, neuroprotective, cardioprotective and anti-ageing activity.

In Part III, Potential Alternative Therapeutic Dietary Supplements, Chapter 5 deals with synbiotics promoting gastrointestinal (GI) health. The metabolic processes of various bacteria and the interactions with dietary inputs impact GI tract health and have systemic influences. The concept of nutritionally using a prebiotic and probiotic in a synbiotic relationship to increase the relative number of beneficial bacteria in the gut is a new and promising area of investigation. Synbiotics may be useful in treating some skin ailments, chronic kidney disease, diarrhoea and inflammatory bowel disease. Chapter 6 describes that nature is an attractive source of new therapeutic compounds with tremendous chemical diversity. Exploitation of microorganisms are being employed for the large scale production of a variety of biochemicals ranging from alcohol to antibiotics and processing of foods and feeds. Microorganisms have a great potential as nutraceuticals and can be used to combat diseases such as protein energy malnutrition, anaemia, diarrhoea, cancer, obesity, ulcerative colitis, Crohn's disease, irritable bowel syndrome and gluten therapy resistant celiac. Chapter 7 describes phytochemicals of nutraceutical importance from cactus and their role in human health. Cacti have been used by ancient civilizations to cure diseases and heal wounds. Cactus cladodes, fruits and flowers have been traditionally used as natural medicines in several countries. Cactus products may be efficiently used as a source of several phytochemicals of nutraceutical importance.

In Part IV, Importance and Benefits of Dietary Phytopharmaceuticals, Chapter 8 deals with the role of omega 3 and omega 6 fatty acids in human health. Foods must supply two essential fatty acids such as linoleic acid and α-linolenic acid, which accomplish fundamental and highly specific physiological roles in humans and are involved in protection from cardiovascular disease, nervous tissue, retina function, seminal glands, inflammatory process, immunity, etc. Chapter 9 deals with glucosinolates present in cruciferous vegetables, which are considered as one of the most significant biologically active phytochemicals with anticancer properties. Consumption of plants of *Brassica* species provides protection against carcinogenesis, mutagenesis and other forms of toxicity of electrophiles and reactive oxygen species. Chapter 10 describes phytoestrogens, which can structurally or functionally mimic mammalian oestrogens and show potential benefits for human health, serving as potential alternatives to the synthetic selective oestrogen receptor modulators currently being used in hormone replacement therapy. Chapter 11 describes phytosterols and their healthy effects. They compete with cholesterol in the intestine for uptake, and aid in the elimination of cholesterol from the body. They are found to exhibit anti-inflammatory, antineoplastic, antipyretic and immunomodulating activity. Chapter 12 deals with the chemistry and health benefits of carotenoids, which comprise carotenes and oxycarotenoids as two main groups of fat-soluble pigments, widely distributed in nature. Carotenes along with xanthophylls, astaxanthin, lycopene and lutein seem to offer protection against lung, colorectal, breast, uterine and prostate cancers. They help to prevent heart disease, and supplementation along with vitamin C and E reduces the risk of developing diabetes and to fight against Alzheimer's disease.

In Part V, Antioxidant Phytonutrients and their Therapeutic Values, Chapter 13 describes phenolic acids as natural antioxidants for reducing lipid oxidation, extending the shelf life of edible fats and oils, replacing synthetic phenolic antioxidants. They are quite common in plants and contribute to the taste and flavour characteristics of many spices. Their antioxidant activity is related to their mechanism of trapping free radicals and their potency is related to their chemical structure. Chapter 14 explains the role of antioxidant polyphenols in nutraceuticals and human health. Polyphenols are considered to be the most effective antioxidants; they can also intensify the activity of other antioxidants. Antioxidants may be of significant importance to offer protection against various degenerative diseases such as cancer, diabetes mellitus, inflammatory diseases, neurodegenerative disorders and ageing. Natural polyphenols afford protection against various stress-induced toxicities through modulating intercellular cascades which inhibit inflammatory molecule synthesis, the formation of free radicals, nuclear damage and induce antioxidant enzyme expression. Chapter 15 deals with the use of antioxidant phytochemicals in cancer chemoprevention. *In vitro* and *in vivo* studies show their potency as preventive and therapeutic agents for various stages and types of cancer. There are several obstacles for the effective use of these phytochemicals for their medicinal values. The proven phytochemicals such as epigallocatechin-gallate (EGCG), curcumin, silibinin, resveratrol and genistein show less bioavailability and durability *in vivo*. Chapter 16 describes antioxidants, their roles and plant sources. Excessive amounts of free radicals are thought to be related to the development of conditions such as heart and liver disease, cancers, arthritis and accelerated ageing. Plants produce an impressive array of antioxidant compounds, which includes carotenoids, flavonoids, tocopherols, tocotrienols, cinnamic acids, benzoic acids, folic acid and ascorbic acid etc. Antioxidants present in the diet enter the blood and are delivered to the cells directly to protect them from damage by free radicals.

In Part VI, Potential Traditional and Novel Food Interventions, Chapter 17 deals with phytochemicals of nutraceutical importance from *Curcuma longa* and their role in human health. *Curcuma longa* is used as a spice, colouring matter and preservative and has a wide range of medicinal and pharmacological activities. It exhibits anti-inflammatory, antioxidant, antibacterial, antiparasitic, nematocidal, anti-human immunodeficiency virus, antispasmodic, antimalarial and anticarcinogenic activities. Chapter 18 considers the phytochemistry of plants used in traditional medicine. There is an increasing interest in natural plant products as a source of biologically active phytopharmaceuticals and an urgent need to develop new clinical drugs. This is a timely review of the latest advances and trends in a field that is becoming commercially significant in the pharmaceutical industry. Chapter 19 deals with vitamins, minerals, their roles and plant sources. These are essential for proper functioning of the human body and provide medicinal benefits. They work individually as well as synergistically. Vitamins and minerals are also required to perform specific cellular functions, boost the immune system and support growth and development. Chapter 20 covers the newly emerging field of nutrigenomics: nurturing of genotype and role in human health. The influence of genetic variation on nutrition by correlating gene expression or single-nucleotide polymorphism (SNP) with a nutrient's absorption, metabolism, elimination or biological effects and to develop rational means to optimize nutrition, with respect to subject's genotype is known as nutrigenomics. It is the application of high-throughput genomic tools in nutrition research to provide methods and tools for disease preventing and health promoting phytochemicals/phytonutrients that match their lifestyles, cultures and genetics, which is determined by the specific demands of genetic signature and perfectly balances the macro- and micronutrient needs. Nutrigenomics is the emerging face of nutrition and phytonutrients that provide the necessary stepping stones to achieve the ambitious goal of optimizing an individual's health via nutritional intervention.

We would like to thank the contributing authors for their sincere and dedicated efforts, generosity and patience. Editors are grateful to Dr Ashok K. Chauhan, Founder President and Mr Atul Chauhan, Chancellor, Amity University Uttar Pradesh, Noida, India for the encouragement, support and valuable guidance.

Dhan Prakash
Girish Sharma

1 Phytochemicals of Nutraceutical Importance: Do They Defend Against Diseases?

Girish Sharma,[1]* Dhan Prakash[2] and Charu Gupta[2]

[1]Amity Center for Cancer Epidemiology & Cancer Research and Amity Institute of Biotechnology; [2]Amity Institute for Herbal Research & Studies, Amity University Uttar Pradesh, Noida, India

1.1 Introduction

The word 'nutraceuticals', coined by Dr Stephen de Felice, is derived from the words 'nutrition' and 'pharmaceutical', and is a food or food product that provides health and medical benefits, including the prevention and treatment of disease (Biesalski, 2001). A potential nutraceutical is one that holds a promise of a particular health or medical benefit; such a potential nutraceutical only becomes an established one after there are sufficient clinical data to demonstrate such a benefit (Pandey *et al.*, 2010). Therefore, a nutraceutical is exhibited to have a physiological benefit or provide protection against chronic disease. Such products may range from isolated nutrients, dietary supplements and specific diets to genetically engineered foods, herbal products, and processed foods such as cereals, soups and beverages. Their bioactive ingredients, the phytochemicals, sustain or promote health and occur at the crossroads of the food and pharmaceutical industries. Such substances may range from isolated nutrients, dietary supplements and specific diets to genetically engineered designer foods, herbal products, processed foods and beverages (Kalra, 2003; Prakash *et al.*, 2004). Chemically the nutraceuticals may be classified as isoprenoid derivatives (terpenoids, carotenoids, saponins, tocotrienols, tocopherols, terpenes), phenolic compounds (cumarins, tannins, lignins, anthocyanins, isoflavones, flavonones, flavonoids), carbohydrate derivatives (ascorbic acid, oligosaccharides, non-starch polysaccharides), fatty acid and structural lipids (n-3 PUFA, CLA, MUFA, sphingolipids, lecithins), amino acid derivatives (amino acids, allyl-S compounds, capsaicinoids, isothiocyanates, indoles, folate, choline), microbes (probiotics, prebiotics) and minerals (Ca, Zn, Cu, K, Se) (Sharma, 2009). They play a crucial role in maintaining optimal immune response, such that deficient or excessive intakes can have negative impacts on health. Around the world, governing bodies have accepted nutraceuticals as possible nutraceutical therapy in mainstream medical education and health. The healthcare industry demonstrated the shift of a growing population from medical treatment of cancer towards non-prescription nutraceuticals as self-medication in cancer management and prevention.

* E-mail: sharmagi03@gmail.com

The growing awareness of nutraceutical benefits and shift of healthcare economics in favour of nutraceuticals brought nutraceutical medicine into the spotlight of government health policy on the systematic use of nutraceuticals in prevention and/or control of various chronic diseases (Sharma, 2009).

The recent notion of 'customized' or 'personalized' medicine and diet is being advocated widely to the field of nutrition that can be used to delay the onset of disease and to sustain optimum human health (Dijsselbloem *et al.*, 2004; Kaput and Rodriguez, 2004). Dietary intake of phytochemicals may promote health benefits, protecting against chronic degenerative disorders, such as cancer, cardiovascular and neurodegenerative diseases. The majority of foods, such as whole grains, beans, fruits, vegetables and herbs, contain phytochemicals (Table 1.1). Among these, fruits and vegetables are significant sources of phytochemicals. These phytochemicals, either alone or in combination, have tremendous therapeutic potential in curing various ailments. Phytochemicals with nutraceutical properties present in food are of enormous significance due to their beneficial effects on human health since they offer protection against numerous diseases or disorders such as cancers, coronary heart disease, diabetes, high blood pressure, inflammation, microbial, viral and parasitic infections, psychotic diseases, spasmodic conditions, ulcers, etc. (Fig. 1.1). The National Cancer Institute has emphasized alternative methods of cancer prevention as public awareness by focusing mainly on lifestyle, eating habits, prevention and control care measures (Sharma, 2009). The major nutraceuticals were reviewed and reported as vitamins and minerals, phytochemicals. The vitamins A, B_6, B_{12}, D, E, folate have been reported as anticancer, immune-protective and reducing cancer risk in the population at risk of cancer and individuals who used self-medication (Holick, 2008; Milner, 2008; Zhang *et al.*, 2008).

Epidemiological and animal studies suggest that the regular consumption of fruits, vegetables and whole grains reduces the risk of chronic diseases associated with oxidative damage (Kris-Etherton *et al.*, 2002; Scalbert *et al.*, 2005; Cieslik *et al.*, 2006). Carotenoids, tocopherols, ascorbates, lipoic acids and polyphenols are strong natural antioxidants with free radical scavenging activity. Endogenous antioxidant enzymes such as superoxide dismutase (SOD), catalase, glutathione peroxidase, glutathione reductase, minerals such as Se, Mn, Cu, Zn, vitamins A, C and E, carotenoids, limonoids and polyphenols exert synergistic actions in scavenging free radicals. Synthetic antioxidants such as butylated hydroxy anisole (BHA) and butylated hydroxy toluene (BHT) play a useful role in the food and pharmaceutical industries (Kondratyuk and Pezzuto, 2004). The natural antioxidant system is mainly classified into two categories, namely *in vitro* and *in vivo* antioxidants.

The majority of the achievement of nutraceuticals is based on self-prescription and own individual experiences. However, it is difficult to realize the phenomenal benefits of nutraceuticals unless controlled clinical trials support the evidence and facts of nutraceutical preventive therapeutic efficacy (Sharma, 2009). This chapter summarizes the evidence for protective and health-beneficial effects of phytochemicals, which have the potential of being incorporated into foods or food supplements as nutraceuticals, or into pharmaceuticals, and to propose implications of the explosion in information for the future development, discovery and use of phytochemicals as nutraceuticals. Although nutraceuticals have significant promise in the promotion of human health and disease prevention, health professionals, nutritionists and regulatory toxicologists should strategically work to plan appropriate regulation to provide the ultimate health and therapeutic benefits to mankind. In this context, long-term clinical studies would be required to scientifically validate the nutraceuticals in various medical conditions. The interaction of nutraceuticals with food and drugs is another area that should be taken into consideration. The effect of different processing methods on the biological availability and effectiveness of nutraceuticals remains to be determined. Similar to drugs, there should also be stringent regulatory controls for nutraceuticals.

Table 1.1. Phytochemicals of nutraceutical importance, their sources and health benefits.

Phytochemicals	Source plant	Health benefits
α-Linolenic acid (ALA)	Flaxseed	Cancer preventive, reduce risk of coronary heart disease
Allicin	Garlic, onion	Antibacterial, anticancer, antifungal, anti-inflammatory, chemopreventive, hepatoprotective, hypolipidaemic, hypotensive and neuroprotective
Anthocyanins	Blackberry, cherry, orange, purple maize, raspberry, red grape	Anti-allergic, anti-inflammatory, antioxidants and pigments
Apigenin	Apple, artichoke, basil, celery, cherry, grape, nuts, parsley	Anti-inflammatory, antioxidant, antispasmodic, chemopreventive, induce apoptosis and inhibits breast and ovarian cancers
Caffeic acid	Artichoke, pear, basil, oregano	Anti-inflammatory, antifatigue and antistress properties
Carotene	Carrots, leafy greens and red, orange and yellow vegetables, pumpkin	Anticarcinogenic, enhances release of immunogenic cytokines IL-1 and TNF-alpha, provide cornea protection against UV light, stimulate DNA repair enzymes
Catechins	Tea	Antioxidant, CNS stimulant and diuretic
Curcumin	Turmeric	Antihypertensive, anti-inflammatory, antioxidant and cancer preventive
Diosgenin	Fenugreek seeds	Hypolipidaemic
Ellagic acid	Cranberry, grape, pecan, pomegranate, raspberry, strawberry, walnut	Anticancer and antioxidant
Ferulic acid	Oats, rice, orange, pineapple, groundnut	Protects against cancer, bone degeneration, menopausal symptoms (hot flushes)
Gallic acid	Tea, mango, strawberry, soy	Cytotoxic and antioxidative activities, antileukemic, antioxidant, anticancer, antineoplastic, anti-inflammatory, antidiabetic
Genistein	Lucerne sprouts, red clover, chickpea, groundnut, soybean	Acts as a phytoestrogen, antioxidant, anticancer agent, heart health and helps people with metabolic syndrome
Lutein	Kale, spinach, red pepper, mango, papaya, kiwi, peach, squash, honeydew melon, plum, avocado	Absorbs damaging blue light, protects against colon cancer

Continued

Table 1.1. Continued.

Phytochemicals	Source plant	Health benefits
Lycopene	Apricot, papaya, pink guava, tomato, watermelon	Lowers risk of atherosclerosis and prostate cancer
Momorbicin	Karela (bitter gourd)	Antidiabetic
Myristicin	Nutmeg	Hypolipidaemic
Piperine	Pepper	Aeromatic, analgesic, hepatoprotective and stomachic
Quercetin	Red onion, buckwheat, red grape, green tea, apple skin	Strong antioxidant, reduces LDL oxidation, vasodilator and blood thinner
Resveratrol	Blueberry, groundnut, red grape and red wine	Antioxidant, prevents ageing, cancer, diabetes and heart diseases
Rutin	Asparagus, buckwheat and citrus fruits	Strengthens capillary walls
Silymarin	Milk thistle (*Silybum marianum*)	Protects from UVB-induced carcinogenesis and hepatoprotective
Stigmasterol	Soybean	Anticancer, hypolipidaemic, prevention of osteoporosis
Sulforaphane, glucosinolates	Broccoli sprouts, cabbage, cauliflower, collards, cruciferous vegetables, kale, radish, turnip	Antioxidant, prevents DNA damage, reduces risk of breast and prostate cancers
Ursolic acid	Apple, basil, cranberry, lavender, oregano, rosemary	Anti-inflammatory, antimicrobial and antitumour
Withaferin, withanolides	*Withania somnifera*	Anticancer and immunomodulator
Zingiberene	Ginger	Antibacterial, antifungal, carminative and in treatment of dizziness

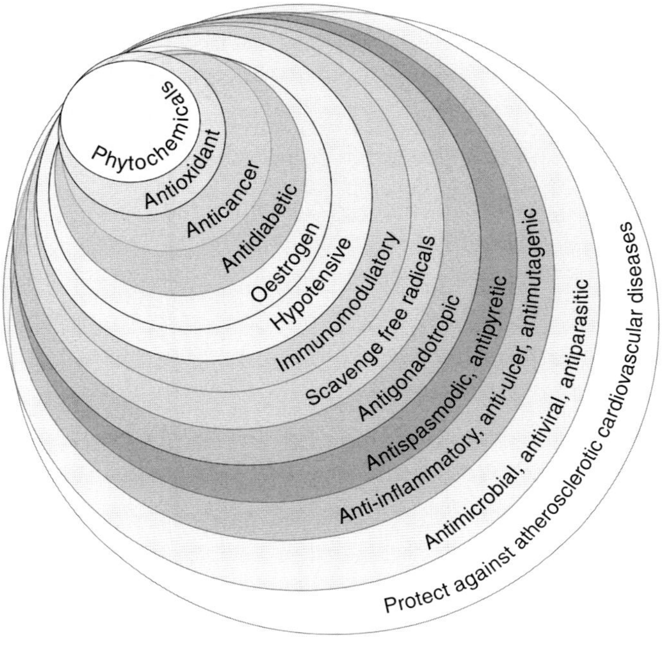

Fig. 1.1. Some important therapeutic properties of phytochemicals.

1.2 Phytochemicals and Their Health Benefits

1.2.1 Polyphenols

Polyphenols are naturally occurring compounds found largely in fruits, vegetables, cereals and beverages. Legumes and chocolate also contribute to the polyphenolic intake. These molecules are secondary metabolites of plants and are generally involved in defence against ultraviolet radiation or aggression by pathogens. Basic researches and epidemiological studies have shown the inverse association between risk of degenerative diseases and intake of a diet rich in polyphenols. The epidemiological studies provide convincing evidence that a diet rich in antioxidants is associated with a lower incidence of degenerative diseases. The major sources of dietary polyphenols are cereals, legumes (barley, maize, nuts, oats, rice, sorghum, wheat, beans and pulses), oilseeds (rapeseed, canola, flaxseed and olive seeds), fruits, vegetables and beverages (fruit juices, tea, coffee, cocoa, beer and wine) (Kaul and Kapoor, 2001; Scalbert *et al.*, 2005; Cieslik *et al.*, 2006; Katalinic *et al.*, 2006; Prakash and Kumar, 2011). Fruits such as apple, grape, pear, cherry and various berries contain up to 200–300 mg polyphenols 100 g^{-1} fresh weights. Similarly, a glass of red wine or a cup of coffee or tea contains about 100 mg polyphenols. Their total dietary intake may be about 1 g day^{-1}, which is about ten times higher than that of vitamin C and 100 times higher than those of vitamin E and carotenoids (Packer and Weber, 2001; Scalbert *et al.*, 2005).

Plant polyphenols are secondary metabolites that are broadly distributed in higher plants. Their unique characteristics are water solubility, intermolecular complexation and antioxidant properties. They are classified as condensed proanthocyanidins, galloyl and hexahydroxydiphenoyl esters and derivatives, or tannins. Polyphenols historically have been considered as antinutrients by nutritionists, because some, e.g. tannins, have adverse effects such as decreasing the activities of digestive enzymes, energy, protein and amino acid availabilities, mineral uptake and having other toxic effects. Detection of the antioxidant activities of many polyphenols has reunited opinion toward the health benefits provided by many of these compounds. The most important dietary phenolics are the phenolic acids (including hydroxybenzoic and hydroxycinnamic acids), polyphenols (hydrolysable and condensed tannins) and flavonoids, the latter being the most studied group. Phenols protect plants from oxidative damage. They have also been studied extensively as antioxidant protectants for human beings and play a beneficial role in reducing the risk of coronary heart disease, diabetes, hypertension and some types of cancer (Gee and Johnson, 2001; Willcox *et al.*, 2004; Arts and Hollman, 2005; Andjelkovic *et al.*, 2006).

The chief constituents of tea polyphenols are flavonols (catechin, epicatechin, catechin-gallate and epigallocatechin-gallate), flavonols (quercetin, kaempferol and their glycosides), flavones (vitexin, isovitexin) and phenolic acids (gallic acid, chlorogenic acid). They constitute up to 30% of the dry weight of green leaves and from 9 to 10% of the dry weight of black tea leaves. Ferulic acid is associated with dietary fibre linked with hemicellulose of the cell wall by means of ester bonds. Caffeic acid in the form of caffeoyl esters and cumaric acids are common in apples, pears and grapes. Additionally, apples and pears are rich in chlorogenic acid and grapes in gallic acid. Apples contain high levels of quercetin among fruits. Grain-derived products are especially significant in human diet as they have higher concentration of phenolic acids in the outer layers of kernel that constitute the bran. Most of the phenolic acid derivatives are hydrolysable tannins and are usually esterified with glucose. Citrus fruits are major sources of flavonones and hesperidin is found in abundance (120–250 mg l^{-1}) in orange juice.

Quercetin occurs in its glycosylated form as rutin in fruits and vegetables and onions are a particularly rich source (Anagnostopoulou *et al.*, 2006; Prakash *et al.*, 2007a; Singh *et al.*, 2009). Anthocyanins are pigments of fruits such as cherries, plums, strawberries, raspberries, blackberries and red currant (Table 1.1) and their content varies from 0.15 to 4.5 mg g^{-1} in fresh berries. Occurrence of some of the flavonoids is restricted to a few foodstuffs; e.g. the main source of isoflavonoids is soy, which contain ~1 mg g^{-1} of genistein and daidzein that have received considerable attention due to their suggested role in prevention of cancer and osteoporosis. People who consume traditional diets rich in soy and tea rarely experience breast, uterus and prostate cancer. Although there is a range of potentially anti-mutagenic fruits, vegetables and cereals, their intake is generally below the level essential to protect from various mutagens (Dillard and German, 2000; Prakash *et al.*, 2004). Extracts from *Silybum marianum* have been used for centuries in folk medicine for the treatment of liver disorders. Silibinin, the main flavolignan occurring in the flavonoids mixture silymarin of this plant, had shown positive effects on the liver. Besides being hepatoprotective, silibinin has been extensively evidenced to induce apoptosis, reduce and/or inhibit cell proliferation and tumour angiogenesis in human lung, bladder and prostate cancer models (Sharma *et al.*, 2003; Singh *et al.*, 2003, 2004, 2008a, b). Kolaviron from seeds of *Garcinia kolu* and hispidulin from *Buccuris frimeru* have also been reported as hepatoprotective (Kris-Etherton *et al.*, 2002; Cai *et al.*, 2004).

Flavonoids

Flavonoids comprise the most common group of plant polyphenols. Flavonoids are a subclass of plant phenols, which includes the minor flavonoids (flavanones and dihydroflavonols), flavones and flavonols. Flavonols are the most ubiquitous flavonoids in food. Quercetin and kaempferol are the main representatives of this group. They are generally present at relatively low concentrations of about 15–30 mg kg^{-1} fresh weight. Onions, curly kale, leeks, broccoli and blueberries are rich sources of flavonols. Flavanones are found in tomatoes and certain aromatic plants such as mint (*Mentha piperita*), but they are present in high concentrations only in citrus fruits. The main flavanones are naringenin in grapefruit, hesperetin in oranges and eriodictyol in lemons. A vast amount of recent literature proposes that the stilbenes provide beneficial health effects (Pandey and Rizvi, 2009). Recent studies indicate that like other polyphenols, stilbenes also show direct antioxidant activity, but due to comparatively dynamic beneficial effects stilbenes get superiority over the other polyphenols. One of the best studied, naturally occurring polyphenol stilbenes is resveratrol (3,4′,5-trihydroxystilbene). Resveratrol is found largely in grapes and red wine that is made from these grapes. Resveratrol is well known for its anticarcinogenic, anti-inflammatory actions. Recently, evidence suggests that stilbenes may act as a signalling molecule within tissues and cells to modulate the expression of genes and proteins (Dore, 2005).

Flavonoids are present in most plant tissues and often in vacuoles (Croteau *et al.*, 2000). Among the biological activities of flavonoids are action against free radicals, free radical-mediated cellular signalling, inflammation, allergies, platelet aggregation, microbes, ulcers, viruses, tumours and hepatotoxins. Proposed mechanisms by which they provide health benefits, in addition to being direct chemical protectants, involve modulatory effects on a variety of metabolic and signalling enzymes. Flavonoids have been shown to block the angiotensin-converting enzyme (ACE) that raises blood pressure; they inhibit cyclooxygenase, which forms prostaglandins; and they block enzymes that produce oestrogen. The implications of these *in vitro* inhibitory actions are that certain flavonoids could prevent platelet aggregation, reducing heart disease and thrombosis; and inhibit oestrogen synthase, which binds oestrogen to receptors in several tissues, thus decreasing the risk of oestrogen-related cancers. Bioactive properties such as free radical scavenging, inhibition of hydrolytic and oxidative enzymes, anti-inflammatory and antiviral (Hodek *et al.*, 2002) action of flavonoids is known. Antiproliferative effects, such as cancers, cardiovascular and inflammatory diseases of dietary flavonoids are recognized. Scavenging activity of hydroxyl radicals, superoxide anion radicals and lipid peroxy radicals signifies the health promoting functions of flavonoids (Kumar and Andy, 2012). The major sources of flavonoid intake are tea (61%), onions (13%) and apples (10%), the other sources include cherry, tomato, broccoli, black grapes and blueberries. There is an inverse association between flavonoid intake and coronary heart disease mortality. Flavonoids in regularly consumed foods appeared to reduce the risk of death from coronary heart disease. Whereas flavonoid intake has been associated with reduced risk of death from coronary heart disease, some flavonoids have been reported to be mutagenic as well (Miller and Larrea, 2002). The capacity of flavonoids to act as antioxidants depends upon their molecular structure. The position of hydroxyl groups and other features in the chemical structure of flavonoids are important for their antioxidant and free radical-scavenging activities. Quercetin, the most abundant dietary flavonol, is a potent antioxidant because it has all the right structural features for free radical-scavenging activity (Kumar and Andy, 2012). Luteolin has anti-inflammatory, antimutagenic and antibacterial activities. Apigenin suppressed 12-O-tetradecanoylphorbol-3-acetate (TPA)-mediated tumour promotion of mouse skin, similar to curcumin, a dietary pigmented polyphenol, possibly through suppression of protein kinase C activity and nuclear oncogene expression (Hasima and Aggarwal, 2012). Apigenin is antibacterial, anti-inflammatory, diuretic, hypotensive, and also promotes smooth muscle relaxation. Myricetin, a hexahydroxyflavone, exhibits antibacterial activity and has antigonadotropic activity, but apparently is not a mutagen. The flavonol kaempferol, which is widely found in the diet, has anti-inflammatory and antibacterial activities and is directly mutagenic. Quercetin, the most common flavonoid in higher plants, seems to contribute to the mutagenicity of kaempferol in the presence of microsomal metabolizing systems. Quercetin inhibits a number of enzymes, inhibits smooth muscle contraction and proliferation of rat lymphocytes. Although it is anti-inflammatory, antibacterial, antiviral and antihepatotoxic, it exhibits mutagenic activity and allergenic properties (Prakash and Gupta, 2009). Major sources of catechins are grapes, berries, cocoa and green tea. Tea contains considerable amounts of gallic acid esters, such as epicatechin, epicatechin-gallate and epigallocatechin-gallate (EGCG). Numerous studies have suggested that these components provide protective benefits by their free radical-scavenging ability and their inhibition of eicosanoid synthesis and platelet aggregation. Green tea provides protection against prostate cancer (Nichenametla *et al.*, 2006). In wines, catechins and procyanidins are involved in the astringency sensation. Catechin is one of the major phenolics in grapes and red wines, and it is considered to be responsible for part of the protective effect of red wine against atherosclerotic cardiovascular disease.

Isoflavonoids

These form another subclass of the phenolic phytonutrients. Isoflavonoids are produced almost exclusively by the members of the *Fabaceae* (*Leguminosae*) family. Their main sources in foods are soy cheese, soy flour, soybean and tofu. Soybeans are an unusually concentrated source of isoflavones, including genistein and daidzein, and soy is the major source of dietary isoflavones. The isoflavones of soy have received considerable attention owing to their binding to the oestrogen receptor class of compounds, thus representing an activity of a number of phytochemicals termed phytoestrogens. Genistein inhibits the growth of most hormone-dependent and independent cancer cells *in vitro*, including colonic cancer cells. Isoflavones have received considerable attention as potentially preventing and treating cancer and osteoporosis (Ko *et al.*, 2010). In mice, dietary soybean components inhibited the growth of experimental prostate cancer and altered tumour biomarkers associated with angiogenesis. Although the epidemiological data suggest that soy potentially decreases the risk of breast and prostate cancer, the evidence that soy exerts a protective effect against colonic cancer is limited.

Antioxidant and antiproliferative properties of isoflavones offer additional, important mechanisms for their protection against many prevalent chronic diseases (Messina *et al.*, 2004; Zeng *et al.*, 2004). Cellular damage resulting from oxidative stress is believed to be a major contributor to the aetiology of cardiovascular disease through the oxidation of LDL, and cancer by causing DNA strand breaks that may lead to mutations (Giles and Wei, 1997; Patel *et al.*, 2001). Exciting mechanistic results that emerged recently showed that the isoflavone genistein from soy selectively bound the beta-oestrogen receptor and reduced binding to the alpha-receptor 20-fold.

Anthocyanidins

These are water-soluble flavonoids that are aglycones of anthocyanins. These compounds are among the principal pigments in fruits and flowers (Prakash *et al.*, 2011). The colour of these pigments is influenced by pH and metal ion complexes. Anthocyanidins are antioxidants *in vitro*, and might be expected to have antioxidative and antimutagenic properties *in vivo*. Although they have been found to have potent antioxidant activity, these compounds did not prevent hydrogen peroxide-induced oxidation of DNA bases in HT29 clone 19A cells.

Anthocyanins usually appear red in leaf cells, but depending on their chemical nature and concentration, the vacuolar pH and interactions with other pigments, they can result in pink, purple, blue, orange, brown and even black leaf colours (Schwinn and Davies, 2004; Andersen and Jordheim, 2006; Hatier and Gould, 2007). Many of the published articles on plant defensive coloration have assumed red foliage to be the outcome of the production of anthocyanins, this despite the fact that other pigments – carotenoids, apo-carotenoids, betalains, condensed tannins, quinones and phytomelanins – can also contribute to plant vermilion (Davies, 2004).

1.2.2 Phytoestrogens

These are non-steroidal phytochemicals quite similar in structure and function to gonadal oestrogen hormone. They offer an alternative therapy for hormone replacement therapy (HRT) with beneficial effects on the cardiovascular system and may even alleviate menopausal symptoms. They are potential alternatives to the synthetic selective oestrogen receptor modulators (SERMs), which are currently applied in HRT. They have antioxidant effects due to their polyphenolic nature, anticarcinogenic, modulation of steroid metabolism or of detoxification enzymes, interference with calcium-transport and favourable effects on lipid and lipoprotein profiles

(Morabito *et al.*, 2002; Prakash and Gupta, 2011). On the basis of chemical structure, phytoestrogens can be classified as flavonoids, isoflavonoids, coumestans, stilbenes and lignans. They occur in either plants or their seeds. Soybean is rich in isoflavones, whereas the soy sprout is a potent source of coumestrol, the major coumestan.

Flavonoids have similar structure to oestrogen and have the capacity to exert both oestrogenic and anti-oestrogenic effects and provide possible protection against bone loss and heart diseases. The precursors of these substances are widespread in the plant kingdom, but mainly found in *Leguminosae* and are especially abundant in soybean and its products, legumes, berries, whole grains and cereals. They share structural features with oestrogen, in the sense that the presence of particular hydroxyl groups that can be positioned in a stereochemical alignment virtually identical to that of oestrogen. Populations in China, Japan, Taiwan and Korea are estimated to consume high quantities of isoflavones and women of these countries complain of fewer incidences of osteoporosis and related health problems, especially hot flushes, cardiovascular diseases, lower incidence of hormone-dependent breast and uterine cancers (Mense *et al.*, 2008; Dip *et al.*, 2009; Sakamoto *et al.*, 2010). The main dietary source of phytoestrogenic stilbenes is resveratrol from red wine and groundnuts. Although there are two isomers of resveratrol, *cis* and *trans*, only the *trans* form has been reported to be oestrogenic. It is found only in the skin of red grapes; in green grapes and white wine very low levels of *trans*-resveratrol are found (Fremont, 2000). The main dietary sources of coumestans are sprouted legumes such as soy and lucerne; however, low levels have been reported in Brussels sprouts and spinach. Clover and soybean sprouts are reported to have its highest concentrations. The term lignan is used for a diverse class of phenylpropanoid dimers and oligomers. Secoisolariciresinol and matairesinol are two lignan dimers that are not oestrogenic by themselves but readily convert to the mammalian lignans, enterodiol and enterolactone, respectively, which are oestrogenic. These are of great interest because of their oestrogenic, anticarcinogenic, antiviral, antifungal and antioxidant activities (Cornwell *et al.*, 2004).

The phytolignans are found in high amounts in flaxseed, asparagus, whole grains, vegetables and tea. Fruits also have low levels with the exception of strawberries and cranberries. In humans, after consumption of plants rich in isoflavones and lignans, enzymatic metabolic conversions occur in the gut, by microflora, and the mammalian lignans are readily absorbed (Cos *et al.*, 2003).

1.2.3 Terpenoids

The terpenes, also known as isoprenoids, form the largest class of phytonutrients in green foods and grains. These compounds are found in higher plants, mosses, liverworts, algae and lichens, as well as in insects, microbes or marine organisms. Terpenoids are derived from a common biosynthetic pathway based on mevalonate as parent, and are named terpenoids, terpenes or isoprenoids, with the subgroup of steroids among them as a class (Tholl, 2006; Bohlmann and Keeling, 2008). Their importance to plants relates to their necessity to fix carbon through photosynthetic reactions using photosensitizing pigments. Animals have evolved to utilize these compounds for hormonal and growth regulatory functions (vitamin A) and, as it is now being understood, the presence of these molecules in animal tissues also provides a measure of protection from certain diseases, especially those related to chronic damage and growth deregulation.

The diverse functional roles of some of the terpenoids are characterized as hormones (gibberellins), photosynthetic pigments (phytol, carotenoids), electron carriers (ubiquinone, plastoquinone), and mediators of polysaccharide assembly, as well as communication and defence mechanisms (Langenheim, 1994). Several biological actions have been reported for diterpenes including antibacterial, antifungal, anti-inflammatory, antileishmanial, cytotoxic and antitumour activities (Singh *et al.*, 1999). Currently, a broad range of biological responses can be elicited in humans through various terpenoids that are applicable to human health care (Paduch *et al.*, 2007). Different terpenoid molecules have antimicrobial, antifungal, antiparasitic, antiviral, anti-allergenic, antispasmodic, antihyperglycaemic, anti-inflammatory, chemotherapeutic and immunomodulatory properties (Hammer *et al.*, 2003; Wagner and Elmadfa, 2003; Paduch *et al.*, 2007). Terpenes are also used as skin penetration enhancers as well as natural insecticides, and can be of use as protective substances in storing agriculture products (Lee *et al.*, 2003). Terpenes have a unique antioxidant activity in their interaction with free radicals. They react with free radicals by partitioning themselves into fatty membranes by virtue of their long carbon side chain. The most studied terpene antioxidants are the tocotrienols and tocopherols. They are found naturally in whole grains and have effects on cancer cells. The tocotrienols are effective apoptotic inducers for human breast cancer cells. The impact of a diet of fruits, vegetables and grains on reduction of cancer risk may be explained by the actions of terpenes *in vivo* (Ikeda *et al.*, 2002; Prakash and Gupta, 2009; Prakash and Kumar, 2011).

1.2.4 Carotenoids

Carotenoids are highly pigmented, yellow, orange and red, are present in fruits and vegetables, and when consumed by birds are incorporated into the yolk of eggs. Carotenoids comprise two types of molecules, carotenes and xanthophylls. All carotenoids possess a polyisoprenoid structure, a long conjugated chain of double bond and a near bilateral symmetry around the central double bond, as common chemical features (Britton, 1995). Due to the presence of the conjugated double bonds, carotenoids can undergo isomerization to *cis-trans* isomers. Although the *trans* isomers are more common in foods and are more stable, very little is known about the biological significance of carotenoid isomerization in human health. Carotenes are tissue specific in their biological activity and beta-carotene has vitamin A activity. Based on epidemiological studies a positive link is suggested between higher dietary intake and tissue concentrations of carotenoids and lower risk of chronic diseases (Agarwal and Rao, 2000; Johnson, 2002; Elliott, 2005). β-carotene and lycopene have been shown to be inversely related to the risk of cardiovascular diseases and certain cancers (Johnson, 2002; Ribaya-Mercado and Blumberg, 2004). Lutein protects against uterine, prostate, breast, colorectal and lung cancers. They may also protect against risk of digestive tract cancer. The xanthophyll types of carotenoids offer protection to other antioxidants, and they may exhibit tissue-specific protection. Zeaxanthin, cryptoxanthin and astaxanthin are members of the xanthophyll group (Prakash *et al.*, 2004; Stahl, 2005). The antioxidant properties of carotenoids have been suggested as being the main mechanism by which they afford their beneficial effects. Recent studies are also showing that carotenoids may mediate their effects via other mechanisms such as gap junction communication, cell growth regulation, modulating gene expression, immune response and as modulators of Phase I and II drug metabolizing enzymes (Astrog, 1997; Bertram, 1999; Jewell and O'Brien, 1999; Paiva and Russell, 1999). Although the antioxidant properties of some carotenoids have been studied, most other mechanisms such as their provitamin A activity, immune, endocrine and metabolic activities, and their role in cell cycle regulation, apoptosis and cell differentiation are also under intense scientific scrutiny. Future areas of research include their bioavailability, metabolism, mechanisms of action and safety.

1.2.5 Limonoids

These are terpenes present in citrus fruit. Limonoids, with diverse structures and broad range of bioactivities, have been an attraction for both natural product and synthesis chemists. Limonoids are unique highly oxygenated triterpenoid compounds long recognized as significant biologically active natural compounds. Citrus limonoids appear in large amounts in citrus juice and citrus tissues as water-soluble limonoid glucosides or in seeds as water-insoluble limonoid aglycones (Ozaki *et al.*, 1995). Several citrus limonoids have recently been subjected to anticancer screen procedures utilizing laboratory animals and human breast cancer cells in culture. In mice, it was found that five limonoid aglycones (limonin, nomilin, obacunone, isoobacunoic acid, ichangin) induced significant amounts of glutathione-S-transferase (GST) in the liver and intestinal mucosa (Lam *et al.*, 1994). GST is a major detoxifying enzyme system, which catalyses the conjugation of glutathione with many potentially carcinogenic compounds which are highly electrophilic in nature. A study of the inhibitory effects of two limonoid aglycones (limonin and nomilin) on the formation of benzo[a]pyrene-induced neoplasia in the forestomach of ICR/Ha mice showed that incidence of tumours could be reduced by more than 50% at 10 mg dose given three times every 2 days (Lam and Hasegawa, 1989). The experimental results described above indicate that citrus limonoids may provide substantial anticancer action. The compounds have been shown to be free of toxic effects in animal models, so potential exists for use of limonoids against human cancer in either the natural fruit, in citrus fortified with limonoids, or in purified forms of specific limonoids. They provide chemotherapeutic activity by inhibiting Phase I enzymes and inducing Phase II detoxification enzymes in the liver. D-Limonene, the commonest monocyclic monoterpene, found within orange peel oil, inhibits pancreatic carcinogenesis induced in experimental models and also provides protection to lung tissue (Prakash *et al.*, 2004; Stahl, 2005). Although the initial studies are very promising, they have been conducted primarily with *in vitro* cell culture and animal models. Thus, research is needed to determine whether the limonoids may be useful in preventing or treating cancer in humans. The first step is to assess the bioavailability of the compounds for humans – are they absorbed after ingestion, do they appear in the blood and tissues, and for how long. If limonoid compounds are found to be bioavailable, further human studies will be needed to assess the effects of limonoid ingestion on biomarkers related to cancer.

1.2.6 Phytosterols

These are another important terpene subclass. The primary sources of phytosterols are vegetables, nuts, fruits and seeds. Seeds contain an average of 120 mg of plant sterols 100 g^{-1} wet weight; vegetables contain 20 mg 100 g^{-1} wet weight and fruits about 15 mg 100 g^{-1} wet weight. Sitosterol, campesterol and stigmasterol are most abundant in nature comprising 65%, 30% and 3% of dietary phytosterol intake (John *et al.*, 2007). Two sterol molecules that are synthesized by plants are beta-sitosterol and its glycoside. In animals, these two molecules exhibit anti-inflammatory, antineoplastic, antipyretic and immunomodulating activity. Phytosterols were reported to block inflammatory enzymes, for example by modifying the prostaglandin pathways in a way that protected platelets. Phytosterols compete with cholesterol in the intestine for uptake, and aid in the elimination of cholesterol from the body. In the intestine, plant sterols are initially solubilized into a micelle form. These micelles interact with brush border cells and are transferred into enterocytes. Plant sterols are esterified within the enterocyte, assembled into chylomicrons and secreted into the lymphatics. They are excreted via the biliary system. The non-esterified phytosterols are transported back into the intestinal lumen by sterolin (1 and 2) pumps containing the ATP binding cassette (ABC) proteins encoded by the genes *ABCG5* and *ABCG8*. These are expressed in the mucosal cells and the canalicular membrane, and they re-secrete sterols, especially absorbed plant sterols, back into the intestinal lumen and from the liver into bile (von Bergmann *et al.*, 2005). Saturated phytosterols appear to be more effective than unsaturated compounds in decreasing cholesterol concentrations in the body. Their actions reduce serum or plasma total cholesterol and low-density lipoprotein (LDL) cholesterol. Competition with cholesterol for absorption from the intestine is not unexpected as the structure of plant sterols is similar to that of cholesterol. In mammals, concentrations of plasma phytosterol are low because of their poor absorption from the intestine and their faster excretion from liver, and metabolism to bile acids, compared with cholesterol (Dillard and German, 2000). Available animal studies suggest that phytosterols reduce atherosclerosis in the Apo-E deficient mouse model. Human studies are mixed, and do not prove or disprove an increase in atherosclerotic risk that can be clearly related to serum phytosterol levels. It is reassuring that vegetarians who consume considerable plant sterols are at decreased risk of ASCVD, but it is impossible to separate the effects of phytosterol excess from animal fat reduction in this population (John *et al.*, 2007).

1.2.7 Glucosinolates

Glucosinolates are present in cruciferous vegetables, and are activators of liver detoxification enzymes. These chemicals are responsible for the pungent aroma and bitter flavour of cruciferous vegetables. Consumption of cruciferous vegetables offers a phytochemical strategy for providing protection against carcinogenesis, mutagenesis and other forms of toxicity of electrophiles and reactive forms of oxygen. The sprouts of certain crucifers, including broccoli and cauliflower, contain higher amounts of glucoraphanin (the glucosinolate of sulforaphane) than do the corresponding mature plants. Crucifer sprouts may protect against the risk of cancer more effectively than the same quantity of mature vegetables of the same variety (Cartea and Velasco, 2008; Traka and Mithen, 2009). During food preparation, chewing and digestion, the glucosinolates in cruciferous vegetables are broken down to form biologically active compounds such as indoles, nitriles, thiocyanates and isothiocyanates (Hayes *et al.*, 2008). Indole-3-carbinol (an indole) and sulforaphane (an isothiocyanate) have been most frequently examined for their anticancer effects. Epidemiological studies indicate that consumption of brassica vegetables is associated with a reduced incidence of cancers at a number of sites including the lung, stomach, colon and rectum (Conaway *et al.*, 2001). Glucosinolates, the thioglucosides, present in brassica vegetables are thought to contribute to this phenomenon. Dietary glucosinolates have been reported to block formation of endogenous or exogenous carcinogens for preventing initiation of carcinogenesis (Vig *et al.*, 2009). The mechanism of the protective effects is thought to involve the modulation of carcinogen metabolism by the induction of Phase 2 detoxification enzymes and inhibition of Phase 1 carcinogen-activating enzymes, thereby possibly influencing several processes related to chemical carcinogenesis, e.g. the metabolism, DNA binding and mutagenic activity of pro-mutagens. A reducing effect on tumour formation has been shown in rats and mice, and studies carried out in humans using high but realistic human consumption amounts of indoles and brassica vegetables have shown putative positive effects on health. Indole-3-carbinol is a glucosinolate metabolite that inhibits organ-site carcinogenesis in rodent models. Its preventive effect on human mammary carcinogenesis may be due in part to its ability to regulate cell cycle progression, increase the formation of antiproliferative oestradiol metabolite and induce cellular apoptosis (Dillard and German, 2000; Cartea and Velasco, 2008; Traka and Mithen, 2009).

1.2.8 Fibres

Most plant foods in their native state contain indigestible residues that used to be classified as crude fibre but are currently classified as dietary fibre (DF) and also as non-starch polysaccharides (NSP). Dietary fibre is not a single entity but consists of a wide range of complex polysaccharides such as cellulose, gums, mucilages, hemicellulose and lignins with different chemical, physiochemical and physiological properties (Narasinga Rao, 2003).These NSP in foods have been shown to be useful in reducing blood glucose levels in diabetes, in reducing blood cholesterol levels for treatment of cardiovascular disease and also in preventing bowel cancer (Schnecman, 1989). The disease-preventing potential of DF will depend upon the proportion and actual quantities of different polysaccharide components present in a given food (Narasinga Rao, 1988). Dietary fibre components exert their beneficial effects mostly by way of their swelling properties, and by increasing transit time in the small intestine. Consequently, they reduce the rate of release of glucose and its absorption, thus helping in the management of certain types of diabetes (e.g. non-insulin-dependent diabetes mellitus). DF components also bind bile salts, thereby promoting cholesterol excretion from the body and thus reducing blood cholesterol levels, and food toxins in the gut to reduce their toxicity. They can also have some adverse nutritional effects by binding dietary calcium, magnesium, zinc and iron, thereby reducing their bioavailability (Narasinga Rao, 2003).

Although dietary meat and fat intake have a positive relation to the incidence of colon cancer, DF has been associated with alterations of the colonic environment that protect against colorectal diseases. Fibre may also provide protection by increasing faecal bulk, which dilutes the increased colonic bile acid concentrations that occur with a high-fat diet. Short chain fatty acids, including butyric acid, and dietary sugarbeet fibre also suppressed cholesterol synthesis in a rat liver and intestine model. Different DFs have markedly diverse cancer protective effects, and the differences may be related to the differential bacterial fermentation of fibre in the colon to short-chain fatty acids, especially butyric acid. Butyric acid induces growth arrest, differentiation and apoptosis of colonic epithelial cells and tumour cells *in vitro*. Butyric acid in the colon also appears to influence the on-going process of apoptosis within the mucosa. The potential for fermentation of fibre to butyric acid and its derivatives is of substantial interest. Its enrichment through food products, such as fibre and starch, may emerge as a molecular-based strategy that provides significant health benefits (Dillard and German, 2000; Packer and Weber, 2001).

1.2.9 Polysaccharides

Polysaccharides widely exist in plants, microorganisms, algae and animals, are essential biomacromoleules in life activities and play important roles in cell–cell communication, cell adhesion and molecular recognition in the immune system (Dwek, 1996). Recently, some bioactive polysaccharides isolated from natural sources have attracted much attention in the field of biochemistry and pharmacology; in particular, plant polysaccharides have shown diverse biological activities such as wound healing, enhancement of the reticulo-endothelial system, stimulation of the immune system, treatment of tumours and effects on the haematopoietic system (Schmidgall *et al.*, 2000). In folk medicine, plants containing polysaccharides have been used as hypoglycaemic (Bnouham *et al.*, 2006; Lopez, 2007) and anti-inflammatory treatments (Atherton, 2002). Traditionally, polysaccharides are used as thickening, emulsifying and stabilizing agents. But, nowadays, a huge market in healthy compounds has appeared with the production of oligo- or monosaccharide syrups, using physical methods or controlled enzymatic degradation of polysaccharides (e.g. starch). Some of them possess interesting biological properties, e.g. oligodextrins (anti-ulcer agents, lowering serum cholesterol in low saturated fat diet) and fructo-oligosaccharides (prebiotics, dietary fibres, stimulate mineral absorption, enhance defence mechanism) (Lopez, 2007).

1.2.10 Saponins

Saponins are secondary plant metabolites that occur in a wide range of plant species (Hostettmann and Marston, 1995). They are stored in plant cells as inactive precursors but are readily converted into biologically active antibiotics by plant enzymes in response to pathogen attack. These compounds can also be regarded as 'preformed', since the plant enzymes that activate them are already present in healthy plant tissues (Osbourn, 1996). The natural role of saponins in plants is thought to be protection against attack by pathogens and pests (Morrissey and Osbourn, 1999). These molecules also have considerable commercial value and are processed as drugs and medicines, foaming agents, sweeteners, taste modifiers and cosmetics (Hostettmann and Marston, 1995). Saponins are glycosylated compounds that are widely distributed in the plant kingdom and can be divided into three major groups: a triterpenoid, a steroid, or a steroidal glycoalkoloid. Triterpenoid saponins are found primarily in dicotyledonous plants but also in some monocots, whereas steroid saponins occur mainly in monocots, such as the *Liliaceae, Droscoraceae* and *Agavaceae* and in certain dicots, such as foxglove (Hostettmann and Marston, 1995). Oats (*Avena* spp.) are unusual because they contain both triterpenoid and steroid saponins (Price *et al.*, 1987). Steroidal glycoalkaloids are found primarily in members of the family *Solanaceae*, which includes potato and tomato. The major saponin in tomato is the steroidal glycoalkaloid α-tomatine. The α-tomatine is present in healthy plants in its biologically active form. The levels of this saponin are particularly high in the leaves, flowers and green fruits of tomato. It is assumed that α-tomatine is present in tomato leaves in the concentration around 1 mM, which is sufficient to inhibit the growth of many non-pathogens of tomato. Therefore it would be expected that this molecule could protect the tomato leaves from fungal pathogens (Mert-Turk, 2006).

1.3 Role of Phytochemicals in Health and Diseases

Epidemiological evidence with respect to cancer and cardiovascular disease cogently suggests that phytochemicals may play a significant part in protection against the development of these diseases. This association has been drawn from the strong correlation that exists between a high dietary intake of fruit and vegetables and a reduction in the incidence of these diseases, which has led nutritionists to investigate the components in fruits and vegetables (phytochemicals) that may confer this protection. Experimental evidence that phytochemicals influence many cellular mechanisms that may optimize health has highlighted the need to identify clearly which effects may be of greater health significance (Dreosti, 2000).

The rapid growth of apparent health foods, now frequently defined by the industry as nutraceuticals, have enormously impacted the consumers. The respective health benefits of nutraceuticals are based on science and ethics, for health claims for functional foods, and presence of certain phytochemicals (Fig. 1.1). They are constituents of plants and have certain pharmacological and/or physiological effects in the ethno-medical treatment of various disorders. Traditionally, natural plant products have been the source for the search for new drugs by pharmaceutical companies. Phytochemicals play an important role in human health as antioxidants, antibacterial, antifungal, anti-inflammatory, anti-allergic, antispasmodic, chemopreventive, hepatoprotective, hypolipidaemic, neuroprotective, hypotensive agents, and help in preventing ageing, diabetes, osteoporosis, cancer and heart diseases, induce apoptosis, diuretic, CNS stimulant, analgesic, protects from UVB-induced carcinogenesis, immunomodulator and carminative (Dillard and German, 2000; Packer and Weber, 2001; Prakash and Gupta, 2009).

Capsaicin, the pungent ingredient present in red pepper and ginger, has anticarcinogenic and antimutagenic effects. Curcumin, another polyphenolic phytochemical, acts as an anti-inflammatory and cancer preventive drug. In a study, tumour volumes in mice treated with genistein, dietary soy phytochemical concentrate, at 1%, or dietary soy protein isolate were decreased 40, 48 or 37%, respectively, as compared with the controls. Genistein (5,7,4′-trihydroxyisoflavone) is one of two major isoflavonoids in soy. In human breast cancer cells in culture, genistein has antiproliferative effects on mitogen-stimulated growth (Dixon and Ferreira, 2002; Prakash *et al.*, 2007b). Soy isoflavonoid conjugates have chemopreventive activity in carcinogen-induced rat models of breast cancer.

Osteoporosis is related to multiple factors including ageing, hormone deficiency and diet. Most of the studies suggest that phytoestrogens are somewhat effective in maintaining bone mineral density (BMD) in post-menopausal women and to alleviate osteoporosis and associated disorders. Evidence from several human studies demonstrates that certain dietary phytoestrogens can produce oestrogenic effects in post-menopausal women, including oestrogen-like effects on vaginal cytology and reductions in hot flushes. In post-menopausal women, cardiovascular diseases (CVDs) are one of the leading causes of death in the USA and Europe. Isoflavonoids or soy products/soy protein and flaxseed have the ability to lower total and LDL cholesterols and raise HDL cholesterol resulting in reduced risk of CVDs. There is evidence to support the hypothesis that phytoestrogen consumption contributes to the lower incidence of CVDs in Asian countries and in vegetarians and that they may also be cardioprotective (Cherdshewasart *et al.*, 2009; Al-Azzawi and Wahab, 2010; Prakash and Gupta, 2011).

A large number of epidemiological studies had shown that people who consume high amounts of isoflavonoids (phytoestrogens) in their diets have lower incidences of various types of cancers including breast, prostate and colon cancer. A high plasma concentration of the mammalian lignan, enterolactone, is correlated with a reduced risk of breast cancer. Similar correlations have also been found between dietary intakes of isoflavonoids and lignans and thyroid, ovarian, and breast cancers in pre- and post-menopausal women. The incidence of hormone-dependent tumours is lower in Asia and eastern Europe where consumption of phytoestrogens is higher than western countries and amongst vegetarians. Breast, ovarian, prostate and colon cancer show a negative correlation with phytoestrogen intake when compared with mortality rates due to cancer. The epidemiological, animal and cell-line data suggest that phytoestrogens may play a protective role against the development of prostate and breast cancer. It has been reported that increased consumption of beans, lentils and peas, tomatoes and dried fruits was associated with significantly decreased prostate cancer risk (Cherdshewasart *et al.*, 2009; Al-Azzawi and Wahab, 2010; Prakash and Gupta, 2011).

Diets rich in phytonutrients may supply a variety of phytoestrogens such as isoflavones, resveratrol, lignans etc., capable of producing a range of pharmacological effects. In females, life is affected by a variety of oestrogen-related conditions such as osteoporosis, cognitive and cardiovascular decline, increased risk of breast cancer and other symptoms that decrease the overall quality of life. Phytoestrogens appear to have physiological effects in humans, with the most supportive data being related to the effects of soy protein supplements on lipids and lipoproteins and on vascular function. Therefore, post-menopausal women who have the greatest breast cancer risk should be encouraged to increase their phytoestrogen intake (Cherdshewasart *et al.*, 2009; Al-Azzawi and Wahab, 2010; Prakash and Gupta, 2011).

1.4 Conclusions

The use of medicinal plants by indigenous people to treat different ailments has a long history. Recently, the scientific data supported the nutritional and medical importance of phytonutrients/phytochemicals for the prevention and treatment of several diseases. The 'novel' nutraceuticals of plant origin may evolve to be considered a vital aspect of dietary disease-preventive food components. Careful studies are necessary on the various phytochemicals for their roles in the prevention of chronic degenerative diseases. The resurgence of interest in these compounds will eventually lead to much-needed information on structure–function relationships.

The ever-widening choice of food ingredients makes it possible for food designers to provide food choices that meet the public's expressed desire for healthy food. Other aspects of determining the role of phytochemicals in functional foods include consumer attitudes, any competitive advantage for manufacturers producing functional foods and identification of those areas of research needed to produce foods with the desired health effects. The future of nutraceuticals of both plant and animal origin holds exciting opportunities for the food industry to create novel food products. The food industry will need to persuade investors of the potential for monetary rewards to be gained by investing in the value of nutraceuticals, and it will need to market the products so as to capture the interest of and, perhaps most important, to please the tastes of consumers.

References

Agarwal, S. and Rao, A.V. (2000) Carotenoids and chronic diseases. *Drug Metabolism and Drug Interactions* 17, 189–210.

Al-Azzawi, F. and Wahab, M. (2010) Effectiveness of phytoestrogens in climacteric medicine. *Annals of the New York Academy of Sciences* 1205, 262–267.

Anagnostopoulou, M.A., Kefalas, P., Papageorgiou, V.P., Assimopoulou, A.N. and Boskou, D. (2006) Radical scavenging activity of various extracts and fractions of sweet orange peel (*Citrus sinensis*). *Food Chemistry* 94, 19–25.

Andersen, O.M. and Jordheim, M. (2006) The anthocyanins. In: Anderson, O.M. and Markham, K.R. (eds) *Flavonoids: Chemistry, Biochemistry, and Applications*. CRC Press, Boca Raton, Florida, pp. 471–553.

Andjelkovic, M., Camp, J.V., Meulenaer, B.D., Depaemelaere, G., Socaciu, C., Verloo, M. and Verhe, R. (2006) Iron-chelation properties of phenolic acids bearing catechol and galloyl groups. *Food Chemistry* 98, 23–31.

Arts, I. and Hollman, P. (2005) Polyphenols and disease risk in epidemiologic studies. *American Journal of Clinical Nutrition* 81, 317S–325S.

Astrog, P. (1997) Food carotenoids and cancer prevention: an overview of current research. *Trends in Food Science and Technology* 8, 406–413.

Atherton, P. (2002) *Aloe vera*: magic or medicine with antidiabetic potential. *Journal of Ethnopharmacology* 81, 81–100.

Bertram, J.S. (1999) Carotenoids and gene regulation. *Nutrition Reviews* 57, 182–191.

Biesalski, H.K. (2001) Nutraceuticals: the link between nutrition and medicine. In: Kramer, K., Hoppe, P.P. and Packer, L. (eds) *Nutraceuticals in Health and Disease Prevention*. Marcel Deckker, New York, pp. 1-26.

Bnouham, M., Ziyyat, A., Mekhfi, H., Tahri, A. and Legssyer, A. (2006) Medicinal plants with potential antidiabetic activity – a review of ten years of herbal medicine research. *International Journal of Diabetes and Metabolism* 14, 1–25.

Bohlmann, J. and Keeling, C.I. (2008) Terpenoid biomaterials. *Plant Journal* 54, 656–669.

Britton, G. (1995) Structure and properties of carotenoids in relation to function. *The Journal of the Federation of American Societies for Experimental Biology* 9, 1551–1558.

Cai, Y., Luo, Q., Sun, M. and Corke, H. (2004) Antioxidant activity and phenolic compounds of 112 traditional medicinal plants associated with anticancer. *Life Sciences* 74, 2157–2184.

Cartea, M.E. and Velasco, P. (2008) Glucosinolates in Brassica foods: Bioavailability in food and significance for human health. *Phytochemistry Reviews* 7, 213–229.

Cherdshewasart, W., Sutjit, W., Pulcharoen, K. and Chulasiri, M. (2009) The mutagenic and antimutagenic effects of the traditional phytoestrogen-rich herbs, *Pueraria mirifica* and *Pueraria lobata*. *Brazilian Journal of Medical and Biological Research* 42, 816–823.

Cieslik, E., Greda, A. and Adamus, W. (2006) Contents of polyphenols in fruits and vegetables. *Food Chemistry* 94, 135–142.

Conaway, C.C., Getachun, S.M., Liebes, L.L., Pusateri, D.J., Tophan, D.K.W., Botero-Omary, M. and Chung, F.L. (2001) Disposition of glucosinolates and sulphoraphanes in human after ingestion of steam and fresh broccoli. *Nutrition and Cancer* 38, 168–178.

Cornwell, T., Cohick, W. and Raskin, I. (2004) Dietary phytoestrogens and health. *Phytochemistry* 65, 995–1016.

Cos, P., De Bruyne, T., Apers, S., Vanden Berghe, D., Pieters, L. and Vlietinck, A.J. (2003) Phytoestrogens: Recent developments. *Planta Medica* 69, 589–599.

Croteau, R., Kutchan, T.M. and Lewis, N.G. (2000) Natural products (secondary metabolites). In: Buchman, V.V., Gruissem, W. and Jones, R.L. (eds) *Biochemistry and Molecular Biology of Plants*. American Society of Plant Physiologists, Rockville, Maryland, pp. 1250–1318.

Davies, K.M. (2004) Important rare plant pigments. In: Davies, K.M. (ed.) *Plant Pigments and their Manipulation. Annual Plant Reviews*, Vol. 14. Blackwell Publishing, Oxford, UK, pp. 214–247.

Dijsselbloem, N., Vanden Berghe, W., De Naeyer, A. and Haegeman, G. (2004) Soy isoflavone phytopharmaceuticals in interleukin-6 affections. Multi-purpose nutraceuticals at the crossroad of hormone replacement, anti-cancer and anti-inflammatory therapy. *Biochemical Pharmacology* 68, 1171–1185.

Dillard, C.J. and German, J.B. (2000) Review Phytochemicals: nutraceuticals and human health. *Journal of the Science of Food and Agriculture* 80, 1744–1756.

Dip, R., Lenz, S. and Gmuender, H. (2009) Pleiotropic combinatorial transcriptomes of human breast cancer cells exposed to mixtures of dietary phytoestrogens. *Food Chemistry and Toxicology* 47, 787–795.

Dixon, R.A. and Ferreira, D. (2002) Molecules of interest: Genistein. *Phytochemistry* 60, 205–211.

Dore, S. (2005) Unique properties of polyphenol stilbenes in the brain: more than direct antioxidant actions; gene/protein regulatory activity. *Neurosignals* 14, 61–70.

Dreosti, I.E. (2000) Recommended dietary intake levels for phytochemicals: Feasible or fanciful? *Asia Pacific Journal of Clinical Nutrition* 9, S119–S122.

Dwek, R. (1996) Glycobiology: Towards understanding the function of sugars. *Chemical Review* 96, 683–720.

Elliott, R. (2005) Mechanisms of genomic and non-genomic actions of carotenoids. *Biochimica et Biophysica Acta* 1740, 147–154.

Fremont, L. (2000) Biological effects of resveratrol. *Life Sciences* 66, 663–673.

Gee, J.M. and Johnson, I.T. (2001) Polyphenolic compounds: Interactions with the gut and implications for human health. *Current Medicinal Chemistry* 8, 1245–1255.

Giles, D. and Wei, H. (1997) Effect of structurally related flavones/isoflavones on hydrogen peroxide production and oxidative DNA damage in phorbol ester-stimulated HL-60 cells. *Nutrition and Cancer* 29, 77–82.

Hammer, K.A., Carson, C.F. and Riley, T.V. (2003) Antifungal activity of the components of *Melaleuca alternifolia* (tea tree) oil. *Journal of Applied Microbiology* 95, 853–860.

Hasima, N. and Aggarwal, B.B. (2012) Cancer-linked targets modulated by curcumin. *International Journal of Biochemistry and Molecular Biology* 3, 328–351.

Hatier, J.H. and Gould, K.S. (2007) Black coloration in leaves of *Ophiopogon planiscapus* 'Nigrescens'. Leaf optics, chromaticity, and internal light gradients. *Functional Plant Biology* 34, 130–138.

Hayes, J.D., Kelleher, M.O. and Eggleston, I.M. (2008) The cancer chemopreventive actions of phytochemicals derived from glucosinolates. *European Journal of Nutrition* 47, 73–88.

Hodek, P., Trebil, P. and Stiborova, M. (2002) Flavonoids- potent and versatile biologically active compounds interacting with cytochromes. *Chemico Biological Interactions* 139, 1–21.

Holick, M.F. (2008) Vitamin D and sunlight: strategies for cancer prevention and other health benefits. *Clinical Journal of the American Society of Nephrology* 3, 1548–1554.

Hostettmann, K.A. and Marston, A. (1995) Saponins: Chemistry and pharmacology of natural products. Cambridge University Press, Cambridge, UK.

Ikeda, K., Arao, Y., Otsuka, H., Nomoto, S., Horiguchi, H., Kato, S. and Kayama, F. (2002) Terpenoids found in the Umbelliferae family act as agonists/antagonists for ERα and ERβ: Differential transcription activity between ferutinine-liganded ERα and ERβ. *Biochemical and Biophysical Research Communications* 291, 354–360.

Jewell, C. and O'Brien, N.M. (1999) Effect of dietary supplementation with carotenoids on xenobiotic metabolizing enzymes in the liver, lung, kidney and small intestine of the rat. *British Journal of Nutrition* 81, 235–242.

John, S., Sorokin, A.V. and Thompson, P.D. (2007) Phytosterols and vascular disease. *Current Opinion in Lipidology* 18, 35–40.

Johnson, E.J. (2002) The role of carotenoids in human health. *Nutrition in Clinical Care* 5, 47–49.

Kalra, E.K. (2003) Nutraceutical – Definition and Introduction. *American Association of Pharmaceutical Scientists – Pharma Science* 5, 1–2.

Kaput, J. and Rodriguez, R.L. (2004) Nutritional genomics: the next frontier in the postgenomic era. *Physiological Genomics* 16, 166–177.

Katalinic, V., Milos, M., Kulisic, T. and Jukic, M. (2006) Screening of 70 medicinal plant extracts for antioxidant capacity and total phenols. *Food Chemistry* 94, 550–557.

Kaul, C. and Kapoor, H.C. (2001) Antioxidants in fruits and vegetables – the millenniums health. *International Journal of Food Science and Technology* 36, 703–725.

Ko, K.P., Park, S.K. and Park, B. (2010) Isoflavones from Phytoestrogens and Gastric Cancer Risk: A Nested Case-Control Study within the Korean Multicenter Cancer Cohort. *Cancer Epidemiology, Biomarkers and Prevention* 19, 1292–1300.

Kondratyuk, T.P. and Pezzuto, J.M. (2004) Natural Product Polyphenols of Relevance to Human Health. *Pharmaceutical Biology* 42, 46–63.

Kris-Etherton, P., Hecker, K., Bonanome, A., Coval, S., Binkoski, A. and Hilpert, K. (2002) Bioactive compounds in foods: their role in the prevention of cardiovascular disease and cancer. *American Journal of Medicine* 113, 71S–88S.

Kumar, S. and Andy, A. (2012) Health promoting bioactive chemicals from *Brassica*. *International Food Research Journal* 19, 141–152.

Lam, L.K.T. and Hasegawa, S. (1989) Inhibition of benzo[a]pyrene-induced forestomach neoplasia in mice by citrus limonoids. *Nutrition and Cancer* 12, 43–47.

Lam, L.K.T., Zhang, J. and Hasegawa, S. (1994) Citrus limonoid reduction of chemically induced tumorigenesis. *Food Technology* 48, 104–108.

Langenheim, J.H. (1994) Higher plant terpenoids: a phytocentric overview of their ecological roles. *Journal of Chemical Ecology* 20, 1223–1280.

Lee, S., Peterson, C.J. and Coats, J.R. (2003) Fumigation toxicity of monoterpenoids to several stored product insects. *Journal of Stored Products Research* 39, 77–85.

Lopez, J.L. (2007) Use of *Opuntia* Cactus as a Hypoglycemic. *Nutrition Bytes* 12, 2.

Mense, S.M., Hei, T.K. and Ganju, R.K. (2008) Phytoestrogens and breast cancer prevention: Possible mechanisms of action. *Environmental Health Perspectives* 116, 426–433.

Mert-Turk, F. (2006) Saponins versus plant fungal pathogens. *Journal of Cell and Molecular Biology* 5, 13–17.

Messina, M., Erdman, Jr, J. and Setchell, K.D.R. (2004) Introduction to and perspectives from the Fifth International Symposium on the role of soy in preventing and treating chronic disease. *Journal of Nutrition* 134, 1205S–1206S.

Miller, N.J. and Larrea, M.B.R. (2002) Flavonoids and other plant phenols in the diet: their significance as antioxidants. *Journal of Nutritional and Environmental Medicine* 12, 39–51.

Milner, J.A. (2008) Nutrition and cancer: essential elements for a roadmap. *Cancer Letters* 269, 189–198.

Morabito, N., Crisafulli, A., Vergara, C., Gaudio, A., Lasco, A., Frisina, N., D'Anna, R., Corrado, F., Pizzoleo, M.A., Cincotta, M., Altavilla, D., Ientile, R. and Squadrito, F. (2002) Effects of genistein and hormone-replacement therapy on bone loss in early postmenopausal women: a randomized double-blind placebo-controlled study. *Journal of Bone and Mineral Research* 17, 1904–1912.

Morrissey, J.P. and Osbourn, A.E. (1999) Fungal resistance to plant antibiotics as a mechanism of pathogenesis. *Microbiology and Molecular Biology Reviews* 63, 708–724.

Narasinga Rao, B.S. (1988) Dietary fibre in Indian diets and its nutritional significance. *Bulletin of Nutrition Foundation of India* 9, 1–5.

Narasinga Rao, B.S. (2003) Bioactive phytochemicals in Indian foods and their potential in health promotion and disease prevention. *Asia Pacific Journal of Clinical Nutrition* 12, 9–22.

Nichenametla, S.N., Taruscio, T.G., Barney, D.L. and Exon, J.H. (2006) A review of the effects and mechanism of polyphenolics in cancer. *Critical Reviews in Food Science and Nutrition* 46, 161–183.

Osbourn, A.E. (1996) Preformed antimicrobial compounds and plant defense against fungal attack. *The Plant Cell* 8, 1821–1831.

Ozaki, Y., Ayano, S., Inaba, N., Miyake, M., Berhow, M.A. and Hasegawa, S. (1995) Limonoid glucosides in fruit, juice and processing by-products of Satsuma Mandarin (*Citrus unshiu* Marcov.). *Journal of Food Science* 60, 186–189.

Packer, L. and Weber, S.U. (2001) *The Role of Vitamin E in the Emerging Field of Nutraceuticals*. Marcel Dekker, New York, pp. 27–43.

Paduch, R., Kandefer-Szerszen, M., Trytek, M. and Fiedurek, J. (2007) Terpenes: substances useful in human healthcare. *Archivum Immunologiae et Therapia Experimentalis* 55, 315–327.

Paiva, S. and Russell, R. (1999) Beta carotene and other carotenoids as antioxidants. *Journal of the American College of Nutrition* 18, 426–433.

Pandey, K.B. and Rizvi, S.I. (2009) Current understanding of dietary polyphenols and their role in health and disease. *Current Nutrition and Food Science* 5, 249–263.

Pandey, M., Verma, R.K. and Saraf, S.A. (2010) Nutraceuticals: new era of medicine and health. *Asian Journal of Pharmaceutical and Clinical Research* 3, 11–15.

Patel, R.P., Boersma, B.J., Crawford, J.H., Hogg, N., Kirk, M., Kalyanaraman, B., Parks, D.A., Barnes, S. and Darley-Usmar, V. (2001) Antioxidant mechanisms of isoflavones in lipid systems: paradoxical effects of peroxyl radical scavenging. *Free Radical Biology and Medicine* 31, 1570–1581.

Prakash, D. and Gupta, C. (2011) Role of Phytoestrogens as Nutraceuticals in Human health. *Pharmacologyonline* 1, 510–523.

Prakash, D. and Gupta, K.R. (2009) The antioxidant phytochemicals of nutraceutical importance. *The Open Nutraceuticals Journal* 2, 20–35.

Prakash, D. and Kumar, N. (2011) Cost Effective Natural Antioxidants. In: Watson, R.R., Gerald, J.K. and Preedy, V.R. (eds) *Nutrients, Dietary Supplements and Nutriceuticals*. Humana Press, Springer, USA, pp. 163–188.

Prakash, D., Dhakarey, R. and Mishra, A. (2004) Carotenoids: the phytochemicals of nutraceutical importance. *Indian Journal of Agricultural Biochemistry* 17, 1–8.

Prakash, D., Singh, B.N. and Upadhyay, G. (2007a) Antioxidant and free radical scavenging activities of phenols from onion (*Allium cepa*). *Food Chemistry* 102, 1389–1393.

Prakash, D., Upadhyay, G., Singh, B.N. and Singh, H.B. (2007b) Antioxidant and free radical-scavenging activities of seeds and agri-wastes of some varieties of soybean (*Glycine max*). *Food Chemistry* 104, 783–790.

Prakash, D., Upadhyay, G., Pushpangadan, P. and Gupta, C. (2011) Antioxidant and Free Radical Scavenging Activities of Some Fruits. *Journal of Complementary and Integrative Medicine* 8, 1–19.

Price, K.R., Johnson, I.T. and Fenwick, G.R. (1987) The chemistry and biological significance of saponins in food and feeding stufs. *Critical Reviews in Food Science and Nutrition* 26, 127–133.

Ribaya-Mercado, J.D. and Blumberg, J.B. (2004) Lutein and zeaxanthin and their potential roles in disease prevention. *Journal of the American College of Nutrition* 23, 567S–587S.

Sakamoto, T., Horiguchi, H. and Oguma, E. (2010) Effects of diverse dietary phytoestrogens on cell growth, cell cycle and apoptosis in estrogen-receptor-positive breast cancer cells. *Journal of Nutritional Biochemistry* 21, 856–864.

Scalbert, A., Manach, C., Morand, C. and Remesy, C. (2005) Dietary polyphenols and the prevention of diseases. *Critical Reviews in Food Science and Nutrition* 45, 287–306.

Schmidgall, J., Schnetz, E. and Hensel, A. (2000) Evidence for bioadhesive effects of polysaccharides and polysaccharide-containing herbs in an *ex vivo* bio adhesion assay on buccal membranes. *Planta Medica* 66, 48–53.

Schnecman, B.O. (1989) Dietary fibre: scientific status summary. *Food Technology* 43, 133–139.

Schwinn, K.E. and Davies, K.M. (2004) Flavonoids. In: Davies, K.M. (ed.) *Plant Pigments and their Manipulation.* Annual Plant Reviews, Vol. 14. Blackwell Publishing, Oxford, UK, pp. 92–149.

Sharma, G., Singh, R.P., Chan, D.C.F. and Agarwal, R. (2003) Silibinin induces growth inhibition and apoptotic cell death in human lung carcinoma cells. *Anticancer Research* 23, 2649–2655.

Sharma, R. (2009) Nutraceuticals and nutraceutical supplementation criteria in cancer: a literature survey. *The Open Nutraceuticals Journal* 2, 92–106.

Singh, B.N., Singh, B.R., Singh, R.L., Prakash, D., Singh, D.P., Sharma, B.K., Upadhyay, G. and Singh, H.B. (2009) Polyphenolics from various extracts/fractions of red onion (*Allium cepa*) peel with potential antioxidant and antimutagenic activities. *Food and Chemical Toxicology* 47, 1161–1167.

Singh, M., Pal, M. and Sharma, R.P. (1999) Biological activity of the labdane diterpenes. *Planta Medica* 65, 2–8.

Singh, R.P., Sharma, G., Sivanandhan, D., Agarwal, C. and Agarwal, R. (2003) Suppression of advanced human prostate tumor growth in athymic mice by silibinin feeding is associated with reduced cell proliferation, increased apoptosis and inhibition of angiogenesis. *Cancer Epidemiology, Biomarkers and Prevention* 12, 933–939.

Singh, R.P., Mallikarjuna, G.U., Sharma, G., Sivanandhan, D., Chan, D.C.F., Agarwal, C. and Agarwal, R. (2004) Silibinin enhances therapeutic efficacy and reduces toxicity of doxorubicin in lung cancer. *Clinical Cancer Research* 10, 8641–8647.

Singh, R.P., Raina, K., Sharma, G. and Agarwal, R. (2008a) Silibinin inhibits established prostate tumor growth, progression, invasion, and metastasis and suppresses tumor angiogenesis and Epithelial-Mesenchymal Transition in Transgenic Adenocarcinoma of the Mouse Prostate Model Mice. *Clinical Cancer Research* 14, 7773–7780.

Singh, R.P., Tyagi, A., Sharma, G., Mohan, S. and Agarwal, R. (2008b) Oral silibinin inhibits *in vivo* human bladder tumor xenograft growth involving down regulation of survivin. *Clinical Cancer Research* 14, 300–308.

Stahl, W. (2005) Bioactivity and protective effects of natural carotenoids. *Biochimica et Biophysica Acta* 1740, 101–107.

Tholl, D. (2006) Terpene synthases and the regulation, diversity and biological roles of terpene metabolism. *Current Opinion Plant Biology* 9, 297–304.

Traka, M. and Mithen, R. (2009) Glucosinolates, isothiocyanates and human health. *Phytochemistry Reviews* 8, 269–282.

Vig, A.P., Rampal, G., Singh, T.S. and Arora, S. (2009) Bioprotective effects of glucosinolates – a review. *LWT Food Science and Technology* 42, 1561–1572.

von Bergmann, K., Sudhop, T. and Lutjohann, D. (2005) Cholesterol and plant sterol absorption: recent insights. *American Journal of Cardiology* 96, 10D–14D.

Wagner, K.H. and Elmadfa, I. (2003) Biological relevance of terpenoids. Overview focusing on mono-, di-, and tetraterpenes. *Annals of Nutrition and Metabolism* 47, 95–106.

Willcox, J.K., Ash, S.L. and Catignani, G.L. (2004) Antioxidants and prevention of chronic diseases. *Critical Reviews in Food Science and Nutrition* 44, 275–295.

Zeng, H., Chen, Q. and Zhao, B. (2004) Genistein ameliorates β-amyloid peptide (25-35)-induced hippocampal neuronal apoptosis. *Free Radical Biology and Medicine* 36, 180–188.

Zhang, S.M., Cook, N.R., Albert, C.M., Gaziano, J.M., Buring, J.E. and Manson, J.E. (2008) Effect of combined folic acid, vitamin B6, and vitamin B12 on cancer risk in women: a randomized trial. *Journal of American Medical Association* 300, 2012–2021.

2 Use of Phytochemicals as Adjuncts to Conventional Therapies for Chronic Kidney Disease

Ken Wojcikowski[1] and Glenda C. Gobe[2]*

[1]*Southern Cross University, Lismore, Australia;* [2]*Centre for Kidney Disease Research, The University of Queensland, Brisbane, Australia*

2.1 Introduction

Chronic kidney disease (CKD) is one of the most common chronic diseases in developed and developing nations, and is often the precursor to end-stage kidney disease (ESKD), the stage at which patients rely on kidney replacement therapies, such as dialysis or renal transplant, to survive (Levey *et al.*, 2011). The National Kidney Foundation's Kidney Disease Outcomes Quality Initiative (K/DOQI) defines CKD as either kidney damage or a decreased glomerular filtration rate (GFR) of less than 60 ml min^{-1} 1.73 m^{-2} for 3 or more months (Eknoyan *et al.*, 2001). However, the term CKD represents a continuum of chronic change in the kidney (Pannu *et al.*, 2011). This continuum is characterized, clinically, by five stages of CKD that depend on GFR and extent of kidney injury (Table 2.1). Many people develop CKD after incomplete recovery from single or multiple episodes of acute kidney injury caused by, for example, ischaemia-reperfusion, nephrotoxic drugs, environmental toxins, radiation or other cancer therapy, or kidney stones (Venkatachalam *et al.*, 2010). The cause and outcome of the initial CKD are often overwhelmed by non-specific tissue changes of chronic progression, including chronic inflammation, tubulo-interstitial fibrosis, glomerulosclerosis and vascular loss. The development of CKD and ESKD is exacerbated by an increasingly ageing population in developed countries, with ageing being one of the key independent risk factors for CKD (Karamouzis *et al.*, 2008).

CKD and ESKD populations are vulnerable to complications and co-morbid conditions that adversely affect quality of life and longevity (Soni *et al.*, 2010). There is an upward trend of obesity in Western societies, diabetes and the metabolic syndrome, hypertension, anaemia and its complications, dialysis and its complications, uraemia-related diseases, and high mortality rates from cardiovascular disease. Since the introduction of angiotensin converting enzyme inhibitors (ACEi) in the early 1980s and angiotensin receptor blockers (ARB) in the mid-1990s as antihypertensive and antifibrotic therapies, pharmacologic blockade of the renin-angiotensin-aldosterone system has become one of the most widespread therapeutic approaches for the management of CKD (Hoogwerf, 2010). However, the incidence of CKD continues to increase. There is a need to

* E-mail: g.gobe@uq.edu.au

Table 2.1. Classification and description of the different stages of chronic kidney disease.

Stage	eGFR[a]	Description
1	90 ml min^{-1}	Normal renal function but abnormal urine findings, or structural abnormalities, or a genetic trait indicating kidney disease
2	60–89 ml min^{-1}	Mildly reduced renal function and other findings (as for stage 1) indicate kidney disease
3A	45–59 ml min^{-1}	Moderately reduced kidney function
3B	30–44 ml min^{-1}	
4	15–29 ml min^{-1}	Severely reduced kidney function
5	<15 ml min^{-1} or on dialysis	Very severe, or end-stage kidney failure (sometimes called established renal failure)

[a]Measured using the MDRD formula (Twomey and Reynolds, 2006). MDRD = Modification of Diet in Renal Disease. All estimated glomerular filtration rate (eGFR) values are normalized to an average surface area of 1.73 m^2.

identify new therapies that will decrease its incidence and slow its progression (Fig. 2.1). Complementary and alternative herbal medicines, and individual phytochemicals derived from them, may provide the key to increasing the benefits of conventional medicines. Equally important is the fact that they may have their own inherent toxicity, or have interactions with drugs prescribed for kidney disease (Wojcikowski *et al.*, 2004a). In this chapter, we report on the potential benefits of the use of phytochemicals for CKD, especially referring to their use as adjuncts to conventional therapies. Some useful medicinal herbs are also discussed.

2.2 CKD as an Inflammatory Disease

CKD and ESKD are often associated with chronic low-grade systemic inflammation, characterized by increased expression of inflammatory markers (Silverstein, 2009). For example, CKD patients have increased serum levels of C-reactive protein, a well-recognized biomarker of inflammation. In some cases, the pro-inflammatory state that exists in patients with CKD and ESKD has developed after an initial normal inflammatory response fails to abate. In other cases, the low-grade systemic inflammation may be a result of contributors to very low birth weight and continuing poor nutrition (Hughson *et al.*, 2008). In either case, the prolonged inflammatory response contributes to CKD progression. Traditional mediators of chronic inflammation in CKD and ESKD patients include hyperglycaemia and advanced glycation end-products (AGE) in diabetes, and atherosclerosis and lipopolysaccharides in hypertension and cardiovascular disease. Molecular mediators include pro-inflammatory cytokines and growth factors such as tumour necrosis factor-α (TNFα) and some of the interleukins (IL) such as IL-1β, IL-6 and IL-13 (Asmis *et al.*, 2006). These mediators interact with vascular endothelial cells and intrinsic renal cells, but they also accumulate in ESKD patients because of decreased renal clearance. Increased oxidative stress also contributes to the progression of CKD (Small *et al.*, 2012), and the close link between inflammatory cells and oxidative stress provides one of the most realistic causative mechanisms, and targets for treatment, for CKD.

In diabetes associated with the metabolic syndrome, the origin of the inflammatory response may, in fact, be adipose tissue (Calle and Fernandez, 2012). Typically, visceral adipose tissue is often found in close vicinity to the kidneys. Apart from energy storage, adipocytes also produce cytokines including IL-1β, IL-6, TNFα, monocyte chemoattractant protein-1 (MCP1) and adipokines such as leptin and adiponectin. Macrophages found in adipose tissue also produce pro-inflammatory cytokines and produce reactive oxygen species (ROS). In addition, other immune cells such as T-cells, natural killer cells, mast cells and eosinophils are known to invade adipose tissue (Silverstein, 2009). The interaction and communication between

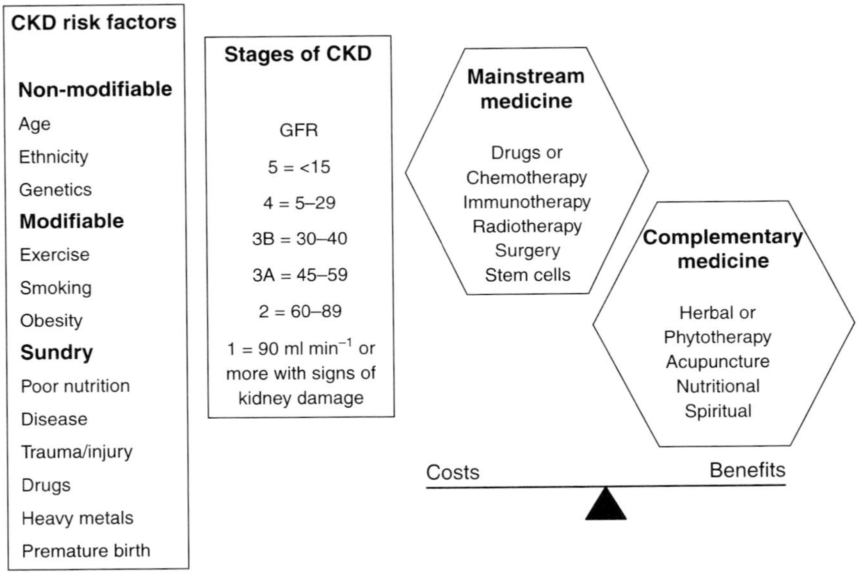

Fig. 2.1. Finding a balance for treating chronic kidney disease. Chronic kidney disease (CKD) has many non-modifiable, modifiable and other contributing factors throughout our lives. CKD is then graded in stages from 1 to 5, according to an estimated glomerular filtration rate (GFR), which is calculated in ml min^{-1} 1.73 m^{-2} average surface area. Examples are given of types of conventional and complementary therapies for disease. CKD is treated traditionally with conventional medicines, but a balance between benefit and cost should be sought for inclusion of alternative therapies, like phytotherapies, to complement the conventional therapies used for CKD.

immune cells and adjacent adipocytes may amplify inflammation in obesity.

Another important link with inflammation as a mechanism for CKD is the increase in serum uric acid levels (hyperuricaemia) seen when kidneys fail (Badve *et al.*, 2011). Hyperuricaemia has harmful systemic effects, and it is also closely linked with the development of cardiovascular disease. Retention of uremic toxins promotes inflammation and oxidative stress by priming acute inflammatory cells such as neutrophils and activating pro-inflammatory and pro-fibrotic cytokines. Additionally, uric acid synthesis can promote oxidative stress directly through the activity of xanthine oxidoreductase (Grootveld *et al.*, 1987).

2.3 Oxidative Stress in CKD

Chronic diseases of the kidney possess various commonalities, which can be linked through pathways controlled by oxidative stress (Vlassara *et al.*, 2009a, b; Small *et al.*, 2012). As mentioned previously, the links with oxidative stress as an endogenous driver of CKD are obvious when one recognizes the close association between oxidative stress and inflammation in the kidney. Many studies investigating antioxidant treatments in CKD patients show a reduction in oxidative stress and many show improved renal function (Small *et al.*, 2012). Therapies such as exercise training for CKD patients has improved inflammation and oxidative stress biomarkers, muscle strength and function, and decreased blood pressure (Howden *et al.*, 2012). Kakuda *et al.* (2012) used a statin (pitavastatin) in healthy volunteers to investigate oxidative stress parameters. Their data indicate pitavastatin affects renal outcomes in both lipid status-dependent and- independent manners, by reducing oxidative stress parameters.

The loss of renal mass during CKD can have other effects on oxidative stress development, for example loss of a vital source of

L-arginine, a precursor for nitric oxide (NO). NO activity is vital for regulation of the vascular endothelium (Mendoza *et al.*, 2008). Decreased endothelial function has the potential to exacerbate CKD development because of vascular loss. This complex picture of cause and effect is common in CKD pathogenesis. There is now also a novel link in the pathogenesis of oxidative stress-induced CKD through a functional mitochondrial angiotensin system (Abadir *et al.*, 2011). Angiotensin type II receptors were co-localized with angiotensin on the inner mitochondrial membrane of human mononuclear cells and mouse renal tubular cells. This system was found to modulate mitochondrial NO production and respiration.

The benefits of phytochemicals as antioxidants for CKD patients have received limited attention. Phytochemicals are often proven antioxidants that can scavenge excessive, damaging, free radicals arising from normal metabolic processes (Wojcikowski *et al.*, 2007). They may also have indirect antioxidant effects through induction of endogenous protective enzymes. We believe that more and improved studies are needed to develop evidence-based guidelines for use of phytochemicals in individuals with CKD, either as adjunct or alternative treatments.

2.4 Modulation of CKD With Phytochemicals

The identification of natural substances that can prevent or delay the development of CKD, and are well-tolerated as a therapy, is of fundamental importance to developing treatment regimens for this disease. In other diseases with close links with CKD, such as type 2 diabetes mellitus, phytochemicals are known to have a curative effect (Leiherer *et al.*, 2012). Plants have great therapeutic potential and their use in preclinical and clinical trials is increasing. The following sections first highlight some promising experimental and clinical data on use of phytochemicals for CKD and closely related chronic diseases such as diabetes and cardiovascular disease, then provide data on some developing aspects of phytotherapeuticals.

2.4.1 Curcumin

Curcumin is the principal curcuminoid of the Indian curry spice turmeric, and is widely used as an anti-inflammatory and antioxidant renoprotective agent, with most work showing potent antioxidant properties (Osawa, 2007). As well as augmenting free radical scavenging activity, curcumin also induced antioxidative enzymes and detoxification enzymes. The major metabolite tetrahydrocurcumin appears to have highest antioxidant activity. Curcumin and its metabolites also improved creatinine and urea clearance in instances of chronic renal pathologies and protected against chronic renal allograft nephropathy, a potential cause of graft loss in the long term.

A considerable amount of preclinical research has been carried out by Ghosh *et al.* (2009, 2010, 2012), in the main using the 5/6th nephrectomy model of CKD in rats, an experimental model of chronic renal failure. In 2009, they investigated the effectiveness of curcumin (75 mg kg^{-1}) versus the ACEi enalapril (10 mg kg^{-1}) in reducing development of CKD (Ghosh *et al.*, 2009). They analysed expression of the inflammatory agents TNFα and nuclear factor-κB (NF-κB) in comparison with the nephrectomized group without treatment. Renal dysfunction, for CKD, was measured by elevated blood urea nitrogen, plasma creatinine, proteinuria, segmental glomerulosclerosis, and tubular dilatation and atrophy. These parameters were significantly reduced by curcumin and enalapril treatment. The nephrectomized animals had significantly higher plasma and kidney TNFα, and NF-κB activation and macrophage infiltration in the kidney, changes that were effectively reduced by curcumin and enalapril treatment. In 2012, to better mimic the scenario for kidney disease in humans, they used a similar model but began curcumin and enalapril therapy when proteinuria was established (Ghosh *et al.*, 2012). They then investigated cyclooxygenase (COX) and phospholipase expression in the kidney, as well as the cytokines TNFα and IL-1β. COX, which has isoforms of COX1 and COX2, is the key enzyme of prostaglandin biosynthesis. They found that curcumin, by antagonizing the inflammatory cytokines TNFα and IL-1β, could significantly reduce both phospholipase 2 and COX1 and 2. Their work in 2010 also looked at the effects of curcumin treatment on attenuating left ventricular hypertrophy, which is frequent in patients with ESKD following chronic renal failure (Ghosh *et al.*, 2010). The nephrectomized rats showed a significant hypertrophic response and increased diameter of the inferior vena cava at inspiration, which was inhibited by treatment with curcumin or enalapril. The signalling pathway molecules critically involved in the hypertrophic response (glycogen synthase kinase-3β phosphorylation, β-catenin expression, calcineurin and several kinases) were all reduced by curcumin and enalapril. One of the frustrating aspects of these reports was that the authors did not report on therapy with enalapril plus curcumin, a regimen more likely to be used by nephrologists in the clinic, than curcumin as a replacement for enalapril. However, considering the safety of curcumin, the studies should facilitate future preclinical and clinical trials.

Using the blood from CKD patients before dialysis, a recent *ex vivo* investigation determined the effect of curcumin, bovine colostrum and fish oil on inflammatory cytokine and tissue factor procoagulant activity (Shing *et al.*, 2011). The peripheral blood mononuclear cells (PBMCs) from these patients, and age- and sex-matched healthy controls, were cultured alone and with low and high doses of the nutritional compounds, with and without lipopolysaccharide. The introduction of lipopolysaccharide represents a model of endotoxin injury, linked to worsening outcome of CKD patients on dialysis. Curcumin decreased secretion of IL-6 and IL-1β and it was more effective than colostrum at decreasing the procoagulant activity of PBMCs in the CKD and control groups. However, the production of pro-inflammatory C-reactive protein, MCP-1, IL-6, and IL-1β by PBMCs was inhibited most by fish oil in the CKD group.

2.4.2 Resveratrol

Resveratrol is a polyphenol phytoalexin that, among other sources, occurs in grapes (Neves *et al.*, 2012). Interest in resveratrol has increased because of its reported antioxidant, anti-inflammatory and 'anti-ageing' benefits. The compound has poor bioavailability as it is rapidly metabolized and excreted, and so the drug delivery systems have to stabilize resveratrol and enhance its bioavailability. The potential benefits of resveratrol delivery on increased oxidative stress and endothelial dysfunction (atherosclerosis) seen in CKD patients was discussed by Caimi *et al.* (2004). Atherosclerosis development is accelerated in chronic renal failure and is the major cause of death in this clinical condition, and this review promotes the idea of red wine or resveratrol usage to decrease the progression of atherosclerosis.

A preclinical trial investigating the renoprotective effect of resveratrol used the 5/6th nephrectomy model of CKD in rats (Chander and Chopra, 2006). Resveratrol (5 mg kg^{-1}) was administered daily by gavage for 12 weeks, with and without nitro-L-arginine methyl ester (L-NAME) (10 mg kg^{-1}), an agent that alters NO synthesis. Proteinuria, hypertension, renal function, glomerulosclerosis and urinary excretion of NO metabolites were analysed. Treatment of animals with resveratrol significantly attenuated the increase in systolic blood pressure, preserved the normal renal function, reduced the urinary protein excretion, increased the urinary excretion of NO metabolites and prevented glomerulosclerosis. Co-administration of animals with L-nitro-arginine methyl ester along with resveratrol prevented the protection observed with resveratrol. The findings suggest that resveratrol exerts its renoprotective effect against CKD through a NO pathway. Damage to the renal vascular component in diabetic nephropathy is one of the main causes of ESKD. Resveratrol was also found to attenuate high glucose-induced endothelial cell apoptosis in diabetic nephropathy, and oxidative stress, with the outcome of reduced renal dysfunction in diabetic rats (Sharma *et al.*, 2006).

However, some resveratrol studies have been disappointing and further work will need to be performed to determine the optimal therapeutic window. For example, the streptozotocin-induced diabetic rat model of CKD was used to investigate changes in COX1 and COX2 mRNA and protein level, with and without resveratrol (Yar *et al.*, 2010). These prostaglandin-synthesizing enzymes, especially COX2, are induced in inflammatory disease such as diabetes mellitus. Resveratrol (10 mg kg^{-1}) was administered intraperitoneally for 4 weeks after the induction of chronic diabetes. It had no significant effect on COX1 and COX2 mRNA and protein levels, thus no significant anti-inflammatory action in CKD. Dey *et al.* (2009) found that resveratrol had a protective effect against indomethacin-induced gastric ulcers at a lower dose (2 mg kg^{-1}) and a contra-indicative effect at a higher dose (10 mg kg^{-1}) in mice. The lower dose maintained normal COX1 levels, allowed angiogenesis and aided healing. The higher dose reduced COX1, thereby significantly reducing prostaglandin E2 synthesis, a change that delayed healing of the ulcers.

2.4.3 Capsaicin

Skin disorders associated with CKD can markedly affect a patient's quality of life and are a considerable cause of morbidity (Kuypers, 2009). Uremic pruritus, which is frequently encountered in patients with CKD, is considered to be an inflammatory systemic disease rather than a local skin disorder. Murphy *et al.* (2009) also describe 'renal itch' as a localized or generalized itch where there is no primary skin disease and no systemic or psychological dysfunction that might cause pruritus, but they state that it does not arise from raised serum urea levels. Thus, the aetiology of renal itch is unclear. Dialysis does not reduce the pruritus. Whilst the definitive treatment for renal itch remains renal transplantation, other treatment options are being sought. There is established and emerging evidence to suggest that topical capsaicin cream may be effective for renal itch (Breneman *et al.*, 1992; Tarng *et al.*, 1996; Weisshaar *et al.*, 2003).

2.4.4 Quercetin

Interventions for controlling development and progression of CKD, such as controlling hypertension, are being investigated. In a study by Peng *et al.* (2012), naringenin (a flavanon), catechin (a flavanol), quercetin (a flavonol) and rutin (a flavonol rutinoside) were used in rats in a model of CKD. Results indicated quercetin to be the most effective therapeutic candidate with respect to CKD (improved serum creatinine, glomerular amyloidosis, collagen deposition, and expressions of TNFα, cleaved caspase-3 as a measure of apoptosis, transcription factors and serum insulin) and hypertension and cardiac pathologies (haematocrit and erythrocyte depletion in bone marrow, aortic calcification). However, quercetin only partially restored GFR and blood urea nitrogen, uric acid, albuminuria, serum cholesterol and triglyceride, and only partially ameliorated oxidative stress (malondialdehyde, superoxide dismutase as biomarkers). Quercetin was completely effective in ameliorating apoptosis (reduced caspase-3 cleavage), but only partially effective in suppressing pro-apoptotic proteins Bax and Bad, and restoring anti-apoptotic Bcl-2.

2.4.5 Genistein

Intracellular levels of the inflammatory cytokines TNFα, IL-6 and IL-10 in monocytes are indistinguishable between haemodialysis patients and healthy controls. However, monocytes from haemodialysis patients are selectively primed for enhanced TNFα secretion in response to lipopolysaccharides. The selective inhibition of monocyte TNFα production by the isoflavone genistein may explain an anti-inflammatory action of this phytochemical observed in experimental animals. Asmis *et al.* (2006) incubated whole blood and isolated mononuclear cells from haemodialysis patients and healthy control subjects with genistein and stimulated with lipopolysaccharides. These induced a robust TNFα response in both whole blood and monocytes, and an increase in IL-6 in whole blood. Genistein did not inhibit IL-6 formation and did not alter basal TNFα, but did block lipopolysaccharide-induced TNFα formation, ultimately resulting in lower TNFα levels than controls.

2.4.6 Combination phytotherapies

It is unlikely, given the complexity of CKD pathogenesis, that a single phytochemical will be effective in slowing or stopping the disease progression, and multiple phytochemicals may be needed. There is growing evidence that dietary phytoestrogens have a beneficial role in CKD. Velasquez and Bhathena (2001) reviewed recent findings from dietary intervention using combination phytochemicals. This group had found that consumption of soy-based protein rich in isoflavones and flaxseed rich in lignans retarded the development and progression of CKD (Velasquez *et al.*, 2003). Wang *et al.* (2012) compared more than 500 Stage 3 CKD patients with primary glomerulonephritis treated with Traditional Chinese Medicine (TCM), the ACEi benazepril, or TCM combined with benazepril. The TCM contained one of four treatments, depending on the patient's needs: (i) replenishing qi and blood decoction with addition and subtraction based on the classic Dang Gui Bu Xue decoction: *Astragalus membranaceus*, *Pesudostellariae* spp., *Angelica sinensis*, Fructus ligustri lucidi; (ii) promoting blood flow decoction for the treatment of blood stasis in the kidney with addition and subtraction based on the classic Xia Yu Xue decoction: *Salvia miltiorrhizae*, *Semen persicae*, *Herba centellae*, Rhizoma rhei; (iii) expel wind-evil and remove wetness decoction for the treatment of wind-dampness interfering in the kidney with addition and subtraction based on the classic Fang Ji Huang Qi decoction: *Tripterygium wilfordii*, *Stephania tetrandrae*, *Euonymi ramulus*; and (iv) clearing heat and dissipating dampness decoction for the treatment of patterns of endoretention of damp heat with addition and subtraction based on the classic Tu Fu Ling decoction: *Polygoni cuspidate*, *Coptidis*, *Smilacis glabrae* and *Serissa serissoides*. Patients were followed up for 24 weeks and GFR, serum creatinine, proteinuria and haemoglobin monitored. The primary endpoint was the time to 50% of the increased serum creatinine, ESKD, or death. TCM, with and without benazapril, maintained or improved renal function, whilst benazapril by itself only reduced proteinuria. Side effects in the TCM group (dry cough, hyperkalaemia) were the lowest in these groups. This suggests that benazepril combined with TCM may have a synergistic advantage in the treatment of CKD.

2.5 Other Medicinal Herbs in Development as Therapies for CKD

2.5.1 *Astragalus membranaceus* and *Angelica sinensis*

A plant extract with promising antifibrotic activity, derived from the roots of *Astragalus membranaceus* and *Angelica sinensis* (A&A), has been identified (Wojcikowski *et al.*, 2010). These plants have been used in TCM as potent renal and cardiovascular therapies for hundreds of years. A model of renal fibrosis (unilateral ureteral obstruction/UUO) was used in rats with aqueous-ethanol extract (A&A), or the ACEi enalapril (in drinking water), or with a combination of both treatments. Enalapril or A&A, individually, were antifibrotic. The pro-fibrotic growth factor, transforming growth factor-beta1 (TGFβ1) and pro-inflammatory TNFα were reduced, along with myofibroblast activation, collagen deposition, macrophage accumulation and tubular cell apoptosis. Importantly, the combination of the two treatments was significantly more effective than enalapril alone in reducing renal fibrosis. Other studies investigating the mechanisms by which A&A exert their antifibrotic effects have suggested A&A may reduce pro-fibrotic growth factors such as TGFβ1 (Zhao *et al.*, 2004), osteopontin (Zhao *et al.*, 2002), ROS and c-Jun N-terminal kinase activity (Wang *et al.*, 2004). A&A also appears to enhance microcirculation of the kidneys by decreasing arginine vasopressin, enhancing the expression of the vasoactivator NO, and enhancing the expression and activity of NO synthase (Meng *et al.*, 2007).

2.5.2 Rhubarb (*Rheum* spp.)

There is some scientific evidence to support the traditional Chinese use of the roots of rhubarb for treatment of disorders involving inflammation, hypertension, hyperlipidaemia and renal failure (Wojcikowski *et al.*, 2004b, 2006). A series of experiments was carried out to test the efficacy of rhubarb extract on rats with adenine-induced renal failure. The aqueous extract of rhubarb, administered orally to rats after the induction of renal failure, lowered blood urea nitrogen and serum creatinine in a dose-dependent manner when compared to the controls. Zhang and El Nahas (1996) tested the efficacy of rhubarb in a subtotal nephrectomy model of CKD with or without rhubarb. Rats consuming rhubarb extract in their drinking water (750 mg kg^{-1} day^{-1}) had significantly less proteinuria and less glomerulosclerosis than rats without rhubarb extract. A few, basically flawed, clinical trials have been performed to determine the efficacy of rhubarb. The results can only be used to support further clinical trials. In a small study of 30 patients with moderate to severe CKD, the combination of Captopril (25 mg, three times daily) and rhubarb extract (6–9 g day^{-1}) induced a non-significant improvement in renal function by normalizing blood urea nitrogen and serum creatinine (Zhang *et al.*, 1990). An uncontrolled observational study of 50 inpatients suffering from CKD was also carried out (Kang *et al.*, 1993). The therapy was complex and included small doses of diuretics and hypertension pills for 3 months, but the main therapy was a decoction of 10 g rhubarb, 20 g dandelion (*Taraxacum officinale*) and 30 g oyster shell, administered orally or by retention enema. In the 1–3 year follow-up, blood urea nitrogen of 37 patients dropped from an average of 35 to 17.56 mmol l^{-1}. The 13 remaining cases needed to be switched to full conventional therapy, seven with successful dialysis treatment and six died from complications of renal failure. Diarrhoea was also a complication of the therapy.

2.5.3 *Salvia miltiorrhiza* root and magnesium lithospermate B

Yokozawa *et al.* (1989) isolated and identified some of the active compounds from the root of *Salvia miltiorrhiza*. The compound magnesium lithospermate B (MLB), a tetramer of caffeic acid, was found to be the most effective constituent, the action at a dose of 20 mg kg^{-1} corresponding to that of the aqueous whole root extract at a dose of 300 mg kg^{-1}. A single i.p. dose administered to rats with adenine-induced renal failure significantly increased the GFR in a dose-dependent manner. The same researchers had previously found that root extracts of *S. miltiorrhizae* (100 mg kg^{-1} day^{-1}) in rats with adenine-induced renal failure increased GFR by 50% and renal blood flow by 40% when compared to the rats with no treatment. The mechanism by which MLB acts in renal failure may include protection from oxidant injury. Restoration of superoxide dismutase and catalase activities, and inhibition of ROS, was found with MLB treatment.

2.5.4 Polyherb Sairei-to

Polyherb Sairei-to (ST) is a combination of 12 herbs that has been traditionally been used in Japan for renal diseases. ST ameliorated renal damage and reduced urinary N-acetyl-beta-D-glucosaminidase and protein excretion in an animal model of gentamicin nephrotoxicity in which rats were fed with or without a diet containing 2.5% ST. These benefits may have been due to the antioxidant actions of ST, given that gentamicin exerts renal tubular toxicity via enhanced generation of superoxide anion and the hydroxyl radical. ST has similarly proved to be beneficial in a subtotal nephrectomy model of CKD, whereby animals administered 2.5% ST in their food had lower blood pressure, reduced renal damage, decreased protein excretion and greater efferent arteriolar dilation when compared to the untreated rats. However, Satoh *et al.* (1995) found no benefit of ST, alone or in combination with enalapril, on survival following subtotal renal ablation in male Wistar rats.

2.6 Oxidation – Molecular Target for Phytochemicals

The use of phytochemicals as a source of antioxidants to combat oxidation warrants attention. Wojcikowsji *et al.* (2007) recently compared the *in vitro* antioxidant capacity of 55 medicinal herbs, and prioritized them for preclinical studies of the value of herbal therapies in the treatment of renal disorders. They used the oxygen radical absorbance capacity (ORAC) method and a sequential multisolvent extraction process. The herbs were selected for their traditional use in kidney or urinary system disorders, or because they have attracted the attention of recent investigations into renal pathologies. Twelve of the 55 herbs had ORAC levels comparable to plants known to have high antioxidant properties. The highest radical-scavenging activity was found in *Olea europaea* (olive leaf), *Cimicifuga racemosa* (black cohosh), *Rheum palmatum* (rhubarb), *Glycyrrhiza glabra* (liquorice) and *Scutellaria lateriflora* (Virginia skullcap). These authors concluded that the antioxidant capacity of many of the herbs studied may, at least in part, be responsible for their reputation as being protective of organs of the urinary system.

In a recent review (Firuzi *et al.*, 2011), the efficiency of antioxidants in preventing and treating various human diseases was reported. Apart from a few therapies, including some phytochemicals, they record little acceptance of antioxidants for clinical use. Ones that have been used are: edaravone (for ischaemic stroke in Japan); N-acetylcysteine (for acetaminophen toxicity); alfa-lipoic acid (for diabetic neuropathy); and some flavonoids, such as micronized purified flavonoid fraction (diosmin and hesperidin) and oxerutins (for chronic venous insufficiency) as well as baicalein and catechins (for osteoarthritis). One problem they discuss is the reliance of clinical trials on antioxidant vitamin supplements, and the disappointing outcomes from clinical trials. They indicate a need for trials of more disease-specific, target-directed, highly bioavailable antioxidants.

Diabetes mellitus is characterized by hyperglycaemia, lipidaemia and oxidative stress and predisposes affected individuals to long-term complications afflicting the eyes, skin, kidneys, nerves and blood vessels (Elosta *et al.*, 2012). Increased protein glycation and the subsequent build-up of tissue AGE contribute towards the pathogenesis of diabetic complications. Glycation-derived free radicals can damage proteins, lipids and nucleic acids and contribute towards oxidative stress in diabetes. There is interest in compounds with antiglycation activity as they may offer therapeutic potential in delaying or preventing the onset of diabetic complications. Data for 42 plants/constituents studied for antiglycation activity was presented, and some commonly used medicinal plants that possess antiglycation activity are described including their active ingredients, mechanism of action and therapeutic potential.

2.7 The Growing Problem of Anaemia in CKD Patients

CKD is closely linked with excessive cardiovascular disease (CVD) and increased morbidity and mortality (Eknoyan, 2001). Anaemia is common among CKD patients and greatly contributes to adverse patient outcome, with low haemoglobin levels in such patients increasing risk for progression of CKD and associated CVD. Clinical trials of anaemia treatment with erythropoiesis stimulating agents (ESA), such as recombinant human erythropoietin and its analogues, have demonstrated improved quality of life but have not demonstrated improved CKD and CVD outcome. In some trials, treatment with ESA was associated with worse outcomes, such as increased thrombosis, and in a few cases there appeared to be increased risk of cancer development. The use of ESA for treatment of anaemia in CKD is common but not without risks to patient health (Nangaku and Eckardt, 2006), and an alternative or adjunct therapy for anaemia in CKD is needed.

Some phytochemicals have been reported as having erythropoietin-like actions. Zheng *et al.* (2010) reported that danggui buxue tang (DBT), a Chinese medicinal decoction that is commonly used as a haematopoietic medicine to treating woman menopausal irregularity, contains *Astragalus membranaceus* and *Angelica sinensis*. Pharmacological results indicate that DBT can stimulate the production of erythropoietin in cultured kidney cells, via the hypoxia-inducible factor-1α pathway. Other similar reports are available.

Nakamoto *et al.* (2008) reported that juzentaiho-to (TJ-48), a mixture of extracts from ten medicinal herbs that has been used traditionally to treat patients with anaemia, anorexia, or fatigue, was effective in improving erythropoietin-resistant anaemia in ESKD patients. However, this effect was at least in part due to the anti-inflammatory effect of TJ-48 in patients on haemodialysis.

2.8 Nephrotoxicity of Phytochemicals

Nephrotoxicity is a potential complication of any human disease therapy. As for other drugs and chemicals, herbal extracts may exert renal toxicity through the inherent filtering and concentrating function of the kidney and also from localized renal toxicity. Thus, it is important to compile information regarding the potential toxicity of all medicinal herbs. The most well known of the nephrotoxic herbal treatments was one involving *Aristolochia* species (Wojcikowski *et al.*, 2004a). In 1991, a number of relatively young women presented with renal failure. All had attended the same private clinic in Belgium and had ingested a weight-reducing formula containing a mixture of several drugs and powdered extracts of Chinese herbs, including the herb *Stephania tetrandra*. It was later discovered that *S. tetrandra* had been inadvertently replaced by *Aristolochia fangchi* by the manufacturers of the weight-reducing formula. *Aristolochia fangchi* contains aristolochic acid, a plant alkaloid that is nephrotoxic and carcinogenic in humans and animals. This alkaloid was the most likely cause of the renal failure, and later development of urothelial-cell atypia and carcinoma (De Broe, 2012). Oxidative damage may have been involved in the toxicity, with the outcome of cell necrosis and apoptosis, and possibly renal cell mutations. Other examples of toxic herbal therapies are reviewed by Wojcikowski *et al.* (2004a, 2009). If herbal therapies are found to have some degree of toxicity, the risks must be weighed against the benefits and decisions made regarding their continued availability. Strict controls on the presence of adulterants within herbal medicines, labelling of dosages and contraindications and manufacturing techniques must be maintained to ensure the safety of those consuming herbal medicines. Adverse interactions between phytotherapies and conventional drugs have also been recorded. This is not unusual considering a high level of herbal use throughout the world. In a questionnaire of patient usage, 43% took garlic, 32% ginkgo biloba, 30% St John's wort, 18% ephedre, 12% echinacea and 10% aloe. Although many of these produced no unwanted outcome, St John's wort repeatedly led to lowering of levels of the immunosuppressive drug cyclosporine in transplant patients (Ernst, 2002). The success of organ transplantations was thus endangered and hospital costs increased.

2.9 Summary on the Benefits of Phytochemicals for CKD

The rise in CKD is associated with an increasingly ageing population, changing diet and lifestyles in developed and developing nations, increasing obesity and smoking, and increasing incidence of inflammation. Slowing the progression of CKD throughout the world is a major challenge for basic science and clinical researchers. Phytochemicals have some proven or suggested roles in reducing the incidence and progression of CKD. There is now an emphasis on regular assessment of renal health in patients over 50 years of age. This has led to the increased prescription of lipid-lowering drugs like the statins. Rigorous investigation, with clinical trials, may indicate phytotherapy supplements that have positive outcome in minimizing progression of CKD, through modulation of the causative mechanisms of the disease, like oxidative stress. Toxicity studies, preliminary *in vivo* preclinical studies and clinical trials are necessary to help decide on which combinations of phytotherapies are appropriate for use alone or with conventional therapies. Such experiments are currently being performed in our laboratories, and should be encouraged and published in the scientific community.

References

Abadir, P.M., Foster, D.B., Crow, M., Cooke, C.A., Rucker, J.J., Jain, A., Smith, B.J., Burks, T.N., Cohn, R.D., Fedarko, N.S., Carey, R.M., O'Rourke, B. and Walston, J.D. (2011) Identification and characterization of a functional mitochondrial angiotensin system. *Proceedings of the National Academy of Science USA* 108, 14849–14854.

Asmis, R., Stevens, J., Begley, J.G., Grimes, B., Van Zant, G. and Fanti, P. (2006) The isoflavone genistein inhibits LPS-stimulated TNFalpha, but not IL-6 expression in monocytes from hemodialysis patients and healthy subjects. *Clinical Nephrology* 65, 267–275.

Badve, S.V., Brown, F., Hawley, C.M., Johnson, D.W., Kanellis, J., Rangan, G.K. and Perkovic, V. (2011) Challenges of conducting a trial of uric-acid-lowering therapy in CKD. *Nature Review Nephrology* 7, 295–300.

Breneman, D.L., Cardone, J.S., Blumsack, R.F., Lather, R.M., Searle, E.A. and Pollack, V.E. (1992) Topical capsaicin for treatment of hemodialysis-related pruritus. *Journal of the American Academy of Dermatology* 26, 91–94.

Caimi, G., Carollo, C. and Lo Presti, R. (2004) Chronic renal failure: oxidative stress, endothelial dysfunction and wine. *Clinical Nephrology* 62, 331–335.

Calle, M.C. and Fernandez, M.L. (2012) Inflammation and type 2 diabetes. *Diabetes Metabolism* 38, 183–191.

Chander, V. and Chopra, K. (2006) Possible role of nitric oxide in the protective effect of resveratrol in 5/6th nephrectomized rats. *Journal of Surgical Research* 133, 129–135.

De Broe, M.E. (2012) Chinese herbs nephropathy and Balkan endemic nephropathy: toward a single entity, aristolochic acid nephropathy. *Kidney International* 81, 513–515.

Dey, A., Guha, P., Chattopadhyay, S. and Bandyopadhyay, S.K. (2009) Biphasic activity of resveratrol on indomethacin-induced gastric ulcers. *Biochemica Biophysica Research Communications* 381, 90–95.

Eknoyan, G. (2001) The importance of early treatment of the anaemia of chronic kidney disease. *Nephrology Dialysis Transplantation* 16(Suppl. 5), 45–49.

Eknoyan, G., Levin, N.W., Eschbach, J.W., Golper, T.A., Owen, W.F. Jr, Schwab, S. and Steinberg, E.P. (2001) Continuous quality improvement: DOQI becomes K/DOQI and is updated. National Kidney Foundation's Dialysis Outcomes Quality Initiative. *American Journal of Kidney Diseases* 37, 179–194.

Elosta, A., Ghous, T. and Ahmed, N. (2012) Natural products as anti-glycation agents: possible therapeutic potential for diabetic complications. *Current Diabetes Review* 8, 92–108.

Ernst, E. (2002) St John's Wort supplements endanger the success of organ transplantation. *Archives of Surgery* 137, 316–319.

Firuzi, O., Miri, R., Tavakkoli, M. and Saso, L. (2011) Antioxidant therapy: current status and future prospects. *Current Medicinal Chemistry* 18, 3871–3888.

Ghosh, S.S., Massey, H.D., Krieg, R., Fazelbhoy, Z.A., Ghosh, S., Sica, D.A., Fakhry, I. and Gehr, T.W. (2009) Curcumin ameliorates renal failure in 5/6 nephrectomized rats: role of inflammation. *American Journal of Physiology Renal Physiology* 296, F1146–F1157.

Ghosh, S.S., Salloum, F.N., Abbate, A., Krieg, R., Sica, D.A., Gehr, T.W. and Kukreja, R.C. (2010) Curcumin prevents cardiac remodeling secondary to chronic renal failure through deactivation of hypertrophic signaling in rats. *American Journal of Physiology Circulation Physiology* 299, H975–H984.

Ghosh, S.S., Krieg, R., Massey, H.D., Sica, D.A., Fakhry, I., Ghosh, S. and Gehr, T.W. (2012) Curcumin and enalapril ameliorate renal failure by antagonizing inflammation in 5/6 nephrectomized rats: role of phospholipase and cyclooxygenase. *American Journal of Physiology Renal Physiology* 302, F439–F454.

Grootveld, M., Halliwell, B. and Moorhouse, C.P. (1987) Action of uric acid, allopurinol and oxypurinol on the myeloperoxidase-derived oxidant hypochlorous acid. *Free Radical Research Communication* 4, 69–76.

Hoogwerf, B.J. (2010) Renin-angiotensin system blockade and cardiovascular and renal protection. *American Journal of Cardiology* 105(Suppl. 1), 30A–35A.

Howden, E.J., Fassett, R.G., Isbel, N.M. and Coombes, J.S. (2012) Exercise training in chronic kidney disease patients. *Sports Medicine* 42, 473–488.

Hughson, M.D., Gobe, G.C., Hoy, W.E., Manning, R.D. Jr, Douglas-Denton, R. and Bertram, J.F. (2008) Associations of glomerular number and birth weight with clinicopathological features of African Americans and whites. *American Journal of Kidney Diseases* 52, 18–28.

Kakuda, H., Kanasaki, K., Koya, D. and Takekoshi, N. (2013) The administration of pitavastatin augments creatinine clearance associated with reduction in oxidative stress parameters: acute and early effects. *Clinical and Experimental Nephrology* 17, 240–247.

Kang, Z., Bi, Z., Ji, W., Zhao, C. and Xie, Y. (1993) Observation of therapeutic effect in 50 cases of chronic renal failure treated with rhubarb and adjuvant drugs. *Journal of Traditional Chinese Medicine* 13, 249–252.

Karamouzis, I., Sarafidis, P.A., Karamouzis, M., Iliadis, S., Haidich, A.B., Sioulis, A., Triantos, A., Vavatsi-Christaki, N. and Grekas, D.M. (2008) Increase in oxidative stress but not in antioxidant capacity with advancing stages of chronic kidney disease. *American Journal of Nephrology* 28, 397–404.

Kuypers, D.R. (2009) Skin problems in chronic kidney disease. *Nature Clinical Practice Nephrology* 5, 157–170.

Leiherer, A., Mundlein, A. and Drexel, H. (2012) Phytochemicals and their impact on adipose tissue inflammation and diabetes. *Vascular Pharmacology* Epub http://dx.doi.org/10.1016/j.vph.2012.09.002

Levey, A.S., de Jong, P.E., Coresh, J., El Nahas, M., Astor, B.C., Matsushita, K., Gansevoort, R.T., Kasiske, B.L. and Eckardt, K.U. (2011) The definition, classification, and prognosis of chronic kidney disease: a KDIGO Controversies Conference report. *Kidney International* 80, 17–28.

Mendoza, M.G., Castillo-Henkel, C., Medina-Santillan, R., Jarillo Luna, R.A., Robles, H.V., Romo, E., Rios, A. and Escalante, B. (2008) Kidney damage after renal ablation is worsened in endothelial nitric oxide synthase -/- mice and improved by combined administration of L-arginine and antioxidants. *Nephrology* (Carlton) 13, 218–227.

Meng, L., Qu, L., Tang, J., Cai, S.-Q., Wang, H. and Li, X. (2007) A combination of Chinese herbs, *Astragalus membranaceus* var. *mongholicus* and *Angelica sinensis*, enhanced nitric oxide production in obstructed rat kidney. *Vascular Pharmacology* 47, 174–183.

Murphy, E.L., Murtagh, F.E., Carey, I. and Sheerin, N.S. (2009) Understanding symptoms in patients with advanced chronic kidney disease managed without dialysis: use of a short patient-completed assessment tool. *Nephron Clinical Practice* 111, c74–c80.

Nakamoto, H., Mimura, T. and Honda, N. (2008) Orally administrated Juzen-taiho-to/TJ-48 ameliorates erythropoietin (rHuEPO)-resistant anemia in patients on hemodialysis. *Hemodialysis International* 12(Suppl. 2), S9–S14.

Nangaku, M. and Eckardt, K.U. (2006) Pathogenesis of renal anemia. *Seminars in Nephrology* 26, 261–268.

Neves, A.R., Lucio, M., Lima, J.L. and Reis, S. (2012) Resveratrol in medicinal chemistry: a critical review of its pharmacokinetics, drug-delivery, and membrane interactions. *Current Medicinal Chemistry* 19, 1663–1681.

Osawa, T. (2007) Nephroprotective and hepatoprotective effects of curcuminoids. *Advances in Experimental Medicine and Biology* 595, 407–423.

Pannu, N. and Hemmelgarn, B.; Alberta Kidney Disease Network (2011) The acute kidney injury to chronic kidney disease continuum: comment on 'The magnitude of acute serum creatinine increase after cardiac surgery and the risk of chronic kidney disease, progression of kidney disease, and death'. *Archives of Internal Medicine* 171, 233–234.

Peng, C.C., Hsieh, C.L., Ker, Y.B., Wang, H.Y., Chen, K.C. and Peng, R.Y. (2012) Selected nutraceutic screening by therapeutic effects on doxorubicin-induced chronic kidney disease. *Molecular Nutrutrition and Food Research* 56, 1541–1558.

Satoh, S., Kaneko, T., Omori, S., Sugimura, J., Ujiie, T., Fujioka, T. and Kubo, T. (1995) The effect of enalapril and sairei-to on survival-time for the rat with subtotal nephrectomy. *Nihon Jinzo Gakkai Shi* 37, 112–118.

Sharma, S., Anjaneyulu, M., Kulkarni, S.K. and Chopra, K. (2006) Resveratrol, a polyphenolic phytoalexin, attenuates diabetic nephropathy in rats. *Pharmacology* 76, 69–75.

Shing, C.M., Adams, M.J., Fassett, R.G. and Coombes, J.S. (2011) Nutritional compounds influence tissue factor expression and inflammation of chronic kidney disease patients in vitro. *Nutrition* 27, 967–972.

Silverstein, D.M. (2009) Inflammation in chronic kidney disease: role in the progression of renal and cardiovascular disease. *Pediatric Nephrology* 24, 1445–1452.

Small, D.M., Coombes, J.S., Bennett, N., Johnson, D.W. and Gobe, G.C. (2012) Oxidative stress, anti-oxidant therapies and chronic kidney disease. *Nephrology (Carlton)* 17, 311–321.

Soni, R.K., Weisbord, S.D. and Unruh, M.L. (2010) Health-related quality of life outcomes in chronic kidney disease. *Current Opinion in Nephrology and Hypertension* 19, 153–159.

Tarng, D.C., Cho, Y.L., Liu, H.N. and Huang, T.P. (1996) Hemodialysis-related pruritus: a double-blind, placebo-controlled, cross-over study of capsaicin 0.025% cream. *Nephron* 72, 617–622.

Twomey, P.J. and Reynolds, T.M. (2006) The MDRD formula and validation. *Quarterly Journal of Medicine* 99, 804–805.

Velasquez, M.T. and Bhathena, S.J. (2001) Dietary phytoestrogens: a possible role in renal disease protection. *American Journal of Kidney Diseases* 37, 1056–1068.

Velasquez, M.T., Bhathena, S.J., Ranich, T., Schwartz, A.M., Kardon, D.E., Ali, A.A., Haudenschild, C.C. and Hansen, C.T. (2003) Dietary flaxseed meal reduces proteinuria and ameliorates nephropathy in an animal model of type II diabetes mellitus. *Kidney International* 64, 2100–2107.

Venkatachalam, M.A., Griffin, K.A., Lan, R., Geng, H., Saikumar, P. and Bidani, A.K. (2010) Acute kidney injury: a springboard for progression in chronic kidney disease. *American Journal of Physiology Renal Physiology* 298, F1078–F1094.

Vlassara, H., Torreggiani, M., Post, J.B., Zheng, F., Uribarri, J. and Striker, G.E. (2009a) Role of oxidants/inflammation in declining renal function in chronic kidney disease and normal aging. *Kidney International* 114, S3–S11.

Vlassara, H., Uribarri, J., Ferrucci, L., Cai, W., Torreggiani, M., Post, J.B., Zheng, F. and Striker, G.E. (2009b) Identifying advanced glycation end products as a major source of oxidants in aging: implications for the management and/or prevention of reduced renal function in elderly persons. *Seminars in Nephrology* 29, 594–603.

Wang, H., Li, J., Yu, L., Yani, Z. and Ding, W. (2004) Antifibrotic effect of the Chinese herbs, *Astragalus mongholicus* and *Angelica sinensis*, in a rat model of chronic puromycin aminonucleoside nephrosis. *Life Sciences* 74, 1645–1658.

Wang, Y.J., He, L.Q., Sun, W., Lu, Y., Wang, X.Q., Zhang, P.Q., Wei, L.B., Cao, S.L., Yang, N.Z., Ma, H.Z., Gao, J., Li, P., Tao, X.J., Yuan, F.H., Li, J., Yao, C. and Liu, X. (2012) Optimized project of traditional Chinese medicine in treating chronic kidney disease stage 3: a multicenter double-blinded randomized controlled trial. *Journal of Ethnopharmacology* 139, 757–764.

Weisshaar, E., Dunker, N. and Gollnick, H. (2003) Topical capsaicin therapy in humans with hemodialysis-related pruritus. *Neuroscience Letters* 345, 192–194.

Wojcikowski, K., Johnson, D.W. and Gobe, G. (2004a) Medicinal herbal extracts – renal friend or foe? Part one: the toxicities of medicinal herbs. *Nephrology* (Carlton) 9, 313–318.

Wojcikowski, K., Johnson, D.W. and Gobe, G. (2004b) Medicinal herbal extracts-renal friend or foe? Part two: herbal extracts with potential renal benefits. *Nephrology* (Carlton) 9, 400–405.

Wojcikowski, K., Johnson, D.W. and Gobe, G.C. (2006) Herbs or natural substances as complementary therapies for chronic kidney disease: ideas for future studies. *Journal of Laboratory and Clinical Medicine* 147, 160–166.

Wojcikowski, K., Stevenson, L., Leach, D., Wohlmuth, H. and Gobe, G. (2007) Anti-oxidant capacity of 55 medicinal herbs traditionally used to treat the urinary system: a comparison using a sequential three-solvent extraction process. *Journal of Alternative and Complementary Medicine* 13, 103–109.

Wojcikowski, K., Wohlmuth, H., Johnson, D.W., Rolfe, M. and Gobe, G. (2009) An *in vitro* investigation of herbs traditionally used for kidney and urinary system disorders: Potential therapeutic and toxic effects. *Nephrology* (Carlton) 14, 70–79.

Wojcikowski, K., Wohlmuth, H., Johnson, D.W. and Gobe, G.C. (2010) Effect of *Astragalus membranaceus* and *Angelica sinensis* combined with enalapril in rats with obstructive uropathy. *Phytotherapy Research* 24, 875–884.

Yar, A.S., Menevse, S., Alp, E., Helvacioglu, F. and Take, G. (2010) The effects of resveratrol on cyclooxygenase-1 and cyclooxygenase-2 mRNA and protein levels in diabetic rat kidneys. *Molecular Biology Reports* 37, 2323–2331.

Yokozawa, T., Chung, H.Y., Oura, H., Nonaka, G. and Nishioka, I. (1989) Isolation of a renal function-facilitating constituent from the Oriental drug, salviae miltiorrhizae radix. *Nihon Jinzo Gakkai Shi* 31, 1091–1098.

Zhang, G. and El Nahas, A.M. (1996) The effect of rhubarb extract on experimental renal fibrosis. *Nephrology Dialysis Transplantation* 11, 186–190.

Zhang, J.H., Li, L.S. and Zhang, M. (1990) Clinical effects of rheum and captopril on preventing progression of chronic renal failure. *Chinese Medical Journal* 103, 788–793.

Zhao, J.-R., Qu, L. and Li, X.-M. (2004) Preventive and therapeutic effects of astragalus and angelica mixture on renal tubulointerstitial fibrosis after unilateral ureteral obstruction in rats. *Beijing da Xue Xue Bao* 36, 119–123.

Zhao, Y.-N., Li, J.-Z. and Yu, L. (2002) Effect of astragalus-angelica mixture on osteopontin expression in rats with chronic nephrosclerosis. *Chinese Journal of Integrative Traditional Western Medicine* 22, 613–617.

Zheng, K.Y., Choi, R.C., Xie, H.Q., Cheung, A.W., Guo, A.J., Leung, K.W., Chen, V.P., Bi, C.W., Zhu, K.Y., Chan, G.K., Fu, Q., Lau, D.T., Dong, T.T., Zhao, K.J. and Tsim, K.W. (2010) The expression of erythropoietin triggered by danggui buxue tang, a Chinese herbal decoction prepared from radix *Astragali* and radix *Angelicae Sinensis*, is mediated by the hypoxia-inducible factor in cultured HEK293T cells. *Journal of Ethnopharmacology* 132, 259–267.

3 Natural Products in the Prevention of Cancer: Investigating Clues in Traditional Diets for Potential Modern-Day Cures

Vondina Moseley, Rebecca Knackstedt* and Michael J. Wargovich
Department of Cellular and Molecular Pharmacology and Experimental Therapeutics, Hollings Cancer Center, Medical University of South Carolina, Charleston, USA

3.1 Introduction

Globally, a process known as 'acculturation' is occurring and may explain the rapid and alarming increase in the prevalence of diseases related to chronic inflammation. Acculturation refers to the process by which a racial or ethnic group adopts the lifestyle of its host group (Satia *et al.*, 2001; Page, 2006). Dietary acculturation refers to the process by which groups exploring new lifestyles exclude traditional foods in favour of newly introduced foods (Satia, 2010). This usually refers to a group that has left its homeland and has immigrated to a new country; however, it is also relevant for areas infiltrated by other ethnic or racial groups or practices. Acculturation has been encouraged by the demographic and socio-economic changes seen globally, which have brought about a decrease in food prices and increased access to different foods occurring simultaneously with a shift to lifestyles with less physical activity (Vorster *et al.*, 1999; Popkin, 2004, 2006, 2009; Astrup *et al.*, 2008). Rapid urbanization is accelerating this process (Solomons and Gross, 1995), especially in lower and middle income countries that, as a consequence, are witnessing a shift toward obesity and non-communicable diseases, such as various cancers (Popkin, 2004, 2006, 2009).

3.2 Overall Trends of Westernization and Acculturation

There have been numerous epidemiological studies characterizing the effects of acculturation on disease prevalence. The majority of these studies have observed countries or groups that have left behind their traditional diet in favour of a more 'Western' diet. As a general rule, 'Westernization', in respect to diet, refers to a way of eating that has a high energy density with a high content of fat, sugary drinks and *trans*-fatty acids with a low intake of fibre-rich foods (Astrup *et al.*, 2008). These trends in eating are allowing for a shift in the traditional patterns of disease evidenced globally from a historically high prevalence of infections and nutrient deficiencies, to now high rates of heart disease and cancer (Omran, 1971; Popkin, 1994; Posner *et al.*, 1994). In fact, some countries that have often

* E-mail: rew27@musc.edu

suffered from nutritional deficiencies with a high proportion of the population being underweight, are now suffering from high rates of obesity (Satia, 2010). The predication globally in relation to disease prevalence is alarming if this trend continues. It is predicted that diabetes will be a pandemic by 2030 (Wild *et al.*, 2004) and that 1.56 billion people will have hypertension by 2025 (Kearney *et al.*, 2005).

3.3 Country-Specific Evidence

Countries around the world have been adopting a Western diet along with the prevalence for Western diseases. Below are examples of some traditional diets and their disease-fighting constituents that are slowly being forgotten.

3.4 Okinawa

The population of Okinawa, Japan, is renowned for its remarkable longevity, which has been likely due to the traditional caloric restriction of this region. Even though the diet of this area was calorically low, it was nutrient dense with ample vitamins, minerals and phytonutrients in the form of antioxidants and flavonoids (Willcox *et al.*, 2007). However, the ability of this regional diet to extend lifespan is now disappearing as younger generations adopt more Western styles of eating. In fact, the longer life expectancy now only applies to generations born before World War II (Gavrilova and Gavrilov, 2012). Before the war, a traditional Okinawan diet consisted of tofu, sweet potatoes, rice, vegetables, soybeans and very little meat. This type of diet is very anti-inflammatory (Willcox *et al.*, 2009). The ample supply of antioxidants could be one reason Okinawa demonstrates a low rate of chronic diseases as multiple disease states, such as coronary heart disease, cancer, stroke, diabetes, rheumatoid arthritis, have been linked to inflammation (Willcox *et al.*, 2009). Shortly after World War II, the Okinawan diet changed to no longer being low in calories but to include greater amounts of meat. This has resulted in a decreased life expectancy and the rise of cardiovascular disease (Gavrilova and Gavrilov, 2012).

3.5 Inuit and Remote Alaskan Populations

The traditional diet consumed by Inuit and remote Alaskan populations had a very high content of fat from marine sources. Marine fat sources are rich in n-3 fatty acids and vitamin D (Receveur *et al.*, 1997), which is crucial at this high latitude where sunlight intensity is not adequate to stimulate vitamin D production in the skin. Despite the high fat content of this diet, this region has historically demonstrated a low prevalence of chronic disease. This is likely due to the healthy aspects of the type of fat consumed (Adler *et al.*, 1994; Parkinson *et al.*, 1994; Ebbesson *et al.*, 2005). Since the 1930s, the import of Western foods has increased and has led to a decrease in hunting and fishing lifestyles (Bjerregaard, 2004). In fact, commercial foods now account for up to 76% of caloric intake for some people in western Alaska (Johnson *et al.*, 2009, 2012). Thus, less of the fat in this region is being consumed from marine sources.

This dietary shift has resulted in a decrease in n-3 fatty acids (Deutch *et al.*, 2006), eicosapentaenoic acid (EPA) and docosahexanoic acid (DHA) and vitamin D in the diet along with an increase in saturated fat consumption (Bersamin *et al.*, 2008). This diet now closely resembles the diet consumed in the continental USA (Nobmann *et al.*, 2005; Risica *et al.*, 2005; Bersamin *et al.*, 2006). Along with this dietary shift has come an increase in the prevalence of cardiovascular disease and co-morbidities such as obesity (Rith-Najarian *et al.*, 2002; McLaughlin *et al.*, 2004; Munch-Andersen *et al.*, 2012). In fact, Alaskan native people now have the highest increase of diabetes in North America (Bjerregaard *et al.*, 2004). A recent survey found that traditional foods were no longer in the top ten foods consumed in this region, which could explain the new prevalence of vitamin D deficiency (Sharma *et al.*, 2009, 2010; Erber *et al.*, 2010;

Hopping *et al.*, 2010). However, individuals who have maintained a traditional diet continue to demonstrate lower triglyceride levels and higher high-density lipoprotein (HDL) levels, both of which would lead to improved glucose tolerance (Mensink *et al.*, 1992; Adler *et al.*, 1994; Hu *et al.*, 2002; Bersamin *et al.*, 2008), healthier lipid profiles (Nobmann *et al.*, 1999) and less obesity (Murphy *et al.*, 1995). Also, it was found that consuming one traditional food per day was adequate to increase vitamin D levels in this largely deficient population (Kuhnlein, 2003; Kuhnlein and Receveur, 2007).

3.6 Grenada, Caribbean

The diet traditionally consumed on this island largely consisted of fish. A marine-dominant diet like this has large amounts of omega-3s, EPA and DHA, all of which have been linked to decreased cardiovascular disease (Kris-Etherton *et al.*, 2002). Not surprisingly, this region has demonstrated a prevalence of cardiovascular disease even lower than that predicted by other risk factors. However, dietary patterns are shifting to include more red meat, poultry and fried foods, which is predicted to increase the prevalence of cardiovascular disease, especially in women (Block *et al.*, 2012).

3.7 China

The Chinese diet has traditionally been low in fat and high in carbohydrates (Campbell and Junshi, 1994). The low amounts of cholesterol and fat consumed traditionally resulted in a low cardiovascular mortality rate (Critchley *et al.*, 2004). However, beginning in the 1950s, the diet in this region began to include fewer carbohydrates, fruits and vegetables and more overall calories, meat, alcohol and fat (Drewnowski and Popkin, 1997; Zhang *et al.*, 2008). This Westernization has resulted in an increase in the prevalence of obesity and hypercholesterolaemia (Yang *et al.*, 1999; Yu *et al.*, 2000; Cheng, 2004; Gu *et al.*, 2005; Jiang *et al.*, 2007; Li *et al.*, 2007; Zhang *et al.*, 2008). In fact, the prevalence of obesity in adults has increased by 97% in the past decade (Liu *et al.*, 2007). As obesity is associated with cardiovascular disease, it is no surprise that in 2005, one-third of all deaths were due to heart disease and this is expected to double by 2020.

3.8 Japan

The traditional Japanese diet included fish, rice with salt and soybeans in various forms along with a low amount of fat (Drewnowski and Popkin, 1997; Sonoda *et al.*, 2004). The prevalence of soybeans ensured a high consumption of dietary isoflavones (Adlercreutz *et al.*, 1993). However, after World War II, Japan experienced accelerated economic growth (Drewnowski and Popkin, 1997), which was accompanied by dietary changes. The Japanese diet now is rich in sugar and fat (Lands *et al.*, 1990). From the 1940s to the 1980s, the fat in this diet increased three-fold due to the introduction of oil fat, meat, chicken, milk and eggs. This was accompanied by a reduction in healthy fats from foods such as sardines and mackerel (Drewnowski and Popkin, 1997). This has resulted in an increase in serum total cholesterol for Japanese children and adults and a new prevalence of chronic diseases that were not historically demonstrated. For example, the incidence of prostate cancer (Sonoda *et al.*, 2004), type 2 diabetes (Yoon *et al.*, 2006), breast cancer (Porter, 2008) and obesity have all been attributed to the Westernization of the Japanese diet.

3.9 Mediterranean

Although it is difficult to characterize a traditional Mediterranean diet due to diversified eating patterns, as a whole, the eating style of this region has been linked to health and longevity. Despite the varied foodstuffs found in this region, certain foods are ubiquitous, such as medium unsaturated fatty acids (MUFA) from olive oil (Menotti *et al.*, 1999; Tunstall-Pedoe *et al.*, 1999), vegetables, fruits, nuts, cereal, seafood and legumes (Trichopoulou *et al.*, 2009). The healthy aspects of this diet are apparent in the number of books available that tout the Mediterranean diet as being ideal for weight loss and heart disease prevention. The Mediterranean diet has been linked to a reduced risk for cardiovascular mortality, low cancer incidence and mortality (Trichopoulou *et al.*, 2003), a reduced risk of heart attack, lower heart disease biomarkers (Panagiotakos *et al.*, 2006) and a beneficial effect on the incidence of Alzheimer's disease and dementia (Scarmeas *et al.*, 2006). Unfortunately, like many other traditional diets, it is no longer being consumed in the region in which it originated (Bonaccio *et al.*, 2012).

3.10 Phytochemical Depletion and Risk for Cancer and Other Chronic Diseases

Acculturation, or the adaption of a migrant population to the habits and mores of their new homeland, figures prominently in the concept known as 'phytochemical depletion'. As populations adopt the dietary patterns of their new homeland, especially in the transit from the developing world to the developed, a casualty of this process is the decrease in the intake of anti-inflammatory fruits and vegetables. It is illustrative to examine what has happened in a country with an epidemic of chronic inflammatory disease since early in the past century: the USA.

In the early 1900s, life in the USA was primarily agrarian due to the vegetable and fruit growing of early pioneers and the kitchen garden was a frequent fixture in rural America. People ate locally due to availability. After World War II, intake of fruits and vegetables started to decline due to their scarcity during the war years. This was accompanied by the introduction of preserved, processed and 'fast' foods. In the 1950s, demographic shifts were well underway from rural areas to cities with populations no longer dependent on farm produce or local gardens for their daily meals (Goldman, 2003). As a consequence, the American diet changed markedly. Cereal grains, meat, poultry and shellfish displaced fruit and vegetable intake. With this shift, a change in the patterns of eating ensued. Since the 1950s, portion sizes have increased and have been accompanied by a greater tendency to eat out. This increase in consumption occurred at the same time as a decrease in the time spent in physical activity.

To illustrate the concept of phytochemical depletion and its role in chronic inflammatory disease prevalence, an examination of fruit and vegetable intake in recent years in the USA is revealing. The US Department of Health and Human Services, in its Healthy People 2010 report, advised that the average daily intake of servings of fruits and vegetables be increased to three or more servings. Only 32% of adults consume two or more servings a day and only 26% of adults consume greater than three servings despite massive campaigns to raise awareness by the federal government and other health agencies (Erovic *et al.*, 2010). According to a recent analysis by the Centers for Disease Control and Prevention (CDC) using data from the Behavioral Risk Factor Surveillance System, the US population is falling well short of the recommendations for fruit and vegetable intake. When one examines fruit and vegetable intake across the last several decades in the USA, it is clear that consumption has flatlined from 1994 to 2004 according to data from the National Cancer Institute (Services, 2012). When examining intakes in younger people, the trends are even more alarming. Between 1994 and 2004, a study from the University of Minnesota showed that fruit and vegetable intake dropped almost a serving among boys and girls transiting from early to middle adolescence and in youths entering late adolescence (Larson *et al.*, 2007). Adequate to robust levels of fruits and vegetables in the diet are associated with lower risk of mouth, pharynx, larynx, oesophagus, stomach, lung and colon cancers. In the constellation of inflammatory-driven diseases, people with robust intakes of fruits and vegetables are at lowered risk for diabetes, heart disease and high blood pressure.

Despite the high availability of fruits and vegetables in this country, it is disconcerting to see no positive movement to increase consumption of these foods with disease-preventing compounds. To understand the unintended consequences of not consuming these healthy foods, one can examine populations that have immigrated to the USA and undergone the process of acculturation. One of the fastest growing Asian-American groups in this country are Korean immigrants. In a study of 486 Korean Americans in the New York City area, it was found that members of this population who underwent rapid acculturation consumed more sweets, grains, cereals, certain meats and soft drinks as compared to those who were slower to acculturate (Kim and Chan, 2004). Another association with diet change, acculturation and chronic inflammatory disease can be observed in the Latino population where it is observed that length of time spent in the USA is proportional to poor health outcomes (Hazuda *et al.*, 1991). Of great interest is a study of current refugees from war-torn Liberia and Somalia who have immigrated to the USA. These refugees reported that in their homeland, greater than 92% of their caloric intake came from vegetable foods as compared to their diet in the USA where only 72% of the caloric intake is from vegetables. A notable claim from these populations is the complaint that the cost of fresh fruits and vegetables is just too high (Patil *et al.*, 2009).

3.11 Phytochemicals Associated with Prevention

The scope of this chapter does not allow for the discussion of all phytochemicals and their abilities to prevent disease. Below are a few highlighted food products rich in health-promoting phytochemicals. Numerous animal studies support the findings described below, but only human studies will be focused on in this chapter.

Cruciferous vegetables, such as broccoli, cabbage and cauliflower, contain glucosinolates (Drewnowski and Gomez-Carneros, 2000), which are broken down into indoles and isothiocyanates (Holst and Williamson, 2004). These active ingredients have shown through epidemiological studies and case-controlled trials that they have the ability to prevent the development of various diseases, especially cancer. A review published in 1996 found that 67% of studies had demonstrated an inverse association between cancer risk and cruciferous vegetable consumption (Verhoeven *et al.*, 1996). The consumption of cruciferous vegetables more than three times per week was associated with a reduction in lung cancer risk in Dutch men and women (Voorrips *et al.*, 2000b) and Finnish men (Neuhouser *et al.*, 2003). Broccoli and Brussels sprouts were able to enhance urinary extraction of a possible carcinogen associated with well-done meat, suggesting that cruciferous vegetables could help eliminate potential carcinogens from the body (Walters *et al.*, 2004). A Dutch study found that adults with high cruciferous vegetable consumption were less likely to develop colon cancer as compared to those with a low intake of these vegetables (Voorrips *et al.*, 2000a). It was found in the USA, Sweden and China, that women diagnosed with breast cancer had consumed less cruciferous vegetables as compared to cancer-free controls (Terry *et al.*, 2001; Fowke *et al.*, 2003; Ambrosone *et al.*, 2004). This correlation was also demonstrated in studies looking at prostate cancer (Jain *et al.*, 1999; Cohen *et al.*, 2000; Kolonel *et al.*, 2000; Joseph *et al.*, 2004).

Green tea contains active ingredients such as catechins, proanthocyanidins and flavonols, which have all been associated with preventing various diseases. The consumption of green tea has been linked to modulating cholesterol, encouraging weight loss and preventing cancer. A meta-analysis of 133 randomized control trials concluded that green tea consumption was able to reduce LDL cholesterol (Hooper *et al.*, 2008). A meta-analysis investigating green tea and weight loss found that green tea catechins plus caffeine were able to promote weight loss and maintenance (Hursel *et al.*, 2009). It has been shown in numerous studies that green tea has the ability to prevent cancer development. The majority of these studies were conducted in Asian countries where green tea consumption is common. A study in the Jiangsu province of China found that alcoholics and cigarette smokers who drank green tea had a reduction in their risk of stomach and oesophageal cancer development (Wang *et al.*, 1999). The protective effects of green tea against stomach and oesophageal cancers were also demonstrated in non-alcoholics and non-smokers (Sasazuki *et al.*, 2008). Green tea consumption has been associated with a reduced risk of breast cancer (Dai *et al.*, 2010), epithelial ovarian cancer (Larsson and Wolk, 2005) and prostate cancer (Jian *et al.*, 2004; Bettuzzi *et al.*, 2006). A Chinese study found that women with established ovarian cancer who drank green tea actually had a longer survival than women with ovarian cancer who did not consume green tea (Zhang *et al.*, 2004).

Soy products have been associated with the ability to prevent numerous diseases. One meta-analysis found that soy protein was able to reduce cholesterol (Anderson *et al.*, 1995) and this was supported by another study, which found soy was able to reduce cholesterol even in healthy individuals with a normal cholesterol level prior to intervention (Cassidy *et al.*, 1995). This hypocholesterolaemic effect might be one of the reasons that the consumption of phytoestrogens, such as those found in soy products, has been linked to lower rates of cardiovascular disease (Keys *et al.*, 1984). Since the first study to demonstrate that genistein, an active ingredient in soy products, was able to increase nitric oxide dilation (Walker *et al.*, 2001), more than 50 trials have been released investigating the efficacy of other food products with isoflavones (Howes *et al.*, 2006; Lethaby *et al.*, 2007). The increase in nitric oxide dilation would also serve to reduce cardiovascular disease. As phytoestrogens are one of the active ingredients in soy, studies have examined correlations between soy consumption and oestrogen-related symptoms and diseases, such as side effects associated with menopause and cancers due to high levels of oestrogen. A case-control study found a reduction in breast cancer risk due to phytoestrogen consumption for both pre- and post-menopausal women (Ingram *et al.*, 1997). The phytoestrogens in soy have also shown to reduce menopause-associated hot flushes (Lock, 1991). Soy was first linked to improving bone density in post-menopausal women in 1998 (Potter *et al.*, 1998). Since then, more than 30 similar trials have been published (Ma *et al.*, 2008; Liu *et al.*, 2009). One study focused on post-menopausal women with high cholesterol and found that women taking a soy protein with high isoflavone concentration had significant increases in bone mineral density and bone mineral content in their lumbar spines as compared with women taking a soy protein with moderate isoflavone content or casein non-fat dry milk (Potter *et al.*, 1998). Soy has also shown the ability to decrease the risk of cancer development in men. Asian populations have a low incidence of prostate cancer and this has been linked to phytoestrogen consumption (Adlercreutz *et al.*, 1991). This population also demonstrates lower levels of 5α-reductase activity, which has also been linked to prostate cancer promotion and progression (Ross *et al.*, 1992). Interestingly, Japanese men who migrate to America develop the same prostate cancer incidence within one or two generations as Americans, thus ruling out genetic predispositions to cancer development (Kolonel *et al.*, 1985).

Few human studies have been conducted on the ability of ginger to prevent disease development or progression. Ginger is typically associated with reducing nausea. This is likely due to its role in accelerating gastric emptying and antral contractions (Wu *et al.*, 2008). This mechanism is similar to that of other pharmaceutics, such as metoclopramide, which are used to treat nausea (Dennehy, 2011). In ten randomized control trials, ginger was compared to a placebo, vitamin B6 or dimenhydrinate in its ability to reduce nausea. Ginger was found to be more effective than placebo (Borrelli *et al.*, 2005; Ozgoli *et al.*, 2009), equal or more efficacious as compared to vitamin B6, and equal in efficacy to dimenhydrinate (Fischer-Rasmussen *et al.*, 1991; Borrelli *et al.*, 2005; Chittumma *et al.*, 2007; Pongrojpaw *et al.*, 2007; Ensiyeh and Sakineh, 2009; Ozgoli *et al.*, 2009). Although ginger is most known for its antinausea effects, it has also been investigated for its ability to reduce coronary artery disease. For example, in patients with coronary artery disease, ginger powder was shown to significantly reduce platelet aggregation (Bordia *et al.*, 1997).

Garlic has been studied for its abilities to modulate cholesterol, improve heart disease and prevent cancer development. Two reviews concluded that garlic is able to reduce total cholesterol and triglycerides (Warshafsky *et al.*, 1993; Silagy and Neil, 1994). A double-blind placebo-controlled randomized study, which occurred after the above review was published, found that garlic powder was able to decrease both total and LDL cholesterol (Sobenin *et al.*, 2010). Another study supported these findings and found that garlic could also increase in HDL cholesterol in individuals with hypercholesterolaemia (Durak *et al.*, 2004; Mahmoodi *et al.*, 2006; Sobenin *et al.*, 2008). Hypertensive adults taking oily macerate of garlic experienced reductions in total and LDL cholesterol as well as in their triacylglyceride levels (Duda *et al.*, 2008). Garlic has been shown to be heart-healthy due to its ability to improve cardiovascular parameters. The intake of only one clove of garlic daily for 16 weeks was able to reduce serum thromboxane B2 by 80% (Ali and Thomson, 1995). A randomized, placebo-controlled, cross-over study demonstrated that aged garlic extract was able to improve brachial artery flow-mediated endothelium-dependent dilation (Williams *et al.*, 2005). Garlic oil over a 2 month period improved both systolic and diastolic blood pressure in individuals with hypertension (Dhawan and Jain, 2004). Garlic oil was also shown to reduce the oxidation of LDL, a precursor step to atherosclerosis, and reduce 8-iso-prostaglandin F2 alpha levels while decreasing both systolic and diastolic blood pressure in hypertensive patients (Dhawan and Jain, 2004). A decrease in oxidized LDL by garlic was supported in at least two other studies (Duda *et al.*, 2008; Budoff *et al.*, 2009). Data also suggest that garlic intake can reduce cancer risk. Those with a high intake of garlic have shown to have a decreased risk of colorectal, stomach and gastric cancer development (Fleischauer and Arab, 2001; Li *et al.*, 2004; Millen *et al.*, 2007). High dose garlic was also able to reduce the size and number of colon adenomas in individuals with colorectal adenomas or precancerous lesions (Tanaka *et al.*, 2006).

Various forms of pomegranate have been used in clinical trials to test its effect on chronic illnesses. The illnesses that pomegranate or some form of the fruit or extracts affect range from cancer to heart disease and blood pressure. In a clinical trial where the 46 male volunteers had recurrent prostate cancer, 35% of them showed a significant decrease in PSA (prostate specific antigen) when treated with 8 ounces of pomegranate juice (Pantuck *et al.*, 2006). Pomegranate is a heart-healthy fruit with many beneficial cardiovascular effects. Pomegranate juice intake is associated with a reduction in myocardial ischaemia and improved myocardial perfusion in patients with coronary heart disease (Sumner *et al.*, 2005). Pomegranate juice consumption also reduced systolic blood pressure as related to hypertension (Stowe, 2011). Another pilot study involving 22 type-2 diabetic patients treated with 40 g of concentrated pomegranate juice for 8 weeks showed significantly decreased levels of total cholesterol (Esmaillzadeh *et al.*, 2006). Clinical research involving pomegranate has also shown promise for dietary weight management. A pomegranate seed oil composite has been found to reduce body weight, waist circumference and liver fat content in obese, non-diabetic women with non-alcoholic fatty liver disease as well as in women with a normal liver (Abidov *et al.*, 2010).

Much of the research done on chilli is with its active compound, capsaicin. Capsaicin is what gives the chilli peppers their spice. The research surrounding chilli peppers and their preventative properties is limited to its use as a protective agent against ulcers. Capsaicin proved to be protective against gastroduodenal mucosal injury in response to aspirin in healthy individuals (Yeoh *et al.*, 1995). In low concentrations, applied intragastrically, capsaicin was protective against ethanol- and indomethacin-induced gastric injury in a prospective study of 84 healthy volunteers (Mozsik *et al.*, 2005).

The clinical applications of assessing fruit for use in disease prevention separates citrus fruits from all others based on the presence of citrus flavones. The citrus flavones are commonly lumped together, but they include hesperidin, quercitin, tangeritin and rutin. The antioxidant properties of these compounds are what make citrus fruits a target for dietary prevention. Citrus flavones, found in grapefruit, tangerines, oranges and lemons are shown in epidemiological studies to be inversely associated with the risk for coronary heart disease (Joshipura *et al.*, 2001). A high intake of citrus fruits is also inversely associated with breast cancer risk. When the consumption of all fruits is taken into consideration, a similar association is not seen (Bao *et al.*, 2012). Citrus fruit intake is associated with decreased risk for heart disease and breast cancer; however, some clinical studies involving citrus flavones are not conclusive. Some studies report a decrease in cholesterol with an increased intake of citrus fruits while others do not. Specifically, grapefruit improved the blood pressure of those who ate grapefruit every day for 6 weeks but it had no effect on weight or lipid profiles (Dow *et al.*, 2012). More research is needed in order to make a conclusive argument as to what effect citrus fruits have on cholesterol. Ultimately, citrus fruits do show promise as preventive agents for chronic disease.

3.12 Concluding Remarks

The world is awash in acculturation. As the economies of developing nations improve, a headlong rush to adopting lifestyles of prosperous nations has been associated with a replacement of anti-inflammatory and cancer-preventive elements from plant foods with energy dense pro-inflammatory factors. Current projections are the global risk for cancer and other chronic inflammatory diseases will heavily impact populations in the developing world and, as in the USA, make unprecedented demands on already overtaxed healthcare systems. The only reasonable remedy is to underscore the importance of research showing that phytochemicals are a major deterrent to pro-inflammatory disease. Traditional diets should not be abandoned and we must learn from other cultures, adopt dietary practices associated with lesser risk, and incorporate new scientific knowledge into altering current dietary practices, beginning at an early age of life.

References

Abidov, M., Ramazanov, Z., Seifulla, R. and Grachev, S. (2010) The effects of Xanthigen in the weight management of obese premenopausal women with non-alcoholic fatty liver disease and normal liver fat. *Diabetes Obesity and Metabolism* 12, 72–81.

Adler, A.I., Boyko, E.J., Schraer, C.D. and Murphy, N.J. (1994) Lower prevalence of impaired glucose tolerance and diabetes associated with daily seal oil or salmon consumption among Alaska Natives. *Diabetes Care* 17, 1498–1501.

Adlercreutz, H., Honjo, H., Higashi, A., Fotsis, T., Hamalainen, E., Hasegawa, T. and Okada, H. (1991) Urinary excretion of lignans and isoflavonoid phytoestrogens in Japanese men and women consuming a traditional Japanese diet. *American Journal of Clinical Nutrition* 54, 1093–1100.

Adlercreutz, H., Markkanen, H. and Watanabe, S. (1993) Plasma concentrations of phyto-oestrogens in Japanese men. *Lancet* 342, 1209–1210.

Ali, M. and Thomson, M. (1995) Consumption of a garlic clove a day could be beneficial in preventing thrombosis. *Prostaglandins Leukotrience and Essential Fatty Acids* 53, 211–212.

Ambrosone, C.B., McCann, S.E., Freudenheim, J.L., Marshall, J.R., Zhang, Y. and Shields, P.G. (2004) Breast cancer risk in premenopausal women is inversely associated with consumption of broccoli, a source of isothiocyanates, but is not modified by GST genotype. *Journal of Nutrition* 134, 1134–1138.

Anderson, J.W., Johnstone, B.M. and Cook-Newell, M.E. (1995) Meta-analysis of the effects of soy protein intake on serum lipids. *New England Journal of Medicine* 333, 276–282.

Astrup, A., Dyerberg, J., Selleck, M. and Stender, S. (2008) Nutrition transition and its relationship to the development of obesity and related chronic diseases. *Obesity Review* 9(Suppl. 1), 48–52.

Bao, P.P., Shu, X.O., Zheng, Y., Cai, H., Ruan, Z.X., Gu, K. and Lu, W. (2012) Fruit, vegetable, and animal food intake and breast cancer risk by hormone receptor status. *Nutrition and Cancer* 64, 806–819.

Bersamin, A., Luick, B.R., Ruppert, E., Stern, J.S. and Zidenberg-Cherr, S. (2006) Diet quality among Yup'ik Eskimos living in rural communities is low: the Center for Alaska Native Health Research Pilot Study. *Journal of the American Dietetic Assocation* 106, 1055–1063.

Bersamin, A., Luick, B.R., King, I.B., Stern, J.S. and Zidenberg-Cherr, S. (2008) Westernizing diets influence fat intake, red blood cell fatty acid composition, and health in remote Alaskan Native communities in the center for Alaska Native health study. *Journal of the American Dietetic Assocation* 108, 266–273.

Bettuzzi, S., Brausi, M., Rizzi, F., Castagnetti, G., Peracchia, G. and Corti, A. (2006) Chemoprevention of human prostate cancer by oral administration of green tea catechins in volunteers with high-grade prostate intraepithelial neoplasia: a preliminary report from a one-year proof-of-principle study. *Cancer Research* 66, 1234–1240.

Bjerregaard, P. (2004) Public health research and practice in Greenland. *International Journal of Circumpolar Health* 63, 210–211.

Bjerregaard, P., Young, T.K., Dewailly, E. and Ebbesson, S.O. (2004) Indigenous health in the Arctic: an overview of the circumpolar Inuit population. *Scandanavian Journal of Public Health* 32, 390–395.

Block, R.C., Dozier, A.M., Hazel-Fernandez, L., Guido, J.J. and Pearson, T.A. (2012) An epidemiologic transition of cardiovascular disease risk in carriacou and petite martinique, grenada: the Grenada heart project, 2005-2007. *Preventing Chronic Disease* 9, E90.

Bonaccio, M., Iacoviello, L., de Gaetano, G. and Moli-Sani, I. (2012) The Mediterranean diet: the reasons for a success. *Thrombosis Research* 129, 401–404.

Bordia, A., Verma, S.K. and Srivastava, K.C. (1997) Effect of ginger (*Zingiber officinale* Rosc.) and fenugreek (*Trigonella foenumgraecum* L.) on blood lipids, blood sugar and platelet aggregation in patients with coronary artery disease. *Prostaglandins Leukotrience and Essential Fatty Acids* 56, 379–384.

Borrelli, F., Capasso, R., Aviello, G., Pittler, M.H. and Izzo, A.A. (2005) Effectiveness and safety of ginger in the treatment of pregnancy-induced nausea and vomiting. *Obstetrics and Gynecology* 105, 849–856.

Budoff, M.J., Ahmadi, N., Gul, K.M., Liu, S.T., Flores, F.R., Tiano, J., Miller, E. and Tsimikas, S. (2009) Aged garlic extract supplemented with B vitamins, folic acid and L-arginine retards the progression of subclinical atherosclerosis: a randomized clinical trial. *Preventive Medicine* 49, 101–107.

Campbell, T.C. and Junshi, C. (1994) Diet and chronic degenerative diseases: perspectives from China. *American Journal of Clinical Nutrition* 59, 1153S–1161S.

Cassidy, A., Bingham, S. and Setchell, K. (1995) Biological effects of isoflavones in young women: importance of the chemical composition of soyabean products. *British Journal of Nutrition* 74, 587–601.

Cheng, T.O. (2004) The current state of cardiology in China. *International Journal of Cardiology* 96, 425–439.

Chittumma, P., Kaewkiattikun, K. and Wiriyasiriwach, B. (2007) Comparison of the effectiveness of ginger and vitamin B6 for treatment of nausea and vomiting in early pregnancy: a randomized double-blind controlled trial. *Journal of the Medical Association of Thailand* 90, 15–20.

Cohen, J.H., Kristal, A.R. and Stanford, J.L. (2000) Fruit and vegetable intakes and prostate cancer risk. *Journal of the National Cancer Institute* 92, 61–68.

Critchley, J., Liu, J., Zhao, D., Wei, W. and Capewell, S. (2004) Explaining the increase in coronary heart disease mortality in Beijing between 1984 and 1999. *Circulation* 110, 1236–1244.

Dai, Q., Shu, X.O., Li, H., Yang, G., Shrubsole, M.J., Cai, H., Ji, B., Wen, W., Franke, A., Gao, Y. and Zheng, W. (2010) Is green tea drinking associated with a later onset of breast cancer? *Annual Epidemiology* 20, 74–81.

Dennehy, C. (2011) Omega-3 fatty acids and ginger in maternal health: pharmacology, efficacy, and safety. *Journal of Midwifery and Womens Health* 56, 584–590.

Deutch, B., Dyerberg, J., Pedersen, H.S., Asmund, G., Moller, P. and Hansen, J.C. (2006) Dietary composition and contaminants in north Greenland, in the 1970s and 2004. *Science of the Total Environment* 370, 372–381.

Dhawan, V. and Jain, S. (2004) Effect of garlic supplementation on oxidized low density lipoproteins and lipid peroxidation in patients of essential hypertension. *Molecular and Cellular Biochemistry* 266, 109–115.

Dow, C.A., Going, S.B., Chow, H.H., Patil, B.S. and Thomson, C.A. (2012) The effects of daily consumption of grapefruit on body weight, lipids, and blood pressure in healthy, overweight adults. *Metabolism* 61, 1026–1035.

Drewnowski, A. and Gomez-Carneros, C. (2000) Bitter taste, phytonutrients, and the consumer: a review. *American Journal of Clinical Nutrition* 72, 1424–1435.

Drewnowski, A. and Popkin, B.M. (1997) The nutrition transition: new trends in the global diet. *Nutrition Reviews* 55, 31–43.

Duda, G., Suliburska, J. and Pupek-Musialik, D. (2008) Effects of short-term garlic supplementation on lipid metabolism and antioxidant status in hypertensive adults. *Pharmacological Reports* 60, 163–170.

Durak, I., Kavutcu, M., Aytac, B., Avci, A., Devrim, E., Ozbek, H. and Ozturk, H.S. (2004) Effects of garlic extract consumption on blood lipid and oxidant/antioxidant parameters in humans with high blood cholesterol. *Journal of Nutritional Biochemistry* 15, 373–377.

Ebbesson, S.O., Risica, P.M., Ebbesson, L.O., Kennish, J.M. and Tejero, M.E. (2005) Omega-3 fatty acids improve glucose tolerance and components of the metabolic syndrome in Alaskan Eskimos: the Alaska Siberia project. *Internal Journal of Circumpolar Health* 64, 396–408.

Ensiyeh, J. and Sakineh, M.A. (2009) Comparing ginger and vitamin B6 for the treatment of nausea and vomiting in pregnancy: a randomised controlled trial. *Midwifery* 25, 649–653.

Erber, E., Hopping, B.N., Beck, L., Sheehy, T., De Roose, E. and Sharma, S. (2010) Assessment of dietary adequacy in a remote Inuvialuit population. *Journal of Human Nutrition and Dietetics* 23(Suppl. 1), 35–42.

Erovic, B.M., Schopper, C., Pammer, J., Vormittag, L., Maleki, A., Brunner, M., Heiduschka, G., Grasl, M. and Thurnher, D. (2010) Multimodal treatment of patients with minor salivary gland cancer in the case of recurrent disease. *Head and Neck* 32, 1167–1172.

Esmaillzadeh, A., Tahbaz, F., Gaieni, I., Alavi-Majd, H. and Azadbakht, L. (2006) Cholesterol-lowering effect of concentrated pomegranate juice consumption in type II diabetic patients with hyperlipidemia. *International Journal for Vitamin and Nutrition Research* 76, 147–151.

Fischer-Rasmussen, W., Kjaer, S.K., Dahl, C. and Asping, U. (1991) Ginger treatment of hyperemesis gravidarum. *European Journal of Obstetrics and Gynecology and Reproductive Biology* 38, 19–24.

Fleischauer, A.T. and Arab, L. (2001) Garlic and cancer: a critical review of the epidemiologic literature. *Journal of Nutrition* 131, 1032S–1040S.

Fowke, J.H., Chung, F.L., Jin, F., Qi, D., Cai, Q., Conaway, C., Cheng, J.R., Shu, X.O., Gao, Y.T. and Zheng, W. (2003) Urinary isothiocyanate levels, brassica, and human breast cancer. *Cancer Research* 63, 3980–3986.

Gavrilova, N.S. and Gavrilov, L.A. (2012) Comments on dietary restriction, okinawa diet and longevity. *Gerontology* 58, 221–223.

Goldman, I.L. (2003) Recognition of fruits and vegetables as healthful: vitamins and phytonutrients. *HortTechnology* 13, 252–258.

Gu, D., Reynolds, K., Wu, X., Chen, J., Duan, X., Reynolds, R.F., Whelton, P.K. and He, J. (2005) Prevalence of the metabolic syndrome and overweight among adults in China. *Lancet* 365, 1398–1405.

Hazuda, H.P., Mitchell, B.D., Haffner, S.M. and Stern, M.P. (1991) Obesity in Mexican American subgroups: findings from the San Antonio Heart Study. *American Journal of Clinical Nutrition* 53, 1529S–1534S.

Holst, B. and Williamson, G. (2004) A critical review of the bioavailability of glucosinolates and related compounds. *Natural Products Reports* 21, 425–447.

Hooper, L., Kroon, P.A., Rimm, E.B., Cohn, J.S., Harvey, I., Le Cornu, K.A., Ryder, J., Hall, W. and Cassidy, A. (2008) Flavonoids, flavonoid-rich foods, and cardiovascular risk: a meta-analysis of randomized controlled trials. *American Journal of Clinical Nutrition* 88, 38–50.

Hopping, B.N., Mead, E., Erber, E., Sheehy, C., Roache, C. and Sharma, S. (2010) Dietary adequacy of Inuit in the Canadian Arctic. *Journal of Human Nutrition and Diet* 23(Suppl. 1), 27–34.

Howes, L.G., Howes, J.B. and Knight, D.C. (2006) Isoflavone therapy for menopausal flushes: a systematic review and meta-analysis. *Maturitas* 55, 203–211.

Hu, F.B., Bronner, L., Willett, W.C., Stampfer, M.J., Rexrode, K.M., Albert, C.M., Hunter, D. and Manson, J.E. (2002) Fish and omega-3 fatty acid intake and risk of coronary heart disease in women. *Journal of the American Medical Association* 287, 1815–1821.

Hursel, R., Viechtbauer, W. and Westerterp-Plantenga, M.S. (2009) The effects of green tea on weight loss and weight maintenance: a meta-analysis. *International Journal of Obesity (London)* 33, 956–961.

Ingram, D., Sanders, K., Kolybaba, M. and Lopez, D. (1997) Case-control study of phyto-oestrogens and breast cancer. *Lancet* 350, 990–994.

Jain, M.G., Hislop, G.T., Howe, G.R. and Ghadirian, P. (1999) Plant foods, antioxidants, and prostate cancer risk: findings from case-control studies in Canada. *Nutrition and Cancer* 34, 173–184.

Jian, L., Xie, L.P., Lee, A.H. and Binns, C.W. (2004) Protective effect of green tea against prostate cancer: a case-control study in southeast China. *International Journal of Cancer* 108, 130–135.

Jiang, B.Q., Zhong, P.H., Cheng, X.B., Yang, X.L., Yang, J. and Cao, Y.F. (2007) Investigation of health and nutrition status of middle-aged and old residents in the urban district of Chongqing. *Asia Pacific Journal of Clinical Nutrition* 16(Suppl. 1), 17–21.

Johnson, J.S., Nobmann, E.D., Asay, E. and Lanier, A.P. (2009) Dietary intake of Alaska Native people in two regions and implications for health: the Alaska Native Dietary and Subsistence Food Assessment Project. *International Journal of Circumpolar Health* 68, 109–122.

Johnson, J.S., Nobmann, E.D. and Asay, E. (2012) Factors related to fruit, vegetable and traditional food consumption which may affect health among Alaska Native People in Western Alaska. *International Journal of Circumpolar Health* 71, 17345.

Joseph, M.A., Moysich, K.B., Freudenheim, J.L., Shields, P.G., Bowman, E.D., Zhang, Y., Marshall, J.R. and Ambrosone, C.B. (2004) Cruciferous vegetables, genetic polymorphisms in glutathione S-transferases M1 and T1, and prostate cancer risk. *Nutrition and Cancer* 50, 206–213.

Joshipura, K.J., Hu, F.B., Manson, J.E., Stampfer, M.J., Rimm, E.B., Speizer, F.E., Colditz, G., Rosner, B., Spiegelman, D. and Willett, W.C. (2001) The effect of fruit and vegetable intake on risk for coronary heart disease. *Annals of Internal Medicine* 134, 1106–1114.

Kearney, P.M., Whelton, M., Reynolds, K., Muntner, P., Whelton, P.K. and He, J. (2005) Global burden of hypertension: analysis of worldwide data. *Lancet* 365, 217–223.

Keys, A., Menotti, A., Aravanis, C., Blackburn, H., Djordevic, B.S., Buzina, R., Dontas, A., Fidanza, F., Karvonen, M., Kimura, N., Mohacek, I., Nedeljkovic, S., Puddu, V., Punsar, S., Taylor, H., Conti, S., Kromhout, D. and Toshima, H. (1984) The seven countries study: 2,289 deaths in 15 years. *Preventive Medicine* 13, 141–154.

Kim, J. and Chan, M.M. (2004) Acculturation and dietary habits of Korean Americans. *British Journal of Nutrition* 91, 469–478.

Kolonel, L.N., Hankin, J.H. and Nomura, A.M. (1985) Multiethnic studies of diet, nutrition, and cancer in Hawaii. *Princess Takamatsu Symposia* 16, 29–40.

Kolonel, L.N., Hankin, J.H., Whittemore, A.S., Wu, A.H., Gallagher, R.P., Wilkens, L.R., John, E.M., Howe, G.R., Dreon, D.M., West, D.W. and Paffenbarger, R.S., Jr (2000) Vegetables, fruits, legumes and prostate cancer: a multiethnic case-control study. *Cancer Epidemiology Biomarkers and Prevention* 9, 795–804.

Kris-Etherton, P.M., Harris, W.S. and Appel, L.J. (2002) Fish consumption, fish oil, omega-3 fatty acids, and cardiovascular disease. *Circulation* 106, 2747–2757.

Kuhnlein, H.V. (2003) Promoting the nutritional and cultural benefits of traditional food systems of Indigenous People. *Forum of Nutrition* 56, 222–223.

Kuhnlein, H.V. and Receveur, O. (2007) Local cultural animal food contributes high levels of nutrients for Arctic Canadian Indigenous adults and children. *Journal of Nutrition* 137, 1110–1114.

Lands, W.E., Hamazaki, T., Yamazaki, K., Okuyama, H., Sakai, K., Goto, Y. and Hubbard, V.S. (1990) Changing dietary patterns. *American Journal of Clinical Nutrition* 51, 991–993.

Larson, N.I., Neumark-Sztainer, D., Hannan, P.J. and Story, M. (2007) Trends in adolescent fruit and vegetable consumption, 1999-2004: project EAT. *American Journal of Preventative Medicine* 32, 147–150.

Larsson, S.C. and Wolk, A. (2005) Tea consumption and ovarian cancer risk in a population-based cohort. *Archives of Internal Medicine* 165, 2683–2686.

Lethaby, A.E., Brown, J., Marjoribanks, J., Kronenberg, F., Roberts, H. and Eden, J. (2007) Phytoestrogens for vasomotor menopausal symptoms. *Cochrane Database System Review*, CD001395.

Li, H., Li, H.Q., Wang, Y., Xu, H.X., Fan, W.T., Wang, M.L., Sun, P.H. and Xie, X.Y. (2004) An intervention study to prevent gastric cancer by micro-selenium and large dose of allitridum. *China Medical Journal (England)* 117, 1155–1160.

Li, Y., Zhai, F., Yang, X., Schouten, E.G., Hu, X., He, Y., Luan, D. and Ma, G. (2007) Determinants of childhood overweight and obesity in China. *British Journal of Nutrition* 97, 210–215.

Liu, J., Ho, S.C., Su, Y.X., Chen, W.Q., Zhang, C.X. and Chen, Y.M. (2009) Effect of long-term intervention of soy isoflavones on bone mineral density in women: a meta-analysis of randomized controlled trials. *Bone* 44, 948–953.

Liu, M., Wu, B., Wang, W.Z., Lee, L.M., Zhang, S.H. and Kong, L.Z. (2007) Stroke in China: epidemiology, prevention, and management strategies. *Lancet Neurology* 6, 456–464.

Lock, M. (1991) Contested meanings of the menopause. *Lancet* 337, 1270–1272.

Ma, D.F., Qin, L.Q., Wang, P.Y. and Katoh, R. (2008) Soy isoflavone intake increases bone mineral density in the spine of menopausal women: meta-analysis of randomized controlled trials. *Clinical Nutrition* 27, 57–64.

Mahmoodi, M., Islami, M.R., Asadi Karam, G.R., Khaksari, M., Sahebghadam Lotfi, A., Hajizadeh, M.R. and Mirzaee, M.R. (2006) Study of the effects of raw garlic consumption on the level of lipids and other blood biochemical factors in hyperlipidemic individuals. *Pakistan Journal of Pharmaceutical Science* 19, 295–298.

McLaughlin, J.B., Middaugh, J.P., Utermohle, C.J., Asay, E.D., Fenaughty, A.M. and Eberhart-Phillips, J.E. (2004) Changing patterns of risk factors and mortality for coronary heart disease among Alaska Natives, 1979-2002. *Journal of the American Medical Association* 291, 2545–2546.

Menotti, A., Kromhout, D., Blackburn, H., Fidanza, F., Buzina, R. and Nissinen, A. (1999) Food intake patterns and 25-year mortality from coronary heart disease: cross-cultural correlations in the Seven Countries Study. The Seven Countries Study Research Group. *European Journal of Epidemiology* 15, 507–515.

Mensink, R.P. and Katan, M.B. (1992) Effect of dietary fatty acids on serum lipids and lipoproteins. A meta-analysis of 27 trials. *Arteriosclerosis, Thrombosis and Vascular Biology* 12, 911–919.

Millen, A.E., Subar, A.F., Graubard, B.I., Peters, U., Hayes, R.B., Weissfeld, J.L., Yokochi, L.A. and Ziegler, R.G. (2007) Fruit and vegetable intake and prevalence of colorectal adenoma in a cancer screening trial. *American Journal of Clinical Nutrition* 86, 1754–1764.

Mozsik, G., Szolcsanyi, J. and Racz, I. (2005) Gastroprotection induced by capsaicin in healthy human subjects. *World Journal of Gastroenterology* 11, 5180–5184.

Munch-Andersen, T., Olsen, D.B., Sondergaard, H., Daugaard, J.R., Bysted, A., Christensen, D.L., Saltin, B. and Helge, J.W. (2012) Metabolic profile in two physically active Inuit groups consuming either a western or a traditional Inuit diet. *International Journal of Circumpolar Health* 71, 17342.

Murphy, N.J., Schraer, C.D., Thiele, M.C., Boyko, E.J., Bulkow, L.R., Doty, B.J. and Lanier, A.P. (1995) Dietary change and obesity associated with glucose intolerance in Alaska Natives. *Journal of American Dietary Association* 95, 676–682.

Neuhouser, M.L., Patterson, R.E., Thornquist, M.D., Omenn, G.S., King, I.B. and Goodman, G.E. (2003) Fruits and vegetables are associated with lower lung cancer risk only in the placebo arm of the beta-carotene and retinol efficacy trial (CARET). *Cancer Epidemiology Biomarkers and Prevention* 12, 350–358.

Nobmann, E.D., Ebbesson, S.O., White, R.G., Bulkow, L.R. and Schraer, C.D. (1999) Associations between dietary factors and plasma lipids related to cardiovascular disease among Siberian Yupiks of Alaska. *International Journal of Circumpolar Health* 58, 254–271.

Nobmann, E.D., Ponce, R., Mattil, C., Devereux, R., Dyke, B., Ebbesson, S.O., Laston, S., Robbins, D., Romenesko, T., Ruotolo, G., Wenger, C.R. and Howard, B.V. (2005) Dietary intakes vary with age among Eskimo adults of Northwest Alaska in the GOCADAN study, 2000-2003. *Journal of Nutrition* 135, 856–862.

Omran, A.R. (1971) The epidemiologic transition. A theory of the epidemiology of population change. *Milbank Memorial Fund Quarterly* 49, 509–538.

Ozgoli, G., Goli, M. and Simbar, M. (2009) Effects of ginger capsules on pregnancy, nausea, and vomiting. *Journal of Alternative and Complementary Medicine* 15, 243–246.

Page, R.L. (2006) Acculturation in Mexican immigrants: a concept analysis. *Journal of Holistic Nursing* 24, 270–278; quiz 279–281.

Panagiotakos, D.B., Arapi, S., Pitsavos, C., Antonoulas, A., Mantas, Y., Zombolos, S. and Stefanadis, C. (2006) The relationship between adherence to the Mediterranean diet and the severity and short-term prognosis of acute coronary syndromes (ACS): The Greek Study of ACS (The GREECS). *Nutrition* 22, 722–730.

Pantuck, A.J., Leppert, J.T., Zomorodian, N., Aronson, W., Hong, J., Barnard, R.J., Seeram, N., Liker, H., Wang, H., Elashoff, R., Heber, D., Aviram, M., Ignarro, L. and Belldegrun, A. (2006) Phase II study of pomegranate juice for men with rising prostate-specific antigen following surgery or radiation for prostate cancer. *Clinical Cancer Research* 12, 4018–4026.

Parkinson, A.J., Cruz, A.L., Heyward, W.L., Bulkow, L.R., Hall, D., Barstaed, L. and Connor, W.E. (1994) Elevated concentrations of plasma omega-3 polyunsaturated fatty acids among Alaskan Eskimos. *American Journal of Clinical Nutrition* 59, 384–388.

Patil, C.L., Hadley, C. and Nahayo, P.D. (2009) Unpacking dietary acculturation among new Americans: results from formative research with African refugees. *Journal of Immigrant Minority Health* 11, 342–358.

Pongrojpaw, D., Somprasit, C. and Chanthasenanont, A. (2007) A randomized comparison of ginger and dimenhydrinate in the treatment of nausea and vomiting in pregnancy. *Journal of the Medical Association of Thailand* 90, 1703–1709.

Popkin, B.M. (1994) The nutrition transition in low-income countries: an emerging crisis. *Nutrition Reviews* 52, 285–298.

Popkin, B.M. (2004) The nutrition transition: an overview of world patterns of change. *Nutrition Reviews* 62, S140–143.

Popkin, B.M. (2006) Global nutrition dynamics: the world is shifting rapidly toward a diet linked with non-communicable diseases. *American Journal of Clinical Nutrition* 84, 289–298.

Popkin, B.M. (2009) Global changes in diet and activity patterns as drivers of the nutrition transition. *Nestle Nutrition Workshop Series. Paediatric Programme* 63, 1–10; discussion 10–14, 259–268.

Porter, P. (2008) 'Westernizing'women's risks? Breast cancer in lower-income countries. *New England Journal of Medicine* 358, 213–216.

Posner, B.M., Franz, M. and Quatromoni, P. (1994) Nutrition and the global risk for chronic diseases: the INTERHEALTH nutrition initiative. The INTERHEALTH Steering Committee. *Nutrition Reviews* 52, 201–207.

Potter, S.M., Baum, J.A., Teng, H., Stillman, R.J., Shay, N.F. and Erdman, J.W., Jr (1998) Soy protein and isoflavones: their effects on blood lipids and bone density in postmenopausal women. *American Journal of Clinical Nutrition* 68, 1375S–1379S.

Receveur, O., Boulay, M. and Kuhnlein, H.V. (1997) Decreasing traditional food use affects diet quality for adult Dene/Metis in 16 communities of the Canadian Northwest Territories. *Journal of Nutrition* 127, 2179–2186.

Risica, P.M., Nobmann, E.D., Caulfield, L.E., Schraer, C. and Ebbesson, S.O. (2005) Springtime macronutrient intake of Alaska natives of the Bering Straits Region: the Alaska Siberia Project. *International Journal of Circumpolar Health* 64, 222–233.

Rith-Najarian, S.J., Gohdes, D.M., Shields, R., Skipper, B., Moore, K.R., Tolbert, B., Raymer, T. and Acton, K.J. (2002) Regional variation in cardiovascular disease risk factors among American Indians and Alaska Natives with diabetes. *Diabetes Care* 25, 279–283.

Ross, R.K., Bernstein, L., Lobo, R.A., Shimizu, H., Stanczyk, F.Z., Pike, M.C. and Henderson, B.E. (1992) 5-alpha-reductase activity and risk of prostate cancer among Japanese and US white and black males. *Lancet* 339, 887–889.

Sasazuki, S., Inoue, M., Miura, T., Iwasaki, M. and Tsugane, S. (2008) Plasma tea polyphenols and gastric cancer risk: a case-control study nested in a large population-based prospective study in Japan. *Cancer Epidemiology Biomarkers and Prevention* 17, 343–351.

Satia, J.A. (2010) Dietary acculturation and the nutrition transition: an overview. *Applied Physiology, Nutrition, and Metabolism* 35, 219–223.

Satia, J.A., Patterson, R.E., Kristal, A.R., Hislop, T.G., Yasui, Y. and Taylor, V.M. (2001) Development of scales to measure dietary acculturation among Chinese-Americans and Chinese-Canadians. *Journal of the American Dietetic Association* 101, 548–553.

Scarmeas, N., Stern, Y., Tang, M.X., Mayeux, R. and Luchsinger, J.A. (2006) Mediterranean diet and risk for Alzheimer's disease. *Annals of Neurology* 59, 912–921.

Services, D. o. H. a. H. (2012) Cancer Trends Progress Report 2011/2012. Update, National Cancer Institute, NIH, DHHS, Bethesda, Maryland, August 2012. Available at: http://progressreport.cancer.gov (Accessed 14 April 2013).

Sharma, S., De Roose, E., Cao, X., Pokiak, A., Gittelsohn, J. and Corriveau, A. (2009) Dietary intake in a population undergoing a rapid transition in diet and lifestyle: the Inuvialuit in the Northwest Territories of Arctic Canada. *Canadian Journal of Public Health* 100, 442–448.

Sharma, S., Cao, X., Roache, C., Buchan, A., Reid, R. and Gittelsohn, J. (2010) Assessing dietary intake in a population undergoing a rapid transition in diet and lifestyle: the Arctic Inuit in Nunavut, Canada. *British Journal of Nutrition* 103, 749–759.

Silagy, C. and Neil, A. (1994) Garlic as a lipid lowering agent – a meta-analysis. *Journal of the Royal College of Physicians of London* 28, 39–45.

Sobenin, I.A., Andrianova, I.V., Demidova, O.N., Gorchakova, T. and Orekhov, A.N. (2008) Lipid-lowering effects of time-released garlic powder tablets in double-blinded placebo-controlled randomized study. *Journal of Athererosclerosis and Thrombosis* 15, 334–338.

Sobenin, I.A., Pryanishnikov, V.V., Kunnova, L.M., Rabinovich, Y.A., Martirosyan, D.M. and Orekhov, A.N. (2010) The effects of time-released garlic powder tablets on multifunctional cardiovascular risk in patients with coronary artery disease. *Lipids in Health and Disease* 9, 119.

Solomons, N.W. and Gross, R. (1995) Urban nutrition in developing countries. *Nutrition Reviews* 53, 90–95.

Sonoda, T., Nagata, Y., Mori, M., Miyanaga, N., Takashima, N., Okumura, K., Goto, K., Naito, S., Fujimoto, K., Hirao, Y., Takahashi, A., Tsukamoto, T., Fujioka, T. and Akaza, H. (2004) A case-control study of diet and prostate cancer in Japan: possible protective effect of traditional Japanese diet. *Cancer Science* 95, 238–242.

Stowe, C.B. (2011) The effects of pomegranate juice consumption on blood pressure and cardiovascular health. *Complemental Therapies in Clinical Practice* 17, 113–115.

Sumner, M.D., Elliott-Eller, M., Weidner, G., Daubenmier, J.J., Chew, M.H., Marlin, R., Raisin C.J. and Ornish, D. (2005) Effects of pomegranate juice consumption on myocardial perfusion in patients with coronary heart disease. *American Journal of Cardiology* 96, 810–814.

Tanaka, S., Haruma, K., Yoshihara, M., Kajiyama, G., Kira, K., Amagase, H. and Chayama, K. (2006) Aged garlic extract has potential suppressive effect on colorectal adenomas in humans. *Journal of Nutrition* 136, 821S–826S.

Terry, P., Wolk, A., Persson, I. and Magnusson, C. (2001) Brassica vegetables and breast cancer risk. *Journal of the American Medical Association* 285, 2975–2977.

Trichopoulou, A., Costacou, T., Bamia, C. and Trichopoulos, D. (2003) Adherence to a Mediterranean diet and survival in a Greek population. *New England Journal of Medicine* 348, 2599–2608.

Trichopoulou, A., Bamia, C. and Trichopoulos, D. (2009) Anatomy of health effects of Mediterranean diet: Greek EPIC prospective cohort study. *British Medical Journal* 339, 26–29.

Tunstall-Pedoe, H., Kuulasmaa, K., Mahonen, M., Tolonen, H., Ruokokoski, E. and Amouyel, P. (1999) Contribution of trends in survival and coronary-event rates to changes in coronary heart disease mortality: 10-year results from 37 WHO MONICA project populations. Monitoring trends and determinants in cardiovascular disease. *Lancet* 353, 1547–1557.

Verhoeven, D.T., Goldbohm, R.A., van Poppel, G., Verhagen, H. and van den Brandt, P.A. (1996) Epidemiological studies on brassica vegetables and cancer risk. *Cancer Epidemiology Biomarkers and Prevention* 5, 733–748.

Voorrips, L.E., Goldbohm, R.A., van Poppel, G., Sturmans, F., Hermus, R.J. and van den Brandt, P.A. (2000a) Vegetable and fruit consumption and risks of colon and rectal cancer in a prospective cohort study: The Netherlands Cohort Study on Diet and Cancer. *American Journal of Epidemiology* 152, 1081–1092.

Voorrips, L.E., Goldbohm, R.A., Verhoeven, D.T., van Poppel, G.A., Sturmans, F., Hermus, R.J. and van den Brandt, P.A. (2000b) Vegetable and fruit consumption and lung cancer risk in the Netherlands Cohort Study on diet and cancer. *Cancer Causes Control* 11, 101–115.

Vorster, H.H., Bourne, L.T., Venter, C.S. and Oosthuizen, W. (1999) Contribution of nutrition to the health transition in developing countries: a framework for research and intervention. *Nutrition Reviews* 57, 341–349.

Walker, H.A., Dean, T.S., Sanders, T.A., Jackson, G., Ritter, J.M. and Chowienczyk, P.J. (2001) The phytoestrogen genistein produces acute nitric oxide-dependent dilation of human forearm vasculature with similar potency to 17beta-estradiol. *Circulation* 103, 258–262.

Walters, D.G., Young, P.J., Agus, C., Knize, M.G., Boobis, A.R., Gooderham, N.J. and Lake, B.G. (2004) Cruciferous vegetable consumption alters the metabolism of the dietary carcinogen 2-amino-1-methyl-6-phenylimidazo[4,5-b]pyridine (PhIP) in humans. *Carcinogenesis* 25, 1659–1669.

Wang, M., Guo, C. and Li, M. (1999) A case-control study on the dietary risk factors of upper digestive tract cancer. *Zhonghua Liu Xing Bing Xue Za Zhi* 20, 95–97.

Warshafsky, S., Kamer, R.S. and Sivak, S.L. (1993) Effect of garlic on total serum cholesterol. A meta-analysis. *Annals of Internal Medicine* 119, 599–605.

Wild, S., Roglic, G., Green, A., Sicree, R. and King, H. (2004) Global prevalence of diabetes: estimates for the year 2000 and projections for 2030. *Diabetes Care* 27, 1047–1053.

Willcox, B.J., Willcox, D.C., Todoriki, H., Fujiyoshi, A., Yano, K., He, Q., Curb, J.D. and Suzuki, M. (2007) Caloric restriction, the traditional Okinawan diet, and healthy aging: the diet of the world's longest-lived people and its potential impact on morbidity and life span. *Annals of the New York Academy of Science* 1114, 434–455.

Willcox, D.C., Willcox, B.J., Todoriki, H. and Suzuki, M. (2009) The Okinawan diet: health implications of a low-calorie, nutrient-dense, antioxidant-rich dietary pattern low in glycemic load. *Journal of the American College of Nutrition* 28(Suppl.), 500S–516S.

Williams, M.J., Sutherland, W.H., McCormick, M.P., Yeoman, D.J. and de Jong, S.A. (2005) Aged garlic extract improves endothelial function in men with coronary artery disease. *Phytotherapy Research* 19, 314–319.

Wu, K.L., Rayner, C.K., Chuah, S.K., Changchien, C.S., Lu, S.N., Chiu, Y.C., Chiu, K.W. and Lee, C.M. (2008) Effects of ginger on gastric emptying and motility in healthy humans. *European Journal of Gastroenterology and Hepatology* 20, 436–440.

Yang, G., Fan, L., Tan, J., Qi, G., Zhang, Y., Samet, J.M., Taylor, C.E., Becker, K. and Xu, J. (1999) Smoking in China: findings of the 1996 National Prevalence Survey. *Journal of the American Medical Association* 282, 1247–1253.

Yeoh, K.G., Kang, J.Y., Yap, I., Guan, R., Tan, C.C., Wee, A. and Teng, C.H. (1995) Chili protects against aspirin-induced gastroduodenal mucosal injury in humans. *Digestive Diseases and Sciences* 40, 580–583.

Yoon, K.H., Lee, J.H., Kim, J.W., Cho, J.H., Choi, Y.H., Ko, S.H. and Son, H.Y. (2006) Epidemic obesity and type 2 diabetes in Asia. *Lancet* 368, 1681–1688.

Yu, Z., Nissinen, A., Vartiainen, E., Song, G., Guo, Z. and Tian, H. (2000) Changes in cardiovascular risk factors in different socioeconomic groups: seven year trends in a Chinese urban population. *Journal of Epidemiology and Community Health* 54, 692–696.

Zhang, M., Lee, A.H., Binns, C.W. and Xie, X. (2004) Green tea consumption enhances survival of epithelial ovarian cancer. *International Journal of Cancer* 112, 465–469.

Zhang, X.H., Lu, Z.L. and Liu, L. (2008) Coronary heart disease in China. *Heart* 94, 1126–1131.

4 Resveratrol: A Chemo-Preventative Agent with Diverse Applications

Charu Gupta,[1] Girish Sharma[2] and Daniel Chan[3*]
[1]*Amity Institute of Herbal Research & Studies;* [2]*Amity Institute of Biotechnology, Amity University-Uttar Pradesh, Noida, India;* [3]*University of Colorado Denver, Division of Medical Oncology, Aurora, Colorado, USA*

4.1 Introduction

With the rapid advances made over the last two decades in biomedical research, there has been an unprecedented interest in unravelling the magical properties of some commonly used natural products. Consequently, a wide variety of natural products are under scrutiny for their clinical potential, both in terms of disease prevention and treatment. Resveratrol, a phytoalexin, found in grapes and wines, exhibits a wide range of pharmacological properties and plays an important role in the prevention of human cardiovascular disease. The history of resveratrol dates back to the early 1990s when epidemiological investigations showed that the incidence of myocardial infarction in France is only one-third of that in the USA, despite the presence of equally prevalent cardiovascular risk factors, exemplified by high fat intake, lack of exercise and heavy cigarette smoking. This phenomenon, commonly referred to as the 'French paradox' (Alarcón de la Lastra *et al.*, 2006), has been attributed to the regular consumption of red wine at meals by the French and further implies that mild or moderate consumption of red wine could reduce the mortality and morbidity of atherosclerosis and coronary heart disease (CHD) (Das and Maulik, 2006). This beneficial effect is increasingly attributed to the polyphenol resveratrol that is present in red wine. A volume of 1 l red wine contains 1.5–3 mg resveratrol.

Resveratrol (Fig. 4.1) is a natural compound belonging to stilbenes and is made only by certain plants classified as spermatophytes. It is a phytoalexin, an antimicrobial compound synthesized by plants in response to injury and infection. In particular, it is made in response to fungal infection and exposure to UV light and ozone.

Resveratrol (3,5,4′-trihydroxystilbene) was first isolated from the roots of white hellebore (*Veratrum grandiflorum* O. Loes) in 1940 (Takaoka, 1940) and later in 1963 from the roots of *Polygonum cuspidatum*, a plant used in traditional Chinese and Japanese medicine (Nonomura *et al.*, 1963). Later on, in 1976, resveratrol was discovered in grapes, and in 1992 resveratrol was also identified in wine.

Resveratrol is *trans*-stilbene that undergoes isomerization under ultraviolet (UV) radiation. It is the *trans* form of resveratrol that has been shown to display a much broader spectrum of pharmacological activity than its *cis* isomer. Stilbenes, in particular

* Email: Dan.Chan@ucdenver.edu

 Phytochemicals of Nutraceutical Importance (eds D. Prakash and G. Sharma)

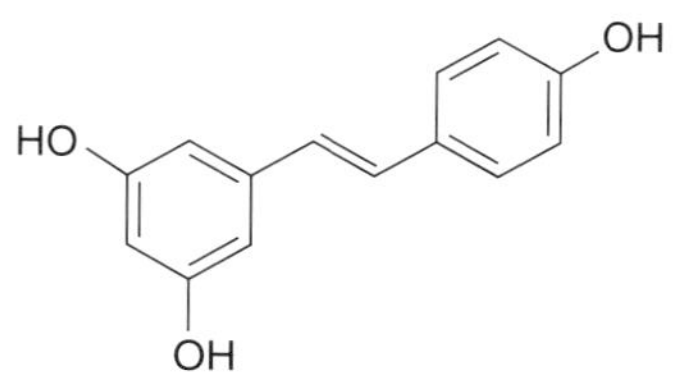

Fig. 4.1. Chemical structure of resveratrol.

trans-resveratrol and its glucoside, are widely reported to be beneficial to health and possess antioxidative, anticarcinogenic and antitumour properties (Burns *et al.*, 2002). *Trans*-resveratrol is synthesized naturally by several plants in response to pathogen infection, traumatic damage, UV irradiation and other stresses. The accumulation of resveratrol in plant cells increases in response to fungal infections (biotic stress) or other physical stresses such as UV radiation, ultrasound, wounding such as slicing as well as in response to invasion of chemicals such as hydrogen peroxide, paraquat etc. (Burns *et al.*, 2002).

This molecule also plays a major role in both cancer prevention and therapy. It was found that low to moderate consumption of red wine has a relatively greater benefit than other alcoholic beverages in the prevention of atherosclerosis and CHD.

Resveratrol is a natural compound made by plants. It is a phytoalexin, made by plants in stress conditions and in response to pathogen attack. It is found in considerable concentrations in grapes, groundnuts etc. In the diet, the major source is found in red wine. Resveratrol is made in grape skins, but not in the flesh, thus it is found in small quantities in white wine, and proportionally more in rosé wines with the highest concentration in red wines. However, the concentration of resveratrol, even in red wines, can differ greatly, depending on the location and country where the grapes were grown, the soil properties, the cultivation and wine-making methods etc. There have been many recent studies of red wine consumption and health. Studies show drinking red wine regularly may have health benefits. However, blood levels of resveratrol are extremely low (nanomolar range) even after one glass of wine, while its metabolites are much higher (micromolar range). This has caused dispute as to whether resveratrol can be present in high enough concentrations to be beneficial by ingestion. However, others have reported that resveratrol can be active even at micromolar concentrations.

4.2 Occurrence/sources

The richest natural sources of resveratrol are dark grape extracts (*Vitis vinifera*) and giant knotweed (*Polygonum cuspidatum*, a perennial shrub). It is also found in abundance in labrusca and muscadine grapes. It is also present in other plants such as *Eucalyptus*, spruce and lily and in foods such as mulberries, groundnuts, blueberries, strawberries, hops and their products (Valenzano *et al.*, 2006; Wang *et al.*, 2007). It also occurs in grape vines, roots, seeds and stalks, but its highest concentration is in the skin, which contains 50–100 $\mu g\ g^{-1}$ (Burns *et al.*, 2002). Resveratrol is a phytoalexin, a class of antibiotic compounds produced as a part of a plant's defence system against disease. For example, in response to an invading fungus, resveratrol is synthesized from p-coumaroyl CoA and malonyl CoA (Soleas *et al.*, 1997). Since fungal infections are more common in cooler climates, grapes grown in cooler climates have a higher concentration of resveratrol (Kopp, 1998). The total resveratrol content of different wines and foods was described by Romero-Perez *et al.* (1996, 1999), Sobolev and Cole (1999), Sanders *et al.* (2000), Burns *et al.* (2002) and Moreno-Labanda *et al.* (2004) and are given in Table 4.1.

4.3 Chemical Structure of Resveratrol

4.3.1 Synonyms

There are various synonyms for resveratrol based on their chemical structure:

- (E)-resveratrol;
- (E)-5-(2-(4-hydroxyphenyl)ethenyl)-1,3-benzenediol;

Table 4.1. Resveratrol content of some foods.

Sources	Resveratrol
Wine	0.32–15.35 μg g^{-1}
Peanut butter	0.02–0.98 μg g^{-1}
Groundnut	0.01–0.07 μg g^{-1}
Green groundnut	0.19–0.72 μg g^{-1}
Polygonum cuspidatum	296–377 μg g^{-1}
Green grape	0.02–0.32 μg g^{-1}
Black grape	0.95–1.88 μg g^{-1}
Raisin	0.0005–0.003 μg g^{-1}
Grape juice – black	Trace–0.09 μg g^{-1}
Grape juice – green	Trace–0.01 μg g^{-1}
White wine (Spanish)	0.05–1.80 mg l^{-1}
Rosé wine (Spanish)	0.43–3.52 mg l^{-1}
Red wine (Spanish)	1.92–12.59 mg l^{-1}
Red wine (global)	1.98–7.13 mg l^{-1}
Red grape juice (Spanish)	1.14–8.69 mg l^{-1}

- (E)-5-(p-hydroxystyryl) resorcinol;
- (E)-5-(2-(4-hydroxyphenyl)ethenyl)-1,3-benzenediol;
- 3,4′,5-stilbenetriol;
- 3,4′,5-trihydroxystilbene;
- *trans*-3,4′,5-trihydroxystilbene.

4.4 Classification

Resveratrol is classified according to its isomers into the following categories.

4.4.1 Oxyresveratrol

An isomer of hydroxylated resveratrol, oxyresveratrol is an excellent inhibitor of the enzyme tyrosinase. Tyrosinase is the enzyme responsible for colouring of skin, hair, eyes in animals and also for the browning of fruits and vegetables. Oxyresveratrol exhibits potent inhibitory activity of tyrosinase with IC_{50} value of 1.2 μM on mushroom tyrosinase activity, which is many times stronger than kojic acid, a de-pigmenting agent, used as a cosmetic material with skin-whitening effect and the medical agent for hyper-pigmentation disorders (Kim *et al.*, 2002). A neuroprotective effect of oxyresveratrol in cultured rat cortical neurons has also been observed in recent studies (Ban *et al.*, 2006).

4.4.2 Gnetol

Gnetol is a positional isomer of oxyresveratrol. It occurs naturally in *Gnetum*, a special group of gymnosperm plants. Gnetol has also been found to have tyrosinase inhibition and was found to be a stronger inhibitor of murine tyrosinase activity (IC_{50}, 4.5 μM) than a standard inhibitor, kojic acid (IC_{50}, 139 μM). In a recent study, gnetol significantly suppressed melanin biosynthesis in murine B16 melanoma cells. Gnetol is a promising pharmacological or cosmetic agent (Ohguchi *et al.*, 2003).

4.4.3 Monomethyl resveratrol

Monomethyl resveratrol occurs naturally in *Muscari comosum*, a popular plant in southern Italy used in gastronomy (Borgonovo *et al.*, 2008). This compound exhibits an apoptosis-inducing activity against sensitive and resistant leukaemia cells and also acts as an antioxidant.

4.4.4 Pterostilbene (dimethyl resveratrol)

Pterostilbene is a dimethylated analogue of resveratrol. It naturally occurs in *Pterocarpus marsupium*, which possesses antidiabetic activity. Pterostilbene significantly reduces the plasma glucose levels and its effect is comparable to metformin (Alarcón de la Castra *et al.*, 2006).

4.4.5 Trimethyl resveratrol

The *cis* isomer of per-methylated resveratrol (trimethylated) is found to occur in more than five different plants. This compound exhibits anticancer properties. It especially inhibits ornithine decarboxylase, an enzyme involved in the rate-determining step in the synthesis of polyamines.

4.5 Physiological Effects

Resveratrol produces various physiological effects. At low concentrations that normally occur in food, resveratrol has been shown to exert neuroprotective effects (De Ruvo *et al.*, 2000), as well as beneficial effects on the cardiovascular system (Pace-Asciak *et al.*, 1995; Pendurthi *et al.*, 1999). These effects are mostly attributed to its antioxidant properties. More recently, resveratrol has been evaluated for its health benefits in other medical areas, such as anticancer activity when administered at higher, non-physiological doses. Researchers believe that 500 mg daily is the minimum amount of resveratrol needed to protect against cancer. A glass of red wine contains approximately 640 μg of resveratrol. In these conditions, resveratrol inhibits the proliferation and induces apoptotic cell death in multiple cancer cell types *in vitro* (Hsieh and Wu, 1999; Pozo-Guisado *et al.*, 2002; Kim *et al.*, 2004); moreover, in animal models of cancer, resveratrol has been shown to inhibit angiogenesis and delay tumour growth (Tseng *et al.*, 2004), impede carcinogenesis (Gusman *et al.*, 2001) and reduce experimental metastasis (Busquets *et al.*, 2006). Resveratrol acts on the process of carcinogenesis by affecting the tumour initiation, promotion and progression phases and suppresses the final steps of carcinogenesis, i.e. angiogenesis and metastasis. It is also able to activate apoptosis, arrest the cell cycle or inhibit kinase pathways. Most noticeable biological activities are antithrombogenic, anti-inflammatory, cardioprotective, neuroprotective, anti-ageing and cancer preventive and therapeutic activities.

Of greatest interest is its ability as chemoprevention and antimutagen. It is shown to inhibit cellular processes involved in the initiation and progression of carcinogenesis. It lowers platelet aggregation and thrombosis, and potentially decreases cardiovascular disease by several mechanisms. It is extremely effective in reducing skin tumours in mouse models. It can act as a phytoestrogen, a plant compound that in humans can mimic the hormone oestrogen. It is a concentration dependent oestrogen/anti-oestrogen. Even in less healthy diets/diets high in fat, resveratrol can have benefits. It intercalates into the lipid membrane of cells and has been shown to help control cholesterol levels.

Resveratrol has two forms, *cis* and *trans*. When isolated from plants, it is only found in the *trans* form. *Trans*-resveratrol can also act as a preventive agent against important pathologies, i.e. vascular diseases, cancers, viral infection or neurodegenerative processes. It is produced in huge amounts in grapevine skin in response to infection by *Botrytis cinerea*. This production of resveratrol blocks the proliferation of the pathogen, thereby acting as a natural antibiotic. It plays a role in inhibiting many cellular pathways associated with cancer and disease, including apoptosis.

4.6 Metabolism and Bioavailability

Although *trans*-resveratrol appears to be well-absorbed by humans when taken orally, its bioavailability is relatively low due to its rapid metabolism and elimination (Ohguchi *et al.*, 2003; Ban *et al.*, 2006). The oral absorption of resveratrol in humans is about 75% and occurs mainly by trans-epithelial diffusion. The major sites of metabolism for resveratrol include the intestine and liver. Deconjugation enzymes such as β-glucuronidase and sulfatase, as well as specific tissue accumulation of resveratrol, may enhance resveratrol efficacy at target sites. Due to extensive metabolism in the intestine and liver, oral bioavailability is less than 1%. Resveratrol metabolites are primarily detected upon oral exposure to *trans*-resveratrol. The bioavailability of resveratrol from grape juice, which contains mostly glucosides of resveratrol (piceid), may be even lower than that of *trans*-resveratrol. A recent study reported that bioavailability of *trans*-resveratrol from red wine did not differ when the wine was consumed with a meal (low- or high-fat) versus on an empty stomach. Resveratrol analogues, such as methylated derivatives with improved bioavailability, may be important in future research (Walle, 2011).

4.7 Role of Resveratrol in Biological Systems

4.7.1 Antioxidants

The antioxidant activity of resveratrol is one of the most beneficial discoveries. Resveratrol is both a free-radical scavenger and a potent antioxidant because of its ability to promote the activities of a variety of antioxidant enzymes. The ability of the polyphenolic compounds to act as antioxidants depends on the redox properties of their phenolic hydroxy groups and the potential for electron delocalization across the chemical structure (Alarcón de la Lastra *et al.*, 2006).

The role of resveratrol as a natural antioxidant was clarified by Zini *et al.* (1999), who suggested three different antioxidant mechanisms: (i) competition with coenzyme Q and, to decrease the oxidative chain complex, the site of ROS generation; (ii) scavenging O_2 free radicals formed in the mitochondria; and (iii) inhibition of LP (lipid peroxidation) induced by Fenton reaction products. Numerous studies have demonstrated the ability of resveratrol to scavenge both O_2 and OH free radicals (Martínez and Moreno, 2000; Orallo *et al.*, 2002; Leonard *et al.*, 2003; Losa, 2003).

By contrast, in a study by Orallo *et al.* (2002) using the enzymatic hypoxanthine oxidase–XO (xanthine oxidase) system, resveratrol neither affected the XO activity nor scavenged O_2 free radicals in rat macrophage extracts.

Resveratrol can maintain the concentration of intracellular antioxidants found in biological systems. For example, stilbene maintained the glutathione content in peripheral blood mononuclear cells isolated *ex vivo* from a healthy human from oxidative damage caused by 2-deoxy-D-ribose (Losa, 2003). In human blood platelets, resveratrol markedly decreased oxidation of thiol groups of proteins in these cells (Olas *et al.*, 2004). Similarly, resveratrol induced an increase in glutathione levels in a concentration-dependent manner in human lymphocytes activated with H_2O_2. In another study, resveratrol increased the amounts of several antioxidant enzymes, including glutathione peroxidase, glutathione S-transferase and glutathione reductase (Yen *et al.*, 2003). The antioxidant potential of resveratrol for the protection of polyunsaturated fatty acids (PUFA) has been described by Lastra and Villegas (2007).

4.7.2 Platelet aggregation inhibitors

Resveratrol has been reported to have antiplatelet activity (Lin *et al.*, 2009); however, the detailed mechanisms have not yet been resolved. In a recent study, protein kinase C inhibitor (PKCI) and resveratrol (RSVL) had an additive effect in inhibiting platelet aggregation and platelet membrane-bound fibrinogen (PFig) content. Furthermore, RSVL (final concentration 50 μM) remarkably depressed the activity of protein kinase C (PKC) in the membrane of platelets and the percentage of membrane PKC activity in total PKC activity (Yang *et al.*, 2011).

In another study, resveratrol (0.05–0.25 μmol l^{-1}) showed stronger inhibition of platelet aggregation stimulated by collagen (1 μg ml^{-1}) than other agonists (Shen *et al.*, 2007). In yet another study, it was found that resveratrol, at 10–1000 μmol l^{-1} concentration, significantly inhibited platelet aggregation *in vitro* induced by collagen, thrombin and ADP in healthy subjects (Wang *et al.*, 2002).

4.7.3 Enzyme inhibitors

Trans-resveratrol inhibits oxidative enzymes in an animal cell system. It inhibited superoxide dismutase, lipoxygenase, catalase, peroxidase, polyphenol oxidase and 1-aminocyclopropane-1-carboxylic acid oxidase. *Trans*-resveratrol also inhibits lipoxygenase activity more effectively than other lipoxygenase inhibitors, including propyl gallate, ibuprofen, ursolic acid, acetylsalicylic acid, and salicyl-hydroxamic acid (Fan and Matthesis, 2001). The rate of inhibition increases with *trans*-resveratrol concentration. Resveratrol with antioxidant activity inhibits matrix metalloproteinase via modulation of SIRT1 in human fibrosarcoma cells, providing evidence that resveratrol can be a potential candidate for chemoprevention of cancer (Lee and Kim, 2011). In a recent study, resveratrol specifically inhibited inducible nitric oxide synthase (iNOS) induction in muscle through a mechanism involving AMP-activated protein kinase (AMPK) but not deacetylase enzyme (SIRT1) activation. This anti-inflammatory action of resveratrol likely contributes to the therapeutic effect of this plant polyphenol (Centeno-Baez *et al.*, 2011).

In another study, resveratrol inhibited neuronal apoptosis and elevated Ca^{2+}/calmodulin-dependent protein kinase II activity in diabetic mouse retina. It was concluded that resveratrol prevents diabetes-induced RGC death via calmodulin-dependent protein kinase II (CaMKII) down-regulation, implying that resveratrol may have potential therapeutic applications for prevention of diabetes-induced visual dysfunction (Kim *et al.*, 2010).

Resveratrol, a red wine polyphenol, suppresses pancreatic cancer by inhibiting leucotriene A4 hydrolase. It exerts relatively stronger inhibitory effects than bestatin, an established inhibitor of LTA(4)H activity (Oi *et al.*, 2010).

4.7.4 Anticarcinogenic agents

Resveratrol has been shown to inhibit carcinogenesis by affecting various molecular events in the initiation, promotion and progression stages (Aggarwal *et al.*, 2004; Signorelli and Ghidoni, 2005; Delmas *et al.*, 2006; Athar *et al.*, 2007; Bishayee, 2009). The anti-initiation activity of resveratrol has been linked to the suppression of the metabolic activation and/or induction of detoxification of carcinogens through modulation of enzymes involved in either phase I reactions (i.e. cytochrome P450 enzymes (CYP)) or phase II conjugation reactions. A number of *in vitro* studies have shown that resveratrol inhibits CYP1A1 and CYP1A2 enzyme activities (Yueh *et al.*, 2005).

Modulation of enzyme systems involved in carcinogen activation and detoxification could be one of the biochemical mechanisms responsible for the cancer-preventive effect of resveratrol. However, such changes may also affect drug efficacy and toxicity because these enzymes are also responsible for drug metabolism.

Extensive *in vitro* studies revealed multiple intracellular targets of resveratrol, which affect cell growth, inflammation, apoptosis, angiogenesis, and invasion and metastasis. These include tumour suppressors p53 and Rb; cell cycle regulators, cyclins, CDKs, p21WAF1, p27KIP and INK and the checkpoint kinases ATM/ATR; transcription factors NF-κB, AP-1, c-Jun, and c-Fos; angiogenic and metastatic factors, VEGF and matrix metalloprotease 2/9; cyclooxygenases for inflammation; and apoptotic and survival regulators, Bax, Bak, PUMA, Noxa, TRAIL, APAF, survivin, Akt, Bcl2 and Bcl-X(L). In addition to its well-documented antioxidant properties, there is increasing evidence that resveratrol exhibits pro-oxidant activity under certain experimental conditions, causing oxidative DNA damage that may lead to cell cycle arrest or apoptosis (Athar *et al.*, 2009).

A recent study showed for the first time antiproliferative, DNA damaging and apoptotic effects of resveratrol in HNSCC cells independent of Smad4 status, both *in vitro* and *in vivo*, suggesting that more studies are needed to establish its potential usefulness against head and neck squamous cell carcinoma (HNSCC) (Tyagi *et al.*, 2011).

4.7.5 Antineoplastic and phytogenic agents

As early as 1997, researchers proposed to use resveratrol as a cancer-preventive agent. Resveratrol was shown to have anticancer activities in assays representing three major stages of carcinogenesis. It has been shown to inhibit cancer initiation and promotion (Jang *et al.*, 1997). It acts as a selective oestrogen receptor modulator (SERM) and regulates proteins involved in DNA synthesis and cell cycle. Resveratrol also affects the activity of transcriptional factors involved in proliferation and stress responses, such as NF-κB, AP-1 and Egr1 (Signorelli and Ghidoni, 2005).

Researchers at Cook County Hospital (Chicago, Illinois) found that gastric adenocarcinoma cells respond to resveratrol treatment with suppression of DNA synthesis, activation of nitric oxide synthase, induction of apoptosis and inhibition of total PKC and PKC-α activity (Atten *et al.*, 2005).

4.7.6 Resveratrol for arthritis

Arthritis is the inflammation of the joints, and is a chronic disease that results from dysregulation of pro-inflammatory cytokines (e.g. tumour necrosis factor (TNF) and interleukin-1-β (IL-1β)) and pro-inflammatory enzymes that mediate the production of prostaglandins (e.g. cyclooxygenase-2) and leucotrienes (e.g. lipooxygenase), together with the expression of adhesion molecules and matrix metalloproteinases, and hyper-proliferation of synovial fibroblasts. All of these factors are regulated by the activation of the transcription factor nuclear factor-kappa B. Thus, any agent that can suppress the expression of TNFα, IL-1β, cyclooxygenase-2, lipooxygenase, matrix metalloproteinases or adhesion molecules, or suppress the activation of NF-κB, have the potential to cure arthritis (Khanna *et al.*, 2007). Resveratrol was found be an inhibitor or a mediator for some of these compounds in our body. Thus, resveratrol may benefit people suffering from arthritis.

4.7.7 Resveratrol for cardiovascular diseases

Resveratrol is a phytoestrogen, potent antioxidant, reactive oxygen species scavenger and metal chelator (Olas and Wachowicz, 2005). Thus, resveratrol may have benefits for protection of the cardiovascular system against ischaemic-reperfusion injury; it may also protect and maintain the intact endothelium; it exhibits anti-atherosclerotic properties, which inhibits the LDL oxidation, suppress platelet aggregation and exhibits oestrogen-like action (Dong and Ren, 2004; Olas and Wachowicz, 2005) Thus, resveratrol may benefit people at risk of certain cardiovascular conditions.

4.7.8 Resveratrol for diabetes

Studies showed that resveratrol increased lifespan in lower organisms by activating the NAD(+)-dependent histone deacetylase Sirt1. It was also found that resveratrol promoted longevity and improved glucose homeostasis in mice by stimulating the Sirt1-mediated deacetylation of the transcriptional co-activator PGC-1α (Koo and Montminy, 2006). In 2001, resveratrol (5–35 μmol l^{-1}) was found to induce concentration-dependent relaxation of mesenteric arteries pre-constricted with noradrenaline (8 μmol l^{-1}) or KCl (125 mmol l^{-1}) from both lean and dietary-obese rats (Naderali *et al.*, 2001). Hyperglycaemia, a symptom of diabetes mellitus, induces hyper-osmotic responses, including apoptosis, in vascular endothelial cells and leucocytes. Hyper-osmotic shock often leads to apoptotic cell death. Resveratrol was found to attenuate high glucose-induced apoptotic changes by virtue of its antioxidant property. Diabetic nephropathy is a serious vascular complication and one of the main causes of end-stage renal disease. Increased oxidative stress plays an important role in the aetiology of diabetic nephropathy. Treatment with resveratrol significantly attenuated renal dysfunction and oxidative stress in diabetic rats (Sharma *et al.*, 2006).

Most of type-2 diabetes mellitus patients eventually become insulin dependent when insulin secretion by the islets of Langerhans is exhausted. Resveratrol was found to possess hypoglycaemic and hypolipidaemic effects in streptozotocin-induced diabetic rats. In resveratrol-treated diabetic rats, the plasma glucose concentration on day 14 was reduced by 25.3%, and the triglyceride concentration was reduced by 50.2% compared with the placebo-treated rats. In nicotinamide-treated diabetic rats, the plasma glucose concentration on day 14 was reduced only by 20.3% and the triglyceride concentration was reduced by 33%. Resveratrol administration ameliorates common diabetes mellitus symptoms, such as body weight loss, polyphagia and polydipsia. In STZ-nicotinamide diabetes mellitus rats, resveratrol administration significantly decreased insulin secretion and delayed the onset of insulin resistance (Su *et al.*, 2006).

4.7.9 Resveratrol for fatty liver/liver protection

The prevalence of non-alcoholic fatty liver disease is linked to obesity, diabetes mellitus and hyper-triglyceridaemia. Approximately 20% of patients with non-alcoholic fatty liver disease develop cirrhosis. Resveratrol was found to decrease non-alcoholic fatty liver disease severity in rats. This effect was mediated, at least in part, by TNFα inhibition and antioxidant activities. Oral administration of resveratrol (dosage 20 mg kg^{-1} daily for 4 weeks) also remarkably prevented the DMN-induced loss in body and liver weight, and inhibited the elevation of serum alanine transaminase, aspartate transaminase, alkaline phosphatase and bilirubin levels (Lee *et al.*, 2010). Thus, resveratrol may provide liver protection.

4.7.10 Resveratrol for skin

When topically applied, resveratrol cream limited the HSV-1 lesion formation in the skin of mice; resveratrol cream also reduced HSV replication in the vagina of mice and limits extra-vaginal disease. Thus, resveratrol cream may have some potential benefits on skin health, but more studies are needed to support this health benefit claim.

4.7.11 Resveratrol for weight loss

Some studies argue that resveratrol speeds up metabolism to help users burn more calories throughout the day. Thus, resveratrol offers benefits of weight loss. In a study, resveratrol in combination with genistein and quercetin synergistically decreased adipogenesis in murine and human adipocytes. An *in vivo* study showed that phytochemicals including resveratrol in combination with vitamin D prevented weight gain and bone loss in a post-menopausal rat model (Rayalam *et al.*, 2011). In another study of aged ovariectomized female rats, a high-dose treatment (dosage: vitamin D + 400 mg kg^{-1} resveratrol + 2000 mg kg^{-1} quercetin + 1040 mg kg^{-1} genistein) reduced body weight gain and the fat pad weights. This treatment also increased the serum concentration of IGF-1 and the bone mineral content of the femur. Thus, the synergistic effects of a combination of resveratrol with vitamin D may be effective in reducing bone loss and weight gain after menopause (Lai *et al.*, 2011). These studies show that resveratrol may help prevent 'weight gain' under certain conditions. However, there is still no solid evidence to support that resveratrol helps weight loss.

4.7.12 Resveratrol for longevity

Caloric restriction is known to extend the lifespans of a number of species, including mammals (Heilbronn and Ravussin, 2003). In yeast, caloric restriction stimulates the activity of an enzyme known as Sir2 (Lin *et al.*, 2000). Providing resveratrol to yeast increased Sir2 activity in the absence of caloric restriction and extended the replicative lifespan of yeast by 70% (Howitz *et al.*, 2003). Resveratrol feeding also extended the lifespans of worms (*Caenorhabditis elegans*) and fruit flies (*Drosophila melanogaster*) by a similar mechanism (Wood *et al.*, 2004). Additionally, resveratrol dose-dependently increased the lifespan of a vertebrate fish (*Nothobranchius furzeri*) (Valenzano *et al.*, 2006). However, it is not known whether resveratrol will have similar effects in higher animals. A recent study reported that resveratrol extended lifespan of mice on a high-calorie diet such that their lifespan was similar to that of mice fed a standard diet (Baur *et al.*, 2006). Although resveratrol increased the activity of the homologous human enzyme (Sirt1) in the test tube (Howitz *et al.*, 2003), it is not known whether resveratrol can extend the human lifespan. Moreover, the resveratrol concentrations required to increase human Sirt1 activity were considerably higher than concentrations that have been measured in human plasma after oral consumption. Interestingly, a recent ageing study in mice found that a low dose of dietary resveratrol altered gene expression in heart, brain and skeletal muscle similar to that induced by caloric restriction (Barger *et al.*, 2008). Like caloric restriction, resveratrol also blunted the age-related decline in heart function in this study. Clinical trials will be needed to determine if these findings are relevant to humans.

4.7.13 Resveratrol for other conditions

Because microcirculation occlusion and cytokine over-production is involved in many diseases such as acute pancreatitis, resveratrol as a platelet and cytokines inhibitor may have benefits on acute pancreatitis (Ma and Ma, 2005). Resveratrol has been shown to have an immunosuppressive property as well as protective effect on hepatocytes under allograft rejection in a study of Wistar rats (Wu *et al.*, 2005). Resveratrol has been shown to reduce ischaemia-reperfusion (I/R) injury of rat kidney both by antioxidant and anti-inflammatory mechanisms (Bertelli *et al.*, 2002). In a study, researchers administered dosage of 20 mg kg^{-1} day^{-1} of *trans*-resveratrol for 90 days. Compared to a control group, the diameter of the seminiferous tubules was significantly reduced from 437.5 +/– 0.1 μm in the controls to 310.9 +/– 0.1 μm. This decrease was accompanied by a significant increase in tubular density. Sperm counts were significantly greater in the resveratrol-treated rats than in the control group, but sperm quality did not differ (Juan *et al.*, 2005). Some online articles claim that resveratrol helps weight loss, but it also depends on the lifestyle.

4.7.14 Resveratrol as anti-inflammatory

Resveratrol exhibits anti-inflammatory activity through modulation of enzymes and pathways that produce mediators of inflammation and also induction of programmed cell death in activated immune cells. Resveratrol has been shown to produce no adverse effects, even when consumed at high concentrations. Hence, resveratrol possesses the potential to be used as an adjunctive or alternative therapy for cancer and inflammatory diseases (Udenigwe *et al.*, 2008).

4.8 Stability of Resveratrol

Resveratrol is a potent antioxidant, and oxidation may occur during the manufacturing process and/or storage of the finished products. It has long been thought that resveratrol supplements were not very effective in comparison with wine. Recently, some manufacturers have developed a technique to solve the stability issues of resveratrol supplements during manufacturing and storage (Alarcón de la Lastra *et al.*, 2006). Resveratrol possess an overall good stability, but may become unstable after light exposure and only its *trans* isomer has been consistently linked to health beneficial effects in pharmacological and clinical studies, while *cis*-resveratrol had only seen limited success as an antiplatelet agent (Signorelli and Ghidoni, 2005; Szkudelska and Szkudelski, 2010).

4.9 Resveratrol Dosage

Resveratrol has been suggested to be safe at doses equivalent to 500 mg day^{-1} (Alarcón de la Lastra *et al.*, 2006). However, the resveratrol dosage needs to be much higher than that to be effective for therapeutic use. On the other hand, dosages lower than 500 mg day^{-1} are definitely suitable for general support.

4.10 Resveratrol Safety

4.10.1 Resveratrol side effects

Resveratrol is not known to be toxic or cause adverse effects in humans at low doses. Researchers administered rats dosages of 0, 300, 1000, and 3000 mg *trans*-resveratrol kg^{-1} body weight day^{-1} for 4 weeks. Most of the adverse or side effects occurred in the rats administered 3000 mg kg^{-1} body weight day^{-1}. Signs of toxicity included: reduced final body weights and food consumption; elevated BUN, creatinine, alkaline phosphatase, alanine aminotransferase, total bilirubin and albumin; reduced haemoglobin, haematocrit and red cell counts; and increased white cell counts. They also observed renal lesions. No adverse or side effects were observed at 300 mg resveratrol kg^{-1} body weight day^{-1} in rats (Crowell *et al.*, 2004).

However, resveratrol was found to affect the activity of P450 and increase the absorption of certain drugs such as diltiazem. The intake of high resveratrol dosages can offer serious unwanted side effects for patients on certain types of medication. Resveratrol was also found to delay the recovery from gastric ulcer in an animal study. Finally, resveratrol is also known as a metal chelator, with high doses of resveratrol possibly preventing the absorption of copper or other metals. Copper deficiency can cause muscle issues and even heart failure.

4.10.2 Resveratrol–drug interaction

Intake of resveratrol significantly increased the extent of the absorption (AUC) of diltiazem, except for resveratrol at a dose of 0.5 mg kg^{-1}, in a rat study. The increased bioavailability of diltiazem is probably related to the inhibition of both the cytochrome P450 3A4-mediated metabolism and the efflux pump P-glycoprotein in the intestine and/or liver (Hong *et al.*, 2008). Further, resveratrol may slow down the metabolism of some drugs including statins, erectile dysfunction medicine, calcium channel blockers and some immune-system suppressants.

4.10.3 Oestrogen issues

Resveratrol is a phytoestrogen, known to serve as an agonist for the oestrogen receptor, promoting oestrogen-like effects such as neuroprotection and bone growth. Oestrogen has been shown to decrease the turnover of bone, prevent bone loss and increase bone mass. Oestrogen replacement therapy is used to decrease bone loss in post-menopausal women, which presumably acts by increasing osteoblast activity and decreasing the number and activity of osteoclasts. A study evaluated the effects of resveratrol on late stage osteogenesis of MC3T3-E1 cells and the feasibility to develop resveratrol-loaded poly (caprolactone) nanofibres as scaffolds for bone regeneration. The results demonstrated that resveratrol-treated groups irrespective of the concentrations studied showed a significant increase in mineralized matrix deposition compared to the control group at both time points. The results demonstrate that the resveratrol-loaded PCL nanofibre matrix is bioactive and has the potential to serve as an osteogenic biomaterial (Singh *et al.*, 2012).

4.11 Conclusions

There is growing evidence that resveratrol can prevent or delay the onset of cancer, heart disease, ischaemic and chemically induced injuries, diabetes, pathological inflammation and viral infection. These effects are observed despite extremely low bioavailability and rapid clearance from the circulation. By dint of diverse biological activity, resveratrol and related compounds have joined many other promising agents being investigated for their disease-preventive and therapeutic potential. Being a natural constituent of wine, fruits and nuts and the fact that it has no untoward effects on normal cells or tissues, resveratrol is under preclinical scrutiny. Future research will lead to a greater number and variety of pharmacologically useful novel compounds of interest.

References

Aggarwal, B.B., Bhardwaj, A., Aggarwal, R.S., Seeram, N.P., Shishodia, S. and Takada, Y. (2004) Role of resveratrol in prevention and therapy of cancer: preclinical and clinical studies. *Anticancer Research* 24, 2783–2840.

Alarcón de la Lastra, C., Villegas, I. and Martín, A.R. (2006) Red wine consumption and associated health benefits, the resveratrol story. In: Aggarwald, B.B. and Shishodia, S. (eds) *Resveratrol in Health and Disease*. CRC Press, Boca Raton, Florida, pp. 33–56.

Athar, M., Back, J.H., Tang, X., Kim, K.H., Kopelovich, L., Bickers, D.R. and Kim, A.L. (2007) Resveratrol: a review of preclinical studies for human cancer prevention. *Toxicology and Applied Pharmacology* 224, 274–283.

Athar, M., Back, J.H., Kopelovich, L., Bickers, D.R. and Kim, A.L. (2009) Multiple molecular targets of resveratrol: Anti-carcinogenic mechanisms. *Archives of Biochemistry and Biophysics* 486, 95–102.

Atten, M.J., Godoy-Romero, E., Attar, B.M., Milson, T., Zopel, M. and Holian, O. (2005) Resveratrol regulates cellular PKC alpha and delta to inhibit growth and induce apoptosis in gastric cancer cells. *Investigational New Drugs* 23, 111–119.

Ban, J.Y., Jeon, S.Y., Nguyen, T.T., Bae, K., Song, K.S. and Seong, Y.H. (2006) Neuroprotective effect of oxyresveratrol from smilacis chinae rhizome on amyloid beta protein (25-35)-induced neurotoxicity in cultured rat cortical neurons. *Biological and Pharmaceutical Bulletin* 29, 2419–2424.

Barger, J.L., Kayo, T., Vann, J.M., Arias, E.B., Wang, J., Hacker, T.A., Wang, Y., Raederstorff, D., Morrow, J.D., Leeuwenburgh, C., Allison, D.B., Saupe, K.W., Cartee, G.D., Weindruch, R. and Prolla, T.A. (2008) A Low Dose of Dietary Resveratrol Partially Mimics Caloric Restriction and Retards Aging Parameters in Mice. *PLoS ONE* 3, e2264.

Baur, J.A., Pearson, K.J., Price, N.L., Jamieson, H.A., Lerin, C., Kalra, A., Prabhu, V.V., Allard, J.S., Lopez-Lluch, G., Lewis, K., Pistell, P.J., Poosala, S., Becker, K.G., Boss, O., Gwinn, D., Wang, M., Ramaswamy, S., Fishbein, K.W., Spencer, R.G., Lakatta, E.G., Le Couteur, D., Shaw, R.J., Navas, P., Puigserver, P., Ingram, D.K., de Cabo, R. and Sinclair, D.A. (2006) Resveratrol improves health and survival of mice on a high-calorie diet. *Nature* 444, 337–342.

Bertelli, A.A., Migliori, M., Panichi, V., Origlia, C., Das Filippi, D.K. and Giovannini, L. (2002) Resveratrol, a component of wine and grapes, in the prevention of kidney disease. *Annals of New York Academy of Sciences* 957, 230–238.

Bishayee, A. (2009) Cancer prevention and treatment with resveratrol: from rodent studies to clinical trials. *Cancer Prevention Research* 2, 409–418.

Borgonovo, G., Caimi, S., Morini, G., Scaglioni, L. and Bassoli, A. (2008) Taste-active compounds in a traditional Italian food: 'Lampascioni'. *Chemistry & Biodiversity* 5, 1184–1194.

Burns, J., Yokota, T., Ashihara, H., Lean, M.E. and Crozier, A. (2002) Plant foods and herbal sources of resveratrol. *Journal of Agricultural and Food Chemistry* 50, 3337–3340.

Busquets, S., Ametller, E., Fuster, G., Olivan, M., Raab, V., Argiles, J.M. and Lopez-Soriano, F.J. (2006) Resveratrol, a natural diphenol, reduces metastatic growth in an experimental cancer model. *Cancer Letters* 245, 144–148.

Centeno-Baez, C., Dallaire, P. and Marette, A. (2011) Resveratrol inhibition of inducible nitric oxide synthase in skeletal muscle involves AMPK but not SIRT1. *American Journal of Physiology* 301, E922–E930.

Crowell, J.A., Korytko, P.J., Morrissey, R.L., Booth, T.D. and Levine, B.S. (2004) Resveratrol-associated renal toxicity. *The Journal of Toxicological Sciences* 82, 614–619.

Das, D.K. and Maulik, N. (2006) Resveratrol in cardio-protection: a therapeutic promise of alternative medicine. *Molecular Interventions* 6, 36–47.

De Ruvo, C., Amodio, R., Algeri, S., Martelli, N., Intilangelo, A., D'Ancona, G.M. and Esposito, E. (2000) Nutritional antioxidants as anti-degenerative agents. *International Journal of Developmental Neuroscience* 18, 359–366.

Delmas, D., Lancon, A., Colin, D., Jannin, B. and Latruffe, N. (2006) Resveratrol as a chemopreventive agent: a promising molecule for fighting cancer. *Current Drug Targets* 7, 423–442.

Dong, H.H. and Ren, H.L. (2004) New progression in the study of protective properties of resveratrol in anti-cardiovascular disease. *Bratisl Lek Listy* 105, 225–229.

Fan, X. and Matthesis, J.P. (2001) Inhibition of Oxidative and Anti-oxidative enzymes by Trans-Resveratrol. *Journal of Food Science* 66, 200–203.

Gusman, J., Malonne, H. and Atassi, G. (2001) A reappraisal of the potential chemopreventive and chemotherapeutic properties of resveratrol. *Carcinogenesis* 22, 1111–1117.

Heilbronn, L.K. and Ravussin, E. (2003) Calorie restriction and aging: review of the literature and implications for studies in humans. *American Journal of Clinical Nutrition* 78, 361–369.

Hong, S.P., Choi, D.H. and Choi, J.S. (2008) Effects of resveratrol on the pharmacokinetics of diltiazem and its major metabolite, desacetyldiltiazem, in rats. *The Journal of Cardiovascular Therapy* Winter 26, 269–275.

Howitz, K.T., Bitterman, K.J., Cohen, H.Y., Lamming, D.W., Lavu, S., Wood, J.G., Zipkin, R.E., Chung P., Kisielewski, A., Zhang, L.L., Scherer, B. and Sinclair, D.A. (2003) Small molecule activators of sirtuins extend *Saccharomyces cerevisiae* lifespan. *Nature* 425, 191–196.

Hsieh, T.C. and Wu, J.M. (1999) Differential effects on growth, cell cycle arrest, and induction of apoptosis by resveratrol in human prostate cancer cell lines. *Experimental Cell Research* 249, 109–115.

Jang, M., Cai, L., Udeani, G.O., Slowing, K.V., Thomas, C.F., Beecher, C.W., Fong, H.H., Farnsworth, N.R., Kinghorn, A.D., Mehta, R.G., Moon, R.C. and Pezzuto, J.M. (1997) Cancer chemopreventive activity of resveratrol, a natural product derived from grapes. *Science* 275, 218–220.

Juan, M.E., González-Pons, E., Munuera, T., Ballester, J., Rodríguez-Gil, J.E. and Planas, J.M. (2005) Trans-Resveratrol, a natural antioxidant from grapes, increases sperm output in healthy rats. *Journal of Nutrition* 135, 757–760.

Khanna, D., Sethi, G., Ahn, K.S., Pandey, M.K., Kunnumakkara, A.B., Sung, B., Aggarwal, A. and Aggarwal, B.B. (2007) Natural products as a gold mine for arthritis treatment. *Current Opinion in Pharmacology* 7, 344–351.

Kim, Y.A., Choi, B.T., Lee, Y.T., Park, D.I., Rhee, S.H., Park, K.Y. and Choi, Y.H. (2004) Resveratrol inhibits cell proliferation and induces apoptosis of human breast carcinoma MCF-7 cells. *Oncology Reports* 11, 441–446.

Kim, Y.H., Kim, Y.S., Kang, S.S., Cho, G.J. and Choi, W.S. (2010) Resveratrol inhibits neuronal apoptosis and elevated Ca^{2+}/calmodulin-dependent protein kinase II activity in diabetic mouse retina. *Diabetes* 59, 1825–1835.

Kim, Y.M., Yun, J., Lee, C.K., Lee, H., Min, K.R. and Kim, Y. (2002) Oxyresveratrol and hydroxystilbene compounds: Inhibitory effect on tyrosinase and mechanism of action. *The Journal of Biological Chemistry* 277, 16340–16344.

Koo, S.H. and Montminy, M. (2006) In vino veritas: a tale of two sirt1s? *Cell* 127, 1109–1122.

Kopp, P. (1998) Resveratrol, a phytoestrogen found in red wine: a possible explanation for the conundrum of the 'French paradox'? *European Journal of Endocrinology* 138, 619–620.

Lai, C.Y., Yang, J.Y., Rayalam, S., Della-Fera, M.A., Ambati, S., Lewis, R.D., Hamrick, M.W., Hartzell, D.L. and Baile, C.A. (2011) Preventing Bone Loss and Weight Gain with Combinations of Vitamin D and Phytochemicals. *Journal of Medicinal Food* 14, 1352–1362.

Lastra, A.C. and Villegas, I. (2007) Resveratrol as an antioxidant and pro-oxidant agent: mechanisms and clinical implications. *Biochemical Society Transactions* 35, 1156–1160.

Lee, E.S., Shin, M.O., Yoon, S. and Moon, J.O. (2010) Resveratrol inhibits dimethylnitrosamine-induced hepatic fibrosis in rats. *Archives of Pharmacalogy Research* 33, 925–932.

Lee, S.J. and Kim, M.M. (2011) Resveratrol with antioxidant activity inhibits matrix metalloproteinase via modulation of SIRT1 in human fibro-sarcoma cells. *Life Sciences* 14, 465–472.

Leonard, S., Xia, C., Jiang, B.H., Stinefelt, B., Klandorf, H., Harris, G.K. and Shi, X. (2003) Resveratrol scavenges reactive oxygen species and effects radical-induced cellular responses. *Biochemical Biophysical Research Communications* 309, 1017–1026.

Lin, K.H., Hsiao, G., Shih, C.M., Chou, D.S. and Sheu, J.R. (2009) Mechanisms of resveratrol-induced platelet apoptosis. *Cardiovascular Research* 83, 575–585.

Lin, S.J., Defossez, P.A. and Guarente, L. (2000) Requirement of NAD and SIR2 for life-span extension by calorie restriction in *Saccharomyces cerevisiae*. *Science* 289, 2126–2128.

Losa, G.A. (2003) Resveratrol modulates apoptosis and oxidation in human blood mononuclear cells. *European Journal of Clinical Investigation* 33, 818–823.

Ma, Z.H. and Ma, Q.Y. (2005) Resveratrol: a medical drug for acute pancreatitis. *World Journal of Gastroenterology* 11, 3171–3174.

Martínez, J. and Moreno, J.J. (2000) Effect of resveratrol, a natural polyphenolic compound, on reactive oxygen species and prostaglandin production. *Biochemical Pharmacology* 59, 865–870.

Moreno-Labanda, J.F., Mallavia, R., Perez-Fons, L., Lizama, V., Saura, D. and Micol, V. (2004) Determination of piceid and resveratrol in Spanish wines deriving from Monastrell (*Vitis vinifera* L.) grape variety. *Journal of Agricultural Food Chemistry* 52, 5396–5403.

Naderali, E.K., Smith, S.L., Doyle, P.J. and Williams, G. (2001) The mechanism of resveratrol-induced vasorelaxation differs in the mesenteric resistance arteries of lean and obese rats. *Clinical Science* 100, 55–60.

Nonomura, S., Kanagawa, H. and Makimoto, A. (1963) Chemical constituents of polygonaceous plants. Studies on the components of Ko-jo-kon (*Polygonum cuspidatum* Sieb. et Zucc.). *Yakugaku Zasshi* 83, 988–990.

Ohguchi, K., Tanaka, T., Iliya, I., Ito, T., Iinuma, M., Matsumoto, K., Akao, Y. and Nozawa, Y. (2003) Gnetol as a potent tyrosinase inhibitor from genus Gnetum. *Bioscience, Biotechnology and Biochemistry* 67, 663–665.

Oi, N., Jeong, C.H., Nadas, J., Cho, Y.Y., Pugliese, A., Bode, A.M. and Dong, Z. (2010) Resveratrol, a red wine polyphenol, suppresses pancreatic cancer by inhibiting leukotriene A4 hydrolase. *Cancer Research* 70, 9755–9764.

Olas, B. and Wachowicz, B. (2005) Resveratrol, a phenolic antioxidant with effects on blood platelet functions. *Platelets* 16, 251–260.

Olas, B., Wachowicz, B., Bald, E. and Glowacki, R. (2004) The Protective effects of resveratrol against changes in blood platelet thiols induced by Platinum compounds. *Journal of Physiology and Pharmacology* 55, 467–476.

Orallo, F., Alvarez, E., Camina, M., Leiro, J.M., Gomez, E. and Fernandez, P. (2002) The possible implication of trans-resveratrol in the cardio-protective effects of long-term moderate wine consumption. *Molecular Pharmacology* 61, 294–302.

Pace-Asciak, C.R., Hahn, S., Diamandis, E.P., Soleas, G. and Goldberg, D.M. (1995) The red wine phenolics trans-resveratrol and quercetin block human platelet aggregation and eicosanoid synthesis: implications for protection against coronary heart disease. *Clinica Chimica Acta* 235, 207–219.

Pendurthi, U.R., Williams, J.T. and Rao, L.V. (1999) Resveratrol, a polyphenolic compound found in wine, inhibits tissue factor expression in vascular cells: a possible mechanism for the cardiovascular benefits associated with moderate consumption of wine. *Arterioscleriosis, Thrombosis & Vascular Biology* 19, 419–426.

Pozo-Guisado, E., Alvarez-Barrientos, A., Mulero-Navarro, S., Santiago-Josefat, B. and Fernandez-Salguero, P.M. (2002) The anti-proliferative activity of resveratrol results in apoptosis in MCF-7 but not in MDAMB-231 human breast cancer cells: cell-specific alteration of the cell cycle. *Biochemical Pharmacology* 64, 1375–1386.

Rayalam, S., Della-Fera, M.A. and Baile, C.A. (2011) Synergism between resveratrol and other phytochemicals: implications for obesity and osteoporosis. *Molecular Nutrition & Food Research* 55, 1177–1185.

Romero-Perez, A.I., Lamuela-Raventos, R.M., Waterhouse, A.L. and de la Torre-Boronat, M.C. (1996) Levels of cis- and trans-resveratrol and their glucosides in white and rosé *Vitis vinifera* wines from Spain. *Journal of Agricultural Food Chemistry* 44, 2124–2128.

Romero-Perez, A.I., Ibern-Gomez, M., Lamuela-Raventos, R.M. and de La Torre-Boronat, M.C. (1999) Piceid, the major resveratrol derivative in grape juices. *Journal of Agricultural Food Chemistry* 47, 1533–1536.

Sanders, T.H., McMichael, R.W. and Hendrix, K.W. (2000) Occurrence of resveratrol in edible peanuts. *Journal of Agricultural Food Chemistry* 48, 1243–1246.

Sharma, S., Anjaneyulu, M., Kulkarni, S.K. and Chopra, K. (2006) Resveratrol, a polyphenolic phytoalexin, attenuates diabetic nephropathy in rats. *Pharmacology* 76, 69–75.

Shen, M.Y., Hsiao, G., Liu, C.L., Fong, T.H., Lin, K.H., Chou, D.S. and Sheu, J.R. (2007) Inhibitory mechanisms of resveratrol in platelet activation: pivotal roles of p38 MAPK and NO/cyclic GMP. *British Journal of Haematology* 139, 475–485.

Signorelli, P. and Ghidoni, R. (2005) Resveratrol as an anticancer nutrient: molecular basis, open questions and promises. *The Journal of Nutritional Biochemistry* 16, 449–466.

Singh, H., James, E., Kan, H.-M. and Lakshmi, S. (2012) Fabrication and Evaluation of Resveratrol Loaded Polymeric Nanofibers. *Journal of Biomaterials and Tissue Engineering* 2, 228–235.

Sobolev, V.S. and Cole, R.J. (1999) Trans-Resveratrol content in commercial peanuts and peanut products. *Journal of Agricultural Food Chemistry* 47, 1435–1439.

Soleas, G.J., Diamandis, E.P. and Goldberg, D.M. (1997) Resveratrol: a molecule whose time has come? And gone? *Clinical Biochemistry* 30, 91–113.

Su, H.C., Hung, L.M. and Chen, J.K. (2006) Resveratrol, a red wine antioxidant, possesses an insulin-like effect in streptozotocin-induced diabetic rats. *American Journal of Physiology – Endocrinology and Metabolism* 290, E1339–E1346.

Szkudelska, K. and Szkudelski, T. (2010) Resveratrol, obesity and diabetes. *European Journal of Pharmacology* 635, 1–8.

Takaoka, M.J. (1940) Of the phenolic substances of white hellebore (*Veratrum grandiflorum* Loes. fil.). *Journal of the Faculty of Science, Hokkaido Imperial University* 3, 1–16.

Tseng, S.H., Lin, S.M., Chen, J.C., Su, Y.H., Huang, H.Y., Chen, C.K., Lin, P.Y. and Chen, Y. (2004) Resveratrol suppresses the angiogenesis and tumor growth of gliomas in rats. *Clinical Cancer Research* 10, 2190–2202.

Tyagi, A., Gu, M., Takahata, T., Frederick, B., Agarwal, C., Siriwardana, S., Agarwal, R. and Sclafani, R.A. (2011) Resveratrol selectively induces DNA damage, independent of Smad4 expression, in its efficacy against human head and neck squamous cell carcinoma. *Clinical Cancer Research* 17, 5402–5411.

Udenigwe, C.C., Ramprasath, V.R., Aluko, R.E. and Jones, P.J. (2008) Potential of resveratrol in anticancer and anti-inflammatory therapy. *Nutrition Reviews* 66, 445–454.

Valenzano, D.R., Terzibasi, E., Genade, T., Cattaneo, A., Domenici, L. and Cellerino, A. (2006) Resveratrol prolongs lifespan and retards the onset of age-related markers in a short-lived vertebrate. *Current Biology* 16, 296–300.

Walle, T. (2011) Bioavailability of resveratrol. *Annals of New York Academy of Sciences* 1215, 9–15.

Wang, S.Y., Chen, C., Wang, C.Y. and Chen, P. (2007) Resveratrol content in strawberry fruit is affected by pre-harvest conditions. *Journal of Agricultural Food Chemistry* 55, 8269–8274.

Wang, Z., Zou, J., Huang, Y., Cao, K., Xu, Y. and Wu, J.M. (2002) Effect of resveratrol on platelet aggregation *in vivo* and *in vitro*. *Chinese Medical Journal* 115, 378–380.

Wood, J.G., Rogina, B., Lavu, S., Howitz, K., Helfand, S.L., Tatar, M. and Sinclair, D. (2004) Sirtuin activators mimic caloric restriction and delay ageing in metazoans. *Nature* 430, 686–689.

Wu, S.L., Yu, L., Meng, K.W., Ma, Z.H. and Pan, C.E. (2005) Resveratrol prolongs allograft survival after liver transplantation in rats. *World Journal of Gastroenterology* 11, 4745–4749.

Yang, Y., Wang, X., Zhang, L., An, H. and Zao, Z. (2011) Inhibitory effects of resveratrol on platelet activation induced by thromboxane a(2) receptor agonist in human platelets. *The American Journal of Chinese Medicine* 39, 145–159.

Yen, G.C., Duh, P.D. and Lin, C.W. (2003) Effects of Resveratrol and 4-hexylresorcinol on Hydrogen Peroxide-induced Oxidative DNA Damage in Human Lymphocytes. *Free Radical Research* 37, 509–514.

Yueh, M.F., Kawahara, M. and Raucy, J. (2005) Cell-based high-throughput bioassays to assess induction and inhibition of CYP1A enzymes. *Toxicology In Vitro* 19, 275–287.

Zini, R., Morin, C., Bertelli, A., Bertelli, A.A. and Tillement, J.P. (1999) Effects of resveratrol on the rat brain respiratory chain. *Drugs under Experimental and Clinical Research* 25, 87–97.

5 Synbiotics: Promoting Gastrointestinal Health

Charu Gupta,[1]* Dhan Prakash,[1] Marcos H. Rostagno[2] and Todd R. Callaway[3]

[1]*Amity Institute for Herbal Research and Studies, Amity University, Noida, India;* [2]*United States Department of Agriculture, Agricultural Research Service, West Lafayette, Indiana, USA;* [3]*United States Department of Agriculture, Agricultural Research Service, College Station, Texas, USA*

5.1 Introduction

For a long time, the gastrointestinal (GI) microbiota has been recognized to play an important role in the maintenance of the health and well-being of the host (Simon and Gorbach, 1984). In addition to promoting normal GI functions and protecting against pathogenic bacteria, the microbiota exerts beneficial effects on systemic metabolism and the immune system (Bengmark, 1998; Nicholson *et al.*, 2012).

The ability to control GI bacterial growth and pathogenic potential essentially depends on the proper function of the microbiota (McCracken and Lorenz, 2001). An imbalance within the GI microbiota with relative predominance of aggressive bacteria and insufficient concentration of protective species has been associated with several inflammatory processes (Fabia *et al.*, 1993; Rusuler-van Embden *et al.*, 1994; Bengmark, 1996; Bosscher *et al.*, 2009; Nicholson *et al.*, 2012).

Intake of probiotics, prebiotics and synbiotics has been demonstrated (Fig. 5.1) to modify the composition of the GI microbiota, restore the microbial balance and therefore have the potential to provide health benefits (Fuller, 1991; Gibson and Roberfroid, 1995; Macfarlane and Cummings, 1999). However, only recently, well designed clinical studies have provided clear evidence of health promoting effects, such as prevention of antibiotic-associated diarrhoea (Andersson *et al.*, 2001), treatment of acute diarrhoea (Majamaa *et al.*, 1995), inflammatory bowel disease (Gionchetti *et al.*, 2000), eradication of *Clostridium difficile* infection (Gorbach *et al.*, 1987) and enhancement of intestinal immunity (Elmer *et al.*, 1996; Gill *et al.*, 2001). Nevertheless, conclusive evidence and in-depth knowledge on the mechanisms involved in the observed effects are still lacking.

There is growing interest in manipulating the GI microbiota to increase the relative numbers of beneficial bacteria. Until recently, this enhancement was typically accomplished by providing supplements consisting of a strain or strains of live beneficial bacteria, referred to as probiotics. Presently, dietary modulation of the GI microbiota has expanded to the use of prebiotics, which are non-digestible food ingredients that beneficially affect the host by selectively

* Email: charumicro@gmail.com

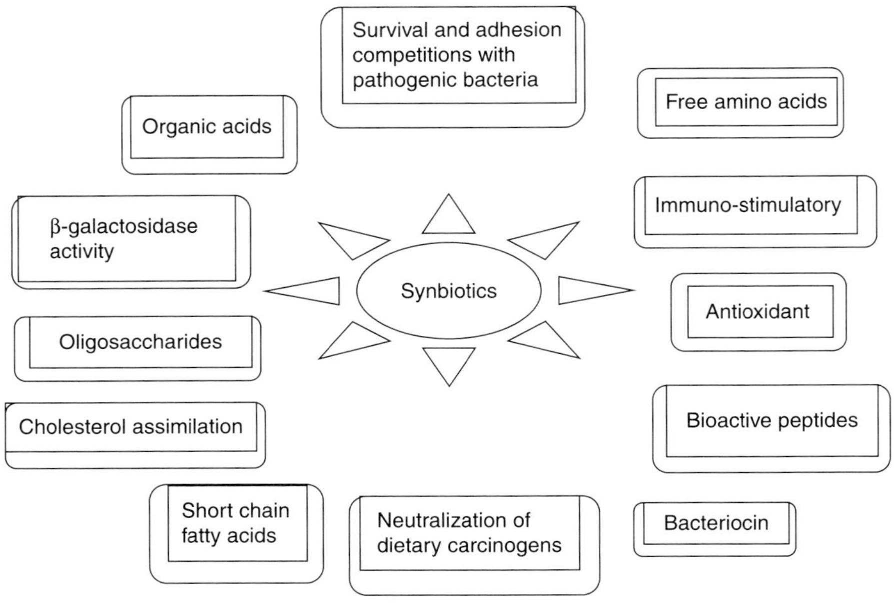

Fig. 5.1. Purported functions of synbiotics.

stimulating the growth and/or activity of beneficial bacteria in the gut and, thus, improve host health. Prebiotics are non-digestible carbohydrates or fibres that resist enzymatic digestion in the upper digestive tract. When the prebiotics reach the large intestine, they serve as a substrate for the resident bacteria for fermentation. Unlike other fibres, prebiotics selectively feed the beneficial bacteria; their positive influence on GI microbiota has been assessed by a number of studies (Van Loo *et al.*, 1999; Gibson *et al.*, 2004; de Vrese and Schrezenmeir, 2008). Many prebiotics are classified as dietary fibre and are not fully digested as they pass through the GI system, so they work in the body like fibre. The use of probiotic strains together with prebiotic substances will provide a combined effect, named 'synbiotic' (Rastall and Maitin, 2002). Synbiotics may help support probiotic bacteria, stimulating their growth and activity in the GI tract and improving the balance of 'good' bacteria (de Vrese and Schrezenmeir, 2008).

As a result, in recent years there has been an increasing demand to select, by means of *in vitro* and *in vivo* approaches, new strains with superior potential probiotic effects (Amalaradjou and Bhunia, 2012). There is a general consensus that probiotic strains should originate from the same host species to which it will be applied, as these bacteria have a greater chance of competing with resident bacteria and of becoming numerically predominant after short intake and to persist in the GI environment for some time after discontinuation of use. The concept of synbiotics has been proposed recently to characterize colonic food with probiotic and prebiotic properties as health-enhancing functional food. Research and development of synbiotic products have been increasingly focusing on evidence of functional benefits, including resistance to infection, antibacterial activity and improved immune status (Gibson and Roberfroid, 1995; Gourbeyre *et al.*, 2011).

5.2 Probiotics

The majority of probiotic microorganisms belong to the genera *Lactobacillus* and *Bifidobacterium*. There are also other genera of bacteria and some yeasts that are widely used (Table 5.1; Baffoni and Biavati, 2008). Lactobacilli and bifidobacteria are Gram-positive lactic acid-producing bacteria that constitute a major part of the normal intestinal microflora

Table 5.1. Microorganisms considered as probiotics (Baffoni and Biavati, 2008).

Lactobacillus	*Bifidobacterium*	*Enterococcus*	*Streptococcus*	*Lactococcus*	*Propionibacterium*	Yeast	Others
L. acidophilus	*B. adolescentis*	*E. faecalis*	*S. thermophilus*	*L. lactis* subsp. *cremoris*	*P. freudenreichii*	*Kluyveromyces lactis*	*Leuconostoc mesenteroides*
L. brevis	*B. animalis*	*E. faecium*		*L. lactis* subsp. *lactis*	*P. freudenreichii* subsp. *shermanii*	*Saccharomyces boulardii*	*Pediococcus acidilactici*
L. casei	*B. bifidum*				*P. jensenii*	*Saccharomyces cerevisiae*	
L. curvatus	*B. breve*						
L. fermentum	*B. infantis*						
L. gasseri	*B. longum*						
L. johnsonii	*B. thermophilum*						
L. reuteri							
L. rhamnosus							
L. salivarius							

in animals and humans. Lactobacilli are Gram-positive, non-spore-forming rods or cocco-bacilli. They have complex nutritional requirements and are strictly fermentative, aero-tolerant or anaerobic, aciduric or acidophilic. Lactobacilli are isolated from a variety of habitats where rich, carbohydrate-containing substrates are available, such as human and animal mucosal membranes, on plants or material of plant origin, sewage and fermented milk products, fermenting or spoiling food. Bifidobacteria constitute a major part of the normal intestinal microflora in humans throughout life. They appear in the faeces a few days after birth and increase in number thereafter. The number of bifidobacteria in the colon of adults is 10^{10}–10^{11} CFU g^{-1}, but this number decreases with age. Bifidobacteria are non-motile, non-spore-forming, Gram-positive rods with varying cell morphology. Most strains are strictly anaerobic.

The desirable properties of probiotic bacteria are discussed in Table 5.2.

5.3 Prebiotics

Prebiotics are predominantly dietary fibres, particularly soluble, also called 'colonic food', consisting of specific carbohydrates. Prebiotics can be defined as: 'a selectively fermented ingredient that allows specific changes, both in the composition and/or activity in the gastrointestinal microflora that confers benefits upon host well-being and health' (Gibson *et al.*, 2004; Roberfroid, 2007; Kelly, 2008).

Prebiotics may have the following characteristics (Gibson and Roberfroid, 1995; Gibson *et al.*, 2004; Roberfroid, 2007; de Vrese and Schrezenmeir, 2008; Kelly, 2008):

- Must pass, almost undamaged and in adequate amount, the digestive processes occurring in the first section of the digestive tract (mouth, stomach and small intestine);
- Must be a nutritional fermentable substrate for intestinal microflora, in order to selectively stimulate the growth and/or metabolism of one or a few bacterial species;
- Should positively change the bacterial flora in favour of the acidophile protective one (bifidobacteria, lactobacilli);
- Should induce systemic or luminal effects that are positive for human health.

Prebiotics are present in many edible plants such as chicory, artichoke, onions, leeks, garlic, asparagus, wheat, bananas, oats, soybeans and other legumes. Many commercial

Table 5.2. Desirable properties of probiotic bacteria (Salminen *et al.*, 1998).

Probiotic strain characteristics	Functional properties
Human origin, if intended for humans	Species-dependent health effects and maintained viability; applicability to functional and clinical foods
Acid and bile stability	Survival in the intestine, maintaining adhesiveness and other colonization properties
Adherence to human intestinal cells and intestinal mucus glycoproteins (mucin)	Immune modulation, competitive exclusion of pathogens
Competitive exclusion and colonization of the human intestinal tract	Multiplication in the intestinal tract, competitive exclusion of pathogens, stimulation of beneficial microflora, immune modulation by contact with gut associated lymphoid tissue
Production of antimicrobial substances	Pathogen activation in the intestine, normalization of the gut flora
Antagonism against cariogenic and pathogenic bacteria	Pathogen exclusion, prevention of pathogen adhesion, normalization of gut flora, normalization of oral microflora
Safety in food and clinical use	Accurate strain identification (genus, species, strain) and characterization, documented safety
Clinically validated and documented health effects	Dose–response data for minimum effective dosage in different products and population groups

prebiotics are obtained from vegetable raw materials, while others are produced by an enzymatic process through the hydrolysis of complex polysaccharides or the transglycosylation of mono- or disaccharides, a beneficial system for mass production starting from simple sugars (sucrose and lactose).

Fructo-oligosaccharides (FOS) have been the most studied prebiotics. FOS occur naturally in several plants (e.g. wheat, bananas, barley, garlic) and are synthesized commercially. Dietary supplementation with FOS positively influences gut health by increasing the concentration of beneficial bacterial populations (bifidobacteria, lactobacilli) and decreasing concentrations of potential pathogens (*Clostridium perfringens*). Increase in the beneficial lactate-producing populations of bifidobacteria and lactobacilli are associated with these health benefits:

- Enhancement of intestinal structure and functions;
- Inhibition of pathogen growth;
- Stimulation of enteric and systemic immune systems;
- Enhanced utilization of indigestible dietary components;
- Treatment of GI disorders;
- Increased mineral absorption;
- Improved stool characteristics.

5.4 Combining Prebiotics and Probiotics: Synbiotics

A synbiotic formulation, consisting of a mixture of the above selected probiotic strains belonging to genera *Lactobacillus* and *Bifidobacterium*, prebiotic ingredients such as oligosaccharides, glutamin, vitamin B_6 and zinc, has been developed. The rationale of formulation is to exploit a complementary probiotic action resulting from the different intrinsic properties of each individual strain and the promotion of bifidobacteria growth due to oligosaccharides. The formulation can be assessed in a nutritional trial aimed at evaluating the ability of the selected strains to survive, grow and persist along the gastrointestinal tract and its efficacy and safety in various gastrointestinal disorders when administered in the final pharmaceutical formulation, in order to follow the FAO/WHO guidelines (FAO, 2006). The most used and already marketed synbiotics are mixtures of oligofructose, FOS, galacto-oligosaccharides (GOS), with probiotic bacterial strains of *L. plantarum*, *L. paracasei*, *L. rhamnosus*, *B. bifidum* or *B. lactis*.

Prebiotics and lactic acid bacteria (LAB) (probiotics) have demonstrated beneficial effects with respect to the function of innate immunity, intestinal barrier function and increased resistance to disease. The gut mucosa and microbiota are intimately linked in the maintenance of a functional interface between the host and the external environment (Henke and Bassler, 2004; Sansonetti, 2004). The hope is that a combined supply of prebiotics and probiotics (synbiotics) shall have synergistic effects in enhancing immunity and facilitating intestinal barrier function.

The term 'defense by diversity' coined by Hill (1999), seems applicable to synbiotic treatment. Natural foods may contain LAB, fibre and prebiotic components. A recent study concluded that combining fibre has more than additive effects on the functions of microbial ecosystems and host immune responses (Peuranen *et al.*, 2004). A recent review suggests that multi-species probiotics may be superior to single-species probiotics in reducing antibiotic-associated diarrhoea, preventing infections (*S. enterica* serovar Typhimurium) and reducing pathogenic colonization (*Escherichia coli*) (Timmerman *et al.*, 2004). The choice of prebiotics and probiotics must be based on scientific evidence, and LAB may have variable effects on immune function and outcome. One consideration is that most LAB have limited abilities to ferment bioactive fibres such as inulin or phlein, variable abilities to adhere to human mucus, low antioxidant capacity, and differences with respect to survival in acid conditions or presence of bile in the gastrointestinal tract. LAB selected for synbiotic studies should be selected for functional activities in the context of a specific combination formulation with prebiotics.

Gibson and Roberfroid (1995) defined synbiotic as 'a mixture of probiotics and prebiotics that beneficially affects the host by improving the survival and implantation of live microbial dietary supplements in the gastrointestinal tract, by selectively stimulating the growth and/or activating the metabolism of one or a limited number of health-promoting bacteria, and thus improving host welfare'. Synbiotic compounds are an alternative opportunity to modulate the intestinal microflora by using pro- and prebiotic together that exploits the synergy between the microorganisms activity and their support for the benefit of the intestinal microflora and consequently of the whole body. Synbiotics have better beneficial effect on intestinal flora than pro- and prebiotics by lowering the pH, promoting growth of potentially protective bifidobacteria and inhibition of potentially pathogenic microorganisms, stabilizing the intestinal environment and releasing short-chain organic acids.

Inulin-type probiotics, FOS or GOS, as well as their synbiotic combination with probiotic bacteria, *L. plantarum, L. paracasei* or *B. bifidum* strains, increased bifidobacteria and lactobacilli and inhibited various human- and animal pathogenic bacterial strains (*Clostridium* sp., *E. coli, Campylobacter jejuni, Enterobacterium* sp., *Salmonella enteritidis* or *S. typhimurium*) (Kanamori *et al.*, 2004). The most used and already marketed synbiotics are the mixtures of oligofructose, FOS, GOS, with probiotic bacterial strains of *L. plantarum, L. paracasei, L. rhamnosus, B. bifidum* or *B. lactis*.

A synbiotic product that contains probiotics in several strains and a prebiotic food source is a great combination. Some products, however, add additional helpful agents to get rid of unwanted organisms in the colon. Products that contain enzymes for the breakdown of the cell wall and cell membrane of *Candida* help get rid of the *Candida* organism much faster. While the probiotic bacteria are busy growing and colonizing the colon, the enzymes break down the *Candida* species and, together, the colon can become healthy again.

5.5 Disturbance of the Gastrointestinal Microbiota

Although the composition of the GI microbiota is considered to be rather stable in healthy individuals, it may be altered by many factors, including exogenous such as antibiotic therapy, excessive hygiene, emotional stress, ageing, travelling, peristaltic disorders, surgical operations, liver or kidney diseases, radiation therapy, chemotherapy, pernicious anaemia, disorders of the immune system, and endogenous factors such as nutrient availability, types of diet, pH value of intestinal lumen, redox potential, diarrhoea, bacterial antagonism, bacterial cooperation, mucin, lysozyme, defensins etc.

Disturbances of the GI microbial ecosystem lead to many different population profiles, including changes of the Firmicutes: Bacteroidetes ratio, remarkable increase in bacterial counts in the small intestine, increase of aerobes (mostly Enterobacteriaceae and streptococci), reduction or disappearance of bifidobacteria and/or an increased incidence of *Clostridium perfringens*. It has been suggested that the loss of indigenous microorganisms implies deregulation of autogenic factors and vacated habitats. Consequently, commensal or transient microorganisms have the chance of occupying these vacant niches. If these microorganisms are potentially pathogenic, the outbreak of an opportunistic infectious disease is quite possible (Jernberg *et al.*, 2010; Sekirov *et al.*, 2010; Jalanka-Tuovinen *et al.*, 2011; Tilg and Kaser, 2011; Candela *et al.*, 2012; Perez-Cobas *et al.*, 2012).

There is increasing evidence of an intricate relationship between the GI tract, its microbiota and the brain. As a consequence, it has been shown that stress can lead to disturbance of the GI microbiota (Tannock, 1983; Sekirov *et al.*, 2010; Aziz *et al.*, 2013). During stress conditions, the number of lactobacilli decreases, whereas the number of potentially pathogenic bacteria increases (e.g. *E. coli*). Stress can be produced by changes in the physical or psychological environment, from which hormonal responses ensue and can affect the production of mucus, which may in turn reduce some groups of beneficial microorganisms associated with it. Moreover, stress causes increased intestinal permeability, which also leads to microbial population changes as well as increased risk of inflammatory processes (Bravo *et al.*, 2012; Collins *et al.*, 2012; Dinan and Cryan, 2012).

5.5.1 Acute diarrhoea in children

Acute diarrhoea in infants is usually caused by infection with rotavirus (Kane *et al.*, 2004). Treatment of such diarrhoea generally consists of oral rehydration solutions to replace lost fluid and electrolytes (Samadi *et al.*, 1983), although these are generally ineffective at reducing the duration of the diarrhoea. It is thought that treatment with probiotics may reduce the time of rotavirus excretion and, therefore, the duration of the illness (Boirivant and Strober, 2007).

Barone *et al.* (2000) found no significant differences in alleviation of symptoms between a single probiotic treatment (*Saccharomyces boulardii*) and two different multi-strain probiotic mixtures. However, the study used a small number of subjects, the probiotic doses were different and there was no placebo control, making it difficult to draw definitive conclusions. A study by Rosenfeldt *et al.* (2002) indicates the efficacy of a mixture of *Lactobacillus rhamnosus* and *L. reuteri* used at high doses (10^{11} CFU; 2 times per day) in treatment of acute diarrhoea in children. By comparison with a control, the probiotic mixture reduced hospital stay and viral shedding. Although the study did not compare the mixture to individual strains, the authors indicated that the combination seems to be no more effective than a single-strain probiotic, *Lactobacillus GG*.

In a double-blind, randomized, placebo-controlled trial with a powdered formula containing *Bifidobacterium bifidum* and *Streptococcus salivarius* subsp. *thermophilus*, Saavedra *et al.* (1994) observed a reduction in incidence, but not in severity or duration, of infant diarrhoea. The much lower dose used may account for the lack of effect on severity and duration, although with the use of different strains, the strain-specific nature of probiotic effect against pathogens would suggest that it is hard to compare the two studies. Grandy *et al.* (2010) observed a reduction in rotavirus diarrhoea in groups taking both single- and multi-strain products. The duration of fever was only reduced in the single-strain group despite the multi-strain product also containing that single strain (*S. boulardii*), suggesting that the strain-specific effect was diminished either by the 1000-fold reduction in the dose of that particular strain, or by the presence of the other probiotic microorganisms in the multi-strain product, possibly outcompeting *S. boulardii* for nutrients or binding sites. Interestingly, duration of vomiting was only reduced in the multi-strain group, despite a lower total dose, indicating that this combination or one of its components was able to produce a beneficial effect that *S. boulardii* alone was unable to provide.

5.5.2 Traveller's diarrhoea

Traveller's diarrhoea (TD) is the most common pathologic condition among travellers and can occur in approximately 20–50% of the subjects during or immediately after a trip to a country with hot and humid weather and inadequate sanitary conditions. In TD cases caused by food contaminated with pathogens, the most frequently occurring are: enterotoxigenic *Escherichia coli*, *Campylobacter jejuni*, *Shighella*, *Salmonella* and some viruses, such as *Rotavirus*, *Calicivirus*, *Enterovirus* or even parasites, such as *Giardia lambdia*, *Entamoeba histolytica*, *Cryptosporidium parvum* and *Cyclospora cayetanensis* (McFarland, 2007; Johnson *et al.*, 2009). In a recent meta-analysis of probiotics for the prevention of TD on randomized, controlled, blinded efficacy trials, several probiotics containing *S. boulardii* and a combination of *L. acidophilus* and *B. bifidum* had significant efficacy (Takahashi *et al.*, 2007; Guslandi, 2008).

There are well founded reasons to believe that probiotics may be a safe and effective strategy to prevent TD, but continued research is needed.

5.5.3 Treatment of antibiotic-associated diarrhoea

One of the major complications of treatment with antimicrobial agents is known as antibiotic-associated diarrhoea (AAD), which occurs in 5–25% of patients (Bergogne-Berezin, 2000). One of the major causes of AAD is infection with *Clostridium difficile*, believed to be responsible for 15–25% of AAD cases (Barbut and Petit, 2001). Three randomized, double-blind, placebo-controlled trials have examined the role of probiotic combinations in the treatment of AAD (Gotz *et al.*, 1979; Plummer *et al.*, 2004; Hickson *et al.*, 2007). Gotz *et al.* (1979), using *L. acidophilus* and *L. bulgaricus*, found a 13% reduction in incidence of ampicillin-related diarrhoea, while Hickson *et al.* (2007), using *L. casei*, *L. bulgaricus* and *S. thermophilus*, found a 22% reduction in incidence in AAD and complete elimination of *C. difficile* toxin in faeces, albeit with a much larger dose of probiotics. In contrast, despite using a larger dose of probiotics, Plummer *et al.* (2004) found no difference between treatment (*L. acidophilus* plus *B. bifidum*) and placebo groups, although the experimental group was found to have a reduced incidence of *C. difficile* toxin. Overall, these studies support the use of multi-strain probiotics in the treatment of infant diarrhoea and AAD, although no definitive information is given on whether mixtures are more effective than single strains.

5.6 Development of Synbiotics

Synbiotic products are completely safe and effective in improving colonic health. Anyone who has concerns about their digestion or who wants to optimize their digestive health would likely find that properly-made synbiotic products are a safe and healthy option.

5.6.1 First-generation prebiotics

First-generation prebiotics are either extracted from plants or manufactured from cheap, readily available sources, generally by means of enzymatic hydrolysis or synthesis reactions. Another approach is enzyme hydrolysis of polysaccharide. Fungal inulinase is used to hydrolyse chicory inulin to oligosaccharides with low monosaccharide contents. Fructo-oligosaccharides and xylo-oligosaccharides are both manufactured by hydrolysis of their parent polysaccharides. Fructo-oligosaccharides can also be manufactured by synthesis from sucrose. Consequently, FOS produced from inulin has reducing activity. The prebiotics, such as galacto-oligosaccharides, lacto-sucrose, isomalto-oligosaccharides (IMO) and some FOS, are manufactured by enzymic glycosyl transfer reactions from cheap sugars such as sucrose and lactose or from starch. All of the sucrose-derived FOS terminate in a non-reducing glucose residue. Ion-exchange chromatography can be used to remove glucose and sucrose (De Vrese and Schrezenmeir, 2008).

5.6.2 Second-generation prebiotics

Two areas of development are being explored in laboratories in Europe at the current time for second-generation prebiotics. Controlled polysaccharide hydrolysis is a commercial manufacturing approach for prebiotics. In this a more controlled partial hydrolysis is carried out in order to achieve control over the molecular weight distribution of the products. Different IMO with average molecular weights up to 12,000 Da can be prepared by controlled partial hydrolysis of dextran and pectins by endodextranase in an enzyme membrane reactor by controlling residence time and ratio of enzyme to substrate. The fractions display good prebiotic fermentation *in vitro* (Kelly, 2008). Targeted prebiotics for probiotics can be developed firstly by screening a wide range of oligosaccharides for their prebiotic attributes, which will provide information about their selectivity towards particular species. Structural diversity and cost-effective manufacture technology for complex oligosaccharides is most important.

The second approach is enzymes-expressed probiotics that can act as synthetic catalysts. These enzymes will produce a mixture of oligosaccharides, which in turn may be more readily metabolized by the producing organism, resulting in higher selectivity, e.g. novel β-galacto-oligosaccharide mixtures have been synthesized from lactose using β-galactosidases from a range of prebiotics.

Synbiotic 2000 consists of a mixture of 10^{10} CFU (or Synbiotic Forte with 10^{11} CFU) of each of four LAB species, including *Pediacoccus pentosaceus*, *Leuconostoc mesenteroides*, *L. paracasei* subsp. *paracasei* and *L. plantarum*, and 2.5 g of each of the four fermentable fibres or prebiotics including β-glucan, inulin, pectin and resistant starch (Medipharm AB, Kågeröd, Sweden, and Des Moines, Iowa, USA). Microbiologists Åsa Ljungh and Torkel Wadström at Lund University developed this multi-component synbiotic formula, which has been extensively used in clinical trials. The choice of LAB for the formulation was finalized after extensive studies of >350 human microbial strains (Kruszewska *et al.*, 2002) and >180 plant microbial strains (Ljungh *et al.*, 2002). Strain selection was based on the ability of LAB to produce bioactive proteins, induce NF-κB signalling, stimulate pro- and anti-inflammatory cytokines, enhance antioxidants, and functionally complement each other. In recent studies both the Synbiotic 2000 Forte and a Probiotic 2000 Forte (no fibre added) containing 10^{11} CFU of each of the four LAB (e.g. 400 billion LAB per dose) have been tested clinically.

5.6.3 Synbiotic foods

Some examples of synbiotic foods with defined health benefits are as follows.

Infant formulae and weaning foods

Bifidogenic factors in milk stimulate the growth of bifidobacteria that result in health benefits to the infant, including a decreased susceptibility to microbial infections. Breast-fed infants' gut is dominated by bifidobacteria while that of formula-fed infants contains a mixed microflora resembling that of an adult. The supplementation of infant milk formula with non-digestible compounds would support growth of bifidobacteria. Hence, it would be of great interest to produce prebiotics with high selectivity towards growth of bifidobacteria that are present in the gut of breast-fed infants as the basis of novel infant food formulations (Rastall and Maitin, 2002).

Functional foods for healthy ageing

Bifidobacterial population decreased markedly in the colon of elderly persons (55–60 years of age) as compared with those of young adults. Species of *Bifidobacterium* are a reasonable target for prebiotics, viz. *B. infantis* and *B. breve* are predominant in infants, whereas *B. adolescentis* and *B. longum* are predominant in adults. Decrease in bifidobacterial numbers results in reduction in resistance to gastrointestinal infections and thus elderly people suffer more with such ailments. The development of a targeted prebiotic that promotes the growth of probiotic strains is able to inhibit gastrointestinal pathogens, viz. *E. coli*, *Salmonella* sp. and *Campylobacter jejuni* more effectively (Guslandi, 2008).

5.7 Safety of Synbiotics

Most probiotics do not have a documented history of safe use, hence safety evaluation is quite necessary. Some of the issues of probiotics concerned to safety are described below.

5.7.1 Antibiotic resistance

Presence of antibiotic resistance-encoding genes must be determined in order to prevent transmission of drug resistance to undesirable organisms. The antibiotic resistance gene, especially vancomycin resistance, should not be unstable plasmid encoded in probiotic organisms as this is one of the last antibiotics used as an effective tool against resistant staphylococci. It is recommended not to use any vancomycin-resistant *Enterococci* as either human or animal probiotics (Macfarlane *et al.*, 2009).

5.7.2 Strain identification

It is not probable that all strains of a genus would confer probiotic health benefits to the host. Proper identification of the organism is desirable by using internationally accepted molecular tools such as DNA-DNA hybridization, 16SrRNA, pulsed field gel electrophoresis (PFGE) or randomly amplified polymorphic DNA (RAPD), or newer systems such as terminal restriction fragment-length polymorphism (T-RFLP) etc., in order to give a correct designation so that it can be easily accessible by researchers. After identification the strain must be deposited in a collection centre so that it can be readily available for workers (Rastall and Maitin, 2002).

5.7.3 Metabolic activities

Certain probiotics are capable of converting food components or biological secretions into secondary metabolites that could be potentially harmful to the host. Hence, these should be assessed for the following parameters.

Biogenic amines

These are produced during degradation of food proteins by certain LAB due to deaminase activity, which is considered as the main factor for the detrimental effects of probiotics. The candidate probiotic can be evaluated for its deaminase activity using Bover-Cid and Holzapfel's method.

Bile salt deconjugation

Bile salts are water soluble end-products of cholesterol metabolism in liver and assist in lipid digestion. They are absorbed actively in the terminal ileum and are subsequently re-secreted, thereby forming an enterohepatic cycle. During bile acid metabolism in microbes, the first step is deconjugation as these are less effective in solubilization of dietary lipids. Further, too early and too much deconjugation, particularly in the upper small intestine, may disturb the lipid digestion and subsequent uptake of fat-soluble vitamins. Primary bile acids can subsequently be dehydroxylated to yield secondary bile acids. The latter are most hydrophobic and toxic to hepatocytes and the gastric and intestinal mucosa, and have been suggested to be cancer promoters and to be involved in the formation of gall stones. Considering the detrimental properties of secondary bile acids, no increase in 7α-dehydroxylase activity can be accepted anywhere in the intestine. Potential probiotics and starters should not exhibit this property (Bengmark, 1998).

D(-)-lactic acid production

Mammalian tissues lack the D-lactate dehydrogenase (DLDH) enzyme to metabolize D(-)-lactic acid. Production of D(-)-lactic acid by probiotic bacteria is also a concern to their use in children, due to D(-)-lactic acidosis. Acidosis is a pathologic condition characterized by neurological alterations.

Others-binding

The binding of probiotics to the mucosal layer is one of the prime selection criteria, as it is more important for immune modulation by competitive exclusion of pathogens. However, binding is also a first step for the pathogenesis. Probiotics adhere to the extracellular matrix (ECM) proteins typically exposed in wound tissue. Pathogens often have affinity for these proteins, which also serve as receptors for invading microbes. Many lactic acid bacteria have been observed to be able to reduce bioavailability of certain toxins by absorption, viz. absorb environmental toxins, mycotoxins, heterocyclic amines etc. Although absorbing these compounds is desirable trait, it is important that such organisms should not be able bind to therapeutic compounds or essential nutrients (Boirivant and Strober, 2007).

- If the strain under evaluation belongs to a species with known haemolytic potential, determination of haemolytic activity is required;
- Assessment of lack of infectivity by a probiotic strain in immunocompromised consumers (add a measure of confidence in the safety of a probiotic);
- Animal and human studies: assessment of side-effects, epidemiological surveillance (post-market) and degradation of mucins must be carried out.

5.8 Evaluation of Synbiotics

5.8.1 Prebiotic characterization

The component that is claimed to have prebiotic attribute(s) must be characterized for source, origin, purity, chemical structure, composition, concentration and amount required to be delivered to the host (Collins *et al.*, 1998).

5.8.2 Functionality evaluation

Correlation of physiological effect and modulation of intestinal microflora should be substantiated based on studies tested in the target host with the final product type along with time framework. A prebiotic can be a fibre but a fibre need not be a prebiotic (De Vrese and Schrezenmeir, 2008).

FAO has recommended the following guidelines for safety evaluation and substantiation of synbiotics:

- If the product has long history of safe usage then it should be considered as GRAS status and thus no need for further human and animal trials. If it is a new candidate, safe levels must be determined;
- Levels of consumption for safe and minimum side effects must be established;
- The product must be free from contaminants and impurities, characterization of contaminant should be done with toxicological studies;
- The prebiotic must not alter the gut microbiota in a way detrimental to the host.

5.9 Applications of Synbiotics

The scientific basis for the development of synbiotics is in their protective role in the host (humans and animals) against colonization of the intestinal tract by non-indigenous microorganisms. The mechanism of synbiotic action is still unknown but different approaches could be developed.

Synbiotics have several health benefits. Some of these applications are the following.

5.9.1 Antimutagenic effect

Mutagenicity means the ability of certain substances to induce genetic mutation, which could prove harmful. Probiotics could exert potent antimutagenic effect. For example, lactobacilli strains in milk (milk cultured with these strains) could reduce the incidence of mutagenicity by binding with harmful chemicals and carcinogens in the gastric juice (Hosno *et al.*, 1997).

5.9.2 Lowering of serum cholesterol

Probiotic bacteria can lower serum cholesterol levels. Certain bacteria found in the gut may break down the bile acids that are secreted into the small intestine. This inhibits re-absorption of bile salts, which in turn leads to reduction of liver cholesterol (liver needs cholesterol to make bile acids). Supposing the gut were more colonized with *Lactobacillus acidophilus* (that has greater propensity to break down bile acids), then obviously serum cholesterol levels could be lowered because there would be more inhibition of re-absorption of bile salts, less cholesterol in the liver (as most of it would be used to make bile acids) and consequently less cholesterol in the serum. For example, it has been found that yoghurt fermented with *L. acidophilus* could reduce LDL cholesterol by 4.5%. But apart from theoretical paradigms it is not clear what level of probiotic dose is appropriate for lowering cholesterol and to what extent hypercholesterolaemic patients could benefit from probiotic therapy (Pereira and Gibson, 2002; Liong and Shah, 2006).

5.9.3 Antihypertensive action

Hypertensive patients could benefit from consuming fermented dairy foods such as fermented milk along with other foods that could possibly lower blood pressure. However, probiotics alone cannot significantly reduce blood pressure on a long-term basis (Liong, 2007).

5.9.4 Benefits to immune system

The immune system produces immunoglobulin A (IgA), an antibody, which protects harmful microbes from binding and penetrating the gut wall. Yoghurt and probiotics like *Lactobacillus casei* are capable of increasing IgA levels by helping to produce more of IgA-producing plasma cells. For instance, people with regular intake of probiotic milk for up to 6 weeks had immune cells with better phagocytic capacity (Schley and Field, 2002).

5.9.5 Antioxidant effects

Normal body metabolism produces free radicals, especially the free radicals from oxygen. If these free radicals are not neutralized quickly enough, they can cause death of cells through oxidation of enzymes, proteins and lipids. Free radical-induced cellular damage can lead to cancer, heart disease and other serious illnesses. Bacteria like *Lactobacillus delbrueckii* ssp. *bulgaricus* and *Streptococcus thermophilus* found in yoghurt can successfully entrap reactive forms of oxygen (hydrogen peroxide and hydroxyl radical). Researchers are working on milk bacteria to be used as an antioxidant food supplement, because milk bacteria can eliminate oxygen free radicals, and also in view of the fact that some lactobacilli have an antioxidant effect in the GI tract (Songisepp *et al.*, 2004).

5.9.6 Synbiotics and *Helicobacter pylori* infection

People with *Helicobacter pylori* bacterium infection could develop peptic ulcer and/or bouts of gastritis together with melena. Potentially this could be a life-threatening illness for the elderly, especially as internal gut haemorrhage could at times lead to a sudden drop in systolic pressure. Probiotics with lactic acid bacteria (for example isolated from yoghurt) can be used to block the growth of *H. pylori* and the complications it could cause. A clinical trial was performed in a school from a low socio-economic area of Santiago, Chile. *Helicobacter pylori*-positive children were randomly distributed into four groups. Children received daily antibiotic treatment (lansoprazole, clarithromycin and amoxicillin) (Ab) for 8 days, '*Saccharomyces boulardii*' (250 mg) plus inulin (5 g) (SbI) daily for 8 weeks, *L. acidophilus* LB (10^9 CFU day^{-1}) (LB), or no treatment (Gotteland *et al.*, 2005). A ^{13}C-urea breath test (13C-UBT) was performed before and after the study, and differences in $^{13}CO_2$ quantities were calculated (DDOB). *Helicobacter pylori* was eradicated in 66, 12, or 6.5% of the children in the Ab, SbI, or LB groups, respectively, while no spontaneous clearance was observed in children not receiving treatment. A moderate but significant difference in DDOB was detected in children receiving SbI (76.31; 95% CI, 711.84 to 70.79), but not LB (+0.70; 95% CI, 75.84 to +7.24). *Helicobacter pylori* infection was eradicated in 12% of synbiotic-treated and 6.5% of probiotic-treated children. Different species of LAB, doses of synbiotics and combinations of antibiotics and synbiotics may yield a wider spectrum of beneficial effects in different disorders. Swiss researchers have found that *L. johnsonii* is beneficial in combating gastritis due to *H. pylori* even if it were to be used alone and not in conjunction with other conventional medicines such as omeprazole or ranitidine. Some researchers believe that *H. pylori* cannot be totally destroyed by probiotics alone, but in any case they can help keep *H. pylori* levels lower (Armuzzi *et al.*, 2001).

5.9.7 Synbiotics and diarrhoea

Diarrhoea, which is a rotavirus infection, is common in children especially under the age of 5, and for this category if probiotics like *Lactobacillus GG*, *L. acidophilus* or *L. acidophilus* combined with *L. bulgaricus* are given it could be helpful in that both the severity and duration of diarrhoea are curbed. Antibiotic-induced diarrhoea can also be taken care of (Cui *et al.*, 2004).

5.9.8 Synbiotics and allergy

It has often been noticed that the incidence of allergy is accompanied by changes in gut microflora. For example, just before an allergy the lactobacilli numbers are found to decrease. Probiotics if given can improve mucosal barrier function and diminish the intensity of allergic effects. For example food allergies linked to milk protein can be reduced by intake of *Lactobacillus GG*. For people susceptible to allergy, probiotics could be a key factor and in the case of young children it can help in the development of the immune system. A synbiotic combination of *L. casei* subsp. *casei* and dextran prevented cedar-pollen-induced onset of nasal and ocular symptoms, cedar pollen-specific IgE responses and elevation of eosinophil counts (Ogawa *et al.*, 2006). In a recent randomized study, children >2 years of age with atopic dermatitis received either a combination of potato starch and *L. rhamnosus* or potato starch alone three times per day for 3 months. Disease scores were reduced with synbiotic treatment from 39.1 to 20.7 (P <0.0001). No differences were observed after 3 months of treatment (P =0.535) (Passeron *et al.*, 2006).

5.9.9 Synbiotics and inflammatory bowel disease (IBD)

Daily rectal instillations with Synbiotic 2000 reconstituted in saline were administered to ten patients with distal colitis during 2 weeks. Synbiotic-treated patients demonstrated dramatic improvements in various disease scores, such as episodes of diarrhoea (initially 2.4, decreased to 0.8), visible blood in stool (2.2 to 0.8), nightly diarrhoea (0.5 to 0), urgency (1.9 to 1.0) and stool consistency (1.1 to 0.8) (Pathmakanthan *et al.*, 2002). Two patients reported significant bloating, but other adverse or side effects were not reported. In another study (Furrie *et al.*, 2005), nine patients with active ulcerative colitis received a synbiotic composed of freeze-dried *Bifidobacterium longum* (4×10^{11} CFU) and a prebiotic FOS/inulin mixture (6 g) daily for 4 weeks. Nine patients received a placebo consisting of powdered maltodextrose (6 g day^{-1}). The quantities of intestinal bifidobacteria were increased 42-fold compared to 4.6-fold in the placebo group. The sigmoidoscopy score, based on clinical assessment of disease activity (Baron *et al.*, 1964), decreased by an average of 1.3 units compared to an increase of 0.58 units in the placebo group (P = 0.06). The bowel habit index scores decreased by 20.4% in the synbiotic group and the scores increased by 70.4% in the placebo group. The prebiotic and probiotic trials reveal the importance of the intestinal environment as a potent regulator of IBD activity.

5.9.10 Synbiotics and irritable bowel syndrome (IBS)

The effects of twice-daily consumption of a probiotic fruit drink, ProViva (Skånemejerier, Malmo, Sweden), containing *L. plantarum* 299v (6×10^7 CFU/drink) or a placebo for 4 weeks were studied in a controlled study including 40 patients (Nobaek *et al.*, 2000). The vast majority (95% of LAB-treated versus 15% of the placebo-treated patients) of individuals in the probiotic consumption group reported general improvement. A total of 20 of 20 patients in the LAB-supplemented group and 11 of 20 patients in the placebo group (P = 0.0012) reported resolution of abdominal pain. A similar study, using the same formula, was performed with patients who received the treatment for 4 weeks. A significant enhancement of LAB composition in probiotics-supplemented patients was described. Flatulence was rapidly and significantly reduced in the LAB-treated group, but no difference in bloating was reported between the groups (Sen *et al.*, 2001). In an another study, 68 patients with IBS were treated for 12 weeks with a vitamin- and plant fibre-enriched diet containing either live or heat-inactivated LAB including 10^9 CFU each of *L. acidophilus, L. helveticus* and *Bifidobacterium* spp. (Tsuchiya *et al.*, 2004). Eighty per cent and 40% of the patients, respectively, reported significant improvements in pain, bloating, constipation and bowel habits (P <0.01).

5.10 Advances in Synbiotic Foods from Traditional Use to the Modern Application in the Medical Field

Synbiotics met a growing scientific interest in the context of the so-called functional foods and/or nutraceuticals, attractive and imaginitive names that reveal their important implications for the human health. Several studies reveal evidence that supports the positive impact of synbiotics on intestinal microflora of pre-term and term infants and in adults, on immunonutritional parameters and on prevention of eczema, in particular atopic eczema (Rastall and Maitin, 2002; Bartosch *et al.*, 2005; Casiraghi *et al.*, 2007; Kukkonen *et al.*, 2007; Panigrahi *et al.*, 2008). It has been also reported that synbiotics administered to newborn infants seem to increase resistance to respiratory infections during the first 2 years of life, reduce the incidence and severity of respiratory diseases during the cold season, decrease the incidence of septic complications in patients with severe systemic inflammatory response syndrome and are safe (Kukkonen *et al.*, 2008; Pregliasco *et al.*, 2008; Shimizu *et al.*, 2009). Leyer *et al.* (2009) reported that daily dietary probiotic supplementation for 6 months was a safe effective way to reduce fever, rhinorrhoea and cough incidence and duration and antibiotic prescription incidence, as well as the number of missed school days attributable to illness, for children 3–5 years of age. An interesting reported activity, that must however be confirmed, is Crohn's disease remission (Fujimori *et al.*, 2007). Moreover, some studies suggest that synbiotic therapy could prove more effective in the treatment of ulcerative colitis (UC) than therapies limited to probiotics or prebiotics; C-reactive protein decreased significantly only with synbiotic therapy and patients with UC on synbiotic therapy experienced greater quality-of life changes than patients on probiotic or prebiotic treatment, suggesting that synbiotic therapy may have a synergistic effect in the treatment of UC (Fujimori *et al.*, 2007; Kanauchi *et al.*, 2009; Macfarlane *et al.*, 2009).

Some other experimental and clinical studies support the fact that, in critically ill patients, early enteral nutrition enriched with synbiotics should restore the balance of microbial communities in a beneficial way with positive effects on intestinal permeability and bacterial translocation and may reduce systemic inflammation, improve the immunological status of the intestinal mucosa and help to prevent infections (Manzanares and Hardy, 2008). A positive effect of synbiotics has been noted also in multiple trauma patients and in patients with high-risk operations (Rayes *et al.*, 2009).

5.11 Conclusions

While probiotics have to compete with already established bacterial communities, prebiotics have the advantage of targeting bacteria already present in the GI tract. However, even though prebiotics may seem as a potentially more efficient and practical way of manipulating the GI microbiota, if for any reason the target bacteria are absent, due to any of several potential disturbance factors, then prebiotics alone are not likely to be effective. Therefore, synbiotics, combinations of prebiotic and probiotic, may be useful and possibly more efficient in several conditions. However, even as increasing evidence becomes available, further well-designed research is still critically needed to definitively prove the effectiveness of synbiotics to promote GI health.

References

Amalaradjou, M.A.R. and Bhunia, A.K. (2012) Modern approaches in probiotics research to control foodborne pathogens. *Advances in Food and Nutrition Research* 67, 185–239.

Andersson, H., Asp, N.-G., Bruce, A., Roos, S., Wadstrom, T. and Wold, A. (2001) Health effects of probiotics and prebiotics: a literature review on human studies. *Scandinavian Journal of Nutrition* 45, 58–75.

Armuzzi, A., Cremonini, F., Ojetti, V., Bartolozzi, F., Canducci, F., Candelli, M., Santarelli, L., Cammarota, G., De Lorenzo, A., Pola, P., Gasbarrini, G. and Gasbarrini, A. (2001) Effect of *Lactobacillus GG* supplementation on antibiotic-associated gastrointestinal side effects during *Helicobacter pylori* eradication therapy: a pilot study. *Digestion* 63, 1–7.

Aziz, Q., Dore, J., Emmanuel, A., Guarner, F. and Quigley, E.M.M. (2013) Gut microbiota and gastrointestinal health: current concepts and future directions. *Neurogastroenterology & Motility* 25, 4–15.

Baffoni, L. and Biavati, B. (2008) Ecologia microbica dell'apparato digerente. In: Biavati, B. and Sorlini, C. (eds) *Microbiologia Agroambientale*. Casa Editrice Ambrosiana, Milan, pp. 147–162.

Barbut, F. and Petit, J.C. (2001) Epidemiology of *Clostridium difficile* associated infections. *Clinical Microbiology and Infection* 7, 405–410.

Baron, J.H., Connell, A.M. and Lennard-Jones, J.E. (1964) Variation between observers in describing mucosal appearances in proctocolitis. *British Medical Journal* 1, 89–92.

Barone, C., Pettinato, R., Avola, E., Alberti, A., Greco, D., Failla, P. and Romano, C. (2000) Comparison of three probiotics in the treatment of acute diarrhoea in mentally retarded children. *Minerva Pediatrica* 52, 161–165.

Bartosch, S., Woodmansey, E.J., Paterson, J.C.M., McMurdo, M.E.T. and Macfarlane, G.T. (2005) Microbiological effects of consuming a synbiotic containing *Bifidobacterium bifidum, Bifidobacterium lactis*, and oligofructose in elderly persons, determined by real-time polymerase chain reaction and counting of viable bacteria. *Clinical Infectious Diseases* 40, 28–37.

Bengmark, S. (1996) Econutrition and health maintenance. A new concept to prevent GI inflammation, ulceration and sepsis. *Clinical Nutrition* 15, 1–10.

Bengmark, S. (1998) Ecological control of the gastrointestinal tract. The role of the probiotic flora. *Gut* 42, 2–7.

Bergogne-Berezin, E. (2000) Treatment and prevention of antibiotic associated diarrhoea. *International Journal of Antimicrobial Agents* 16, 521–526.

Boirivant, M. and Strober, W. (2007) The mechanism of action of probiotics. *Current Opinion in Gastroenterology* 23, 679–692.

Bosscher, D., Breynaert, A., Pieters, L. and Hermans, N. (2009) Food-based strategies to modulate the composition of the intestinal microbiota and their associated health effects. *Journal of Physiology and Pharmacology* 60(Suppl. 1), 5–11.

Bravo, J.A., Julio-Pieper, M., Forsythe, P., Kunze, W., Dinan, T.G., Bienenstock, J. and Cryan, J.F. (2012) Communication between gastrointestinal bacteria and the nervous system. *Current Opinion in Pharmacology* 12, 667–672.

Candela, M., Biagi, E., Maccaferri, S., Turroni, S. and Brigidi, P. (2012) Intestinal microbiota is a plastic factor responding to environmental changes. *Trends in Microbiology* 20, 385–391.

Casiraghi, M.C., Canzi, E., Zanchi, R., Donati, E. and Villa, L. (2007) Effects of a synbiotic milk product on human intestinal ecosystem. *Journal of Applied Microbiology* 103, 499–506.

Collins, J.K., Thronton, G. and O'Sullivan, G.O. (1998) Selection of probiotic strains for human applications. *International Dairy Journal* 8, 487–490.

Collins, S.M., Surette, M. and Bercik, P. (2012) The interplay between the intestinal microbiota and the brain. *Nature Reviews Microbiology* 10, 735–742.

Cui, H.H., Chen, C.L., Wang, J.D., Yang, Y.J., Sun, Y., Wang, Y.D. and Lai, Z.S. (2004) The effects of *Bifidobacterium* on the intestinal mucosa of the patients with ulcerative colitis. *Zhonghua Neike Zazhi* 10, 1521–1525.

De Vrese, M. and Schrezenmeir, J. (2008) Probiotics, prebiotics, and synbiotics. *Advances in Biochemical Engineering/Biotechnology* 111, 1–66.

Dinan, T.G. and Cryan, J.F. (2012) Regulation of the stress response by the gut microbiota: Implications for psychoneuroendocrinology. *Psychoneuroendocrinology* 37, 1369–1378.

Elmer, G.W., Surawicz, C.M. and McFarland, L.V. (1996) Biotherapeutic agents. A neglected modality for the treatment and prevention of selected intestinal and vaginal infections. *The Journal of American Medical Association* 75, 870–876.

Fabia, R., Ae'Rajab, A., Johansson, M.L., Andersson, R., Willen, R., Jeppsson, B., Molin, G. and Bengmark, S. (1993) Impairment of bacterial flora in human ulcerative colitis and experimental colitis in the rat. *Digestion* 54, 248–255.

Food Agriculture Organization (FAO) (2006) Probiotics in food. Health and nutritional properties and guidelines for evaluation. Available at: ftp://ftp.fao.org/docrep/fao/009/a0512e/a0512e00.pdf (accessed 15 April 2013).

Fujimori, S., Tatsuguchi, A., Gudis, K., Kishida, T., Mitsui, K., Ehara, A., Kobayashi, T., Sekita, Y., Seo, T. and Sakamoto, C. (2007) High dose probiotic and prebiotic co-therapy for remission induction of active Crohn's disease. *Journal of Gastroenterology and Hepatology* 22, 1199–1204.

Fuller, R. (1991) Probiotics in human medicines. *Gut* 32, 439–442.

Furrie, E., Macfarlane, S., Kennedy, A., Cummings, J.H., Walsh, S.V., O'Neil, D.A. and Macfarlane, G.T. (2005) Synbiotic therapy (*Bifidobacterium longum*/Synergy 1) initiates resolution of inflammation in patients with active ulcerative colitis: a randomized controlled pilot trial. *Gut* 54, 242–249.
Gibson, G.R. and Roberfroid, M.B. (1995) Dietary modulation of the human colonic microbiota: introducing the concept of prebiotics. *Journal of Nutrition* 125, 1401–1412.
Gibson, G.R., Probert, H.M., van Loo, J.A.E., Rastall, R.A. and Roberfroid, M.B. (2004) Dietary modulation of the human colonic microbiota: updating the concept of prebiotics. *Nutrition Research Reviews* 17, 259–275.
Gill, H.S., Cross, M.L., Rutherfurd, K.J. and Gopal, P.K. (2001) Dietary probiotic supplementation to enhance cellular immunity in the elderly. *British Journal of Biomedical Science* 57, 94–96.
Gionchetti, Rizzello, F., Venturi, A., Brigidi, P., Matteuzzi, D., Bazzocchi, G., Poggioli, G., Miglioli, M. and Campieri, M. (2000) Oral bacteriotherapy as maintenance treatment in patients with chronic pouchitis: a double-blind placebo controlled trial. *Gastroenterology* 19, 305–309.
Gorbach, S.L., Chang, T.W. and Goldin, B. (1987) Successful treatment of relapsing *Clostridium difficile* colitis with *Lactobacillus GG*. *Lancet* 26, 1519.
Gotteland, M., Poliak, L., Cruchet, S. and Brunser, O. (2005) Effect of regular ingestion of *Saccharomyces boulardii* plus inulin or *Lactobacillus acidophilus* LB in children colonized by *Helicobacter pylori*. *Acta Paediatrica* 94, 1747–1751.
Gotz, V., Romankiewicz, J.A., Moss, J. and Murray, H.W. (1979) Prophylaxis against ampicillin-associated diarrhoea with a *Lactobacillus* preparation. *American Journal of Hospital Pharmacy* 36, 754–757.
Gourbeyre, P., Denery, S. and Bodinier, M. (2011) Probiotics, prebiotics, and synbiotics: impact on the gut immune system and allergic reactions. *Journal of Leukocyte Biology* 89, 685–695.
Grandy, G., Medina, M., Soria, R., Teran, C.G. and Araya, M. (2010) Probiotics in the treatment of acute rotavirus diarrhoea. A randomized, double-blind, controlled trial using two different probiotic preparations in Bolivian children. *BMC Infectious Diseases* 10, 253.
Guslandi, M. (2008) Prevention of traveler's diarrhoea with probiotics. *Journal of Clinical Gastroenterology* 42, 1066.
Henke, J.M. and Bassler, B.L. (2004) Bacterial social engagements. *Trends in Cell Biology* 14, 648–656.
Hickson, M., D'Souza, A.L., Muthu, N., Rogers, T.R., Want, S., Rajkumar, C. and Bulpitt, C.J. (2007) Use of probiotic *Lactobacillus* preparation to prevent diarrhoea associated with antibiotics: randomized double blind placebo controlled trial. *British Medical Journal* 335, 80.
Hill, A.V. (1999) Immunogenetics. Defence by diversity. *Nature* 398, 668–669.
Hosno, A., Kitazawa, H. and Yamaguchi, T. (1997) Anti-mutagenic and antitumor activities of lactic acid bacteria. In: Fuller, R. (ed.) *Probiotics 2: Applications and Practical Aspects*. Chapman & Hall, London, pp. 89–132.
Jalanka-Tuovinen, J., Salonen, A., Nikkila, J., Immonen, O., Kekkonen, R., Lahti, L., Palva, A. and de Vos, W.M. (2011) Intestinal microbiota in healthy adults: temporal analysis reveals individual and common core and relation to intestinal symptoms. *PLoS ONE* 6(7), e23035.
Jernberg, C., Lofmark, S., Edlund, C. and Jansson, J.K. (2010) Long-term impacts of antibiotic exposure on the human intestinal microbiota. *Microbiology* 156, 3216–3223.
Johnson, A.M., Kaushik, R.S., Rotella, N.J. and Hardwidge, P.R. (2009) Entero-toxigenic *Escherichia coli* modulates host intestinal cell membrane asymmetry and metabolic activity. *Archive of Infection and Immunity* 77, 341–347.
Kanamori, Y., Sugiyama, M., Hashizume, K., Yuki, N., Morotomi, M. and Tanaka, R. (2004) Experience of long-term synbiotic therapy in seven short bowel patients with refractory enterocolitis. *Journal of Pediatric Surgery* 39, 1686–1692.
Kanauchi, O., Mitsuyama, K. and Andoh, A. (2009) The therapeutic impact of manipulating microbiota in inflammatory bowel disease. *Current Pharmaceutical Design* 15, 2074–2086.
Kane, E.M., Turcios, R.M., Arvay, M.L., Garcia, S., Bresee, J.S. and Glass, R.I. (2004) The epidemiology of rotavirus diarrhoea in Latin America. Anticipating rotavirus vaccines. *Revista Panamericana de Salud Pública* 16, 371–377.
Kelly, G. (2008) Inulin-type prebiotics – a review: Part I. *Alternative Medicine Review* 13, 315–329.
Kruszewska, D., Lan, J., Lorca, G., Yanagisawa, N., Marklinder, I. and Ljungh, A. (2002) Selection of lactic acid bacteria as probiotic strains by *in vitro* tests. *Microecology Therapy* 29, 37–49.
Kukkonen, K., Savilahti, E., Haahtela, T., Juntunen-Backman, K., Korpela, R., Poussa, T., Tuure, T. and Kuitunen, M. (2007) Probiotics and prebiotic galacto-oligosaccharides in the prevention of allergic diseases: a randomized, double-blind, placebo-controlled trial. *Journal of Allergy and Clinical Immunology* 119, 192–198.

Kukkonen, K., Savilahti, E., Haahtela, T., Juntunen-Backman, K., Korpela, R., Poussa, T., Tuure, T. and Kuitunen, M. (2008) Long-term safety and impact on infection rates of postnatal probiotic and prebiotic (synbiotic) treatment: randomized, double-blind, placebo controlled trial. *Pediatrics* 122, 8–12.

Leyer, G.J., Li, S., Mubasher, M.E., Reifer, C. and Ouwehand, A.C. (2009) Probiotic effects on cold and influenza-like symptom incidence and duration in children. *Pediatrics* 124, 172–179.

Liong, M.T. (2007) Probiotics: A Critical Review of Their Potential Role as Antihypertensive, Immune Modulators, Hypo-cholesterolemics and Peri-menopausal Treatments. *Nutrition Reviews* 65, 1–13.

Liong, M.T. and Shah, N.P. (2006) Effects of a *Lactobacillus casei* Synbiotic on Serum Lipoprotein, Intestinal Microflora and Organic Acids in Rats. *Journal of Dairy Science* 89, 1390–1399.

Ljungh, A., Lan, J. and Yanagisawa, N. (2002) Isolation, selection and characteristics of *Lactobacillus paracasei* subsp. *paracasei* F19. *Microbial Ecology in Health and Disease* 14, 4–6.

Macfarlane, G.T. and Cummings, J.H. (1999) Probiotics and prebiotics: can regulating the activities of intestinal bacteria benefit health? *British Medical Journal* 318, 999–1003.

Macfarlane, S., Steed, H. and Macfarlane, G.T. (2009) Intestinal bacteria and inflammatory bowel disease. *Critical Reviews in Clinical Laboratory Sciences* 46, 25–54.

Majamaa, H., Isolauri, E., Saxelin, M. and Vesikari, T. (1995) Lactic acid bacteria in the treatment of acute rotavirus gastroenteritis. *Journal of Pediatric Gastroenterology and Nutrition* 20, 333–338.

Manzanares, W. and Hardy, G. (2008) The role of prebiotics and synbiotics in critically ill patients. *Current Opinion in Clinical Nutrition and Metabolic Care* 11, 782–789.

McCracken, V.J. and Lorenz, R.G. (2001) The gastrointestinal ecosystem: a precarious alliance among epithelium, immunity and microbiota. *Cell Microbiology* 3, 1–11.

McFarland, L.V. (2007) Meta-analysis of probiotics for the prevention of traveller's diarrhoea. *Travel Medicine and Infectious Disease* 5, 97–105.

Nicholson, J.K., Holmes, E., Kinross, J., Burcelin, R., Gibson, G., Jia, W. and Pettersson, S. (2012) Host-gut microbiota metabolic interactions. *Science* 336, 1262–1267.

Nobaek, S., Johansson, M.L., Molin, G., Ahrne, S. and Jeppsson, B. (2000) Alteration of intestinal microflora is associated with reduction in abdominal bloating and pain in patients with irritable bowel syndrome. *American Journal of Gastroenterology* 95, 1231–1238.

Ogawa, T., Hashikawa, S., Asai, Y., Sakamoto, H., Yasuda, K. and Makimura, Y. (2006) A new synbiotic, *Lactobacillus casei* subsp. *casei* together with dextran, reduces murine and human allergic reaction. *FEMS Immunology and Medical Microbiology* 46, 400–409.

Panigrahi, P., Parida, S., Pradhan, L., Mohapatra, S., Misra, P., Johnson, J., Chaudhry, R., Taylor, S., Hansen, N. and Gewolb, I. (2008) Long-term colonization of a *Lactobacillus plantarum* synbiotic preparation in the neonatal gut. *Journal of Pediatric Gastroenterology and Nutrition* 47, 45–53.

Passeron, T., Lacour, J.P., Fontas, E. and Ortonne, J.P. (2006) Prebiotics and synbiotics: two promising approaches for the treatment of atopic dermatitis in children above 2 years. *Allergy* 61, 431–437.

Pathmakanthan, S., Walsh, M., Bengmark, S., Willemse, P.J.A. and Bardhan, K.D. (2002) Efficacy and tolerability treating acute distal ulcerative colitis with synbiotic enemas: a pilot trial. *Gut* 51, A307.

Pereira, D.I.A. and Gibson, G.R. (2002) Effects of Consumption of Probiotics and Prebiotics on Serum Lipid Levels in Human. *Critical Reviews in Biochemistry and Molecular Biology* 37, 259–281.

Perez-Cobas, A.E., Gosalbes, M.J., Friedrichs, A., Knecht, H., Artacho, A., Eismann, K., Otto, W., Rojo, D., Bargiela, R., von Bergen, M., Neulinger, S.C., Daumer, C., Heinsen, F.A., Latorre, A., Barbas, C., Seifert, J., dos Santos, V.M., Ott, S.J., Ferrer, M. and Moya, A. (2012) Gut microbiota disturbance during antibiotic therapy: a multi-omic approach. *Gut* 1, 1–11.

Peuranen, S., Tiihonen, K., Apajalahti, J., Kettunen, A., Saarinen, M. and Rautonen, N. (2004) Combination of polydextrose and lactitol affects microbial ecosystem and immune responses in rat gastrointestinal tract. *British Journal of Nutrition* 91, 905–914.

Plummer, S., Weaver, M.A., Harris, J.C., Dee, P. and Hunter, J. (2004) *Clostridium difficile* pilot study: effects of probiotic supplementation on the incidence of *C. difficile* diarrhoea. *International Microbiology* 7, 59–62.

Pregliasco, F., Anselmi, G., Fonte, L., Giussani, F., Schieppati, S. and Soletti, L. (2008) A new chance of preventing winter diseases by the administration of symbiotic formulations. *Journal of Clinical Gastroenterology* 42, S224–S233.

Rastall, R.A. and Maitin, V. (2002) Prebiotics and synbiotics: towards the next generation. *Current Opinion in Biotechnology* 13, 490–496.

Rayes, N., Seehofer, D. and Neuhaus, P. (2009) Prebiotics, probiotics, synbiotics in surgery – are they only trendy, truly effective or even dangerous? *Langenbeck's Archives of Surgery* 394, 547–555.

Roberfroid, M. (2007) Prebiotics: the concept revisited. *Journal of Nutrition* 137, 830S–837S.

Rosenfeldt, V., Michaelsen, K.F., Jakobsen, M., Larsen, C.N., Moller, P.L., Pedersen, P., Tvede, M., Weyrehter, H., Valerius, N.H. and Paerregaard, A. (2002) Effect of probiotic *Lactobacillus strains* in young children hospitalized with acute diarrhoea. *The Pediatric Infectious Disease Journal* 21, 411–416.

Rusuler-van Embden, J.G.H., Schouten, W.R. and van Lieshout, L.M.C. (1994) Pouchitis: result of microbial imbalance? *Gut* 35, 658–664.

Saavedra, J.M., Bauman, N.A., Oung, I., Perman, J.A. and Yolken, R.H. (1994) Feeding of *Bifidobacterium bifidum* and *Streptococcus thermophilus* to infants in hospital for prevention of diarrhoea and shedding of rotavirus. *Lancet* 344, 1046–1049.

Salminen, S., Deighton, M.A., Benno, Y. and Gorbach, S.L. (1998) Lactic acid bacteria in health and disease. In: Salminen, S. (ed.) *Lactic Acid Bacteria: Microbiology and Functional Aspects*. Marcel Dekker, New York, pp. 211–253.

Samadi, A.R., Islam, R. and Huq, M.I. (1983) Replacement of intravenous therapy by oral rehydration solution in a large treatment centre for diarrhoea with dehydration. *Bulletin of the World Health Organization* 61, 471–476.

Sansonetti, P.J. (2004) War and peace at mucosal surfaces. *Nature Reviews Immunology* 4, 953–964.

Schley, P.D. and Field, C.J. (2002) The Immune-Enhancing Effects of Dietary Fiber and Prebiotics. *British Journal of Nutrition* 87, 221–230.

Sekirov, I., Russell, S.L., Antunes, L.C.M. and Finlay, B.B. (2010) Gut microbiota in health and disease. *Physiology Reviews* 90, 859–904.

Sen, S., Mullan, M., Parker, T.J., Woolner, J., Tarry, S.A. and Hunter, J.O. (2001) Effects of *Lactobacillus plantarum* 299V on symptoms and colonic fermentation in irritable bowel syndrome (IBS). *Gut* 48, A57.

Shimizu, K., Ogura, H., Goto, M., Asahara, T., Nomoto, K., Morotomi, M., Matsushima, A., Tasaki, O., Fujita, K., Hosotsubo, H., Kuwagata, K., Tanaka, H., Shimazu, T. and Sugimoto, H. (2009) Synbiotics decrease the incidence of septic complications in patients with severe SIRS: a preliminary report. *Digestive Diseases and Sciences* 54, 1071–1078.

Simon, G.L. and Gorbach, S.L. (1984) Intestinal flora in health and disease. *Gastroenterology* 86,174–193.

Songisepp, E., Kulisaar, T., Hütt, P., Elias, P., Brilene, T., Zilmer, M. and Mikelsaar, M. (2004) A New Probiotic Cheese with Anti-oxidative and Antimicrobial. *Journal of Dairy Sciences* 87, 2017–2023.

Takahashi, O., Noguchi, Y., Omata, F., Tokuda, Y. and Fukui, T. (2007) Probiotics in the prevention of traveller's diarrhoea: meta-analysis. *Journal of Clinical Gastroenterology* 41, 336–337.

Tannock, G.W. (1983) The effect of dietary and environmental stress on the gastrointestinal microbiota. In: Hentges, D.J. (ed.) *Human Intestinal Microflora in Health and Disease*. Academic Press, New York, pp. 517–539.

Tilg, H. and Kaser, A. (2011) Gut microbiome, obesity, and metabolic dysfunction. *The Journal of Clinical Investigation* 121, 2126–2132.

Timmerman, H.M., Koning, C.J., Mulder, L., Rombouts, F.M. and Beynen, A.C. (2004) Monostrain, multistrain and multispecies probiotics – a comparison of functionality and efficacy. *International Journal of Food Microbiology* 96, 219–233.

Tsuchiya, J., Barreto, R., Okura, R., Kawakita, S., Fesce, E. and Marotta, F. (2004) Single-blind follow-up study on the effectiveness of a symbiotic preparation in irritable bowel syndrome. *Chinese Journal of Digestive Diseases* 5, 169–174.

Van Loo, J., Cummings, J., Delzenne, N., Englyst, H., Franck, A., Hopkins, M., Kok, N., Macfarlane, G., Newton, D., Quigley, M., Roberfroid, M., van Vliet, T. and van den Heuvel, E. (1999) Functional food properties of non-digestible oligosaccharides: a consensus report from the ENDO project (DGXII AIRII-CT94-1095). *British Journal of Nutrition* 2, 121–132.

6 Nutraceuticals from Microbes

Charu Gupta,[1]* Dhan Prakash,[1] Amar P. Garg[2] and Sneh Gupta[3]
Amity Institute for Herbal Research and Studies, Amity University, Noida, India; [2]Department of Microbiology, CCS University, Meerut, India; [3]Department of Zoology, R.G.P.G. College, Meerut, India

6.1 Introduction

Eating habits, actual trends in production and consumption of food have a health, environmental and social impact. In northern European countries eating is an individual affair, whereas in continental and southern countries consumers attach importance to the social dimension of food and sharing a meal (Bronzwaer, 2008). These nutritional differences contribute greatly to the apparent differences in the health of populations. In spite of the fact that today's consumers are increasingly attentive to food safety, quality and health-related issues, people are still fighting with the diseases of a modern age such as obesity, osteoporosis, cancer, diabetes, allergies and dental problems.

Diet has tremendous implications on gut health. Gut complications, such as ulcerative colitis, Crohn's disease, irritable bowel syndrome and gluten therapy-resistant celiac, result from overgrowth and imbalance of intestinal microbial flora, and are related to one's diet. Over the past few years a number of new food ingredients labelled as being nutraceuticals have been launched on the food and pharmaceutical market. These include components that have a proven beneficial effect on human health (Hugenholtz and Smid, 2002). 'Nutraceutical' is the term used to describe a medicinal or nutritional component that includes a food, plant or naturally occurring material, for the improvement of health, by preventing or treating a disease.

Bacteria, yeast and micro-algae can act as producers or catalysts for the production of food ingredients, enzymes and nutraceuticals (Hugenholtz and Smid, 2002). With the current trend towards natural ingredients, there is renewed interest in microbial flavours, colours and bio-processing using enzymes. Microbial production of substances such as organic acids, enzymes, proteins, vitamins, antibiotics and hydrocolloids also remains important (Dufosse, 2009). Lactic acid bacteria, in particular *Lactococcus lactis*, have been demonstrated to be ideal cell factories for the production of these important nutraceuticals. Development in the genetic engineering of food-grade microorganisms means that the production of certain nutraceuticals can be enhanced or newly induced through over-expression and/or disruption of relevant metabolic genes (Bronzwaer, 2008).

Microorganisms can be used as an adjunct therapy for the diseases like protein

* E-mail: charumicro@gmail.com

energy malnutrition (PEM), anaemia, diarrhoea, cancer, obesity, ulcerative colitis, Crohn's disease, irritable bowel syndrome and gluten therapy-resistant celiac. The potential of microbes as a source of nutraceuticals/dietary supplements/functional foods/food supplements in mitigating various health problems is discussed in the present chapter.

6.2 Microorganisms as Protein Source

Microbial biomass has been eaten by man since times immemorial either directly as food, e.g. mushrooms, or as part of fermented foods. Biomass produced by unicellular and multicellular organisms like bacteria, yeast, filamentous fungi and algae is processed and used as human food or animal feed supplement.

Various bacteria, mould, yeast and algae have been employed for the production of single cell proteins (SCP). The bacteria include *Brevibacterium, Methylophilus methylotrophus, Acromobacter delvaevate, Acinetobacter calcoaceticus, Aeromonas hydrophila, Bacillus megaterium, B. subtilis, Lactobacillus* sp., *Cellulomonas* sp., *Methylomonas methylotrophus, Pseudomonas fluorescens, Rhodopseudomonas capsulata, Flavobacterium* species, *Thermomonospora fusca* and others. Some of the algae used are *Chlorella pyrenoidosa, C. sorokiana, Chondrus crispus, Scenedesmus acutus, Porphyrium* sp. and *Spirulina maxima* (Mahasneh, 1997). The filamentous fungi that have been used include *Chaetomium celluloliticum, Fusarium graminearum, Aspergillus fumigatus, A. niger, A. oryzae, Cephalosporium cichhorniae, Penicillium cyclopium, Rhizopus chinensis, Scytalidium acidophilum, Tricoderma viridae, T. alba* and *Paecilomyces varioti*. Yeasts such as *Candida utilis* (Torula yeast), *C. lipolytica, C. tropicalis, C. novellas, C. intermedia* and *Saccharomyces cerevisiae* are all among the various organisms that have been used for the production of SCP (Becker, 2007). The desired microorganisms to be cultured should be non-pathogenic, must have good nutritional values, must be usable as food and feed, should not contain toxic compounds and production cost should be low. For example Quorn, a leading brand of mycoprotein food product in the UK and Ireland, is extracted from a fungus, *Fusarium venenatum*. It is high in protein and dietary fibre and low in saturated fat and salt.

6.2.1 Nutritional benefits of single cell proteins

SCP basically comprise high proteins, low fats, important minerals, vitamins particularly thiamine, riboflavin, B-complex, glutathione, folic acid and good amino acid composition but deficient in S-containing amino acids. Its nutritive value is comparable with soy-meal and fishmeal. The nutritive and food values depend upon the organism and the substrate upon which it grows. The method of harvesting, drying and processing also has an effect on the nutritive value of the finished product. SCP from yeast and fungi has up to about 50–55% protein. It is rich in lysine but poor in methionine and cysteine. It has good balance of amino acids and rich in B-complex vitamins and more suitable as poultry feed. SCP produced from bacteria have more than 80% protein although they are poor in S-amino acids and have high nucleic acid content (Kurbanoğlu, 2001). Yeast SCPs are playing a greater role in the evolution of aquaculture diets. With excellent nutrient profiles and capacity to be mass-produced economically, SCPs have been added to aquaculture diets as partial replacement for fishmeal (Olvera-Novoa *et al.*, 2002) and for HUFA-fortification of rotifer and *Artemia* (Li and Gatlin, 2003). Some yeast strains with probiotic properties, such as *S. cerevisiae* (Oliva-Teles and Gonçalves, 2001) and *Debaryomyces hansenii* (Tovar *et al.*, 2002) boost larval survival either by colonizing the gut of fish larvae, thus triggering the early maturation of the pancreas, or via the immunostimulating glucans derived from the yeast cell wall (Campa-Córdova *et al.*, 2002; Burgents *et al.*, 2004). However, many of these yeast supplements are deficient in S-amino acids, particularly methionine, which restricts their extensive use as the sole protein source.

6.3 Microbes for Treatment and Prevention of Anaemia

Prevalence of anaemia in developing and third world countries is high because of malnutrition, poor availability of iron and chronic blood loss. The other contributing factors of anaemia are deficiency in folic acid, vitamin B_{12} and iron. Vitamin B_{12}, along with folate, is involved in making the haem molecule, an integral part of haemoglobin. Vitamin B_{12} is important in living beings and it is used to treat pernicious anaemia and peripheral neuritis. Folates perform an important role as cofactors in 1-carbon transfer reactions occurring in purine and pyrimidine biosynthesis. They are also required for efficient DNA replication, repair and methylation. For these key roles in the cellular cycle, tissues with a high cell growth rate or turnover, such as haemopoietic cells and intestinal mucosa, have a high requirement for folate. Their deficiency leads to megaloblastic anaemia resulting in inhibition of DNA synthesis in red blood cell production (Bronzwaer, 2008).

There are some folic acid-producing bacteria and yeast that can be cultured in whey or milk plasma, thereby accumulating high concentrations of folic acid in the medium. The various identified bacteria that produce and enhance the uptake of folic acid are *Lactococcus lactis* subsp. *cremoris, L. lactis* subsp. *lactis, Bifidobacterium adolescentis, B. pseudocatenulatum*; yeasts such as *Candida famata, C. guilliermondii, C. glabrata, Yarrowia lipolytica, Saccharomyces cerevisiae* and *Pichia glucozyma*. The vitamin B_{12}-producing bacteria are *Pseudomonas denitrificans* and *Propionibacterium shermanii* (Kassinen *et al.*, 2007). Application of probiotic bacteria for the prevention of megaloblastic anaemia is a novel scientific approach that involves lactic acid-fermented foods, increases iron absorption by optimization of pH in the digestive tract, activates enzyme phytases, produces organic acids and other digestive enzymes (Scarpignato, 2008).

A strain of probiotic bacteria developed by Swedish firm Probi doubled the absorption of iron from food in women. *Lactobacillus plantarum* 299v (Lp299v) not only helps our digestive system, but also helps in our immune system. It is also good for the heart and helpful in reducing unexpected gas and bloating conditions. It improves bowel moments by making them more normal and regular. It also reduces the negative effects of an antibiotic drug on colonic fermentation (Bemgmark, 1998; Liu *et al.*, 2001).

Several medicines are produced using genetically engineered bacteria or fungi that synthesize the medicine in giant bioreactors. Erythropoietin can be man-made in bioreactors by bacteria and is used to treat anaemia, but the treatment requires frequent, even daily injections. Researchers have purified specialized cells from human blood that normally repair the lining of blood vessels. These cells were also genetically engineered to express erythropoietin. The cells were then mixed with mesenchymal stem cells, which are able to form blood vessels. This mixture was then injected underneath the skin of anaemic mice. The cell mixture spontaneously formed networks of blood vessels underneath the skin. The vessel lining secreted erythropoietin and cured anaemia in mice (Kassinen *et al.*, 2007).

6.4 Microbes for the Prevention or Treatment of Obesity and Cholesterol

Obesity is a significant risk factor for major diseases including type-II diabetes, coronary heart disease, hypertension and certain forms of cancer (Barsh *et al.*, 2000; Kopelman, 2000). Obesity arises when energy intake, principally stored as triglycerides, exceeds energy expenditure (Spiegelman and Flier, 2001; Flier, 2004). Obesity is a complex trait influenced by diet, developmental stage, age, physical activity and genes (Brockmann and Bevova, 2002; Friedman, 2003). Probiotic bacteria are used for the manufacture of a natural remedy for controlling weight gain, preventing obesity, increasing satiety, prolonging satiation, reducing food intake, reducing fat deposition, improving energy metabolism, treating and enhancing insulin sensitivity and treating obesity.

Animal studies have demonstrated the efficacy of some strains of lactic acid bacteria

(LAB) to be able to lower serum cholesterol levels, presumably by breaking down bile in the gut, thus inhibiting its re-absorption (Sanders, 2000). A meta-analysis that included five double-blind trials examining the short-term effects of a yoghurt with probiotic strains on serum cholesterol levels found a minor change (~4% decrease) in total cholesterol concentration and a decrease (~5%) in serum LDL concentration (Agerholm-Larsen *et al.*, 2002). A slightly longer study evaluating the effect of yoghurt with probiotic strains on 29 subjects over 6 months found no statistically significant differences in total serum cholesterol or LDL but a significant increase in serum HDL following treatment (Kiessling *et al.*, 2002).

Lactobacillus (*Lb. sporogenes* and *Lb. acidophilus* NCFB 1748) and *Bifidobacterium* genus representatives have been reported to play a critical role in weight regulation as an anti-obesity effect in experimental models and humans (Mercenier *et al.*, 2002). *Lactobacillus sporogenes* has the ability to lower cholesterol levels. It produces a significant reduction in low density lipoprotein (LDL) levels and a small but significant increase in high density lipoprotein (HDL) cholesterol. *Lactobacillus sporogenes* can be used as a side effect-free alternative to drug therapy in the treatment of high cholesterol and heart disease. It has also been used in the treatment of gut dysbiosis, vaginitis and aphthous stomatitis. Clinical studies have revealed that *Lb. sporogenes* can be successfully implanted in the intestine. *Lactobacillus sporogenes* satisfies the essential requirement of an efficient probiotic. The spores of *Lb. sporogenes* are resistant to heat and other adverse environmental conditions. This property of spore formation by *Lb. sporogenes* is the main characteristic that makes it the probiotic of choice in clinical applications (Mohan *et al.*, 1990). Being sporulated, they germinate under favourable conditions and produce sufficient viable cells, which proliferate and perform vital healthful functions. In addition, *Lb. sporogenes* spores are semi-resident and are slowly excreted out of the body. *Lactobacillus sporogenes* is effective in the form of dietary supplements as well as when added to food products (Montrose and Floch, 2005).

Lactobacillus acidophilus ferments lactose into lactic acid, like many lactic acid bacteria. *Lactobacillus acidophilus*, a homeo-fermentative microorganism, produces only lactic acid. *Lactobacillus acidophilus* has been found helpful in reducing serum cholesterol levels (Anderson and Gilliland, 1999). During digestion, *L. acidophilus* assists in the production of niacin, folic acid and pyridoxine. It also helps in bile deconjugation, separating amino acids from bile acids, which can then be recycled by the body (Gilliland and Speck, 1977). *Lactobacillus acidophilus* may provide additional health benefits, including improved gastrointestinal (GI) function, a boosted immune system, and a decrease in the frequency of vaginal yeast infections and relief from indigestion and diarrhoea (De Roos and Katan, 2000).

A Japanese study (Kadooka *et al.*, 2010) showed a *Lactobacillus* probiotic to reduce abdominal fat by 4.6% and subcutaneous fat by 3.3%. The probiotic milk containing *Lactobacillus gasseri* SBT2055 showed significant decreases in body weight BMI, in waist circumference and in the hips. Previous findings (Hamad *et al.*, 2009) showed the same *Lactobacillus* probiotic to reduce fat levels in animals.

Scientists in Ireland found that another probiotic of the *Lactobacillus* genus could seemingly influence the fat composition of the host (Rosberg-Cody *et al.*, 2011). Researchers engineered a specific strain of *Lactobacillus* to produce a specific kind of fatty acid, t10, c12 CLA. Mice fed with the probiotic showed significant alterations to their fat tissues. The t10, c12 CLA molecule has already been associated with decreased body fat in humans and other animals, as well as having demonstrated ability to inhibit the growth of colon cancer cells.

Other studies on obesity following pregnancy showed that where women had taken probiotics (*Lactobacillus* and *Bifidobacterium* strains) during pregnancy they were less likely to become obese after giving birth. Sousa *et al.* (2008) found that administration of *L. acidophilus* in rats resulted in weight loss, showed increased levels of leptin, a protein found to decrease the appetite and increase the metabolism. Bajzer and Seeley (2006) reported a clear difference in gut microbial populations between obese and lean people, suggesting a link between types of bacteria in the gut and obesity. Delzenne *et al.* (2005) suggested that prebiotics have a capacity to promote satiety, by increasing levels of the satiety hormone, glucagon-like peptide (GLP-1), or by reducing the production of ghrelin, a peptide that triggers the appetite.

6.5 Microbes for Treatment and Prevention of Diarrhoea

6.5.1 Microbes for irritable bowel syndrome

Irritable bowel syndrome (IBS) is the most common functional GI disorder that results in abdominal pain, altered bowel habits and irregular stool characteristics (Lacy and Lee, 2005). *Lactobacillus salivarius* or *Bifidobacterium infantis* found significant improvements in typical IBS symptoms with the administration of probiotics. Commonly reported improvements were reductions in bloating, flatulence, speed of colonic transit and abdominal pain. Intestinal microorganisms play various roles in human health such as complex food digestion, metabolizing drugs, detoxifying toxic compounds, producing essential vitamins and preventing colonization of pathogens. Most of the microorganisms found in the GI tract are anaerobic bacteria, which are uncultivable under standard laboratory conditions. The type and number of bacteria in the GI tract varies depending on age, gender, geographical origin (Mueller *et al*., 2006) and environmental factors, such as diet and dietary supplements (Kajander *et al*., 2009). Firmicutes and Bacteroidetes are the dominant beneficial bacteria present in the normal human GI tract, and the latter was reported in lower numbers in constipation-predominant IBS patients (Rajilić-Stojanović *et al*., 2007).

6.5.2 Microbes for inflammatory bowel disorders and colitis

A protein has been isolated from probiotic bacteria that helps to alleviate inflammatory bowel disorders (IBD). The protein called p40 was found effective in animal models suffering from colitis (colon inflammation). The protein supports intestinal epithelial cell growth and function, and reduces inflammatory responses that can cause intestinal cells to die. Many of the hundreds of bacterial species that live in our gut (known as the 'human microbiome') help to digest certain substances, produce vitamins and fight off more dangerous bacteria. But miscommunication between these bacteria and our gut lining can lead to conditions such as ulcerative colitis and Crohn's disease; e.g. *Lactobacillus rhamnosus* GG (LGG), has been used to prevent intestinal disorders such as IBD and diarrhoea, as well as other conditions such as dermatitis. There are two specific proteins secreted by LGG (called p75 and p40) that are responsible for the bacterium's beneficial effects (Gupta *et al*., 2000). LGG prevents epithelial cells from inflammation-induced apoptosis. The p40 activates epidermal growth factor receptor (EGFR), a protein critical for cell survival and growth. Activation of EGFR protected epithelial cells in two ways: by preventing both apoptosis and inflammation-induced disruption of the 'tight junctions' between epithelial cells, which form a barrier to keep toxic substances and pathogens out of the bloodstream (Lacy and Lee, 2005; Kassinen *et al*., 2007).

Lactic acid bacteria, in particular *Lactococcus lactis*, have been demonstrated to be ideal cell factories for the production of such important nutraceuticals. Development in the genetic engineering of food-grade microorganisms means that the production of certain nutraceuticals can be enhanced or newly induced through over-expression and/or disruption of relevant metabolic genes (Hugenholtz and Smid, 2002). *Lactococcus acidophilus* is primarily considered essential because of its ability to produce vitamin K and lactase. Furthermore, it also has the ability to produce antimicrobial substances like acidolin, acidolphilin, lactocidin and bacteriocin, which could aid with further health improvements in the digestive system (Gupta *et al*., 2000).

6.6 Microbes for Treatment and Prevention of Atopic Dermatitis

Atopic dermatitis (AD) is the most common chronic skin condition in infants and children, with a prevalence of from 10% to 20% (Laughter *et al.*, 2000; Schultz-Larsen, 2002). Geographic location affects the prevalence of this disease, with the highest prevalence in the USA and Europe (Thestrup-Pedersen, 2002). Important factors in the susceptibility to develop AD include a genetic basis (82%) and environmental factors (18%) (Thomsen *et al.*, 2007). In addition, AD has been linked to food hypersensitivity, especially milk and egg proteins (Sicherer and Sampson, 1999). Allergic diseases are associated with an imbalance in the T-lymphocytes type (TH1/TH2) cytokine activation of TH2 cells and with stimulation of IgE and IgA synthesis, leading to allergic reactions (Kruisselbrink *et al.*, 2001; Winkler *et al.*, 2007). Probiotics can inhibit the TH2 response while stimulating the production of TH1 and TH1 cytokines, such as interferon (Isolauri *et al.*, 2001; Winkler *et al.*, 2007). In children with atopic disease, the use of probiotics was associated with an increase in interferon-γ production and inhibition of allergen-induced tumour necrosis factor-α, IgE and several allergy-induced cytokines (Prescott *et al.*, 2005; Flinterman *et al.*, 2007). According to a study that evaluated the impact of a mixture of *Lactobacillus acidophilus* DDS-1 and *Bifidobacterium lactis* UABLA-12 with moderate to severe AD, it was found that probiotic supplementation stabilizes the intestinal barrier function and decreases GI symptoms in children with AD (Rosenfeldt *et al.*, 2004). Therefore, probiotics may present attractive alternatives, given the low probability for the development of adverse effects.

6.6.1 Probiotics for allergies

An increased prevalence of atopic diseases, atopic dermatitis, allergic rhinitis and asthma has been reported (Isolauri, 2004). Evidence suggests that specific strains of probiotics have an effect on inflammatory processes as demonstrated by the reduction of certain local and systemic immune markers and these actions may be mediated via the gut-associated lymphoid tissue (GALT), one of the three intestinal lines of defence (Isolauri, 2004; Miraglia del Giudice and De Luca, 2004). Probiotics may affect the production of inflammation-producing cells and accessibility of allergens, normalize the intestinal microbiota, impacting on the intestinal barrier function and help regulate the secretion of inflammatory mediators (Isolauri, 2004).

6.7 Microbes for Cancer

Administration of microorganisms in the treatment of cancer is less widely known in the scientific community. It goes back more than 100 years when William B. Coley, physician and surgeon of the Memorial (Sloan-Kettering) Hospital, New York, observed that many of his cancer patients showed tumour regression when they were infected with bacterial pathogens. Treatment to eliminate the infections resulted in cancer relapse (Coley, 1893). He developed a treatment modality by making extracts of defined bacteria (e.g. *Streptococcus pyogenes, Serratia marcescens*) called Coley's Toxin, which he administered to shrink tumours in his patients (Coley, 1893; Nauts *et al.*, 1946).

Subsequently, other bacteria have been investigated in an effort to reduce the growth or the size of tumours. The most prominent example was the use of *Mycobacterium bovis* BCG (Bacillus Calmette-Guérin), the vaccine strain in the treatment of defined stages of urinary bladder cancer (Lamm *et al.*, 1980). Several good clinical practices (GCP) performed clinical studies (randomized controlled trials; RCTs) that have shown a clear relationship between the use of *M. bovis* BCG immunoprophylaxis after surgical removal of the tumour and the decreased recurrence rate or the prolonged relapse-free interval of urinary bladder cancer (Morales *et al.*, 1976; Lamm *et al.*, 1980). The mode of action of BCG to induce its antineoplastic effect is suggested to result from its effects on the (local, mucosal) immune system, with mononuclear cells (T-lymphocytes, monocytes) playing a major role (Ratliff *et al.*, 1993). Thus, intravesical instillation of BCG induces a non-specific cystitis, which is accompanied by local production of cytokines and accumulation of inflammatory cells being able to damage malignant cells (Alexandroff *et al.*, 1999). The requirement of live cells of BCG for its anticancer activity is reflected in the fact that monocytes and helper T-lymphocytes type 1 (TH1) are most important for its effectiveness (Thanhäuser *et al.*, 1995) and that high doses of defined vitamins have shown a positive effect on the treatment of bladder cancer in human clinical trials (Lamm *et al.*, 1994).

The ability of bacteria to modulate the immune response to non-related antigens is well documented. *Propionibacterium* species are amongst the most potent immunomodulators stimulating cell populations involved in non-specific resistance (Isenberg *et al.*, 1995). Three species (*Propionibacterium acnes, P. granulosum, P. avidum*) appeared to be of special medical interest and after evaluating the immunoactive potential of a great number of strains, *P. avidum* KP-40 and *P. granulosum* KP-45 were selected for further experimental and clinical studies (Ko *et al.*, 1981). For practical reasons (e.g. cultivation procedure, biological and immunological standardization) *P. avidum* KP-40 was preferably introduced for clinical evaluation, although its immunoactive capacity is absolutely identical to *P. granulosum* KP-45.

Bacteria can be used to make anticancer agents and provide an extra source of lead compounds for the pharmaceutical industry. Genetically engineered strains of *Streptomyces parvulus* are used to produce compounds that selectively inhibit growth of human cancer cells. It naturally produces a compound called borrelidin, which is an effective inhibitor of angiogenesis, a key process in the spread of malignant tumours. Genetically engineered *Escherichia coli* bacteria are used to produce a large quantity of a critical compound that is a precursor to the cancer drug Taxol, originally isolated from the bark of the Pacific yew tree. The tree's bacteria can produce 1000 times more of the precursor, known as taxadiene, than any other engineered microbial strain. Taxol intermediates are taxadiene and taxadien-5-alpha-ol. *Escherichia coli* does not naturally produce taxadiene, but it does synthesize a compound called IPP, which is two steps away from taxadiene. Taxol, also known as paclitaxel, is a powerful cell-division inhibitor commonly used to treat ovarian, lung and breast cancers. Similarly, the bacterium *Streptomyces coelicolor* naturally produces red-coloured alkaloids called prodiginines; its synthetic analogue called GX15-070 is in phase 1 and 2 cancer treatment trials. They can be used to target and kill cancer cells (Alexandroff *et al.*, 1999; Adjei, 2000). Analogues of other prodiginines, such as streptorubin B, are also powerful anticancer drugs. Many pharmaceutical agents have been discovered by screening natural products from a wide range of microorganisms. Rapamycin and its analogues, products of *Streptomyces hygroscopicus*, have potent immunosuppressive activity. They inhibit signalling pathways required for T-cell activation and proliferation. Rapamycin blocks progression of the cell cycle at middle-to-late G1 phase in T cells and B cells, and osteosarcoma and rhabdomyosarcoma cell lines, among others. Geldanamycin is a benzoquinone ansamycin natural fermentation product and inhibits heat-shock protein HSP 90 (Schulte and Neckers, 1998). Some examples of anticancer agents derived from microorganisms are cited in Table 6.1.

Wortmannin (Table 6.1) is a product of the fungus *Talaromyces wortmanni* and inhibits signal transduction pathways by forming a covalent complex with an active-site residue of phosphoinositide-3-kinase (PI3K), inhibiting PI3K activity. Thus, toxins that originally evolved to kill competing microorganisms can have a variety of physiological effects in animals. In many cases, the targets of these compounds are components of signal transduction cascades that are conserved in many species, and that have been considered novel targets for anticancer drug discovery (Adjei, 2000).

Laxaphycins A and B are the products of an unknown marine bacterium that produce cyclic peptides having anticancer effect by increasing polyploidy by putative topoisomerase II alterations. Leptosins C and F (alkaloids) are produced by an unknown marine fungus that inhibits DNA topoisomerase I and II and induces apoptosis. Dehydrothrysiferol, a triterpene, produced by a marine alga, enhances apoptosis induction in oestrogen receptor negative breast cancer cells. Similarly, GA3 polysaccharide is also produced by a marine alga, which causes

Table 6.1. Microorganism-derived anticancer agents.

Compound	Microorganism	Use in cancer
Actinomycin	*Streptomyces* spp.	Sarcoma and germ cell tumours
Bleomycin	*S. verticillus*	Germ cell, cervix and head and neck cancer
Daunomycin	*S. coeruleorubidus*	Leukaemia
Doxorubicin	*S. pneuceticus*	Lymphoma, breast, ovary, lung and sarcomas
Epirubicin	*S. pneuceticus*	Breast cancer
Idarubicin	*S. pneuceticus*	Breast cancer and leukaemia
Mitomycin C	*S. caespitosus*	Gastric, colorectal, anal and lung cancer
Geldanamycin	*S. hygroscopicus*	Experimental
Rapamicin	*S. hygroscopicus*	Experimental
Wortamannin	*Talaromyces wortmanni*	Experimental

inhibition of topoisomerase I and II (Alexandroff *et al.*, 1999; Cinque *et al.*, 2006).

A harmless soil-dwelling bacteria has been discovered that successfully kills cancer cells. The therapy uses *Clostridium sporogenes* – a bacterium that is widespread in the soil. Spores of the bacterium are injected into patients and they only grow in solid tumours, where a specific bacterial enzyme is produced. An anticancer drug is injected separately into the patient in an inactive 'pro-drug' form. When the pro-drug reaches the site of the tumour, the bacterial enzyme activates the drug, allowing it to destroy only the cells in its vicinity – the tumour cells. Thus the bacterial strain specifically targets tumours and can be used as a vehicle to deliver drugs in frontline cancer therapy. The strain is expected to be tested in cancer patients in 2013 (De Roos and Katan, 2000).

A fundamental requirement for any new cancer therapy is the ability to target cancer cells while excluding healthy cells. This therapy naturally fulfils this need. *Clostridia* are an ancient group of bacteria that evolved on the planet before it had an oxygen-rich atmosphere and so they thrive in low oxygen conditions. When *Clostridia* spores are injected into a cancer patient, they will only grow in oxygen-depleted environments, i.e. the centre of solid tumours. This is a totally natural phenomenon, which requires no fundamental alterations and is exquisitely specific. This specificity can be exploited to kill tumour cells leaving the healthy tissue unscathed (Thanhäuser *et al.*, 1995). The treatment is superior to a surgical procedure, especially for patients at high risk or with difficult tumour locations. A successful outcome could lead to its adoption as a frontline therapy for treating solid tumours.

A similar study aims to use viruses and bacteria that are disease-causing, such as measles, botulism, gangrene and common cold, as the basis of new forms of cancer treatment (Ratliff *et al.*, 1993; Pandey *et al.*, 2007). These microbes will be engineered to make an enzyme that can activate cancer pro-drugs – a new generation of therapies that remain inert in the body until activated by the enzymes. Pro-drugs are an attractive alternative to chemotherapy because of their ability to kill cancer cells while leaving healthy cells unharmed (Pandey *et al.*, 2007). The fundamental idea behind cancer therapy is that certain viruses and bacteria can more easily infect a cell in a cancerous tumour than a healthy, human cell (Adjei, 2000; Sakamoto *et al.*, 2010). USFDA has approved two types of vaccines to prevent cancer: vaccines against the hepatitis B virus, which can cause liver cancer, and vaccines against human papillomavirus types 16 and 18, which are responsible for about 70% of cervical cancer cases (Pandey *et al.*, 2007).

6.8 Microbes for Treatment of Gluten Therapy-Resistant Celiac

Celiac disease, an intestinal inflammatory disease with autoimmune features, is caused by oral ingestion of gluten peptides that escape intestinal degradation. These peptides are antigenically presented on HLA-DQ2 or HLA-DQ8, preferentially after de-amidation of certain glutamines by the celiac disease auto-antigen tissue transglutaminase (tTG), eliciting a destructive Th1 T cell response (Koning *et al.*, 2005; Rubio-Tapia and Murray, 2010). Strict elimination of gluten from the diet is the therapy of choice, however, it is difficult to maintain and poses a significant social and financial burden on the patient. Therefore, an additive non-dietary therapy that relieves patients from a highly restricted gluten free diet is much needed (Green and Jabri, 2006; Schuppan *et al.*, 2009). A novel therapeutic approach for celiac disease is the use of enzymes to achieve proteolytic fragmentation of gluten proteins which otherwise escape proteolytic inactivation by gastric, pancreatic and intestinal brush border enzymes into smaller non-immunogenic peptides. The resistance of gluten to digestive proteases is due to a particular primary structure based on their high number of proline (P) and glutamine (Q) residues, and repetitive PQ sequences that are not cleavable by common GI proteases. A number of gluten-degrading enzymes from microbial and cereal sources have been discovered. Prolyl-endopeptidases target the conformationally constrained peptide bonds' C-terminal to proline residues. Prolyl endopeptidases from *Sphingomonas capsulata, Flavobacterium meningosepticum, Myxococcus xanthus* and *Aspergillus niger* have been pursued as drug candidates for enzymatic treatment of gluten in celiac disease (Shan *et al.*, 2004; Stepniak *et al.*, 2006).

A biologically more favourable and likely source for gluten-degrading enzymes would be the microbiome colonizing the human GI tract. It is well recognized that bacteria populating the human body supply the host with numerous functions that are not encoded by the human genome (Gill *et al.*, 2006). It has been found that gluten-degrading bacteria are naturally residing in the oral cavity (Aas *et al.*, 2005; Helmerhorst *et al.*, 2008). The finding of gluten-degrading oral microbes may serve as a novel source for therapeutic gluten-degrading enzymes.

So the novel treatment therapy is through administration of enzyme supplements that are focused on inactivating immunogenic gluten epitopes. Some of the methods are described below.

6.8.1 Oral administration of bacterial endopeptidases

After ingestion, degraded gluten proteins reach the small intestine. However, because of their unusually high proline and glutamine content, especially in immuno-dominant gliadin peptides, gluten is poorly degraded by the enzymes present in the GI tract. Hence, oral enzyme therapy has been suggested as an alternative to the gluten-free diet. Promising enzymes (expressed in various microorganisms) tested are the prolyl oligopeptidases from *Flavobacterium meningosepticum, Sphingomonas capsulata* and *Myxococcus xanthus*. These enzymes are capable of degrading proline-containing peptides that are otherwise resistant to degradation by proteases in the GI tract *in vitro* (Gass *et al.*, 2005; Marti *et al.*, 2005). However, most of these enzymes are irreversibly inactivated in the stomach by pepsin and acidic pH, thus failing to degrade gluten before it reaches the small intestine (Shan *et al.*, 2004). Encapsulation of these prolyl oligopeptidases was proposed in order to protect them from gastric juices (Gass *et al.*, 2005). However, in *ex vivo* study it was observed that only high doses of prolyl oligopeptidase were capable of eliminating the accumulation of immunogenic peptides in the serosal compartment (Marti *et al.*, 2005). A new enzyme, prolyl endoprotease from *Aspergillus niger*, was found to degrade gluten peptides and intact gluten proteins efficiently in the stomach, to such an extent that hardly any traces of gluten reached the duodenal compartment (Mitea *et al.*, 2008). Moreover, the optimum pH of this enzyme is compatible with that found in the stomach and the enzyme is resistant to degradation by pepsin. Finally, prolyl endoprotease from *A. niger* is derived from the food-grade microorganism and is available on an industrial scale. These results indicate that this enzyme might be suitable for oral supplementation to degrade gluten proteins in food before they reach the small intestine (Mitea *et al.*, 2008).

Gass *et al.* (2007) evaluated a new combination therapy, consisting of two gastrically active enzymes that detoxify gluten before its release in the small intestine. They used a glutamine-specific endoprotease (EP-B2;

a cysteine endoprotease from germinating barley seeds) and a prolyl-specific endopeptidase from *Sphingomonas capsulata*, for its ability to digest gluten under gastric conditions. Endoprotease EP-B2 extensively hydrolyses the gluten network in bread into relatively short oligopeptides, whereas prolyl-specific endopeptidase from *Sphingomonas capsulata* rapidly detoxifies oligopeptides after primary proteolysis at internal proline residue level to yield non-toxic metabolites (Gass *et al.*, 2007). A practical advantage of this combination product is that both enzymes are active and stable in the stomach and can therefore be administered as lyophilized powders or simple capsules or tablets.

6.8.2 Pretreatment of whole gluten with bacterial-derived peptidase

An alternative approach to detoxify gluten is represented by the digestion of wheat gluten peptides with bacterial-derived peptidase during food processing and before administration to patients. Lactobacilli (*L. alimentarius* 15M, *L. brevis* 14G, *L. sanfranciscensis* 7A and *L. hilgardii* 51B) showed considerable hydrolysis of albumin, globulin and gliadin fractions during wheat sourdough fermentation. These lactobacilli had the capacity to hydrolyse the 31–43 fragment of A-gliadin *in vitro* and, after hydrolysis, greatly reduced the agglutination of K 562(S) subclone cells of human myelogenous leukaemia origin by a toxic peptic-tryptic digest of gliadins (Cagno *et al.*, 2002). On the basis of these results, and with the goal of decreasing gluten intolerance in humans, the authors investigated a novel bread-making method that used selected lactobacilli to hydrolyse various Pro-rich peptides (Cagno *et al.*, 2004). The different probiotic bacterial strains have their characteristic set of peptidases, which may diverge considerably from each other and have variable substrate specificities (Angelis *et al.*, 2006). It is interesting to underline that probiotics, defined as the viable microorganisms that exhibit a beneficial effect on the health of the host by improving its intestinal microbial balance, could directly modulate the function of epithelial cells (Resta-Lenert and Barrett, 2003; Lindfors *et al.*, 2008). Furthermore, several probiotic bacterial strains are able to protect the epithelium, presumably from various insults, including pathogenic bacteria (Otte and Podolsky, 2004; Cinque *et al.*, 2006) and inflammatory cytokines (Rizzello *et al.*, 2007; Yan *et al.*, 2007). The use of proteases from germinating wheat seeds has also been proposed to create safe cereal products for Crohn's disease patients (Gianfrani *et al.*, 2007; Stenman *et al.*, 2009).

6.9 Microbes for Treatment of Crohn's Disease

Crohn's disease (CD) is a common chronic disorder that affects the GI tract and is believed to develop as a result of an aberrant immune response to intestinal microbes in a genetically susceptible host. CD may involve the small intestine, the large intestine, the rectum, or the mouth (Shan *et al.*, 2004). The infectious pathogens implicated in it are mainly *Escherichia coli, Listeria monocytogenes, Yersinia enterocolitica* and *Mycobacterium avium* subsp. *paratuberculosis* (Rosenfeldt *et al.*, 2004). Infection with a probiotic strain of *E. coli* bacteria could help treat and reduce the negative effects of another *E. coli* infection that may be associated with CD (Sanders, 2000; Schuppan *et al.*, 2009). *Lactobacillus* GG is a safe probiotic bacterium known to transiently colonize the human intestine. It has been found to be useful in treatment of several GI conditions characterized by increased gut permeability. It may improve gut barrier function and clinical status in children with mildly to moderately active, stable CD (Gupta *et al.*, 2000; Sanders, 2000).

6.10 Microbes as Source of Antioxidants

There is increased evidence for the participation of free radicals in the aetiology of various diseases such as cancer, diabetes, cardiovascular diseases, autoimmune disorders, neurodegenerative diseases, ageing etc.

Free radicals can cause a wide range of toxic oxidative reactions leading to the accumulation of lipid peroxides, direct inhibition of mitochondrial respiratory chain enzymes, damage to DNA and proteins, which ultimately leads to cell death. The probiotic bacteria *Streptococcus thermophilus* is reported to have powerful antioxidant activity (AOA), protecting the body from dangerous free radicals and has antitumour activity especially against colon cancer cells (Dock *et al.*, 2004). Extracts from *Penicillium* and *Aspergillus* species, including *Rhizopus oryzae*, were found with good antioxidant activity to protect linoleic acid (Malpur *et al.*, 2006).

6.10.1 *Aspergillus* species are effective producers of AOA compounds

Esaki *et al.* (1999) evaluated the AOA of 30 strains of *Aspergillus* and found that methanol extracts of fermented soybeans (MEFS) prevented oxidation of methyl linoleate. The MEFS of 28 strains had better AOA than the non-fermented soybean. The MEFS obtained from *A. saitoi* had the best AOA and 2,3-dihydroxybenzoic acid was found responsible for the activity. Hayashi *et al.* (1995) also identified this phytochemical in *Penicillium roquefortii* IFO 5956 cultures. The AOA of fermentation products by mould cultures was more than bacterial (*Bacillus natto*) ones in producing antioxidants. Hoppe *et al.* (2002) identified tocopherols as antioxidant from tempeh fermented by *Rhizopus oligosporus*. Gallic acid, a phenolic acid, has been isolated from cultures of *Penicillium* and *Aspergillus* (Sarözlü and Kivanc, 2009). *Eurotium* species have been found to produce several antioxidants (Yoshiaki *et al.*, 2009). *Eurotium chevalieri* IFO 4086 and *E. repens* IFO 4041 produce three antioxidants, dihydroauroglaucin, auroglaucin and flavoglaucin. *Aspergillus chevalieri* also produced all three antioxidants while *Penicillium charlesii* produced flavoglaucin. *Streptomyces* sp. USF-319 produces three radical-scavenging antioxidants, of which one inhibits 5-lipoxygenase (Morimitsu and Hirota, 1996). The antioxidants include mycotrienin II, trienomycin A and trienomycin B, which are ansamycin antibiotics. Atroventin was isolated from *Penicillium paraherquei* and found to have good antioxidant activity (Ishikawa *et al.*, 1991). Carazostatin and 7-demethylnaphterpin are free-radical scavengers isolated from *Streptomyces chromofucus* and *S. prunicolor*, respectively (Kato *et al.*, 1989; Shin-Ya *et al.*, 1991).

Carotenoids are the group of antioxidants that can be synthesized by microorganisms. Nelis and de Leenheer (1991) reported that β-carotene from *Blakeslea trispora* and *Duniella salina*, and lycopene from *B. trispora* and *Streptomyces chrestomyceticus* subsp. *rubescens* are approved for human foods as colourants. Astaxanthin from *Xanthophyllomyces dendrorhous* has been approved for use in fish foods. Astaxanthin and lycopene were found to have excellent singlet oxygen quenching activity (Lee and Min, 1990; Gavrilov *et al.*, 1996). The AOA of astaxanthin was ten times greater than that of lutein, β-carotene, zeaxanthin and canthaxanthan. *Blakeslea trispora* and *X. dendrorhous* are most promising candidates for microbial production of carotenoids (Miki, 1991; Naguib, 2000).

It is widely accepted that a diet with high intake of fruits, vegetables and other nutrient-rich plant foods may reduce the risk of oxidative stress-related diseases (Carlsen *et al.*, 2010). Superoxide dismutases (SODs) are also produced efficiently by many microbial species and aerobic microorganisms like *Corynebacterium glutamicum* represent an excellent source for their production (El Shafey *et al.*, 2008). Cloning techniques have been reported to be used successfully with many corynebacterial genes. Thus it would be interesting to enhance superoxide dismutase production using cloning strategies (Schaaf and Bott, 2007; Vetting *et al.*, 2008; El Shafey *et al.*, 2009).

6.11 Microbes for the Treatment of Diabetes

Diabetes is a common and sometimes fatal disease that occurs when the supply of insulin is insufficient for the body to break down sugar properly. The majority of insulin used by people to manage diabetes is produced using biotechnology. Bacterial cells are genetically modified to produce large quantities of human insulin, which is then purified for therapeutic use. Millions of people worldwide now use Humuline, which is a major brand name for 'human' insulin produced using genetically modified (GM) bacteria (Barsh et al., 2000; Bajzer and Seeley, 2006). Some friendly gut microbes have been engineered to make a specific protein that can help regulate blood sugar in diabetic mice. Although the research is still in the very early stages, the microbes can be grown in yoghurt, and may provide an alternative treatment for people with diabetes (Hossain *et al.*, 2007).

People with type-1 diabetes lack the ability to make insulin, a hormone that triggers muscle and liver cells to take up glucose and store it for energy (Doria *et al.*, 2008).

The researchers have created a strain of non-pathogenic *E. coli* bacteria that produce a protein called GLP-1. This protein triggers cells in the pancreas to make insulin. In other research, scientists fed the engineered bacteria to diabetic mice. After 80 days, the mice went from being diabetic to having normal glucose blood levels. Diabetic mice that were not fed the engineered bacteria still had high blood sugar levels. The promise is that a diabetic could eat yoghurt or drink a smoothie as glucose-responsive insulin therapy rather than relying on insulin injections. Creating bacteria that produce the protein has a number of advantages over using the protein itself as the treatment. The bacteria can secrete just the right amount of the protein in response to conditions in the host that could ultimately minimize the need for self-monitoring and allow the patient's own cells (or the cells of the commensal *E. coli*) to provide the appropriate amount of insulin when needed (Bronzwaer, 2008; Cani *et al.*, 2009).

6.12 Microbes for the Treatment of Allergies

In a mouse model, researchers discovered a highly interesting effect: mice that were very susceptible to neuro-dermatitis developed the disease less frequently when a lysate of certain pathogens was applied on to the skin. The susceptibility of the lysate-treated mice to developing neuro-dermatitis was considerably lower than that of control mice. More detailed analyses showed that the animals had a higher concentration of immune-modulating interleukin-10 and a lower concentration of the pro-inflammatory mediator interferon-γ. Interferon-γ is secreted when the immune system recognizes foreign invaders and fine-tunes the immune system to effectively rid it of foreign intruders. The secretion of interferon-γ causes inflammatory reactions that create a hostile environment for the intruders, thereby initiating their elimination (Schultz-Larsen, 2002; Thomsen *et al.*, 2007). Non-pathogenic and probiotic bacteria do not cause inflammation, but are nevertheless recognized by the human immune system. The elevated production of interleukin-10 in certain immune system cells leads to anti-inflammatory reactions, which in turn creates an active immunological tolerance. This means that the human body is able to learn and tolerate self-peptides and harmless allergens when it is exposed to specific non-pathogenic microorganisms (Thomsen *et al.*, 2007; Yan *et al.*, 2007).

It has also been observed that use of probiotic bacteria or lysates of non-pathogenic microorganisms leads to a significant and permanent improvement in neuro-dermatitis. The application of non-pathogenic or probiotic bacteria to the skin is more effective in reducing the recurrence of diseases following the successful primary treatment of diseases such as neurodermatitis. The bacterium leads to permanent stabilization of the immune system once glucocorticoid treatment has alleviated the acute reaction (Sicherer and Sampson, 1999).

Probiotics are perceived to exert beneficial effects in the prevention and treatment of allergic diseases via modifying the gut ecosystem. The effect of ingestion of fermented milk containing *Lactobacillus paracasei*-33 (LP-33) was observed on patients with perennial allergic rhinitis and it was found that ingestion of LP-33-fortified fermented milk can effectively and safely improve the quality of life of patients with allergic rhinitis, and may serve as an alternative treatment (Sanders, 2000; Wang *et al.*, 2004).

6.13 Microbes as Source of Natural Colours

The production of synthetic colours is economically efficient and technically advanced but many of them cause various hazardous effects on health. Various types of microorganisms such as bacteria, fungi, yeasts and algae may be potential sources of natural colours. The major pigments produced by microbes are red, yellow and blue. Most research has been focused on yellow and red pigment production, such as monascue produced by *Monascus* sp., carotenoids from *Phaffia rhodozyma, Micrococcus roseus, Brevibacterium linens* and *Bradyrhizobium* sp. and xanthomonadin from *Xanthomonas campestris* pathovars (Hayman *et al.*, 1995; Chattopadhyay *et al.*, 2008). Actinorhodine-related blue pigments are produced by *Streptomyces coelicolor* A3(2), mixture of violacein and deoxybiolacein by *Chromobacterium violaceum* and *Janthinobacterium lividum* (Dufosse, 2009). In addition to its application in dyeing fabrics, violacein also exhibits cytotoxic activity in human colon cancer cells, and antileishmanial, anti-ulcerogenic, antiviral, antibiotic, antitumoral and anti-*Trypanosoma cruzi* activities (Joshi and Attri, 2006). The major natural biocolours of microbial origin include the following.

6.13.1 Moulds

Fungal carotenoids have also been recently approved as future food colourants by the European Union for the production of polyketide azaphilone pigments. The main advantages of using colourants from fungal source is that it makes the manufacturer independent of the seasonal supply of raw materials, thus minimizing batch to batch variations. Non-toxigenic fungal strains like *Penicillium* and *Epicoccum* sp. can be used as food colourants (Mapari *et al.*, 2010).

Monascus

Monascus species produce monascus pigments that are used in production of traditional East Asian foods, such as red rice wine and red bean curd. 'Ang-Khak', a traditional fermentation product in China, produced by fermenting rice with *Monascus purpureus*, is ground and its powdered form used as food colourant or spice. The pigments responsible for coloration in *Monascus* are ankaflavine, monascine and monascoavin (yellow), rubropunctatine and monascorubrin (orange) and rubropunctamine and monascorubramine (purple) (Nakanishi, 2006). These pigments are secondary metabolites of *Monascus* fermentation and produced mainly in cell-bound state. The variation in colour is influenced by the culture conditions, in particular pH and the phosphorus and nitrogen source in the substrate (Nelis and de Leenheer, 1991). It is used in processed meat products, marine products, tomato ketchup etc.

Blakeslea trispora

This fungus thrives in symbiosis with tropical plants and many of its strains can produce high levels of carotene. The production of carotene from this mould includes two steps: (i) the glucose, maize steep liquor or whey are used as substrates for aerobic submerged fermentation to produce the biomass; and (ii) the biomass is isolated and transformed into a form suitable for the isolation of carotene. It is then extracted using ethyl acetate and subsequently purified and concentrated. Carotene from *B. trispora* is mainly *trans*-βcarotene with approximately 3% other carotenoids (Marshall and Wilmoth, 1981; Naguib, 2000).

Ashbya gossypii, Candida *sp.*, Bacillus *sp.*

These microbes produce riboflavin (vitamin B_2). It possesses a yellow or yellow-orange colour and is being used as a food colourant and as a nutrient supplement in food products. In the food industry, it is used in baby foods, breakfast cereals, pasta, sauces and processed cheese etc. Various biotechnological processes have been developed for industrial scale production of riboflavin. The riboflavin fermentation could be produced by bacteria *Clostridium acetobutylicum*, yeast *Candida gulliiermundii* or fungi *Ashbya gossypii* (Ozbas and Kutsal, 1986; Koizumi *et al.*, 2000; Stahmann *et al.*, 2000).

6.13.2 Yeast

Xanthophyllomyces dendrorhous (*Phaffia rhodozyma*) yeast is known for the production of astaxanthin pigment. It is widely distributed in nature and is a principal pigment in crustaceans and salmonids. These carotenoid pigments impart an orange-red colour to farm animal species when supplemented in their feeds (Naguib, 2000). The carotenoids produced by the yeast *Rhodotorula* are torulene, torularhodin and carotene (Parajo *et al.*, 1997). It is a common environmental inhabitant and can be cultured from soil, water and air samples.

6.13.3 Bacteria

Multifaceted secondary metabolites are produced by *Serratia marcescens, Pseudomonas magneslorubra, Vibrio psychroerythrous, S. rubidaea, Vibrio gazogenes, Alteromonas rubra, Rugamonas rubra* and Gram-positive actinomycetes such as *Streptoverticillium rubrireticuli* and *Streptomyces longisporus*. The actinomycete *Streptomyces coelicolor* A3 produces a closely related linear tripyrrole, undecylprodigiosin, and a cyclic derivative, butyl-meta-cycloheptyl-prodiginine in a 2:1 ratio (Harris *et al.*, 2004). The red pigment of *S. marcescens* was isolated and named as prodigiosine (Venil and Lakshmanaperumalsamy, 2009), which is a multifaceted secondary metabolite produced by *S. marcescens, V. psychroerythrous, S. rubidaea, V. gazogenes, A. rubra, Lugomonas rubra* and Gram-positive Actinomycetes such as *S. rubrireticuli* and *S. longisporus* (Khanafari *et al.*, 2006). This promising pigment possesses antifungal, immunosuppressive, antiproliferative and anticancer activity (Pandey *et al.*, 2007). The microbial production of carotenoids when compared with extraction from vegetables or chemical synthesis seems to be a better option because of the problems of seasonal and geographic variability. Microbial colours are used in the fish industry to enhance the pink colour of farmed salmon. Microorganisms produce various pigments such as carotenoids, melanins, flavins, quinones, prodigiosins and more specifically monascins, violacin or indigo as mentioned in Table 6.2 (Dufosse, 2009; Venil and Lakshmanaperumalsamy, 2009).

Microorganisms such as *Monascus, Rhodotorula, Bacillus, Achromobacter, Yarrowia* and *Phaffia* produce a large number of pigments. Improvements in stability, safety and solubility can certainly lead to widespread use of microbial pigments in the food industry (Joshi *et al.*, 2003; Joshi and Attri, 2006).

Examples of some important food grade biocolourants that can be produced on a large scale by fermentation and bioprocess engineering are given in Table 6.3 (Chattopadhyay *et al.*, 2008).

6.14 Microbes as Source of Vitamins

Most vitamins are essential for the metabolism of all living organisms, and they are synthesized by microorganisms and plants.

6.14.1 Water-soluble vitamins

Riboflavin (vitamin B_2) and related coenzymes

Riboflavin is produced by both synthetic and fermentation processes. Two closely related ascomycete fungi, *Eremothecium ashbyii* and *Ashbya gossypii*, are mainly used for industrial production (Ozbas and Kutsal, 1986; Stahmann *et al.*, 2000). Yeasts (*Candida flaeri, C. famata* etc.) and bacteria can also be used for practical production. Riboflavin production by genetically engineered *Bacillus subtilis* and *Corynebacterium ammoniagenes* over-express genes of the enzymes involved in riboflavin biosynthesis. Several tonnes of flavin adenine dinucleotides (FADs) are annually produced by chemical synthesis or by microbial transformation. The latter uses flavin mononucleotide (FMN) and adenosine-5b-triphosphate (ATP) as the substrates and *C. ammoniagenes* cells as a source of FMN adenylyltransferase (Koizumi *et al.*, 2000).

Nicotinic acid and nicotinamide

Bacterial nitrilase has been shown to be useful for nicotinic acid production. For example,

Table 6.2. Some metabolites and colours produced by microorganisms.

Microorganism	Metabolites	Colour
Alteromonas rubra	Prodigiosin-like pigment	Red
Ashbya gossypii	Riboflavin	Yellow
Blakeslea trispora	Lycopene	Red
	β-carotene	Yellow-orange
Bradyrhizobium sp.	Canthaxanthin	Orange/dark red
Cordyceps unilateralis	Naphtoquinone	Deep blood red
Corynebacterium insidiosum	Indigoidine	Blue
Dunaliella salina	β-carotene	Orange
Flavobacterium spp.	Zeaxanthin	Yellow
Haematococcus pluvialis	Astaxanthin	Red
Janthinobacterium lividum	Violacein	Purple
Monascus roseus	Canthaxanthin	Orange, pink
Monascus spp.	Ankaflavin	Yellow
	Monascorubramin	Red
	Rubropunctatin	Orange
Pacilomyces farinosus	Anthraquinone	Red
Paecilomyces sinclairii		
Penicillium oxalicum	Anthraquinone	Red
P. purpurogenum		
Phaffia rhodozyma	Astaxanthin	Red
Pseudomonas aeruginosa	Pyocyanin Blue	Green
Rhodotorula spp.	Torularhodin	Orange-red
Rugamonas rubra	Prodigiosin-like pigment	Red
Saccharomyces neoformans	Melanin	Black
Serratia marcescens	Prodigiosin	Red
Serratia rubidaea	Prodigiosin-like pigment	Red
Staphylococcus aureus	Zeaxanthin	Yellow
Streptomyces echinoruber	Rubrolone	Red
Streptoverticillium rubrireticuli	Prodigiosin-like pigment	Red
Vibrio gaogenes	Prodigiosin-like pigment	Red
Xanthophyllomyces dendrorhous	Astaxanthin	Pink-red
Xanthomonas oryzae	Xanthomonadin	Yellow

3-cyanopyridine is almost stoichiometrically converted to nicotinic acid on incubation with the nitrilase over-expressed *Rhodococcus rhodochrous* J1 cells (Nagasawa and Yamada, 1989). The same *R. rhodochrous* enzyme can be used for the production of *p*-aminobenzoic acid from *p*-aminobenzonitrile. Nicotinamide is available from partial hydrolysis of 3-cyanopyridine, which is performed by both chemical and enzymatic processes. The enzymatic process uses nitrile hydratase as the catalyst (Asano *et al.*, 1980). The Co-containing enzyme from *R. rhodochrous* J1 hydrates various kinds of aliphatic and aromatic nitriles to the corresponding amides and has been shown to be useful for the production of useful amides (Yamada and Kobayashi, 1996). The enzymatic process surpasses the chemical process in regard to several points such as stoichiometric conversion of high concentration of the substrate and the quality of the product actually with zero contents of by-products (Shimizu *et al.*, 1997; Shimizu and Kataoka, 1999). Some notable examples of vitamins that are produced by various microorganisms are given in Table 6.4.

6.15 Microbes as Source of Synbiotics

Synbiotics are probiotics including both probiotics and prebiotics. According to WHO,

Table 6.3. Important food-grade biocolourants.

Food grade biocolourants (original source)	Biotechnological source
Monascorubramine (*Monascus purpureus*)	–
Astaxanthin (plants)	Fungus: *Xanthophyllomyces dendrorhous* Algae: *Haematococcus lacustris, H. pluvialis*
Arpink red	Fungus: *Penicillium oxalicum* var. *armeniaca* CCM 8242
β-Carotene (*Daucus carota*)	Fungus: *Blakeslea trispora, Phycomyces blakesleeanus* car *S* mutant Algae: *Dunaliella salina, D. bardwil* GM plant: Golden rice
Riboflavin (milk)	Moulds: *Ashbya gossypii, Eremothecium ashbyii* Yeast: *Candida guilliermondii, Debaryomyces subglobosus* Bacteria: *Clostridium acetobutylicum*
Betanin (*Beta vulgaris*)	Higher yielding plant generated through somaclonal variation, hairy root culture
Canthaxanthin	Algae: *Haematococcus lacustris* Bacteria: *Bradyrhizobium* sp.
Cyanidin and Peonidin (cherry, cranberry)	Higher yielding plant generated through somaclonal variation Cell culture
Lycopene (tomato)	GM fungus: *Fusarium sporotrichioides* GM bacteria: *Erwinia uredovora*
Zeaxanthin (maize)	Bacteria: *Flavobacterium* sp.

Table 6.4. Microbial production of water-soluble vitamins.

Vitamin	Enzyme/Microorganism	Process
Biotin	Fermentation (*Serratia marcescens*)	Fermentation of glucose by a genetically engineered bacterium
	Multiple enzyme system (*Bacillus sphaericus*)	Conversion from diaminopimelic acid using the biotin biosynthesis enzyme system of a mutant of *B. sphaericus*
Nicotinamide	Nitrile hydratase (*Rhodococcus rhodochrous*)	Hydration of 3-cyano-pyridine
Nicotinic acid	Nitrilase (*Rhodococcus rhodochrous*)	Hydrolysis of 3-cyanopyridine to form nicotinic acid and ammonia
Pantothenic acid	Lactono-hydrolase (*Fusarium oxysporum*)	Resolution of D,L-pantolactone to D-pantoic acid and L-pantolactone by stereoselective hydrolysis
Riboflavin	Fermentation (*Eremothecium ashbyii, Ashbya gossypii, Bacillus* sp., etc.)	Fermentation of glucose
Vitamin B_{12}	Fermentation (*Propionibacterium shermanii, Pseudomonas denitrificans*)	Fermentation of glucose
Vitamin C	2,5-diketo-D-gulonic acid reductase (*Corynebacterium* sp.)	Enzymatic conversion of 2,5-diketo-D-gluconate by fermentation to 2-keto-L-gulonic, followed by chemical conversion to L-ascorbic acid

probiotics are live microorganisms, which, when administered in adequate amounts, confer a health benefit on the host. The microorganisms must be alive (they are not alive in products that are pasteurized after fermentation), present in high numbers (generally more than 1 billion per daily ingested dose), with scientifically established human health benefits.

Some probiotics belong to the genera *Lactobacillus* and *Bifidobacterium* and are isolated from dairy products or from human or animal intestinal tracts (Flinterman *et al.*, 2007).

Prebiotics are non-digestible carbohydrates that survive digestion, enter the colon and selectively enhance the growth or activity of the body's own native beneficial bacteria. Prebiotics change the composition of faecal bacteria by: (i) increasing beneficial bacteria – such as lactobacilli and bifidobacteria – that help modulate the activity of the immune system; and (ii) by decreasing organisms such as clostridia and protein-degrading *Bacteroides*, which can produce tumour promoters from metabolism of proteins that escaped digestion in the upper gut (Kadooka *et al.*, 2010). Because prebiotics are non-digestible carbohydrates, they help normalize bowel conditions due to an osmotic effect or other effects on indigenous microbes. Prebiotics are found naturally in some plants or are produced enzymatically from sucrose, and often are used in dietary supplements. Unlike probiotics, which need to be in a viable state to maximize biological activity, prebiotics are not alive and can be formulated into a wide range of food formulations and products prior to cooking (De Roos and Katan, 2000). Synbiotics are one of the most promising approaches to manage correct balance of gut microflora. They also improve survival of bacteria during storage and passage of the upper part of the GI tract, thereby enhancing their health effects in the large intestine. The combined effects of synbiotics can be additive or even synergistic (Resta-Lenert and Barrett, 2003). Some applications of synbiotic foods are in the manufacture of infant formulae and weaning foods (Rizzello *et al.*, 2007). It would be of great interest to produce prebiotics with high selectivity towards growth of bifidobacteria (probiotic bacteria) that are present in the gut of breast-fed infants as the basis of novel infant food formulations (Hugenholtz and Smid, 2002; Li and Gatlin, 2003).

6.15.1 Functional foods for healthy ageing

It is known that the bifidobacterial population decreases markedly in the colon of the elderly person. Species of *Bifidobacterium* like *B. infantis* and *B. breve* are predominant in infants, whereas *B. adolescentis* and *B. longum* are predominant in adults. Decrease in bifidobacterial numbers results in reduction in resistance to GI infections and thus elderly people suffer more with such ailments. The criterion of designing functional foods for healthy ageing is through the development of a targeted prebiotic that promotes the probiotic strains that are able to inhibit GI pathogens, viz. *E. coli*, *Salmonella* sp. and *Campylobacter jejuni* (De Roos and Katan, 2000; Rizzello *et al.*, 2007).

6.16 Microbes for Production of Low Calorie/Natural Sweetener

Xylitol, a natural low calorie sweetener, can be produced from genetically engineered bacteria that utilize hemicellulose in maize fibre and other sources (Vazquez *et al.*, 1998). At USDA-ARS Fermentation Biotechnology Research Unit, scientists used metabolic pathway engineering to re-tool the enzyme-making machinery of *E. coli* bacteria so that they could convert two hemicellulose sugars, xylose and arabinose, into xylitol. Another sweetener, aspartame, found in thousands of products worldwide, can also be created using GM bacteria. Monsanto, the largest biotech corporation in the world, uses GM bacteria to produce aspartame in their US production plants. The process in which aspartame is created involves combining an amino acid phenylalanine with aspartic acid. The bacteria require aspartame for the sole purpose of producing phenylalanine. Monsanto discovered that through genetically altering the bacteria, phenylalanine could be created much more quickly.

6.17 Chalcones, Flavonoids and Stilbenes from Lignin using Microbial Biosynthesis

Microbial biosynthesis of chalcones, stilbenes and flavonoids in engineered microbes offers a production route with greater selectivity and purity than current methods that rely on chemical extraction from plants. In addition, the feedstock for microbial biosynthesis (phenylpropionic acid) can be derived from lignin, an inexpensive agricultural by-product. Microbial biosynthesis of flavonoids and stilbenes also allows for production of novel compounds that are not found in nature. It has been found that consumption of high quantities of isoflavones are related to fewer incidences of osteoporosis and related health problems, especially hot flushes, cardiovascular diseases, lower incidence of hormone-dependent breast and uterine cancer. Isoflavones have also been reported to inhibit angiogenesis, cell cycle progression, aromatase enzyme inhibition, stimulation of sex hormone binding globulin (SHBG) synthesis and digitalis-like activity (Sakamoto *et al.*, 2010; Prakash and Gupta, 2011).

Stilbenes have diverse pharmacological activities, which include cancer prevention, a cholesterol-lowering effect, enhanced insulin sensitivity and increased lifespan. One of the best-characterized stilbenes, resveratrol, has been known as an antioxidant and an anti-ageing compound as well as an anti-inflammatory agent (Prakash and Gupta, 2011).

Chalcones are found to be effective as an anticancer, antiviral, cardiovascular and anti-inflammatory agent. Certain chalcone derivatives are reported to inhibit the polymerization of tubulin to form microtubules and are, therefore, antimitotic agents that can be used as antiguot agents. Chalcone derivatives are also known to inhibit the destruction of myelin sheath in the central nervous system of multiple sclerosis patients and are thus useful in controlling the progressive nature of the disease (Prakash and Gupta, 2011).

The microbes have great significance in nutraceuticals, particularly in the production of resveratrol, piceatannol, naringenin and eriodictyol (cancer prevention antioxidants). Thus it can be safely concluded that microbes have a tremendous potential of being used as nutraceuticals and in designing various disease-targeting preventative functional foods and dietary supplements.

References

Aas, J.A., Paster, B.J., Stokes, L.N., Olsen, I. and Dewhirst, F.E. (2005) Defining the normal bacterial flora of the oral cavity. *Journal of Clinical Microbiology* 43, 5721–5732.

Adjei, A.A. (2000) Signal transduction pathway targets for anticancer drug discovery. *Current Pharmaceutical Design* 6, 361–378.

Agerholm-Larsen, L., Bell, M.L., Grunwald, G.K. and Astrup, A. (2002) The effect of a probiotic milk product on plasma cholesterol: a meta-analysis of short term intervention studies. *European Journal of Clinical Nutrition* 54, 856–860.

Alexandroff, A.B., Jackson A.M., O'Donnell, M.A. and James, K. (1999) BCG immunotherapy of bladder cancer: 20 years on. Lancet 353, 1689–1694.

Anderson, J.W. and Gilliland, S.E. (1999) Effect of fermented milk (yogurt) containing *Lactobacillus acidophilus* L1 on serum cholesterol in hypercholesterolemic humans. *Journal of the American College of Nutrition* 18, 43–50.

Angelis, M.D., Rizzello, C.G., Fasano, A., Clemente, M.G., De Simone, C., Silano, M., De Vincenzi, M., Losito, I. and Gobbetti, M. (2006) VSL#3 probiotic preparation has the capacity to hydrolyze gliadin polypeptides responsible for Celiac Sprue. *Biochimica et Biophysica Acta* 1762, 80–93.

Asano, Y., Tani, Y. and Yamada, H. (1980) A new enzyme 'nitrile hydratase' which degrades acetonitrile in combination with amidase. *Agricultural and Biological Chemistry* 44, 2251–2252.

Bajzer, M. and Seeley, R. (2006) Physiology: obesity and gut flora. *Nature* 444, 1009–1010.

Barsh, G.S., Farooqi, I.S. and O'Rahilly S. (2000) Genetics of body-weight regulation. *Nature* 404, 644–651.

Becker, E.W. (2007) Micro-algae as a source of protein. *Biotechnology Advances* 25, 207–210.

Bemgmark, S. (1998) Ecological control of the gastrointestinal tract: the role of probiotic flora. *Gut* 42, 2–7.

Brockmann, G.A. and Bevova, M.R. (2002) Using mouse models to dissect the genetics of obesity. *Trends in Genetics* 18, 367–376.

Bronzwaer, S. (2008) EFSA scientific forum 'from safe food to healthy diets'. EU risk assessment – Past, present and Future. *Trends in Food Science and Technology* 19, S2–S8.

Burgents, J.E., Burnett, K.G. and Burnett, L.E. (2004) Disease resistance of Pacific white shrimp, *Litopenaeus vannamei*, following the dietary administration of a yeast culture food supplement. *Aquaculture* 231, 1–8.

Cagno, R., Di Angelis, M., De Lavermicocca, P., De Vincenzi, M., Giovannini, C., Faccia, M. and Gobbetti, M. (2002) Proteolysis by sourdough lactic acid bacteria: effects on wheat flour protein fractions and gliadin peptides involved in human cereal intolerance. *Applied and Environmental Microbiology* 68, 623–633.

Cagno, R., Di Angelis, M., De Auricchio, S., Greco, L., Clarke, C., De Vincenzi, M., Giovannini, C., D'Archivio, M., Landolfo, F., Parrilli, G., Minervini, F., Arendt, E. and Gobbetti, M. (2004) Sourdough bread made from wheat and nontoxic flours and started with selected lactobacilli is tolerated in Celiac Sprue patients. *Applied and Environmental Microbiology* 70, 1088–1096.

Campa-Córdova, A.I., Hernández-Saavedra, N.Y., De Philippis, R. and Ascencio, F. (2002) Generation of superoxide anion and SOD activity in haemocytes and muscle of American white shrimp (*Litopenaeus vannamei*) as a response to β-glucan and sulphated polysaccharide. *Fish and Shellfish Immunology* 12, 353–366.

Cani, P.D., Possemiers, S., Van de Wiele, T., Guiot, Y., Everard, A., Rottier, O., Geurts, L., Naslain, D., Neyrinck, A., Lambert, D.M., Muccioli, G.G. and Delzenne, N.M. (2009) Changes in gut microbiota control inflammation in obese mice through a mechanism involving GLP-2-driven improvement of gut permeability. *Gut* 58, 1091–1103.

Carlsen, M.H., Halvorsen, B.L., Holte, K., Bøhn, S.K., Dragland, S., Sampson, L., Willey, C., Senoo, H., Umezono, Y., Sanada, C., Barikmo, I., Berhe, N., Willett, W.C., Phillips, K.M., Jacobs, D.R. and Blomhoff, R. (2010) The total antioxidant content of more than 3100 foods, beverages, spices, herbs and supplements used worldwide. *Nutrition Journal* 9, 1–11.

Chattopadhyay, P., Chatterjee, S. and Sukanta, S.K. (2008) Biotechnological potential of natural food grade biocolourants. *African Journal of Biotechnology* 7, 2972–2985.

Cinque, B., Di Marzio, L., Della Riccia, D., Bizzini, F., Giuliani, M., Fanini, D., De Simone, C. and Cifone, M. (2006) Effect of *Bifidobacterium infantis* on interferon-γ-induced keratinocyte apoptosis: a potential therapeutic approach to skin immune abnormalities. *International Journal of Immunopathology and Pharmacology* 19, 775–786.

Coley, W.B. (1893) The treatment of malignant tumors by repeated inoculations of erysipelas: with a report of ten original cases. *The American Journal of Medical Sciences* 105, 487–511.

De Roos, N.M. and Katan, M.B. (2000) Effects of probiotic bacteria on diarrhea, lipid metabolism, and carcinogenesis: a review of papers published between 1988 and 1998. *American Journal of Clinical Nutrition* 71, 405–411.

Delzenne, N.M., Cani, P.D., Daubioul, C. and Neyrinck, A.M. (2005) Impact of inulin and oligofructose on gastrointestinal peptides. *British Journal of Nutrition* 93, 157–161.

Dock, D.B., Aguilar-Nascimento, J.E. and Latorraca, M.Q. (2004) Probiotics enhance the recovery of gut atrophy in experimental malnutrition. *Biocell* 28, 143–150.

Doria, A., Patti, M.E. and Kahn, C.R. (2008) The emerging genetic architecture of type 2 diabetes. *Cell Metabolism* 8, 186–200.

Dufosse, L. (2009) Pigments. *Microbial Encyclopedia Microbiology* 4, 457–471.

El Shafey, H.M., Ghanem, S., Merkamm, M. and Guyonvarch, A. (2008) *Corynebacterium glutamicum* superoxide dismutase is a manganese strict non-cambialistic enzyme *in vitro*. *Microbiological Research* 163, 80–86.

El Shafey, H.M., Ghanem, S. and Guyonvarch, A. (2009) Cloning of recA gene of *Corynebacterium glutamicum* and phenotypic complementation of *Escherichia coli* recombinant deficient strain. *World Journal of Microbiology and Biotechnology* 25, 367–373.

Esaki, H., Kawakishi, S., Morimitsu, Y. and Osawa, T. (1999) New potent anti-oxidative 0-Dihydroxyisoflavones in fermented Japanese soybean products. *Bioscience, Biotechnology and Biochemistry* 63, 1637–1639.

Flier, J.S. (2004) Obesity wars: molecular progress confronts an expanding epidemic. *Cell* 116, 337–350.

Flinterman, A.E., Knol, E.F., Van Ieperen-van Dijk, A.G., Timmerman, H.M., Knulst, A.C., Bruijnzeel-Koomen, C.A., Pasmans, S.G. and Van Hoffen, E. (2007) Probiotics have a different immunomodulatory potential in vitro versus *ex vivo* upon oral administration in children with food allergy. *International Archives of Allergy and Immunology* 143, 237–244.

Friedman, J.M. (2003) A war on obesity, not the obese. *Science* 299, 856–858.

Gass, J., Ehren, J., Strohmeier, G., Isaacs, I. and Khosla, C. (2005) Fermentation, purification, formulation and pharmacological evaluation of a prolyl endopeptidase from *Myxococcus xanthus*: implications for Celiac Sprue therapy. *Biotechnology and Bioengineering* 92, 674–684.

Gass, J., Bethune, M.T., Siegel, M., Spencer, A. and Khosla C. (2007) Combination enzyme therapy for gastric digestion of dietary gluten in patients with Celiac Sprue. *Gastroenterology* 133, 472–480.

Gavrilov, A., Kiseleva, A., Matushkina, S., Kordyukova, N. and Feofilova, E. (1996) Industrial production of lycopene by a microbiological method. *Applied Biochemistry and Microbiology* 32, 492–494.

Gianfrani, C., Siciliano, R.A., Facchiano, A.M., Camarca, A., Mazzeo, M.F., Costantini, S., Salvati, V.M., Maurano, F., Mazzarella, G., Iaquinto, G., Bergamo, P. and Rossi M. (2007) Transamidation of wheat flour inhibits the response to gliadin of intestinal T cells in celiac disease. *Gastroenterology* 133, 780–789.

Gill, S.R., Pop, M., Deboy, R.T., Eckburg, P.B., Turnbaugh, P.J., Samuel, B.S., Gordon, J.I., Relman, D.A., Fraser-Liggett, C.M. and Nelson, K.E. (2006) Metagenomic analysis of the human distal gut microbiome. *Science* 312, 1355–1359.

Gilliland, S.E. and Speck, M.L. (1977) Antagonistic action of *Lactobacillus acidophilus* toward intestinal and food-borne pathogens in associative cultures. *Journal of Food Protection* 40, 820–823.

Green, P.H. and Jabri, B. (2006) Celiac disease. *Annual Review of Medicine* 57, 207–221.

Gupta, P., Andrew, H., Kirschner, B.S. and Guandalini, S. (2000) Is *Lactobacillus* GG Helpful in Children with Crohn's Disease? Results of a Preliminary, Open-Label Study. *Journal of Pediatric Gastroenterology and Nutrition* 31, 453–457.

Hamad, E.M., Sato, M., Uzu, K., Yoshida, T., Higashi, S., Kawakami, H., Kadooka, Y., Matsuyama, H., Abd El-Gawad, I. and Imaizumi, K. (2009) Milk fermented by *Lactobacillus gasseri* SBT2055 influences adipocyte size via inhibition of dietary fat absorption in Zucker rats. *British Journal of Nutrition* 101, 716–724.

Harris, K.P., Williamson, R., Slater, H., Cox, A., Abbasi, S., Foulds, I., Simonsen, H.T., Leeper, F.J. and Salmond, G.P. (2004) The Serratia gene cluster encoding biosynthesis of the red antibiotic, prodigiosin, shows species and strain dependent genome context variation. *Microbiology* 150, 3547–3560.

Hayashi, K., Suzuki, K., Kawaguchi, M., Nakajima, T., Suzuki, T., Murata, M. and Nakamura, T. (1995) Isolation of an antioxidant from *Penicillium roquefortii* IFO 5956. *Bioscience, Biotechnology and Biochemistry* 59, 319–320.

Hayman, G., Mannarelli, B. and Leathers, T. (1995) Production of carotenoids by *Phaffia rhodozyma* grown on media composed of corn wet-milling co-products. *Journal of Industrial Microbiology* 14, 389–395.

Helmerhorst, E.J., Sun, X., Salih, E. and Oppenheim, F.G. (2008) Identification of Lys-Pro-Gln as a novel cleavage site specificity of saliva-associated proteases. *The Journal of Biological Chemistry* 283, 19957–19966.

Hoppe, U., Bergemann, J., Diembeck, W., Ennen, J., Gohla, S., Harris, I., Jacob, J., Kielholz, J., Mei, W., Pollet, D., Schachtschabel, D., Sauermann, G., Schreiner, V., Stab, F. and Steckel, F. (2002) Lipophilic antioxidants in human sebum and aging. *Free Radical Research* 36, 471–477.

Hossain, P., Kawar, B. and El Nahas, M. (2007) Obesity and diabetes in the developing world – a growing challenge. *New England Journal of Medicine* 356, 213–215.

Hugenholtz, J. and Smid, E.J. (2002) Nutraceutical production with food-grade microorganisms. *Current Opinion in Biotechnology* 13, 497–507.

Isenberg, J., Stoffel, B., Wolters, U., Beuth, J., Stutzer, H., Ko, H.L. and Pichlmaier, H. (1995) Immunostimulation by *Propionibacteria* effects on immune status and anti-neoplastic treatment. *Anticancer Research* 15, 2363–2368.

Ishikawa, Y., Morimoto, K. and Iseki, S.J. (1991) Atrovenetin as a potent antioxidant compound from *Penicillium* species. *Journal of the American Oil Chemists' Society* 68, 666–668.

Isolauri, E. (2004) Dietary modification of atopic disease: use of probiotics in the prevention of atopic dermatitis. *Current Allergy and Asthma Reports* 4, 270–275.

Isolauri, E., Sutas, Y., Kankaanpaa, P., Arvilommi, H. and Salminen, S. (2001) Probiotics: effects on immunity. *The American Journal of Clinical Nutrition* 73 (Suppl.), 444S–450S.

Joshi, V.K. and Attri, D. (2006) Solid state fermentation of apple pomace for the production of value added products. *Natural Product Radiance* 5, 289–296.

Joshi, V.K., Attri, D., Bala, A. and Bhushan, S. (2003) Microbial Pigments. *Indian Journal of Biotechnology* 3, 362–369.

Kadooka, Y., Sato, M., Imaizumi, K., Ogawa, A., Ikuyama, K., Akai, Y., Okano, M., Kagoshima, M. and Tsuchida, T. (2010) Regulation of abdominal adiposity by probiotics (*Lactobacillus gasseri* SBT2055) in adults with obese tendencies in a randomized controlled trial. *European Journal of Clinical Nutrition* 64, 636–643.

Kajander, K., Myllyluoma, E., Kyrönpalo, S., Rasmussen, M., Sipponen, P., Mattila, I., Seppänen-Laakso, T., Vapaatalo, H., Oresic, M. and Korpela, R. (2009) Elevated pro-inflammatory and lipotoxic mucosal lipids characterize irritable bowel syndrome. *World Journal of Gastroenterology* 15, 6068–6074.

Kassinen, A., Krogius-Kurikka, L., Makivuokko, H., Rinttila, T., Paulin, L., Corander, J., Malinen, E., Apajalahti, J. and Palva, A. (2007) The fecal microbiota of irritable bowel syndrome patients differs significantly from that of healthy subjects. *Gastroenterology* 133, 24–33.

Kato, S., Kawai, H., Kawasaki, T., Toda, Y., Urada, T. and Hayagawa, Y. (1989) Studies on free radical scavenging substances from microorganism. *Journal of Antibiotics* 42, 1879–1881.

Khanafari, A., Assadi, M.M. and Fakhr, F.A. (2006) Review of prodigiosin, pigmentation in *Serratia marcescens*. *Online Journal of Biological Sciences* 6, 1–13.

Kiessling, G., Schneider, J. and Jahreis, G. (2002) Long term consumption of fermented dairy products over 6 months increases HDL cholesterol. *European Journal of Clinical Nutrition* 56, 843–849.

Ko, H.L., Roszkowski, W., Jeljaszewicz, J. and Pulverer, G. (1981) Comparative study on the immunostimulating potency of different *Propionibacterium* strains. *Medical Microbiology and Immunology* 170, 1–9.

Koizumi, S., Yonetani, Y., Maruyama, A. and Teshiba, S. (2000) Production of riboflavin by metabolically engineered *Corynebacterium ammoniagenes*. *Applied Microbiology and Biotechnology* 53, 674–679.

Koning, F., Schuppan, D., Cerf-Bensussan, N. and Sollid, L.M. (2005) Patho-mechanisms in celiac disease. *Best Practice and Research Clinical Gastroenterology* 19, 373–387.

Kopelman, P.G. (2000) Obesity as a medical problem. *Nature* 404, 635–643.

Kruisselbrink, A., Heijne Den Bak-Glashouwer, M.J., Havenith, C.E., Thole, J.E. and Janssen, R. (2001) Recombinant *Lactobacillus plantarum* inhibits house dust mite-specific T-cell responses. *Clinical and Experimental Immunology* 126, 2–8.

Kurbanoğlu, E.B. (2001) Production of single-cell protein from ram horn hydrolyzate. *Turkish Journal of Biology* 25, 371–377.

Lacy, B.E. and Lee, R.D. (2005) Irritable bowel syndrome: a syndrome in evolution. *Journal of Clinical Gastroenterology* 39, S230–S242.

Lamm, D.L., Riggs, D.R., Shriver, J.S., Van Gilder, P.F., Rach, J.F. and DeHaven, J.I. (1994) Megadose vitamins in bladder cancer: a double-blind clinical trial. *Journal of Urology* 151, 21–26.

Lamm, D.L., Thor, D.E., Harris, S.C., Reyna, J.A., Stogdill, V.D. and Radwin, H.M. (1980) *Bacillus* Calmette-Guérin immunotherapy of superficial bladder cancer. *Journal of Urology* 124, 38–40.

Laughter, D., Istvan, J.A., Tofte, S.J. and Hanifin, J.M. (2000) The prevalence of atopic dermatitis in Oregon schoolchildren. *Journal of American Academy of Dermatology* 43, 649–655.

Lee, S. and Min, D. (1990) Effects, quenching mechanisms, and kinetics of carotenoids in chlorophyll-sensitized photo-oxidation of soybean oil. *Journal of Agricultural and Food Chemistry* 38, 1630–1634.

Li, P. and Gatlin, D.M. (2003) Evaluation of brewer's yeast (*Saccharomyces cerevisiae*) as a feed supplement for hybrid striped bass (*Morone chrysops* x *M. saxatilis*). *Aquaculture* 219, 681–692.

Lindfors, K., Blomqvist, T., Juuti-Uusitalo, K., Stenman, S., Venäläinen, J., Mäki, M. and Kaukinen, K. (2008) Live probiotic *Bifidobacterium lactis* bacteria inhibit the toxic effects induced by wheat gliadin in epithelial cell culture. *Clinical and Experimental Immunology* 152, 552–558.

Liu, Q., Nobaek, S., Adawi, D., Mao, Y., Wang, M., Molin, G., Ekelund, M. and Jeppsson, B. (2001) Administration of *Lactobacillus plantarum* 299v reduces side-effects of external radiation on colon anastomotic healing in an experimental model. *Colorectal Disease* 3, 245–252.

Mahasneh, I.A. (1997) Production of single cell protein from five strains of the micro-alga *Chlorella* sp. (Chlorophyta). *Cytobiosciences* 90, 153–161.

Malpur, P.P., Shah, A.S. and Juvekar, A.R. (2006) Antioxidant and anti-inflammatory activity of extract obtained from *Aspergillus candidus* MTCC 2202 broth filtrate. *Indian Journal of Experimental Biology* 44, 468–473.

Mapari, S.A.S., Thrane, U. and Meyer, A. (2010) Fungal polyketide azaphilone pigments as future natural food colourants: a review. *Trends in Biotechnology* 28, 300–307.

Marshall, J.H. and Wilmoth, G.J. (1981) Pigments of *Staphylococcus aureus*, a series of triterpenoid carotenoids. *Journal of Bacteriology* 147, 900–913.

Marti, T., Molberg, Ø., Li, Q., Gray, G.M., Khosla, C. and Sollid, L.M. (2005) Prolyl endopeptidase-mediated destruction of T cell epitopes in whole gluten: chemical and immunological characterization. *Journal of Pharmacology and Experimental Therapeutics* 312, 19–26.

Mercenier, A., Pavan, S. and Pot, B. (2002) Probiotics as biotherapeutic agents: present knowledge and future prospects. *Current Pharmaceutical Design* 8, 99–110.

Miki, W. (1991) Biological functions and activities of animal carotenoids. *Pure and Applied Chemistry* 63, 141–146.

Miraglia del Giudice, M. and De Luca, M.G. (2004) The role of probiotics in the clinical management of food allergy and atopic dermatitis. *Journal of Clinical Gastroenterology* 38(Suppl. 2), S84–S85.

Mitea, C., Havenaar, R., Drijfhout, J.W., Edens, L., Dekking, L. and Koning, F. (2008) Efficient degradation of gluten by a prolyl endoprotease in a gastrointestinal model: implications for coeliac disease. *Gut* 57, 25–32.

Mohan, J.C., Arora, R. and Khalilullah, M. (1990) Short term hypolipidemic effects of oral *Lactobacillus sporogenes* therapy in patients with primary dyslipidemias. *Indian Heart Journal* 42, 361–364.

Montrose, D.C. and Floch, M.H. (2005) Probiotics Used in Human Studies. *Journal of Clinical Gastroenterology* 39, 469–484.

Morales, A., Eidinger, D. and Bruce, A.W. (1976) Intracavitory BCG in the treatment of superficial bladder tumours. *Journal of Urology* 116, 180–183.

Morimitsu, Y. and Hirota, A. (1996) Ansamycin antibiotics as free radical scavengers isolated from *Streptomyces* by using the bactericidal action of the hydroxyl radical. *Bioscience, Biotechnology and Biochemistry* 60, 1507–1509.

Mueller, S., Saunier, K., Hanisch, C., Norin, E., Alm, L., Midtvedt, T., Cresci, A., Silvi, S., Orpianesi, C., Verdenelli, M.C., Clavel, T., Koebnick, C., Zunft, H-J.F., Doré, J. and Blaut, M. (2006) Differences in fecal microbiota in different European study populations in relation to age, gender, and country: a cross-sectional study. *Applied and Environmental Microbiology* 72, 1027–1033.

Nagasawa, T. and Yamada, H. (1989) Microbial transformations of nitriles. *Trends in Biotechnology* 7, 153–158.

Naguib, Y. (2000) Antioxidant activities of astaxanthin and related carotenoids. *Journal of Agricultural and Food Chemistry* 48, 1150–1154.

Nakanishi, K. (2006) Studies in microbial and insect natural product chemistry. *Journal of Natural Medicines* 60, 2–20.

Nauts, H.C., Swift, W.E. and Coley, B.L. (1946) The treatment of malignant tumor by bacterial toxins as developed by the late William B Coley, M.D., reviewed in the light of modern research. *Cancer Research* 6, 205–216.

Nelis, H. and de Leenheer, A. (1991) Microbial sources of carotenoid pigments used in foods and feeds. *Journal of Applied Bacteriology* 70, 181–191.

Oliva-Teles, A. and Gonçalves, P. (2001) Partial replacement of fishmeal by brewer's yeast *Saccaromyces cerevisae* in diets for sea bass *Dicentrarchus labrax* juveniles. *Aquaculture* 202, 269–278.

Olvera-Novoa, M.A., Martinez-Palacios, C.A. and Olivera-Castillo, L. (2002) Utilization of torula yeast (*Candida utilis*) as a protein source in diets for tilapia (*Oreochromis mossambicus* Peters) fry. *Aquaculture Nutrition* 8, 257–264.

Otte, J.M. and Podolsky, D.K. (2004) Functional modulation of enterocytes by Gram-positive and Gram-negative microorganisms. *American Journal of Physiology* 286, G613–G626.

Ozbas, T. and Kutsal, T. (1986) Comparative study of riboflavin production from two microorganisms: *Eremotbecium asbbyii* and *Asbbya gossypii*. *Enzyme and Microbial Technology* 10, 593–596.

Pandey, R., Chander, R. and Sainis, K.B. (2007) Prodigiosins: a novel family of immuno-suppressants with anticancer activity. *Indian Journal of Biochemistry and Biophysics* 44, 295–302.

Parajo, J., Santos, V., Vazquez, M. and Cruz, J. (1997) Production of carotenoids by *Xanthophyllomyces dendrorhous* growing on enzymatic hydrolysates of pre-hydrolyzed wood. *Food Chemistry* 60, 347–355.

Prakash, D. and Gupta, C. (2011) Role of Phytoestrogens as Nutraceuticals in Human Health. *Pharmacology Online* 1, 510–523.

Prescott, S.L., Dunstan, J.A., Hale, J., Breckler, L., Lehmann, H., Weston, S. and Richmond, P. (2005) Clinical effects of probiotics are associated with increased interferon-responses in very young children with atopic dermatitis. *Clinical and Experimental Allergy* 35, 1557–1564.

Rajilić-Stojanović, M., Smidt, H. and De Vos, W.M. (2007) Diversity of the human gastrointestinal tract microbiota revisited. Environmental Microbiology 9, 2125–2136.

Ratliff, T.L., Ritchey, J.K., Yuan, J.J., Andrilow, G.L. and Catalona, W.J. (1993) T-cell subsets required for intravesical BCG immunotherapy for bladder cancer. *Journal of Urology* 150, 1018–1023.

Resta-Lenert, S. and Barrett, K.E. (2003) Live probiotics protect intestinal epithelial cells from the effects of infection with entero-invasive *Escherichia coli* (EIEC). *Gut* 52, 988–997.

Rizzello, C.G., De Angelis, M., Di Cagno, R., Camarca, A., Silano, M., Losito, A., De Vincenzi, M., De Bari, M.D., Palmisano, F., Maurano, F., Gianfrani, C. and Gobbetti, M. (2007) Highly efficient gluten degradation by lactobacilli and fungal proteases during food processing: new perspectives for celiac disease. *Applied and Environmental Microbiology* 73, 4499–4507.

Rosberg-Cody, E., Stanton, C., O'Mahony, L., Wall, R., Shanahan, F., Quigley, E.M., Fitzgerald, G.F. and Paul Ross, R. (2011) Recombinant *Lactobacilli* expressing linoleic acid isomerize can modulate the fatty acid composition of host adipose tissue in mice. *Microbiology* 157, 609–615.

Rosenfeldt, V., Benfeldt, E., Valerius, N.H., Paerregaard, A. and Michaelsen, K.F. (2004) Effect of probiotics on gastrointestinal symptoms and small intestinal permeability in children with atopic dermatitis. *Journal of Pediatrics* 145, 612–616.

Rubio-Tapia, A. and Murray, J.A. (2010) Celiac disease. *Current Opinion in Gastroenterology* 26, 116–122.

Sakamoto, T., Horiguchi, H. and Oguma, E. (2010) Effects of diverse dietary phytoestrogens on cell growth, cell cycle and apoptosis in estrogen-receptor-positive breast cancer cells. *Journal of Nutritional Biochemistry* 21, 856–864.

Sanders, M.E. (2000) Considerations for use of probiotic bacteria to modulate human health. *The Journal of Nutrition* 130 (2S Suppl.), 384S–390S.

Sarözlü, N.Y. and Kivanc, M. (2009) Isolation of gallic acid- producing microorganisms and their use in the production of gallic acid from gall nuts and sumac. *African Journal of Biotechnology* 8, 1110–1115.

Scarpignato, C. (2008) NSAID-induced intestinal damage: are luminal bacteria the therapeutic target? *Gut* 57, 145–148.

Schaaf, S. and Bott, M. (2007) Target genes and DNA-binding sites of the response regulator PhoR from *Corynebacterium glutamicum. Journal of Bacteriology* 189, 5002–5011.

Schulte, T.W. and Neckers, L.M. (1998) The benzoquinone ansamycin 17-allylamino-17 demethoxygeldanamycin binds to HSP90 and shares important biologic activities with geldanamycin. *Cancer Chemotherapy and Pharmacology* 42, 273–279.

Schultz-Larsen, F.H.J. (2002) Epidemiology of atopic dermatitis. *Immunology and Allergy Clinics of North America* 22, 1–24.

Schuppan, D., Junker, Y. and Barisani, D. (2009) Celiac disease: from pathogenesis to novel therapies. *Gastroenterology* 137, 1912–1933.

Shan, L., Marti, T., Sollid, L.M., Gray, G.M. and Khosla, C. (2004) Comparative biochemical analysis of three bacterial prolyl endopeptidases: implications for coeliac sprue. *Biochemical Journal* 383, 311–318.

Shimizu, S. and Kataoka, M. (1999) Lactonohydrolase. In: Flickinger, M.C. and Drew, S.W. (eds) *Encyclopedia of Bioprocess Technology: Fermentation, Biocatalysis, and Bio-separation*. Wiley, New York, pp. 1571–1577.

Shimizu, Y., Sakai, M., Umemura, Y. and Ueda, H. (1997) Immuno-histochemical localization of nitric oxide synthase in normal human skin: expression of endothelial-type and inducible-type nitric oxide synthase in keratinocytes. *Journal of Dermatology* 24, 80–87.

Shin-Ya, K., Shimazu, A., Hayakawa, Y. and Seto, H. (1991) 7-Demethylnaphterpin, a new free radical scavenger from *Streptomyces prunicolor. Journal of Antibiotics* 45, 124–125.

Sicherer, S.H. and Sampson, H.A. (1999) Food hypersensitivity and atopic dermatitis: pathophysiology, epidemiology, diagnosis and management. *Journal of Allergy Clinical Immunology* 104, S114–S122.

Sousa, R., Halper, J., Zhang, J., Lewis, S.J. and Li, W. (2008) Effect of *Lactobacillus acidophilus* supernatants on body weight and leptin expression in rats. *BMC Complementary and Alternative Medicine* 8, 5–12.

Spiegelman, B.M. and Flier, J.S. (2001) Obesity and the regulation of energy balance. *Cell* 104, 531–543.

Stahmann, K.P., Revuelta, J.L. and Seulberger, H. (2000) Three biotechnical processes using *Ashbya gossypii, Candida famata* or *Bacillus subtilis* compete with chemical riboflavin production. *Applied Microbiology and Biotechnology* 53, 509–516.

Stenman, S.M., Venäläinen, J.I., Lindfors, K., Auriola, S., Mauriala, T., Kaukovirta-Norja, A., Jantunen, A., Laurila, K., Qiao, S.-W., Ludvig, M.S., Männistö, P.T., Kaukinen, K. and Mäki, M. (2009) Enzymatic detoxification of gluten by germinating wheat proteases: implications for new treatment of celiac disease. *Annals of Medicine* 41, 390–400.

Stepniak, D., Spaenij-Dekking, L., Mitea, C., Moester, M., de Ru, A., Baak-Pablo, R., Van Veelen, P., Edens, L. and Koning, F. (2006) Highly efficient gluten degradation with a newly identified prolyl endoprotease: implications for celiac disease. *American Journal of Physiology – Gastrointestinal and Liver Physiology* 291, G621–629.

Thanhäuser, A., Böhle, A., Schneider, B., Reiling, N., Mattern, T., Ernst, M., Flad, H.D. and Ulmer, A.J. (1995) The induction of Bacillus *Calmette-Guerin*-activated killer cells requires the presence of monocytes and T helper type-1 cells. *Cancer Immunology and Immunotherapy* 40, 103–108.

Thestrup-Pedersen, K. (2002) Treatment principles of atopic dermatitis. *Journal of the European Academy of Dermatology and Venereology* 16, 1–9.

Thomsen, S.F., Ulrik, C.S., Kyvik, K.O., Hjelmborg, J.B., Skadhauge, L.R., Steffensen, I. and Backer, V. (2007) Importance of genetic factors in the etiology of atopic dermatitis: a twin study. *Allergy and Asthma Proceedings* 28, 535–539.

Tovar, D., Zambonino, J., Cahu, C., Gatesoupe, F.J., Vázquez-Juárez, R. and Lésel, R. (2002) Effect of live yeast incorporation in compound diet on digestive enzyme activity in sea bass (*Dicentrarchus labrax*) larvae. *Aquaculture* 204, 113–123.

Vazquez, M., Santos, V. and Parajo, J. (1998) Fed-batch cultures of *Phaffia rhodozyma* in xylose-containing media made from wood hydrolysates. *Food Biotechnology* 12, 43–55.

Venil, C.K. and Lakshmanaperumalsamy, P. (2009) An insightful overview on microbial pigment, prodigiosin. *Electronic Journal of Biology* 5, 49-61.

Vetting, M.W., Frantom, P.A. and Blanchard, J.S. (2008) Structural and enzymatic analysis of MshA from *Corynebacterium glutamicum*: substrate-assisted catalysis. *Journal of Biological Chemistry* 283, 15834–15844.

Wang, M.F., Lin, H.C., Wang, Y.Y. and Hsu, C.H. (2004) Treatment of perennial allergic rhinitis with lactic acid bacteria. *Pediatric Allergy and Immunology* 15, 152–158.

Winkler, P., Ghadimi, D., Schrezenmeir, J. and Kraehenbuhl, J.P. (2007) Molecular and cellular basis of microflora-host interactions. *Journal of Nutrition* 137, 756S–772S.

Yamada, H. and Kobayashi, M. (1996) Nitrile hydratase and its application to industrial production of acrylamide. *Bioscience, Biotechnology, and Biochemistry* 60, 1391–1400.

Yan, F., Cao, H., Cover, T.L., Whitehead, R., Washington, M.K. and Polk, D.B. (2007) Soluble proteins produced by probiotic bacteria regulate intestinal epithelial cell survival and growth. *Gastroenterology* 132, 562–575.

Yoshiaki, M., Chihiro, I., Masataka, I. and Toshihiko, O. (2009) Antioxidants produced by *Eurotium herbariorum* of filamentous fungi used for the manufacture of Karebushi, Dried Bonito (Katsuobushi). *Bioscience, Biotechnology, and Biochemistry* 73, 1323–1327.

7 Phytochemicals of Nutraceutical Importance from Cactus and their Role in Human Health

Mónica Azucena Nazareno*
Universidad Nacional de Santiago del Estero, Argentina

7.1 Introduction

Cacti are xerophyte plants, which present a wide variability in germplasm. These plants have developed phenological, physiological and structural adaptations to survive in arid regions where water is the main limiting factor for vegetal species. Thus, cacti have evolved in harsh environments and hard stressed conditions by developing special physiological traits and distinctive appearances, such as stem morphology, spine presence, succulence, nocturnal stomatal opening (CAM plants) and attractive flowers (Nobel and De la Barrera, 2003).

Cacti belong to the *Cactaceae* family, constituted by more than 1600 species (Gibson and Nobel, 1986; Barthlott and Hunt, 1993). Among them, the *Opuntia* genus comprises about 300 species. They are native to Mesoamerica, although, nowadays, they are globally spread as wild or cultivated species. These plants grow in arid and semi-arid regions of the world where many traditional plants have few possibilities to survive.

Cactus plants can be considered as multipurpose crops, since they provide not only food and feed but they are also sources of health-promoting substances. These properties, known by ancient civilizations, have gained interest among the scientific community in recent years.

7.2 Uses of Cactus in Traditional and Popular Medicine

Cactus plants have been used by ancient civilizations to cure diseases and heal wounds for thousands of years. The origin and history of cacti are closely related to the ancient Mesoamerican civilizations. For over 12,000 years, fresh cactus has been consumed by the natives for its nutritional qualities and healing properties. Cactus cladodes, fruits, seeds and flowers have been used as folk medicines in several countries for centuries. The use of cactus flower petals to treat urological problems is well known in Sicily. Kidney colic treatment with *Opuntia ficus-indica* flowers was mentioned by Pitrè (1896). They are also recognized for the properties of the infusions of dried flowers to prevent prostate cancer. Among the most recognized popular uses of cactus flowers include the depurative and diuretic effects and renal calculus expulsion. Cladodes have also been used in many countries to take advantage of their wound

* E-mail: manazar2004@yahoo.com; nazareno@unse.edu.ar

healing properties, cicatrizing activity, emollient and moisturizing effects. A decoction of plant roots is used in folk medicine for diarrhoea and gonorrhoea. The consumption of cactus fruits and their juices has traditionally been recommended in terms of their diuretic effect, their functions as hypoglycaemic agent, hypocholesterolaemic factor, anti-allergic, analgesic and anti-inflammatory actions and for gastritis relief. Cactus products have also been used to treat indigestion and to alleviate alcohol hangover symptoms (Nefzaoui *et al.*, 2008).

7.3 Cactus Plant as a Source of Phytochemicals

Several reports indicate that a diet rich in fruits and vegetables is related to lower incidences of heart conditions and some types of cancer, suggesting that this kind of diet has positive effects on health (Bazzano *et al.*, 2002). These beneficial effects have been associated to the action of some antioxidant components present in the natural food and not only to their vitamin content (Hertog *et al.*, 1995; Terry *et al.*, 2002).

In today's lifestyle the human body is exposed to the deleterious action of numerous sources of pro-oxidants. Reactive oxygen species (ROS) and free radicals are constantly formed in the human body by normal metabolic function. When the generation of harmful agents greatly exceeds the cell's protective system, serious oxidative stress occurs, and the accumulation of damage will result in pathophysiologic events. Antioxidant phytochemicals are involved in the redox balance of normal physiological functions and against the pathogenesis of various diseases such as neurodegenerative disorders, e.g. Alzheimer's and Parkinson's diseases, heart conditions, cataracts, cancer, inflammatory processes, premature ageing and atherosclerosis, among others (Halliwell and Gutteridge, 1999).

Phytochemicals with antioxidant properties promote a healthy status by protecting against the oxidative damage induced by ROS (Prakash and Gupta, 2009). There are many groups of bioactive compounds such as carotenoids, betalains, flavonoids and other phenolic compounds occurring in nature with these properties. Most of them are derived from plants and widely spread in fruits, vegetables and tea, so they are incorporated frequently as part of a diet. Natural bioactive substances can be used in the food industry to replace synthetic additives, antioxidant and colourants (Nazareno *et al.*, 2011).

Remarkable progress has been made over the past decades in disease prevention with the use of fruit, vegetables and herbs. In recent years there has been a global trend toward the use of natural phytochemicals obtained from plant resources, such as fruit, vegetables, oilseeds and herbs, as antioxidants and functional foods. These plant-derived foods are promising raw materials to obtain bioactive compounds of nutraceutical importance. It is estimated that over 80% of the world population uses medicinal plants for their health care as phytotherapeutic substances and phytomedicines. Phytotherapeutic substances utilise the medicinal properties obtained from plants used for alternative therapies or natural medicine. On the other hand, phytomedicines are drugs, whose active ingredient is a plant extract, developed and standardized according to traditional pharmaceutical forms and with demonstrated biological activity.

Nowadays, there is increasing evidence that the use of these substances may have beneficial effects on consumers' health beyond their nutritive action. The growing demand for nutraceuticals correlates to an increased effort in developing natural products for the prevention or treatment of human diseases. Moreover, the discovery of new compounds obtained from natural sources with high antioxidant activity is a constant challenge for researchers.

In the search for health-promoting substances, the assessment of a relative activity scale is necessary. However, antioxidants present different behaviour depending on the nature of the oxidative species since different mechanisms of action take place for their deactivation, radical scavenging or inhibition of pro-oxidant enzymes. Scientific investigations confirmed that cacti may be efficiently used as a source of several phytochemicals of nutraceutical importance, such as mucilage, fibres, pigments and vitamins. Recent data have revealed high contents of these chemical constituents in fruits, cladodes, seeds and flowers. Figure 7.1 shows typical cactus plants with their corresponding fruits.

The cactus plant can be fully exploited in an integrated manner since its bioactive components can be extracted from different parts of its anatomy: flowers, fruit, cladodes, roots and seeds. According to several studies demonstrating both cactus fruit and cladode yield high values of important nutrients, minerals, vitamins, as well as further antioxidants, the cactus plant appears to be an excellent candidate as a source of phytochemicals of nutraceutical importance as shown in Table 7.1.

7.4 Fruits

There are about 100 cactus species, mainly of the genus *Opuntia*, that yield edible fruits. Their successful cultivation may be achieved in arid lands, where only few plants can survive (Pimienta-Barrios, 1994). Currently, most of the cactus fruits offered in the global market belong to the *Opuntia ficus-indica* species, whose fruits are known as 'cactus pear'. These fruits have as major constituents of their dry matter sugars, fibre, mucilage and pectins, and as minor constituents amino acids and proteins, vitamins, minerals, phenolic compounds and pigments.

Habibi and co-workers investigated the composition of the polysaccharide fraction of the fruit skins, discovered they are composed of galactose and arabinose with 6.3:3.3 ratio, and classified them as arabinogalactane (Habibi *et al.*, 2003, 2004a, b, 2005a). Lipid fraction in fruit skins has been described by Ramadan and Mörsel (2003a). They reported the presence of unsaturated fatty acids, vitamins and sterols.

7.5 Cladodes

In general, cladodes are rich in pectins, mucilages and minerals. Chemical composition of fresh young cladodes has been reported by Sáenz (2002), indicating a 91% (w/w) moisture content and 1.5 and 0.2% for protein and fat contents w/w dry basis, respectively. Ash content represented 1.3%, calcium being about 90% of this component. They also contain 11 mg vitamin C 100 g^{-1} and 30 μg of carotenoids 100 g^{-1}; their fibre content (1.1% fresh weight basis) is comparable to that of spinach.

7.6 Seeds

Seeds are a good source of nutrients as lipids and proteins. Seeds constitute about 10–15% of the edible pulp and are usually discarded as waste after extraction of the pulp. Seed oil represents 7–15% of the seed weight (Ramadan

Fig. 7.1. Different cactus plants with their fruits: (a) *Opuntia ficus-indica;* (b) *Harrisia pomanensis* and (c) *Cereus forbesii*.

Table 7.1. Major functional constituents of the different parts of the cactus plant.

Plant parts	Major phytochemicals	Main phytochemical identified	References
Fruits	Betalains	Indicaxanthin (betaxanthin), betanin, isobetanin (betacyanins)	a
	Polyphenols	Ferulic acid and isorhamnetin glycosides	b
	Vitamin C	Ascorbic acid	b
	Fibre		c
	Mucilage		d
	Minerals	K, Ca, Mg	c, e
	Tocopherols	δ-tocopherol	f
	Phytosterols	β-sitosterol	f
Fruit skin	Polysaccharides	Arabinan-rich polysaccharides	g, h, i
	Lipids	Unsaturated lipids	j
Cladodes or pads	Mucilage and pectins	Polysaccharides	k, l, e, m
	Dietary fibre	Insoluble dietary fibre	n
	Chlorophylls	Chlorophyll-*a*	n
	Minerals	K, Ca, Mg	n
	Flavonoids	Kaempherol, isorhamnetin glycosides	o
	Phenolic compounds	Gallic acid, cumaric acid, 3,4-dihydroxy-benzoic acid, 4-hydroxybenzoic acid, ferulic acid and salicylic acid, iso-quercitrin, isorhamnetin-3-O-glucoside, nicotiflorin, narcissin and rutin	p
Seeds	Lipids	Polyunsaturated fatty acids (linoleic acid), monounsaturated fatty acids (oleic acid)	q, r
	Phytosterols	β-sitosterol	f
	Tocopherols	γ-tocopherol	f
	Polysaccharides	Arabinan-rich polysaccharides	s
Flowers	Flavonoids	Quercetin, isorhamnetin and kaempherol glycosides	t, u, v
	Betalains	Betanin, phyllocactin	w
Root	Flavonoids		x

References: (a) Castellanos-Santiago and Yahia, 2008, (b) Galati *et al.*, 2003, (c) Díaz Medina *et al.*, 2007, (d) Matsuhiro *et al.*, 2006, (e) Sepúlveda *et al.*, 2007, (f) Ramadan and Mörsel, 2003a, (g) Habibi *et al.*, 2004a, (h) Habibi *et al.*, 2004b, (i) Habibi *et al.*, 2005a, (j) Ramadan and Mörsel, 2003b, (k) Trachtenberg and Mayer, 1981, (l) Karawya *et al.*, 1980, (m) Madjdoub *et al.*, 2001, (n) Ayadi *et al.*, 2009, (o) Valente *et al.*, 2010, (p) Guevara-Figueroa *et al.*, 2010, (q) Ennouri *et al.*, 2005, (r) Labuschagne and Hugo, 2010, (s) Habibi *et al.*, 2005b, (t) Clark *et al.*, 1980, (u) Ahmed *et al.*, 2005, (v) De Léo *et al.*, 2010, (w) Piattelli and Imperato, 1969, (x) Alimi *et al.*, 2010.

and Mörsel, 2003b). This oil is constituted by polyunsaturated fatty acids, mainly linoleic acid and monounsaturated fatty acids such as oleic acid (Ennouri *et al.*, 2005; Labuschagne and Hugo, 2010). Other minor compounds found in this valuable oil are β-sitosterol and γ-tocopherol (Ramadan and Mörsel, 2003a). The seed endosperm is rich in arabinan polysaccharides while the main component of the seed coat is D-xylan (Habibi *et al.*, 2005b). Besides being rich in lipids, seeds have been reported to accumulate proanthocyanidins (Nieto, 1987).

7.7 Flowers

Cactus flowers accumulate betalains as well as colourless phenolic compounds (Piattelli and Imperato, 1969; Clark *et al.*, 1980; Ahmed *et al.*, 2005). The chemical composition of *Opuntia ficus-indica* and *Opuntia stricta* flowers extracted at four flowering stage has been studied by Ammar *et al.* (2012). The ethanolic as well as hexane extracts have also been analysed and their corresponding antiradical, antibacterial and antifungal activities have been determined. Phenolic content varies markedly with flowering stage, the maximum level of these active constituents being detected during post-flowering stage. De Léo *et al.* (2010) have reported the chemical profile of methanol extract of *O. ficus-indica* flowers. The volatile fraction composition of three *Opuntia* species (*O. lindheimeri, O. macrohiza* and *O. microdasys*) obtained from aqueous distillation have been reported by Bergaoui *et al.* (2007).

7.8 Functional Phytochemicals of Cacti

7.8.1 Polysaccharides and mucilage

The major components of cladodes are carbohydrate-containing polymers, which consist of a mixture of mucilage and pectins (Karawya *et al.*, 1980; Trachtenberg and Mayer, 1981). The mucilage is a biopolymer present in specialized storage cells or free within cells or intracellular spaces of the chlorenchymatic and parenchymatic tissue of the cladodes (Ting, 1997; Terrazas-Salgado and Mausseth, 2002). Mucilage is a high molecular weight polysaccharide (Cárdenas *et al.*, 1997), which contains arabinose, galactose, xylose, galacturonic acid and rhamnose (Majdoub *et al.*, 2001). The dried mucilage has in average 5.6% moisture, 7.3% protein, 37.3% ash, 1.14% nitrogen, 9.86% calcium and 1.55% potassium (Sepúlveda *et al.*, 2007).

Because of its high ability to retain water it is considered as a hydrocolloid. Hydrocolloids are biopolymers, which can modify the properties of aqueous media by forming colloidal solutions due to their water affinity. As such, they can exhibit thickening, gelling, stabilizing and emulsifying properties, which make them ingredients frequently used in formulated food products. Lyophilized cladodes have significant anti-ulcer effect, a protective effect against gastric lesions as well as anti-inflammatory activity. Galati *et al.* (2001) proposed that *O. ficus-indica* cladodes stimulate a protective response from the gastric mucosa, which prevents the development of ethanol-induced ulcers (preventive treatment) and promotes a faster recovery (curative treatment). The cytoprotective effect of *O. ficus-indica* cladodes has been ascribed to the physicochemical properties of the mucilage (Galati *et al.*, 2001). *Opuntia ficus-indica* cladodes produce cyto-protection by increasing mucus secretion in the gastric mucosa of rats affected by ethanol-induced ulcers (Galati *et al.*, 2002). Cactus cladode administration has been recommended for preventive as well as for curative treatments of gastric ulcers.

The mucilage from fruits of *O. ficus-indica* has been characterized by Matsuhiro *et al.* (2006). The pectins are abundant in fruits and can be used as food or cosmetic additives as a gelling agent and with low calories. Nopal flour is a rich source of dietary fibre, reaching up to 43% on a dry basis (Sáenz-Hernández *et al.*, 2002). Fibres have been recognized for health benefits such as control of obesity, diabetes and cholesterol. Soluble fibres, including mucilages, gums, pectins and hemicelluloses, have been associated with reduced levels of glucose and cholesterol in blood and to regulate intestinal transit (Fernández *et al.*, 1994). Moreover, the insoluble fibres are known for their ability to retain water, ion exchange and promote the absorption of bile salts, minerals and vitamins (Sáenz *et al.*, 2004). The dry cladode powder contains about 43% fibre, 28.5% of which is insoluble. Some functional foods elaborated with cladodes are commercially available in Mexico and other countries such as Morocco and Tunisia. These products are prepared from young cladodes (3–6 months) and are low in insoluble fibre. This powder is used in the production of various foods such as cookies, creams and other desserts (Sáenz, 2002).

7.8.2 Betalains

Betalains are found in only ten families of the *Caryophyllaleae*. Betalains are nitrogen-containing pigments widely used in the food industry. Betalain presence has been reported in different parts of the plants such as in fruits (Wybraniec and Mizrahi, 2002; Stintzing *et al.*, 2005), in roots (Nemzer *et al.*, 2011), as well as in flowers (Stintzing and Carle, 2004). The occurrence of betalains has been ascribed to a great variety of functions (Stintzing and Carle,

2004). Interest in betalains has grown since they have been characterized as good free-radical scavengers (Escribano *et al.*, 1998; Kanner *et al.*, 2001). They are widely used as additives in the food industry because of their natural tinctorial properties and absence of toxicity, even at high concentrations (Schwartz *et al.*, 1983).

According to their chemical structures, betalains are classified into two different groups: (i) betaxanthins, yellow and orange coloured substances; and (ii) betacyanins, red to purple coloured (Hendry and Houghton, 1996). Figure 7.2 shows the typical chemical structure of betaxanthins and betacyanins. They differ only in the chemical group linked to the betalamic acid moiety. Betacyanins are ammonium conjugates of betalamic acid with cyclo-DOPA (e.g. betanin) while betaxanthins are conjugates with amino acids or amines (e.g. indicaxanthin). Betalainic fraction composition has been characterized in cactus pear fruit (Stintzing *et al.*, 2005). A varied composition of betalains has been also found in coloured cactus flowers (Kobayashi *et al.*, 2000).

Evolution of external and internal fruit colours has been studied during ripening of *Opuntia megacantha* fruits showing that the chlorophyll content decreased in the fruit skins and pulps at different rates, while betalain biogenesis progressed faster in the pulp. A good correlation has been observed between the betalain content increase and the antiradical activity variation (Coria-Cayupán *et al.*, 2011). Different coloured fruits of *Opuntia megacantha* and *Opuntia ficus-indica* are shown in Fig. 7.3 as examples of the great variety of hues found in these fruits, caused by the mixtures of different levels of betaxanthins and betacyanins.

In addition to colour, these pigments have other interesting properties from the technological point of view, due to their antioxidant capacities. Moreover, these properties are higher than that of ascorbic acid (Butera *et al.*, 2002; Stintzing *et al.*, 2005). Antioxidant abilities of a group of betalains have been studied by Zacharova and Petrova (1998) and a relation between the chemical structure and their activities has been proposed. In contrast to anthocyanins, betalains are stable in a wider pH range (Stintzing and Carle, 2004). This property makes the latter ideal for their use as food colourants in low acidity products (Stintzing *et al.*, 2001). Due to the wide structural variety, and hence colour diversity, betalains

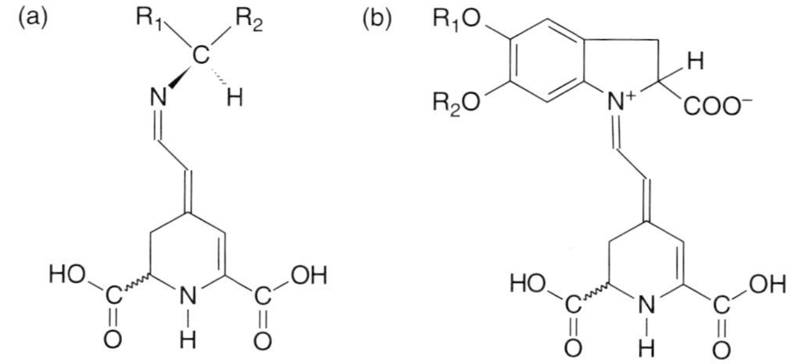

Fig. 7.2. Chemical structure of betalains: (a) betaxanthins and (b) betacyanins.

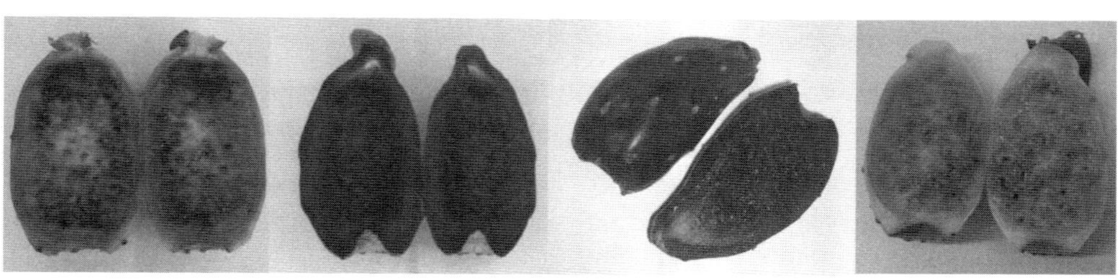

Fig. 7.3. Cactus pears of the *Opuntia* genus.

constitute a very promising source of natural compounds to be used as functional food colourants. Compared with red-beets, cactus pear offers a wider range of colours and, due to its natural character, can be used free of certification (Stintzing *et al.*, 2002).

Betalains have been reported to exert an important role in preventing oxidative damage. The effect of cactus betalains on oxidative stress in humans has been studied by Tesoriere *et al.* (2004). Oxidation of low (LDL) and very low (VLDL) density lipoproteins are crucial steps in atherosclerotic lesion formation. They reported that fruit ingestion produced a decrease of oxidative stress markers, inhibited LDL oxidation and increased the resistance to oxidative haemolysis of red blood cells in *ex vivo* experiments. Budinsky *et al.* (2001) indicated that regular consumption of *Opuntia robusta* reduces oxidative injuries in humans. Betalains have also been reported to increase the resistance to human LDL oxidation (Tesoriere *et al.*, 2003). They are able to bind to LDL and this interaction prevents their oxidation. After cactus pear consumption, LDL extracted from plasma in this condition indicated that 3 h after intake, indica xanthin was incorporated in 98 pmol mg^{-1} of protein diminishing in 5 h and it had completely disappeared in 8 h. Betanin was detected only 3 h after fruit consumption and the lag time of LDL oxidation induced by copper was longer compared with control.

The protective effect of cactus fruit ingestion on the oxidative haemolysis of human red blood cells induced by free radicals has been studied in an *ex vivo* assay. Blood samples were taken from volunteers after ingestion of cactus fruits. Results indicated that cell resistance to haemolysis increased up to 3 h after the fruit intake corresponding to the highest plasma level of betalains (Tesoriere *et al.*, 2005). Methanolic extract containing 1 to 5 mg of *Opuntia ficus-indica* pulp inhibited malondialdehyde (MDA) formation in 4 h incubation of blood cells exposed to organic hydroperoxides as a free radical source.

The main interests of investigations are focused to determine not only the betalain contents and antioxidant effects *in vivo* but also their absorption, since antioxidant efficiency *in vivo* fundamentally depends on its bioavailability. Experiments have demonstrated that red beet betalains have been detected in urine after 2 or 3 h of ingestion, recovering 0.5 to 0.9% of the total intake. Betalains have also been detected in blood after 3 h of ingestion of 500 g fresh cactus pulp. After this period, pigment concentration decreased progressively and almost disappeared in 8 h (Tesoriere *et al.*, 2005). Betalain levels were also measured in human body fluids after intake of cactus pear fruits. Urine recovery after 12 h was about 76% for indicaxanthin and 3.7% for betanin. Pigment peak concentrations were reached in plasma 3 h after ingestion for both of them and completely disappeared 12 h after ingestion.

7.8.3 Flavonoids

Flavonoids are naturally occurring phenolic compounds present in fruits and vegetables and are an integral part of the human diet. Consumption of flavonoid-rich foods is inversely correlated with the risk of coronary heart disease (Hertog *et al.*, 1995). Moreover, the antioxidant and free-radical-scavenging properties of flavonoids have been proposed to contribute to this chemopreventive effect (Geleijnse *et al.*, 2002).

Neuroprotective action of *Opuntia ficus-indica* var. Saboten flavonoid extract has been evaluated against neuronal oxidative injuries in rat cortical cells (Dok-Go *et al.*, 2003). They evaluated their protective effects against oxidative neuronal injuries induced in primary cultured rat cortical cells and their antioxidant activities. They found that quercetin inhibits $H_2O_2^-$ or xanthine (X)/xanthine oxidase (XO)-induced oxidative neuronal cell injury. Moreover, Quercetin-3-methyl-ether potently and dramatically inhibited $H_2O_2^-$ and X/XO-induced neuronal injuries. Kuti (2004) has reported an antioxidative effect due to the major flavonoids found in cactus fruits (quercetin, kaempferol and isorhamnetin). Flavonol derivatives detected in *Opuntia* ssp. have been well reviewed by Stintzing and Carle (2007). Lee *et al.* (2002) found that the flavonoids quercetin, (+)-dihydroquercetin and quercetin-3-methyl-ether, isolated from

Opuntia ficus-indica var. Saboten are also effective in protecting plasmid DNA against the strand breakage induced by hydroxyl radicals. All above mentioned flavonoids markedly inhibited lipid peroxidation and scavenged 1,1-diphenyl-2-picrylhydrazyl free radicals. These results indicate that these are the active constituents in the fruits and cladodes of *Opuntia ficus-indica* var. Saboten. Furthermore, quercetin-3-methyl-ether appears to be the most potent neuroprotectant of the three flavonoids isolated from this plant. Guevara-Figueroa *et al.* (2010) analysed commercial and wild *Opuntia* spp. cladodes from Mexico and evaluated the polyphenolic and flavonoid profiles of fresh nopal and their processed products. The wild morado, tempranillo, blanco and cristalino varieties had the highest total phenolic acid content, while the commercial varieties had the highest total flavonoids. The presence of five major flavonoids (isoquercitrin, isorhamnetin-3-O-glucoside, nicotiflorin, rutin and narcissin) has been observed in all varieties, with nicotiflorin being predominant. Cactus flowers are also rich in active compounds such as flavonoids. Several flavonoids were identified in *Opuntia* flowers by chromatographic methods as quercetin-3-glucoside, quercetin 3-rutinoside and kaempferol-3-glucoside (Clark *et al.*, 1980). Quercetin-3-galactoside (hyperin) and the isorhamnetin 3-rutinoside (narcissin), 3-galactoside and 3-rhamnogalactoside were found in the flowers of *O. lindheimeri* (Rösler *et al.*, 1966). The aqueous ethanolic extract obtained from the fresh cladodes of *Opuntia dillenii* exerts potent radical scavenging activity; this was ascribed to three compounds, opuntioside I, 4-ethoxyl-6-hydroxymethyl-alpha-pyrone and a kaempferol glycoside, isolated and identified from the extract (Qiu *et al.*, 2002).

7.8.4 Vitamin C

Among other nutrients like sugars, cactus fruits are good sources of vitamin C (Kuti, 2004). Cactus fruit has a higher concentration of vitamin C than other common fruits, such as apple, pear, grape and banana. Significant amounts of ascorbic acid are present in *Opuntia ficus-indica*, ranging from 180 to 300 mg kg^{-1}. Cactus fruits have a high antioxidant activity ascribed to the presence of vitamin C, β-carotene, flavonoids and betalains (Galati *et al.*, 2003; Kuti, 2004). Fruit antioxidant activity is twice as high as pears, apples, tomato, bananas, white grapes, and similar to red grapes and grapefruit (Butera *et al.*, 2002).

7.8.5 Vitamin E

Tocopherols, the fat soluble vitamins, are found in the lipid fraction of both the cactus fruit seed and pulp. The vitamin E homologue isoforms gamma- and delta-tocopherol are the main components in seed and pulp oils, respectively, amounting to about 80% of the total vitamin E content. Similar to β-carotene, it is predominant in pulp lipids (Ramadan and Morsel, 2003b).

A comparative study was carried out to evaluate the effect of *O. ficus-indica* intake and vitamin C supplementation in the oxidative status of healthy volunteers. Results showed that the consumption of 250 g fresh fruit pulp as well as its equivalent, 75 mg vitamin C supplementation, produced an increase in vitamin C and E levels. No changes were found in vitamin A content or trolox equivalent antioxidant capacity (TEAC) value in plasma. Differences between both treatments become relevant in oxidative stress biomarker levels. After fruit intake, 8-epi-$PGF_{2\alpha}$ and MDA plasma values decreased in 30 and 75%, respectively, while glutation redox-status relation (GSH:GSSG) in red blood cells was enhanced since the reduced form increased. LDL-conjugated dienes hydroperoxide level was significantly reduced as was LDL oxidation. Vitamin C supplementation does not produce changes in oxidative stress markers, indicating that cactus fruit efficiency is due to antioxidant combination including vitamin C and other co-nutrients such as betalains.

7.8.6 Lipids

Cactus lipids are distributed in the fruit peel, pulp and seeds. Fruit peel contains appreciable amounts of polyunsaturated fatty acids, mainly linoleic acid, as well as other fat-soluble compounds such as sterols, β-carotene and vitamin K_1, the main sterol being β-sitosterol (Ramadan and Mörsel, 2003b). On the other hand, after fruit processing in juice and jam preparation, great amounts of seeds are usually discarded. The cactus pear fruit contains many hard-coated seeds that represent 10–15% of the pulp weight. The fruits contain a large number of seeds although their oil content is relatively low (7–15%). Seeds from 11 commercial cactus pear cultivars were analysed for oil content and fatty acid composition by South African researchers. They reported a palmitic acid content ranging between 11.4 and 15.9%, considerably lower than that of cotton oil. Linoleic acid content varied between 61.4 and 68.9%. The α-linolenic acid of all the cultivars was less than 1%. The oleic acid content varied between 12.4 and 16.5% (lower than that of cotton seed). Unsaturated fatty acids made up about 80% of all fatty acids (Ennouri *et al.*, 2005). Therefore, although the seed oil content is relatively low, the fatty acid composition indicates that it has potential as an oil for the health and cosmetic market (Labuschagne and Hugo, 2010). The seeds can be ground or pressed to obtain the oil as a lucrative part of the plant. It takes approximately 1 t of these tiny seeds to make 1 l of oil. The *Opuntia* seed oil is obtained by cool pressed seeds and some of its main applications are being developed by the cosmetic industry. The seed oil is actually destined for cosmetic product production, and it sells at a very high price as organic oil for anti-ageing and anti-wrinkle purposes.

7.9 Nutraceutical Products

Diverse functional foods are prepared using cactus fruits as ingredients for juices, marmalades, candies, liquors and syrups and are offered as health foods. Cereal bars, dessert preparations, soups and other foods manufactured using dehydrated cladodes and cactus fruits are proposed to take benefits from the medicinal properties of cactus plants. Nowadays, dietary supplements based on dehydrated nopal flour are also commercially available. Several manufactured products are currently available in the nutraceutical market, and consumer interest in these beneficial products is growing globally. Some examples of cactus nutraceutical products are shown in Fig. 7.4.

Fig. 7.4. Cactus products currently available in the global market.

7.10 Conclusions

Cacti can be considered an important source of bioactive substances and excellent candidates for nutraceutical and functional food preparation. Scientific data revealed a high content of some chemical constituents in fruits, cladodes, seeds and flowers, which can add value to cactus products. Additionally, some of their constituents show promising characteristics in terms of functionality as health-promoting substances. Several manufactured products are currently available in the nutraceutical market to promote health benefits from the medicinal properties of cactus.

References

Ahmed, M.S., El Tanbouly, N.D., Islam, W.T., Sleem, A.A. and El Senousy, A.S. (2005) Anti-inflammatory flavonoids from *Opuntia dillenii* (Ker-Gawl) Haw. Flowers growing in Egypt. *Phytotherapy Research* 19(9), 807–809.

Alimi, H., Hfaiedha, N., Bouonia, Z., Hfaiedha, M., Saklyb, M., Zourgui, L. and Rhouma, K.B. (2010) Antioxidant and antiulcerogenic activities of *Opuntia ficus-indica* f. *inermis* root extract in rats. *Phytomedicine* 17, 1120–1126.

Ammar, I., Ennouri, M., Khemakhem, B., Yangui, T. and Attia, H. (2012) Variation in chemical composition and biological activities of two species of *Opuntia* flowers at four stages of flowering. *Industrial Crops and Products* 37, 34–40.

Ayadi, M.A., Abdelmaksoud, W., Ennouri, M. and Hamadi, A. (2009) Cladodes from *Opuntia ficus indica* as a source of dietary fiber: effect on dough characteristics and cake making. *Industrial Crops and Products* 30, 40–47.

Barthlott, W. and Hunt, D.R. (1993) Cactaceae. In: Kubitzki, K. (ed.) *The Families and Genera of Vascular Plants*, Vol. II. Springer, New York, pp, 161–197.

Bazzano, L.A., He, J., Ogden, L.G., Loria, C.M., Vupputuri, S., Myers, L. and Whelton, P.K. (2002) Fruit and vegetable intake and risk of cardiovascular disease in US adults: the first national health and nutrition examination survey epidemiologic follow-up study. *American Journal of Clinical Nutrition* 76, 93–99.

Bergaoui, A., Boughalleb, N., Ben Jannet, H., Harzallah-Shiric, F., El Mahjoub, M. and Mighri, Z. (2007) Chemical composition and antifungal activity of volatiles from three *Opuntia* species growing in Tunisia. *Pakistan Journal of Biological Sciences* 10, 2485–2489.

Budinsky, A., Wolfram, R., Oguogho, A., Elthimiou, Y., Stamatopoulos, Y. and Sinzinger, H. (2001) Regular ingestion of *Opuntia robusta* lowers oxidation injury. *Prostaglandins, Leucotrienes and Essential Fatty Acids* 55, 45–50.

Butera, D., Tesoriere, L., Di Gaudio, F., Bongiorno, A., Allegra, M., Pintaudi, A.M., Kohen, R. and Livrea, M.A. (2002) Antioxidant activities of Sicilian prickly pear (*Opuntia ficus-indica*) fruit extracts and reducing properties of its betalains: betanin and indica-xanthin. *Journal of Agricultural and Food Chemistry* 50, 6895–6901.

Cárdenas, A., Higuera-Ciapara, I. and Goycoolea, F.M. (1997) Rheology and aggregation of cactus (*Opuntia ficus-indica*) mucilage in solutions. *Journal of the Professional Association for Cactus Development* 2, 152–159.

Castellanos-Santiago, E. and Yahia, E.M. (2008) Identification and Quantification of Betalains from the Fruits of 10 Mexican Prickly Pear Cultivars by High-Performance Liquid Chromatography and Electrospray Ionization Mass Spectrometry. *Journal of Agricultural and Food Chemistry* 56, 5758–5764.

Clark, W.D., Brown, G.K. and Mays, R.L. (1980) Flower flavonoids of *Opuntia* subgenus *Cylindropuntia*. *Phytochemistry* 19, 2042–2043.

Coria-Cayupán, Y.S., Ochoa, M.J. and Nazareno, M.A. (2011) Health-Promoting Substances and Antioxidant Properties of *Opuntia* sp. Fruits. Changes in bioactive-compound contents during ripening process. *Food Chemistry* 126, 514–519.

De Léo, M., De Abreu, M.B., Pawlowska, A.M., Cioni, P.L. and Braca, A. (2010) Profiling the chemical content of *Opuntia ficus-indica* flowers by HPLC–PDA–ESI-MS and GC/EIMS analyses. *Phytochemistry Letters* 3, 48–52.

Díaz Medina, E.M., Rodríguez Rodríguez, E.M. and Díaz Romero, C. (2007) Chemical characterization of *Opuntia dillenii* and *Opuntia ficus-indica* fruits. *Food Chemistry* 103, 38–45.

Dok-Go, H., Lee, K.H., Kim, H.J., Lee, E.H., Lee, J., Song, Y.S., Lee, Y.H., Jin, C., Lee, Y.S. and Cho, J. (2003) Neuroprotective effects of antioxidative flavonoids, quercetin, (+)-dihydroquercetin and quercetin 3-methyl ether, isolated from *Opuntia ficus-indica* var. Saboten. *Brain Research* 965(13), 130–136.

Ennouri, M., Bourret, E., Mondolot, L. and Attia, H. (2005) Fatty acid composition and rheological behaviour of prickly pear seed oils. *Food Chemistry* 93, 431–437.

Escribano, J., Pedreño, M.A., García-Carmona, F. and Muñoz, R. (1998) Characterization of the antiradical activity of betalains from *Beta vulgaris* L. roots. *Phytochemical Analysis* 9, 124–127.

Fernández, L.M., Lin, E.C.K., Trejo, A. and McNamara, D.J. (1994) Prickly pear (*Opuntia* sp.) pectin alters hepatic cholesterol metabolism without affecting cholesterol absorption in Guinea pigs fed a hypercholesterolemic diet. *Journal of Nutrition* 124, 817–824.

Galati, E.M., Monforte, M.T., Tripodo, M.M., d'Aquino A. and Mondello, M.R. (2001) Antiulcer activity of *Opuntia ficus-indica* (L.) Mill. (Cactaeceae): ultrastructural study. *Journal of Ethnopharmacology* 76, 1–9.

Galati, E.M., Pergolizzi, S., Miceli, N., Monforte, M.T. and Tripodo, M.M. (2002) Study on the increment of the production of gastric mucus in rats treated with *Opuntia ficusindica* (L.) Mill. Cladodes. *Journal of Ethnopharmacology* 83, 229–233.

Galati, E.M., Mondello, M.R., Giuffrida, D., Dugo, G., Miceli, N., Pergolizzi, S. and Taviano, M.F. (2003) Chemical Characterization and Biological Effects of Sicilian *Opuntia ficus-indica* (L.) Mill. Fruit juice: antioxidant and antiulcerogenic activity. *Journal of Agricultural and Food Chemistry* 51, 4903–4908.

Geleijnse, J.M., Launer, L.J., van der Kurp, D.A.M., Hofman, A. and Witteman, J.C.M. (2002) Inverse association of tea and flavonoidintakes with incidence of myocardial infarction: the Rotterdam Study. *American Journal of Clinical Nutrition* 75, 880–886.

Gibson, A.C. and Nobel P.S. (1986) *The Cactus Primer*. Harvard University Press, Cambridge, Massachusetts, 286 pp.

Guevara-Figueroa, T., Jiménez-Islas, H., Reyes-Escogido, M.L., Mortensen, A., Laursen, B., Lin, L.W., De León-Rodríguez, A., Fomsgaard, I.S. and Barba de la Rosa, A.P. (2010) Proximate composition, phenolic acids, and flavonoids characterization of commercial and wild nopal (*Opuntia* spp.) *Journal of Food Composition Analysis* 23, 525–532.

Habibi, Y., Mahrouz, M. and Vignon, M.R. (2003) Isolation and structure characterization of a (4-O-methyl-D-glucurono)-D-xylan from the skin of *Opuntia ficus-indica* prickly pear fruits. *Journal of Carbohydrate Chemistry* 22, 331–337.

Habibi, Y., Heyraud, A., Mahrouz, M. and Vignon, M.R. (2004a) Structural features of pectic polysaccharides from the skin of *Opuntia ficus-indica* prickly pear fruits. *Carbohydrate Research* 339, 1119–1127.

Habibi, Y., Mahrouz, M., Marais, M.F. and Vignon, M.R. (2004b) An arabinogalactan from the skin of *Opuntia ficus-indica* prickly pear fruits. *Carbohydrate Research* 339, 1201–1205.

Habibi, Y., Mahrouz, M. and Vignon, M.R. (2005a) Isolation and structural characterization of protopectin from the skin of *Opuntia ficus-indica* prickly pear fruits. *Carbohydrate Polymers* 60, 205–213.

Habibi, Y., Mahrouz, M. and Vignon, M.R. (2005b) Arabinan-rich polysaccharides isolated and characterized from the endosperm of the seed of *Opuntia ficus indica* prickly pear fruits. *Carbohydrate Research* 60, 319–329.

Halliwell, B. and Gutteridge, J.M.C. (1999) *Free Radicals in Biology and Medicine*. Oxford University Press, Oxford, UK.

Hendry, G.A.F. and Houghton, J.D. (1996) *Natural Food Colorants*, 2nd edn. Chapman & Hall, Glasgow, UK.

Hertog, G.L., Kromhout, D., Aravanis, C., Blackburn, H., Buzina, R. and Fidanza F. (1995) Flavonoid intake and long term risk of coronary heart disease and cancer in the seven country study. *Archives of Internal Medicine* 155, 381–386.

Kanner, J., Harel, S. and Granit, R. (2001) Betalains – a new class of dietary cationized antioxidants. *Journal of Agricultural and Food Chemistry* 49, 5178–5185.

Karawya, M.S., Wassel, G.M., Baghdadi, H.H. and Ammar, N.M. (1980) Mucilages and pectins of *Opuntia, Tamarindus* and *Cydonia*. *Planta Medica* (Suppl.), 68–75.

Kobayashi, N., Schmidt, J., Nimtz, M., Wray, V. and Schliemann, W. (2000) Betalains from Christmas cactus. *Phytochemistry* 54, 419–426.

Kuti, J.O. (2004) Antioxidant compounds from four *Opuntia* cactus pear fruit varieties. *Food Chemistry* 85, 527–533.

Labuschagne, M.T. and Hugo, A. (2010) Oil Content and Fatty Acid Composition of Cactus Pear Seed Compared with Cotton and Grape Seed. *Journal of Food Biochemistry* 34, 93–100.

Lee, J.C., Kim, H.R., Kim, J. and Jang, Y.S. (2002) Antioxidant property of an ethanol extract of the stem of *Opuntia ficus-indica* var. Saboten. *Journal of Agricultural and Food Chemistry* 50, 6490–6496.

Majdoub, H., Rousdeli, S. and Dertani, A. (2001) Polysaccharides from prickly pear and nopals of *Opuntia ficus-indica*: extraction, characterization and polyelectrolyte behavior. *Polymer International* 50, 552–560.

Matsuhiro, B., Lillo, L.E., Sáenz, C., Urzúa, C.C. and Zárate, O. (2006) Chemical characterization of the mucilage from fruits of *Opuntia ficus-indica*. *Carbohydrate Polymers* 63, 263–267.

Nazareno, M.A., Chaillou, L.L. and González, E.A. (2011) New Insights about Natural Antioxidants in Food. In: Filip, R. (ed.) *Multidisciplinary Approaches on Food Science and Nutrition for the XXI Century*. Editorial Research Signpost, Kerala, India, pp. 91–112.

Nefzaoui, A., Nazareno, M.A. and El Mourid, M. (2008) Review of medicinal uses of cactus. *Cactus net Newsletter* 11, 3–17.

Nemzer, B., Pietrzkowski, Z., Sporna, A., Stalica, P., Threshe, W., Michałowski, T. and Wybraniec, S. (2011) Betalainic and nutritional profiles of pigment-enriched red beet root (*Beta vulgaris* L.) dried extracts. *Food Chemistry* 127, 42–53.

Nieto, M. (1987) Alcaloides de Cactáceas. Estudio de cinco especies argentinas. *Anales de la Asociación Química Argentina* 75, 11–13.

Nobel, P.S. and De la Barrera, E. (2003) Tolerances and acclimation to low and high temperatures for cladodes, fruits and roots of a widely cultivated cactus, *Opuntia ficus-indica*. *New Phytologist* 157, 271–279.

Piattelli, M. and Imperato, F. (1969) Betacyanins of the family Cactaceae. *Phytochemistry* 9, 1503–1507.

Pimienta-Barrios, E. (1994) Prickly pear (*Opuntia* spp.): a valuable fruit crop for the semi-arid lands of Mexico. *Journal of Arid Environments* 28, 1–11.

Pitrè, G. (1896) Medicina Popolare Siciliana. In: Barbèra, G. (ed.) *Medicina Popolare Siciliana*. Biblioteca delle Tradizioni Popolari Siciliane (1949), 2nd edn. Firenze, Italy.

Prakash, D. and Gupta, K.R. (2009) The Antioxidant Phytochemicals of Nutraceutical Importance. *The Open Nutraceuticals Journal* 2, 20–35.

Qiu, Y., Chen, Y., Pei, Y., Matsuda, H. and Yoshikawa, M. (2002) Constituents with radical scavenging effect from *Opuntia dillenii*, Structures of new a-pyrones and flavonol-glycoside. *Chemical and Pharmaceutical Bulletin* 50, 1507–1510.

Ramadan, M.F. and Mörsel, J.T. (2003a) Oil cactus pear (*Opuntia ficus-indica* L.). *Food Chemistry* 82, 339–345.

Ramadan, M.F. and Mörsel, J.T. (2003b) Recovered lipids from prickly pear [*Opuntia ficus-indica* (L.) Mill] peel: a good source of polyunsaturated fatty acids, natural antioxidant vitamins and sterols. *Food Chemistry* 83, 447–456.

Rösler, H., Rösler, U., Mabry, T.J. and Kagan, J. (1966) The flavonoid pigments of *Opuntia lindheimeri*. *Phytochemistry* 5(1), 189–192.

Sáenz, C. (2002) Cactus pear fruit and cladodes: a source of functional components for foods. *Acta Horticulturae* 581, 253–263.

Sáenz, C., Sepúlveda, E. and Matsuhiro, B. (2004) *Opuntia* spp mucilage's: a functional component with industrial perspectives. *Journal of Arid Environments* 57, 275–290.

Sáenz-Hernández, C., Corrales-García, J. and Aquino-Perez, G. (2002) Nopalitos, mucilage, fiber and cochineal. In: Nobel, P. (ed.) *Cacti*. University of California Press, London, UK, pp. 211–234.

Schwartz, S.J., von Elbe, J.H., Pariza, M.W., Goldsworthy, T. and Pilot, H.C. (1983) Inability of red beet betalain pigments to initiate or promote hepatocarcinogenesis. *Food Chemical Toxicology* 21, 531–535.

Sepúlveda, E., Sáenz, C., Aliaga, E. and Aceituno, C. (2007) Extraction and characterization of mucilage in *Opuntia* spp. *Journal of Arid Environments* 68, 534–545.

Stintzing, F.C. and Carle, R. (2004) Functional properties of anthocyanins and betalains in plants, food and in human nutrition. *Trends in Food Science and Technology* 15, 19–38.

Stintzing, F.C. and Carle, R. (2007) Betalains – emerging prospects for food scientists. *Trends in Food Science and Technology* 18, 514–525.

Stintzing, F.C., Schieber, A. and Carle, R. (2001) Phytochemical and nutritional significance of cactus pear. *European Food Research and Technology* 212, 396–407.

Stintzing, F.C., Schieber, A. and Carle, R. (2002) Identification of betalains from yelllow beet (*Beta vulgaris* L.) and cactus pear (*Opuntia ficus-indica* (L.) Mill.) by HPLC-Electrospray Ionization Mass Spectrometry. *Journal of Agricultural and Food Chemistry* 50, 2302–2307.

Stintzing, F.C., Herbach, K.M., Moßhammer, M.R., Carle, R., Yi, W.G., Sellappan, S., Akoh, C.C., Bunch, R. and Felker, P. (2005) Color, betalain pattern, and antioxidant properties of cactus pear (*Opuntia* spp.) clones. *Journal of Agricultural and Food Chemistry* 53, 442–451.

Terrazas-Salgado, T. and Mausseth, J.D. (2002) Shoot anatomy and morphology. In: Nobel, P. (ed.) *Cacti*. University of California Press, London, UK, pp. 23–40.

Terry, P., Lain, M., Miller, A.B., Howe, G.R. and Rohan, T.E. (2002) Dietary carotenoids and risk of breast cancer. *American Journal of Clinical Nutrition* 76, 883–888.

Tesoriere, L., Butera, D., D'Arpa, F., Di Gaudio, M., Allegra, M., Gentile, C. and Livrea, M.A. (2003) Increased resistance to oxidation of betalain-enriched human low density lipoproteins. *Free Radical Research* 37, 689–696.

Tesoriere, L., Butera, D., Pintaudi, A.M., Allegra, M. and Livrea, M.A. (2004) Supplementation with cactus pear (*Opuntia ficus-indica*) fruit decreases oxidative stress in healthy humans: a comparative study with vitamin C. *American Journal of Clinical Nutrition* 80, 391–395.

Tesoriere, L., Butera, D., Allegra, M., Fazzari, M. and Livrea, M.A. (2005) Distribution of betalain pigments in red blood cells after consumption of cactus pear fruits and increased resistance of the cells to *ex vivo* induced oxidative hemolysis in humans. *Journal of Agricultural and Food Chemistry* 53, 1266–1270.

Ting, I. (1997) Carbohydrate Metabolism in Cacti: Gums and Mucilage. *Journal of the Professional Association for Cactus Development* 2, 7–12.

Trachtenberg, S. and Mayer, A.M. (1981) Composition and properties of *Opuntia ficus-indica* mucilage. *Phytochemistry* 20(12), 2665–2668.

Valente, L.M.M., da Paixão, D., do Nascimento, A.C., dos Santos, P.F.P., Scheinvar, L.A., Moura, M.R.L., Tinoco, L.W., Gomes, L.N.F. and da Silva, J.F.M. (2010) Antiradical activity, nutritional potential and flavonoids of the cladodes of *Opuntia monacantha* (Cactaceae). *Food Chemistry* 123, 1127–1131.

Wybraniec, S. and Mizrahi, Y. (2002) Fruit flesh betacyanin pigments in *Hylocereus* cacti. *Journal of Agricultural and Food Chemistry* 50, 6086–6089.

Zakharova, N.S. and Petrova, T.A. (1998) Relationships between the structure and antioxidant activity of certain betalains. *Applied Biochemistry and Microbiology* 34, 182–185.

8 Omega 3 and Omega 6 Fatty Acids in Human Health

Lilia Masson*

Universidad de Chile, Santiago, Chile; Fundación CAPES, Universidad Federal de Rio de Janeiro, Instituto de Nutrición Josué de Castro, Rio de Janeiro, Brasil

8.1 Introduction

Fat and oil structures correspond to triacylglycerol (TAG), in which three fatty acids (FA) are incorporated in one glycerol molecule through an ester union. The distribution of FA in the three positions has great importance from the physiological and technological point of view, such as in digestion, absorption and physical properties such as melting point, solid content and spread properties. Depending on the FA composition, there is either simple or mixed TAG. Three homologous series of FA exist, saturated (SAT), monounsaturated (MUFA; with one double bond in the hydrocarbon chain) and polyunsaturated (PUFA; with two or more double bonds) (Henry, 2009).

On the basis of chain length and number of double bonds, PUFA can either be as long chain polyunsaturated (LCPUFA) or as highly long chain polyunsaturated fatty acids (HUFA). When double bonds are present in the hydrocarbon chain, positional and geometric isomers are formed. These isomers are responsible for important physiological and technological functions (Kraweczk, 2001). MUFA and PUFA are further subdivided according to the position of the first double bond starting from the terminal CH3, then oleic n-9 or ω-9, linoleic (LA) n-6 or ω-6, α-linolenic acid (ALA) n-3 or ω-3 families (Ruiz-Rodriguez *et al.*, 2010).

LA 18:2 9c, 12c, n-6, ALA 18:3 9c, 12c, 15c, n-3 are essential for human organisms as our physiology does not have biological systems to introduce double bonds into the hydrocarbon chain between n-9 position and terminal FA CH3. Therefore our diet must provide these two preformed fatty acids that are necessary to synthesize the LCPUFAs through elongase and Δ5 and Δ6 desaturase, to introduce successive double bonds between n-9 carbon and FA carbon-1 carboxylic group, to synthesize LCPUFAs 20, 22 carbon atoms with three to six double bonds. HUFAs are highly specific and they play an essential physiological role in the human organism (Ferrucci *et al.*, 2006)

It is important to note that a competition in the biosynthesis pathways between LA and ALA through the Δ5 and Δ6 desaturase enzymatic system synthesizes arachidonic acid 20:4 n-6 (AA) and eicosapentaenoic acid 20:5 n-3 (EPA), respectively. These essential fatty acids have special significance in the diet (Simopoulos, 2002).

AA and EPA are the parent molecules for eicosanoid synthesis, which regulates many

* E-mail: masson_lilia@yahoo.es

physiological and important functions in the human organism. From the point of view of human physiology, it is necessary that there should be equilibrium in the synthesis of both parent FA, AA and EPA, which is regulated by the daily dietary intake of LA and ALA (Tapiero *et al.*, 2002). Nowadays, there is a great concern related to the dietary fat consumed by many populations, especially in the occidental world, which presents a great imbalance between its LA and its ALA content. These can have unhealthy effects on populations specifically with regard to non-transmission diseases such as cardiovascular risks (Lands, 2005a).

The role of dietary fats and oils in health is a matter of permanent attention and it continues to be in the first line of biological research. Dietary fats and oils are related with the different physiological roles in human health and development and are associated with cardiovascular risks, inflammatory process, neurological disorders, etc. (Burlingame *et al.*, 2009).

Thus in consideration with the importance of these two essential FAs, this chapter contains more specific information in relation to their healthy implications, metabolism and bioconversion, principal physiological roles, the n-6 n-3 imbalance in dietary sources, daily intake recommendations and common dietary sources in the occidental world. Some special oils are also discussed in the chapter.

8.2 Essential Fatty Acids: Linoleic and α-Linolenic

Until 1929, fats and oils were only considered as the principal and unique source of energetic reserve for humans, but thereafter, the whole scenario changed as fats and oils were found to be a good source of essential fatty acids (EFA) for the animal kingdom. The essential fatty acids identified were linoleic acid (LA) and arachidonic acid (AA), which must be provided daily through the fat intake (Burr and Burr, 1929, 1930). This biological discovery initiated a strong scientific research interest on fats and oils for establishing daily recommendations and physiological functions (FAO, 2010). It was found that the clinical deficiency of essential fatty acids in humans leads to the appearance of characteristic symptoms like dry skin, scale formation, irritation, particularly in the pediatric area, when infants received prolonged parenteral fat-free alimentation. The plasma FA profile showed an increase of eicosatrienoic acid 20:3 n-9 and a great decrease in AA. The relationship between trienic and tetraenic FA increased; this ratio is a biological marker of this deficiency (Caldwell *et al.*, 1972; Paulsrud *et al.*, 1972).

Linoleic acid (LA) was considered as the most important essential fatty acid as the human body could synthesize AA from it. Later ALA was also incorporated as an essential fatty acid (Sinclair *et al.*, 2000). It has been proved that both LA and ALA are essential, since they play an important role in many physiological conditions that require FA with longer carbon chains, belonging to both families of fatty acid, i.e. n-6 and n-3 (Barceló-Coblijn and Murphy, 2009). A critical position about essential FA, definitions and different physiological roles has been discussed by Cunnane (2003).

8.3 Metabolism of LA and ALA, Conversion to HUFAs

8.3.1 General considerations

β-oxidation is the biological mechanism for producing energy in humans from fatty acids. The oxidation rate is related to their structure as short chains oxidize faster than medium and long chain fatty acids (DeLany *et al.*, 2000). Saturated palmitic acid 16:0 is synthesized by '*de novo*' biological mechanism, in a biological process involving successive elongation. Then, new elongation can occur and stearic acid 18:0 is produced followed by the action of enzyme Δ9-desaturase that introduces a double bond at position 9, thus forming oleic acid 18:1 9c, which is not an essential fatty acid (Strawford *et al.*, 2004).

Both LA and ALA are essential fatty acids and they must be provided through the diet.

LA and ALA are metabolized producing two families of HUFAs, ω6 or n-6 derived from LA, ω3 or n-3 derived from ALA. FA in these two families, with terminal structure n-6 and n-3 respectively, have the same initial 9 *cis* structure, however LA has two *cis* double bonds at positions 9 and 12, while ALA 18:3 n-3 or ω3, has three *cis* double bonds in positions 9, 12 and 15 (Sanders, 2009) (Fig. 8.1). These two structural chemical configurations LA and ALA are converted by the enzyme 'elongases' and Δ5 and Δ6 'desaturases', into the HUFAs belonging to each family n-6 and n-3, respectively (Sprecher, 2000). The capacity to metabolize ALA 18:3 n-3 to its longer and polyunsaturated metabolites as 20:5 n-3 EPA, 22:5 n-3 DPA and 22:6 n-3 DHA (HUFAs) is considered an important nutritional and physiological event. In addition, a certain ratio between n-3 and n-6 HUFAs is fundamental for maintaining a good physiological equilibrium (Brenna, 2002; Burdge and Wootton, 2003).

Due to the direct effect of desaturase enzyme, they suffer inhibition by the same unsaturated FA produced; diets high in PUFAs show a tendency to decrease HUFA synthesis through the elongases and Δ5 and Δ6 desaturase enzymatic systems. Probably there exists a narrow regulation mechanism for auto control in the synthesis of HUFAs that have the most labile chemical structure for suffering oxidation *in vitro* and *in vivo*, generating free radicals, which are very aggressive molecules. From the biological point of view, it is correct that HUFA synthesis be auto-controlled at the first level (Blair, 2001).

LA from the n-6 family produces specific fatty acid as dihomo-gamma-linolenic acid 20:3 n-6 (DHGLnA) through the enzymatic elongation system and then arachidonic acid 20:4 n-6 (AA). Both represent the parent molecules for the synthesis of specific eicosanoids. New elongations and unsaturation continues and docosapentaenoic acid 22:5 n-6 is formed (Lands, 2005a). For the n-3 family, the same metabolic route is available, the same enzymatic elongation and Δ5 and Δ6 desaturases work for the synthesis of the n-3 HUFAs through the route stearidonic acid 18:4 n-3, 20:4 n-3, EPA 20:5 n-3, docosapentaenoic acid n-3 DPA 22:5 n-3, followed by 24:5 n-3, 24:6 n-3, finishing with DHA 22:6 n-3 (Sprecher, 2000; Burdge and Wootton, 2002).

Dietary ALA has two different biological routes: a fast energy source through β-oxidation or to enter biochemical pathways to synthesize their HUFAs metabolites. The percentage of ALA involved in β-oxidation is around 30% of the intake value according to other n-3 PUFAs. This is a normal biochemistry process called carbon recycling. It means that carbons coming from β-oxidation can be used through *de novo* synthesis of other FA needed (Cunnane *et al.*, 2003). The conversion of ALA to palmitic, palmitoleic, stearic and oleic acids in men and women has been described by Burdge and Wootton (2003). The priority decision of each one of these physiological alternatives is determined by the specific physiological situations that humans must afford during their different development steps, from newborn until old age. DHA and AA are so important for human development that both are biosynthesized from their 18 carbon precursors in human infants (Salem *et al.*, 1996). It has been reported that gender can have a role in ALA conversion to DHA in young women; this conversion was higher than in men. This result is related to the presence of oestrogens, which could stimulate the Δ6 and elongase enzymatic systems activity (Burdge *et al.*, 2002). Burdge and Wooton (2002) considered a mean ALA intake of 1.5 g day^{-1} for both sexes, the reserve for men would be around for 53 days and for

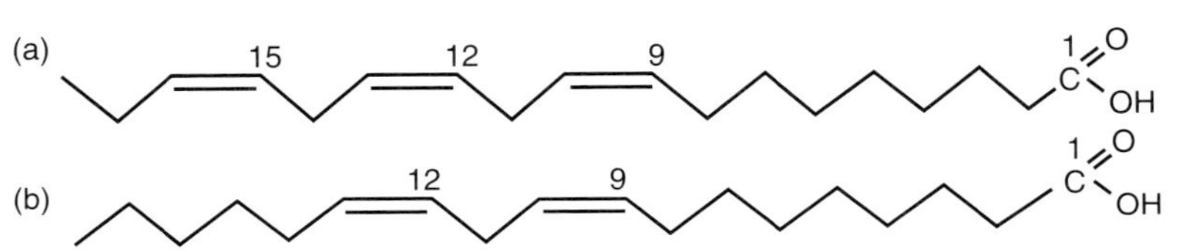

Fig. 8.1. (a) Structure of α-linolenic acid and (b) linoleic acid.

women around 70 days, suggesting a possible gender difference related to ALA metabolism; it is possible that during pregnancy the conversion of ALA to DHA is more efficient (Burdge, 2004). This high synthesis of DHA in pregnant and lactating women has been related with the high demands from the fetus for DHA for the brain, adipose tissue and liver. Vegans show a higher activity of Δ6 and Δ5 desaturases, which probably favours the conversion of ALA to DHA. Its absorption in the human intestine is considered good and its bioavailability is efficient (Burdge and Calder, 2005; Burdge, 2006).

8.3.2 Conversion ALA to HUFAs n-3

Conversion steps n-6 and n-3 are similar but independent. Both use the same enzymatic systems, with each substrate LA and ALA competing between them for the synthesis of their respective HUFAs. As LA is more abundant in the majority of the vegetable oils consumed in the world, LA dietary intake is higher than ALA, and as a biological consequence more HUFAs n-6 are formed related to n-3 HUFAs synthesized (D'Andrea *et al.*, 2002).

The biochemistry pathways for LA and ALA conversion to their respective HUFAs are clearly known in humans but the question is how much ALA is converted to LCPUFAs. A competition exists between n-6 and n-3 families; the step for ALA 18:3 n-3 conversion to 18:4 n-3, stearidonic acid, by the Δ6 desaturase, is considered the limiting reaction, Δ6 desaturase presenting more affinity for ALA than LA (Simopoulos, 2002; Burdge and Calder, 2005). At cellular level, if LA is at a higher concentration than ALA, then the conversion to its next n-6 PUFA metabolite is preferred, confirming that the relationship between dietary LA and ALA influences the conversion to their respective PUFAs metabolites (Barceló-Coblijn and Murphy, 2009). In the conversion of n-3 to DHA, after 24:6 n-3 FA is synthesized, only one β-oxidation cycle is produced, and DHA 22:6 n-3 is formed; it represents the critical point for the metabolic regulation that controls DHA synthesis independent of the previous steps (Sprecher, 2000). Goyens *et al.* (2006) have demonstrated that the total ALA and LA amounts in diets have a higher influence in the ALA conversion to their derivatives than the ratio ALA:LA. Diets supplemented with high amounts of ALA showed low but significant increases in EPA and DPA in blood level; conversion of ALA to DHA in infants was around 1%, very low in adults. The conversion of ALA to LCPUFAs was decreased by high ratios LA/ALA, while n-3 PUFA levels can be improved by increasing ALA or decreasing LA intake (Brenna *et al.*, 2009). Dietary DHA increased blood and tissue levels much more than any n-3 precursor. The bioconversion from ALA to DHA is low; the percentage range is between 0.05% and 4%; the conversion of ALA to EPA is also low, i.e. around 8% (Barceló-Coblijn and Murphy, 2009). Using the stable isotopes procedure, it has been estimated that the conversion percentage from ALA to EPA is 0.2%, to DPA n-3 is 0.13% and to DHA 0.5% (Emken, 2001; Pawlosky *et al.*, 2001). ALA has seven metabolic pathways to finally generate DHA; each one has its own conversion percentage, and each FA formed in this cascade has its own biological function and its own metabolism. LA has only two metabolic pathways to form AA and it is superior to ALA in the habitual occidental diet (Burdge, 2006; Barceló-Coblijn and Murphy, 2009).

In other studies where the diet was modified and supplemented with different amounts of ALA or using special oil rich in this essential fatty acid, some improvements in HUFA levels were observed (Blank *et al.*, 2002; Brenna, 2002; Ghafoorunissa *et al.*, 2002). Austria *et al.* (2008) studied the ALA availability from whole flaxseed, milled flaxseed and flaxseed oil in subjects; the results indicated that milled flaxseed and flaxseed oil delivered significant levels of ALA to the plasma compared to whole flaxseed; some subjects presented adverse gastrointestinal effects with whole flaxseed and flaxseed oil preparation. Artenburg *et al.* (2006), considering that in humans plasma lipids represent the dietary intake of LCPUFAs, studied the effect of diet supplemented with n-3 FA; for EPA the increase was linear; for DHA the maximal plasma response was obtained with a dose of approximately 2 g

DHA day^{-1}; with ALA supplementation EPA increased but not DHA.

In relationship to the conversion efficiency and the ratio LA:ALA in the diet, studies in animals have shown that the best results were obtained with the ratio between LA:ALA 4:1 to 2:1 (Blank *et al.*, 2002). Abbott *et al.* (2010) studied rat skeletal muscle FA phospholipids using 12 moderated fat diets with a wide range of percentages of LA and ALA. The results indicated that the low diet PUFA balance had the greater influence in the FA composition than LA and ALA alone, which means both FA are necessary but in an adequate proportion. On the other hand, the hepatic bioconversion of ALA to EPA and DHA was tested in rats fed with two seed oils with a high content of ALA and a different ratio LA:ALA: chia (*Salvia hispanica*) and rosa mosqueta (*Rosa rubiginosa*) (González-Mañan *et al.*, 2012; Valenzuela *et al.*, 2012). The results indicated that hepatic levels of ALA, EPA and DHA increased, compared with the control oils used; with sunflower and a mix of olive oil with fish oil, the ratio n-6:n-3 decreased with respect to sunflower oil; hepatic damage was also observed and rat liver conversion has also been studied (Fei *et al.*, 2011; Gibson *et al.*, 2012). In rats, DHA synthesis from ALA is inhibited by diets high in PUFAs. The scientists assayed different blends of plant oils in rats with increasing PUFA content and demonstrated that DHA levels were maximum when PUFAs represented less than 2% energy; this means that the present human diet is exceeded in 18 carbon PUFAs (Tu *et al.*, 2010). Gibson *et al.* (2012) reported that the synthesis of HUFA n-3 is more regulated by the level of the substrate than by gene expression (Burdge and Calder, 2005; Barceló-Coblijn and Murphy, 2009).

8.4 Eicosanoids and Docosanoids

Eicosanoids are a group of molecules with 20 carbons derived from three HUFAs: arachidonic acid (AA) 20:4 n-6, dihomogammalinolenic acid (DHGLnA) 20:3 n-6 and EPA 20:5 n-3; the more active eicosanoids are formed from AA and EPA. Two enzymes participate in eicosanoid synthesis: cyclooxygenase forms the prostanoids known as prostaglandins, prostacyclines and thromboxanes; lipooxygenase produces hydroxyperoxytetraenoic acids, which pass to leukotrienes, hydroxyeicosatetraenoic acids and lipoxins (Lands, 2005b). The three parent fatty acids, AA, DHGLnA and EPA, are incorporated in cellular membrane phospholipids normally in the 2 position; they must be liberated by the phospholipase in a competitive way according to their concentration in the respective phospholipid membrane. Normally AA is more abundant than EPA, DHGLnA and DHA, in which case the first option will be the synthesis of eicosanoids from AA; if the diet is high in EPA and DHA, these HUFAs n-3 can inhibit AA eicosanoid synthesis (Lands, 2008). New docosanoids synthesized from DHA called resolvins and protectines have been described by Serhan *et al.* (2002).

The principal biological functions of the eicosanoids from n-6 and n-3 HUFAs are related with AA in pro-inflammatory processes and with EPA and DHA in anti-inflammatory processes, respectively, affecting the inflammatory activity of cytokines. They are also involved in enzymes related with lipid metabolism through the gene activation of fatty acid transport and oxidation (Duplus *et al.*, 2000). Eicosanoids represent the molecules involved in a complex net of cell communication in the animal organism, mediators in the central nervous system with inflammatory process regulation, tissue constriction, relaxation, immune response and platelet aggregation. Some n-3 eicosanoids have the opposite effect related to those n-6 eicosanoids, which requires that at the membrane of the specific tissue, both parent molecules LA and ALA be present in the adequate relationship. This antagonism is the key for getting the expected harmonized physiological response (Bazan, 2007).

8.5 Healthy Biological Effects of n-6 and n-3 Fatty Acids

8.5.1 Cardiovascular disease

This has been the most investigated subject of the physiological role that different FA could play in this non-transmissible disease (CVD) (Lands, 2003b; WHO, 2003). The positive effect of LA in human diet is to decrease LDL-cholesterol, one important biomarker of cardiovascular risk. This effect was first published by Hegsted *et al.* (1965) and Keys *et al.* (1965) and since then it has been demonstrated in numerous publications. It was recommended that the consumption of saturated FA with carbon chain superior to 10 should be decreased due to some saturated FA that increased plasma LDL; and the daily intake of LA n-6 present in high amounts in vegetable oils such as maize, sunflower, safflower, should be increased where ALA is practically absent (Masson and Mella, 1985). These recommendations had a very important impact in the change of consumers' diet through to the present day. The result has been due to an imbalance between LA and ALA that induced a more inflammatory response. High controversy and discussion about this imbalance has been produced in recent years and will be commented on later (Lands, 2003a; Simopoulos, 2006, 2008).

N-3 FA also has a positive biological effect in CVD (Mozaffarian, 2005). It is therefore necessary to differentiate between the beneficial effects of ALA from EPA and DHA. In the case of ALA, the dietary intake of moderate amounts of essential FA has beneficial effects related with prevention of coronary heart disease (Djousse *et al.*, 2001; Hu *et al.*, 2001). Bloedon *et al.* (2008) used flaxseed meal, which has ALA, to test cardiovascular risk factors in humans; the results indicated a modest and short effect lowering LDL-cholesterol. In an another study by Bhathena *et al.* (2003) using flaxseed meal, to test the effect in hypertriglyceridaemia and liver steatosis in an animal model, the results indicated positive effects. Vuksan *et al.* (2007) supplemented conventional therapy with grain of *Salvia hispanica*, a seed with very high ALA content, in individuals with type-2 diabetes; the results indicated an improvement in major and emerging cardiovascular risk factors in type-2 diabetes. Ayerza and Coates (2005) tested chia seed and chia oil in rats and their effects in plasma lipids and fatty acids. The results indicated that chia diets decreased TAG levels, increased HDL-cholesterol and ω-3 FA content in rat serum. Similarly, Chicco *et al.* (2009) demonstrated that chia seed rich in α-linolenic acid reduced adiposity and normalized hypertriacylglycerolaemia and insulin resistance in the dyslipaemic rats studied.

In relation to LCPUFAs n-3, EPA and DHA, it is important to note that when in 1970 marine origin oil appeared as the main source of LCPUFAs n-3, particularly EPA and DHA, ALA took another biological dimension. Bang and Dyerberg (1972) published their results related to an Eskimo population that had a high fat intake, around 40% of total energy. They have very low ALA and thus are more prone to bleeding by accidents; their levels of serum cholesterol, LDL and HDL were more equilibrated than a Danish population. In addition, the dietary cholesterol intake was more than the double the recommended daily intake, due to their high consumption of marine mammals. The explanation was found in the quality of the dietary fat. Marine fats have a predominance of n-3 FA, the most important being 20:5 n-3 (EPA) and 22:6 n-3 (DHA); in addition they contain different n-3 as 22:5 n-3 DPA, stearidonic acid 18:4 n-3, very low ALA, and from n-6 family they contain AA 20:4 n-6 and very low LA (Lands, 2005b; Moreau and Kamal-Eldin, 2009; Rincón-Cervera *et al.*, 2009).

Studies on the protective role of fish lipids in cardiovascular disease proved that fish lipids play an important role in the prevention of cardiovascular diseases and atherogenesis principally by lowering serum triacylglycerol level, along with other beneficial effects (Masson *et al.*, 1990; Gissi, 2008; Strijbosch *et al.*, 2008). EPA is the parent molecule for the synthesis of the parallel AA eicosanoids cascade; thromboxane TXA3 had an opposite effect to TXA2 produced from AA on platelet aggregation, which explained why the Eskimo population suffered more frequent accidents by bleeding than thrombosis. They had a contrary imbalance; n-3 intake was higher than n-6, with their high consumption of marine foods and low sources of LA. It was also shown that the LCPUAs n-3 pre-formed, present in marine origin foods as EPA and DHA, were fundamental for human health and they are part of many important physiological process (Hu *et al.*, 2001).

DHA has its own highly specific physiological and fundamental functions in CVD. It has been concluded that DHA can have anti-inflammatory activity and anti-arrhythmic effect due to the presence of resolvins and protectines (Serhan *et al.*, 2002, 2008). Various cardiologist societies have emphasized their importance and have recommended the consumption of EPA and DHA for the prevention of cardiovascular complications. Perpetual intake is recommended for individuals suffering from myocardial infarct and for preventing sudden death by cardiovascular failure (Gebauer *et al.*, 2006). Therefore it is recommended for all populations to increase their fish intake, especially fatty fish, which are excellent dietary sources of EPA and DHA, such as sardine, anchovy, horse mackerel, trout, salmon, etc. According to production in every country, its consumption has been enhanced and considered in the plans for improving health of the population (Burlingame *et al.*, 2009). Presently, functional foods, enriched in DHA and structured lipids with EPA and DHA, are being developed (Villeneuve *et al.*, 2000).

8.5.2 Pregnancy and lactation

There are two transcendent and important physiological statuses for women, pregnancy and lactation. During these periods there is a mandatory requirement of n-3 HUFAs, mainly pre-formed DHA, for brain formation and development, vision, together with AA n-6 (Brenna *et al.*, 2007). The DHA transfer from the pregnant woman to the fetus via the placenta and for the infant during the lactating period through maternal milk is a point of high relevance for infant development (Uauy *et al.*, 2000a, b). The enrichment of milk formulas with DHA and other foods enriched with EPA and DHA have been developed (FAO, 2010). According to Sanders and Reddy (1992), diet supplementation with ALA is not effective for increasing the DHA breast milk level and vegetarian's breast milk has a low DHA content, therefore their recommended intakes will be commented on separately. In rats, the maternal intake of flaxseed-based diets improved spatial memory to the detriment of growth; the authors recommend caution in encouraging maternal intake of flaxseed during human pregnancy and lactation (Fernandes *et al.*, 2011).

8.5.3 Inflammatory processes

In humans, asthma, inflammatory bowel disease and rheumatoid arthritis are related to AA, EPA and DHA. Eicosanoids and docosenoids play an important physiological role. Special interest has been focused on resolvins and protectins due to their relationship with HUFAs and anti-inflammatory activity (Serhan *et al.*, 2002, 2008; Calder, 2006, 2009; Bazan, 2007).

8.5.4 Brain

Another important physiological role of HUFAs n-3 and n-6 in humans is related to brain development, functions and its composition (Innis, 2000). It has been shown that EFAs are fundamental in early life for structural and functional roles in brain and visual functions; in pre-term infants DHA demand is very high and must be present in infant formulas. The ratio between dietary LA:ALA intake was also evaluated in growth and visual function in term infants (Uauy *et al.*, 2000a, b).

According to Crawford *et al.* (1976), the diet should provide a proportion of n-6:n-3 between 2:1 and 1:1. Brenna *et al.* (2009) have discussed the DHA requirements for brain and ALA conversion. Estimation for adult's brain requirement for HUFAs is around 18 mg of AA and 5 mg of DHA per day. A cross-national comparison of seafood consumption and rates of bipolar disorders is described by Noaghiul and Hibbein (2003). Different disorders can affect brain functions with adverse health consequences for human health as in Alzheimer's disease, macular degeneration, different expressions of depression, etc., where HUFAs can have an important role (Freeman *et al.*, 2006). DHA is the most important brain lipid due to its specific physiological role in vision and the retina, other organs and seminal glands. In pre-term infants with very low weight, serious alterations in the retina have been detected, which can be reverted by supplementing their diet with DHA. All these important physiological events where LCPUFAs are participating in specific physiological and pathological events, seriously affect the human organism (Cunnane, 2000; Sinclair *et al.*, 2000); DHA intake through diet is therefore important (Brenna *et al.*, 2009).

Much research related to brain lipid classes and their fatty acid composition has been carried out in experimental animals. Sjövall *et al.* (2004) studied the spatial imaging of lipids in rat brain tissue applying time-of-flight secondary ion mass spectrometry (TOF-SIMS). The results showed a localization of high cholesterol and phosphatidyl-choline and large variations in the lipid composition in different regions. Kim *et al.* (2011) studied PUFA metabolism in rat brain with dietary n-3 PUFA deprivation; the results indicated that brain maintains the control of DHA concentration. Rapoport *et al.* (2011) published a very interesting review related to DHA incorporated into the brain of rodents and humans, thus developing a method to obtain images. DHA brain content and metabolism is dependent on the diet and the ability of liver to synthesize DHA from circulating ALA. Due to the complex matrix associated with brain lipids, improved analytical methodologies have been developed, coupled and applied to brain lipid analysis in animals such as Matrix-Assisted Laser Desorption Ionization Time-of-Flight Mass Spectrometry and High-Performance TLC (Fuchs *et al.*, 2008). This methodology known as MALDI-TOF-MS has the advantage of minor fragmentation of the sample. It was applied to a crude extract of porcine brain lipids. The principal phospholipid classes were separated and the two FA present were detected; for example in phosphatidyl-choline 16:0/16:0; 16:0/18:1; 18:0/18:1; in phosphatidyl-ethanolamine 16:0/18:1; 18:1/20:4; 18:0/18:1; in sphingomieline 24:1; in galactosyl-ceramide 24:0/18:1 etc. Detailed information about the methods is included in the publication but quantification was not considered. Astigarraga *et al.* (2008) studied brain and liver tissues in rats using MALDI-TOF-MS methodology. They identified a large number of lipid species. Using the novel methodologies in lipid analysis, especially in complex matrix such as brain, lipidomics permits the study of the composition of intact molecular species in biological systems combining different advanced instrumentation techniques as different mass spectrometry procedures, stable isotopes labeling, mass spectrometric imaging, data analysis and bioinformatics, etc. (Wenk, 2005; Postle, 2012).

In addition, docosapentaenoic acid n-3 and n-6 (DPA), also present in fish oils, have been studied for their biochemical properties (Simopoulos, 2008). Other novel physiological roles of HUFAs from fish oils related to their functional effect to modify the membrane lipid rafts, improving flexibility and producing low affinity for cholesterol and saturated acyl chains; the possibility to apply these findings in clinics is described by Shaikh (2012).

8.5.5 Cancer

There are several publications related to n-6, n-3 FA and cancer and it is estimated that cancer can have a relationship with inflammatory processes as a response to oxidative tissue damage (Kushi and Giovanuce, 2002; Espada *et al.*, 2007). The effects of chia oil (*Salvia hispanica*), which is high in ALA, safflower oil (*Carthamus tinctorious*), which is high in LA, and a control diet were compared in mice, in relation to eicosanoids and metastasis of a murine mammary gland adenocarcinoma. The results indicated that chia oil diet inhibited growth and metastasis in the tumour model.

8.6 Dietary Recommendation for n-6 and n-3 Fatty Acids

FAO (2010) contains dietary recommendations for total fat, saturated, *trans-*, polyunsaturated FA, LA, ALA and their metabolites EPA and DHA, for adults, pregnant and lactating women, infants 0–2 years old and children 2–18 years old. The recommendations are expressed as energy percentage. For adults, total fat intake 20–35% (Elmadfa and Kornstainer, 2009), minimum intake for LA and ALA to prevent deficiency is 2.5% plus 0.5%, respectively; minimum total PUFAs for reducing cardiovascular risk is 6%. Another international report is also available in Eurodiet (2000).

All the recommendations are given in percentage of energy for specific and separate fatty acids. Each fat and oil has its own FA composition and own TAG fatty acid distribution, considering that position-2 has also biological significance.

8.7 Dietary Imbalance LA:ALA

The LA:ALA imbalance principally in the occidental diet is an actual controversy. According to Simopoulos (2006, 2008), the occidental diet is high in LA intake compared with ALA intake. The high oilseed production in the world has contributed to this situation. The ratio LA:ALA has increased many times, producing an unfavourable physiological condition for inducing high synthesis of AA and more inflammatory events. In ancient times, the equilibrium in the human diet between n-6 and n-3 PUFAs was around 1:1 (Crawford *et al.*, 1976; Eaton and Konner, 1985). The principal PUFAs in foods are LA and ALA; their HUFA metabolites can be mediators affecting diet-dependent diseases. Ethnic groups in Japan have maintained their diet on n-3 HUFAs as 53% and n-6 HUFAs as 47% in plasma phospholipids. In the individuals of Detroit, this is inverse: 18% n-3 and 82% n-6. Cardiovascular deaths are correlated with these inverted percentages (Lands, 2008).

Pischon *et al.* (2003) compared the intake of EPA and DHA in a group of healthy men and women with inflammatory response in plasma, measuring different biomarkers. EPA + DHA intake in men was around 55–1120 mg day^{-1}, for women it was around 22–471 mg day^{-1}, LA intake in men was around 12 g day^{-1} and in women around 9 g day^{-1}; in both groups ALA intake was around 1 g day^{-1}. The results indicated an inverse relationship between EPA+DHA n-3 intake and inflammatory biomarkers in plasma, combining both types of FA n-6 and n-3 lowest levels of inflammations. Hibbeln *et al.* (2006) estimated a healthy dietary allowance for n-3 LCFAs for the US population. They evaluated LA, AA, ALA, EPA, DPA and DHA content of food commodities in 38 countries and correlated the composition tissue data to each illness according to a model. Fish and seafoods continued to be the best dietary source of EPA+DHA; seed oils, eggs, poultry and pig meat had the greatest amounts of n-6 FA. The n-3 LCPUFA intake in Japan is 0.37% energy corresponding to 750 mg day^{-1} and met the criteria for uniformly protecting >98% of the population worldwide. Ambring *et al.* (2006) investigated Mediterranean diet (MID) in serum PL n-6, n-3 FA content inflammatory endothelial indexes in healthy subjects in the Gotemburg area compared with traditional Sweden diet (OSD). The difference between MID and OSD diet was in the higher fibre, more antioxidants, fat quality, twice amount of n-3 FA, one-half saturated FA, one-half cholesterol, and sterol esters 2 g day^{-1} were included in margarine. Results of MID diet indicated reduced platelets, leucocytes and VECF concentration thus indicating a beneficial influence of MID diet in these inflammatory parameters, which may be linked to the ratio n-6:n-3 FA reduction.

Ramsden *et al.* (2010) critiqued a consumption of at least 5–10% of energy as n-6 PUFA to reduce CHD risk, based on the results of randomized controlled trials (RCT). The authors reviewed all the RCT that increased PUFAs and, after the detailed evaluation, they concluded that n-6 specific PUFA diets tended to increase CHD risks. They recommended decreasing the intake of n-6 PUFA in the diet. Blasbalg *et al.* (2011) evaluated essential FA consumption changes in the USA during the years 1909–1999 and it was found that apparent increased consumption of LA, primarily from soybean oil, decreased EPA and DHA in tissues during the 20th century. In this important debate, Harris *et al.* (2009) presented the position of the American Heart Association (AHA) relating to proposals recommending the reduction of omega-6 PUFA in diet without considering many publications suggesting that higher intakes of n-6 PUFAs reduce CHD risk. The current AHA recommendations for LA are 5–10% total energy. Kriss-Etherton *et al.* (2010) supported the same concept related to maintain higher LA dietary intake, as per AHA recommendations.

The presence of both precursors LA and ALA in adequate proportion is fundamental to reach a harmonized physiological status (Kushi and Giovanuce, 2002; Serhan *et al.*, 2002; Simopoulos, 2002; Lands, 2003b, 2008; Calder, 2006; Bazán, 2007; Sanders, 2009).

8.8 Dietary Sources of LA and ALA

8.8.1 Vegetable sources

As it has been commented above that the occidental diet has changed principally in the LA n-6:ALA n-3 ratio probably from 1:1 to a high predominance of LA, the world production of vegetable oils according to FAO (2010) indicates four species, palm (*Elaeis guineensis*), soybean oil (*Glycine hispida*) canola, rapeseed oil (*Brassica* sp.) and sunflower seed oil (*Helianthus annus*) as principal oil crops. There are other oil sources produced on a minor scale. Their FA composition and LA:ALA ratio are as follows: palm oil is principally saturated, ratio 50:1; soybean oil principally polyunsaturated, ratio 8:1; Canola oil more equilibrated FA composition, highly monounsaturated, ratio 2.5:1; and sunflower seed oil highly polyunsaturated ratio 345:1 (Masson and Mella, 1985). The consumers do not have much option for selecting a more diversified fatty acid composition.

Some alternatives are to stimulate the culture of seed oils with a better FA profile, to combine oils and to search in nature for other better options that can be present in some uncommon seeds (Simopoulos, 2004; Rodriguez-Pérez, 2005; Masson *et al.*, 2008; Moreau and Kamal-Eldin, 2009).

A classification of 80 vegetable oils according to their principal and secondary FA related cholesterol-lipoproteins effect was carried out by Dubois *et al.* (2007). There are three main classes, SAT, MUFA and PUFA, that are further subdivided into sub-classes. In the PUFA class, there are five sub-classes: LA>60%, LA+SAT, LA+MUFA, ALA+MUFA and ALA+LA. This classification gives a wide view of the great diversity in nature for FA composition in seeds and fruits and can help to search special seed oils whose fatty acid composition and ratio LA:ALA is very particular.

There are some special seed oils with more than 7% ALA, whose n-6:n-3 ratio are the following: quinoa (*Chenopodium quinoa* Wild.) ratio 6:1; walnut (*Juglans regia*) ratio 5.5:1; rapsberry (*Rubus idaeus*) ratio 2.4:1; lupine (*Lupinus albus*) ratio 2.4:1; rosa mosqueta (*Rosa rubiginosa*) ratio 1.3:1; linseed or flaxseed (*Linum usitatissimum*) ratio 0.4:1 (Masson and Mella, 1985); perilla (*Perilla frutescens* L.) ratio 0.3:1; hempseed (*Cannabis sativa*) ratio 3:1 (Firestone, 2006; Moreau and Kamal-Eldin, 2009); and chia (*Salvia hispanica*) ratio 0.3:1 (Peiretti and Gai, 2009). Some of these high ALA content seed oils have been assayed by researchers in humans and experimental animals to search their biological behaviour compared with traditional oils. Martin *et al.* (2006) published a selection of different Brazilian foods giving the n-6:n-3 ratio of their lipids.

8.8.2 LA and ALA: animal terrestrial and marine sources

The sources are principally meats from different origins like avian, marine, rivers, lakes, ruminants, herbivores, omnivores, hen eggs, milk and their derivatives (Speedy, 2003). Ruminant animal depot fats are low in LA and much lower in ALA (Masson and Mella, 1985). In fish and seafoods, principally EPA, DPA and DHA are present; amongst fat fishes sardine, anchovy, horse mackerel, caballa, salmon and trout are recognized as the best food sources (Firestone, 2006; Moreau and Kamal-Eldin, 2009). Each species has its own FA profile; in large scale production, the diet is formulated with raw material available in the respective country or region and the fat depots reflect dietary fats (Speedy, 2003; Scollan *et al.*, 2006; Wood *et al.*, 2008). The FA composition can present a wide variation amongst the same food produced by different food industries and the composition depends on the formulation of industrially processed foods (Paeratakul *et al.*, 2003). Non-traditional sources of EPA and DHA are micro-algal, some of which are good sources of these HUFAS (Guil-Guerrero *et al.*, 2000; Andrich *et al.*, 2005; Moreau and Kamal-Eldin, 2009).

8.9 Conclusion

Fats and oils are important macronutrients for human health. They are the only source of energy to survive in extreme physiological conditions and are the dietary source of the parent essential fatty acids linoleic n-6 and α-linolenic n-3. Many important beneficial physiological functions are performed by both essential FA and their long chain polyunsaturated derivatives, principally AA, EPA and DHA and the eicosanoids and docosanoids synthesized from these parent molecules. AA, EPA and DHA also have proper physiological functions in important human tissues, especially AA and DHA in brain development and function. The requirements to maintain equilibrium between dietary intake and tissue storage are not so high, but the important thing is that in the diet equilibrium must exist between both LA and ALA because human physiology was genetically organized in an environmental media where the intake of both FA was equilibrated. This equilibrium has been broken in the last 100 years. The human physiological response has been according to this imbalance producing more inflammation compounds, which can stimulate the induction of non-transmissible diseases among different populations. This problem affects principally occidental countries with the exception of the Mediterranean region. Some action must be taken to advise the population to decrease LA intake, to stimulate the intake of natural sources of ALA and to enhance the consumption of fish and seafoods.

Nature provides the highest biodiversity in food supplies, using their intelligence; men must select the best choices for a healthy life.

References

Abbott, S.K., Else, P.L. and Hulbert, A.J. (2010) Membrane fatty acid composition of rat skeletal muscle is most responsive to the balance of dietary n-3 and n-6 PUFA. *British Journal of Nutrition* 103, 522–529.

Ambring, A., Johansson, M., Axelsen, M., Gan, L.M., Strandvik, B. and Friberg, P. (2006) Mediterranean-inspired diet lowers the ratio of serum phospholipids n-6 to n-3 fatty acids, the number of leukocytes and platelets, and vascular endothelial growth factor in healthy subjects. *American Journal of Clinical Nutrition* 83, 575–581.

Andrich, G., Nesti, U., Venturi, F., Zinnai, A. and Fiorentini, A. (2005) Supercritical fluid extraction of bioactive lipids from the microalga *Nannochloropsis* sp. *European Journal Lipid Science Technology* 107, 381–386.

Artenburg, L.M., Hall, E.B. and Oken, H. (2006) Distribution, interconversion and dose response of n-3 fatty acids in humans. *American Journal of Clinical Nutrition* 83(Suppl.), 1467S–1476S.

Astigarraga, E., Barreda-Gómez, G., Lombardero, L., Fresnedo, O., Castaño, F., Giralt, M.T., Ochoa, B., Rodriguez-Puertas, R. and Fernàndez, J.A. (2008) Profiling and Imaging of Lipids on Brain and Liver Tissue by Matrix-Assisted Laser Desorption/Ionization Mass Spectrometry Using 2. Mercaptobenzothiazole as a Matrix. *Analytical Chemistry* 80, 9105–9114.

Austria, J.A., Richard, M.N., Chahine, M.N., Edel, A.L., Malcolmsom, L.J., Dupasquier, C.M.C. and Pierce, G.N. (2008) Bioavailabity of Alpha-Linolenic Acid in Subjects after Ingestion of Three Different Forms of Flaxseed. *Journal of the American College of Nutrition* 27, 214–221.

Ayerza, R. and Coates, W. (2005) Ground chia seed and chia oil effects on plasma and fatty acids in the rat. *Nutrition Research* 25, 995–1003.

Bang, H. and Dyerberg, J. (1972) Plasma lipids and lipoprotein in Greenlandic west coast Eskimos. *Acta Medica Scandinava* 192, 85–94.

Barceló-Coblijn, G. and Murphy, E.J. (2009) Alpha-linolenic acid and its conversion to longer chain n-3 fatty acids. Benefits for human health and a role in maintaining tissue n-3 fatty acids level. *Progress in Lipid Research* 48, 355–374.

Bazan, N.G. (2007) Omega 3 fatty acids pro-inflammatory signaling and neuroprotection. *Current Opinion Clinical Nutrition Metabolism Care* 10, 136–141.

Bhathena, S.J., Ali, A.A., Haudenschild, C., Latham, P., Ranich, T., Mohamed, A.I., Hansen, C.T. and Velasquez, M.T. (2003) Dietary Flaxseed Meal is More Protective Than Soy Protein Concentrate Against Hypertriglyceridemia and Steatosis of the Liver in an Animal Model of Obesity. *Journal of the American College of Nutrition* 22, 157–164.

Blair, I.A. (2001) Lipid Hydroperoxide–Mediated DNA Damage. *Experimental Gerontology* 36, 1473–1481.

Blank, C., Neumann, M.A., Makrides, M. and Gibson, R.A. (2002) Optimizing DHA levels in by lowering the linoleic acid to α-linolenic acid acid ratio. *Journal of Lipid Research* 43, 1537–1543.

Blasbalg, T.L., Hibbeln, J.R., Ramsden, C.E., Majchrrrzak, S.F. and Rawlings, R.R. (2011) Changes in consumption of omega-3 and omega-6 fatty acids in the United States during the 20th century. *American Journal of Clinical Nutrition* 93, 950–962.

Bloedon, L.T., Bulikai, S., Chittans, J., Cunnane, S.C., Berlin, J.A., Rader, D.J., Philippe, M.D. and Szapary, O. (2008) Flaxeed and Cardiovascular Risk Factors: Resuts from a Double Blind, Randomized, Controlled Clinical Trial. *Journal of the American College of Nutrition* 27, 65–74.

Brenna, J.T. (2002) Efficiency of conversion alpha-linoleic acid to long chain n-3 fatty acids in man. *Current Opinion Clinical Nutrition Metabolism Care* 5, 127–132.

Brenna, J.T., Varamini, B., Jensen, R.J., Diersen-Schade, D.A., Boettcher, J.A. and Arterburn, L.M. (2007) Docosahexaenoic acid and arachidonic acid concentrations in human breast milk worldwide. *American Journal of Clinical Nutrition* 85, 1457–1464.

Brenna, J.T., Salem, N. Jr, Sinclair, A.J. and Cunnane, S.C. (2009) α-Linolenic acid supplementation and conversion to n-3 long chain polyunsaturated fatty acids in humans. *Prostaglandins, Leukotrienes and Essential Fatty Acids* 80, 85–91.

Burdge G.C. (2004) Alpha-linolenic acid metabolism in men and women: nutritional and biological implications. *Current Opinion Clinical Nutrition Metabolism Care* 7, 137–144.

Burdge, G.C. (2006) Metabolism of alpha-linolenic acid in human. *Prostaglandins, Leukotrienes and Essential Fatty Acids* 75, 161–168.

Burdge, G.C. and Calder, P.C. (2005) Review. Conversion of α-linolenic acid to longer-chain polyunsaturated fatty acids in human adults. *Reproductive Nutrition Development* 45, 581–597.

Burdge, G.C. and Wootton, S.A. (2002) Conversion of α-linolenic acid to eicosapentaenoic, docosapentaenoic and docosahexaenoic acids in young women. *British Journal of Nutrition* 88, 411–420.

Burdge, G.C. and Wootton, S.A. (2003) Conversion of α linolenic acid to palmitic, palmitoleic, stearic and oleic acids in men and women. *Prostaglandins, Leukotrienes and Essential Fatty Acids* 69, 283–290.

Burdge, G.C., Jones, A.E. and Wootton, S.A. (2002) Eicosapentaenoic and docosapentae are the principal products of α-linolenic acid metabolim in young women. *British Journal of Nutrition* 88, 355–363.

Burlingame, B., Mishida, C., Uauy, R. and Weissel, R. (2009) Fats and Fatty Acids in Human Nutrition. Joint FAO/WHO. Expert Consultation. *Annals Nutrition Metabolism* 55, 1–3.

Burr, G. and Burr, M. (1929) A new deficiency disease produced by the rigid exclusion of fat from the diet. *Journal of Biological Chem*istry 82, 345–367.

Burr, G. and Burr, M. (1930) On the nature and role of fatty acids essential in nutrition. *Journal of Biological Chemistry* 82, 587.

Calder, P.C. (2006) N-3 polyunsaturated fatty acids inflammation and inflammatory diseases. *American Journal of Clinical Nutrition* 83(Suppl.), 1505S–1519S.

Calder, P.C. (2009) Polyunsaturated fatty acids and inflammatory processes: new twists in an old tale. *Biochimie* 91, 791–795.

Caldwell, M.D., Johnson, H.T. and Othersen, H.B. (1972) Essential fatty acid deficiency in an infant receiving prolonged parenteral alimentation. *Journal of Pediatric* 81, 894–898.

Chicco, A.G., D'Alessandro, M.E., Hein, G.H., Oliva, M.E. and Lombardo, Y.B. (2009) Dietary chia seed (*Salvia hispanica* L.) rich in α-linoleic acid improves adiposity and normalises hypertriacylglyceroaemia and insulin resistance in dyslipaemic rats. *British Journal of Nutrition* 101, 41–50.

Crawford, M.A., Casperd, N.M. and Sinclair, A.J. (1976) The long chain metabolites of linoleic and linolenic acid in liver and brain in hervivores and carnivores. *Comparative Biochemistry Physiology* 54B, 395–401.

Cunnane, S.C. (2000) The Conditional Nature of the Dietary Need for Polyunsaturated: a Proposal to Reclassify 'Essential Fatty acids' as 'Conditionally-Indispensable' or 'Conditionally-Dispensable' Fatty Acids. *British Journal of Nutrition* 84, 803–812.

Cunnane, S.C. (2003) Problems with Essential Fatty Acids: Time for a New Paradigm? *Progress in Lipid Research* 42, 544–568.

Cunnane, S.C., Ryan, M.A., Nadeau C.R., Bazinet, R.P., Musa-Veloso, K. and McCloy, U. (2003) Why is carbon from some polyunsaturated extensively recycled in lipid synthesis? *Lipids* 38, 47–484.

D'Andrea, S., Guillou, H., Jan, S., Catheline, D., Thibault, J.-N., Bouriel, M., Rioux, V. and Bray, G.A. (2002) The same Δ6-desaturase not only acts on 18- but also on 24-carbon fatty acids in very-long chain polyunsaturated fatty acids in humans. *Biochemistry Journal* 364, 49–55.

DeLany, J.P., Windhauser, M.M., Champagne, C.M. and Bray, G.A. (2000) Differencial oxidation of individual dietary fatty acids in humans. *American Journal of Clinical Nutrition* 72, 905–911.

Djousse, L., Pankow, J.H., Eckfeldt, J.S., Folsom, A.R., Hopkins, P.N., Province, M.A., Hong, Y. and Ellison, R.C. (2001) Relation Between Dietary Linolenic Acid and Coronary Artery Disease in the National Heart, Lung and Blood Institute Family Heart Study. *American Journal of Clinical Nutrition* 74, 612–619.

Dubois, V., Breton, S., Linder, M., Fanni, J. and Parmentier, M. (2007) Fatty acid profiles of 80 vegetable oils with regard to their nutritional potential. *European Journal of Lipid Science Technology* 109, 710–732.

Duplus, E., Glorian, M. and Forest, C. (2000) Fatty acid regulation of gene transcription. *Journal of Biological Chemistry* 275, 30749–30752.

Eaton, S.B. and Konner, M. (1985) Paleolithic nutrition. A consideration of its nature and current implications. *New England Journal of Medicine* 312, 283–289.

Elmadfa, I. and Kornstainer, M. (2009) Fats and fatty acids requirements for adults. *Annals Nutrition Metabolism* 55, 57–75.

Emken, E.A. (2001) Stable isotope approach, applications and issues related to polyunsaturated fatty acids metabolism studies. *Lipids* 36, 965–1073.

Espada, C.E., Berra, M.A., Martinez, M.J., Eynard, A.R. and Pasqualini, M.E. (2007) Effect of Chia oil (*Salvia hispanica*) rich in ω-3 fatty acids on the eicosanoid release, apoptosis and T-lymphocytes tumor infiltration in a murine mammary gland adenocarcinoma. *Prostaglandins, Leukotrienes and Essential Fatty Acids* 77, 21–28.

Eurodiet (2000) Eurodiet core report. European dietary guidelines. Available at: http://eurodiet.med.uoc.gr/eurodietcorereport.pdf

FAO (2010) Fats and fatty acids in human nutrition. Report of an expert consultation. FAO, *Food Nutrition Paper* 91, Food and Agriculture Organization of the United Nations, Rome, Italy.

Fei, G., Hyung Wook, K., Miki, I., Iragashi, M., Kiesenwetter, D., Chang, L., Kaizong, M. and Rappaport, S. (2011) Liver conversion of docosahexaenoic and arachidonic acids from their 18-carbon precursors in rats on a DHA-free but α LNA-containing n-3 Pufa adequate diet. *Biochimica et Biophysica Acta* 1811, 484–489.

Fernandes, F.S., Santos de Souza, A., Tavares do Carmo, M.G. and Boaventura, G.T. (2011) Maternal intake of flaxseed-based diet (*Linum usitatissimum*) on hippocampus fatty acids: Implications for growth, locomotor activity and spatial memory. *Nutrition* 27, 1040–1047.

Ferrucci, L., Cherubini, A., Bandinelli, S., Bartali, B., Corsi, A. and Lauretani, F. (2006) Relationship of plasma polyunsaturated fatty acids to circulating inflammatory markers. *Journal of Clinical Endocrinology Metabolism* 91, 439–446.

Firestone, D. (ed.) (2006) *Physical and Chemical Characteristics of Oils, Fat, and Waxes*, 2nd edn. AOCS Press, USA, pp. 3–237.

Freeman, M.P., Hibbeln, J.R., Wisner, K.L., Davis, J.M., Mischoulon, D., Peet, M., Keck, P.E. Jr, Marangell, L.B., Richardson, A.J., Lake, J. and Stoll, A.L. (2006) N-3 fatty acids: evidence basis for treatment and future research in psychiatry. *Journal of Clinical Psychiatry* 67, 1954–1967.

Fuchs, B., Nimptsch, A., Sur, R. and Schiller, J. (2008) Analysis of Brain Lipids by Direct Coupled Matrix-Assisted Laser Desorption Ionization Time-of-Flight Mass spectrometry and High-performance Thin-Layer Chromatography. *Journal of AOAC International* 91, 1227–1236.

Gebauer, S.K., Psote, T.L., Harris, W.-S. and Kris-Etherton, P.M. (2006) n-3 fatty acids dietary recommendations and food sources to achieve essentiality and cardiovascular benefits. *American Journal of Clinical Nutrition* 83, 1526S–1535S.

Ghafoorunissa, A., Vanni, R., Laxmi, R. and Sesikeran, B. (2002) Effects of dietary alpha-linolenic acid from bended oils on biochemical indices of coronary heart disease in Indians. *Lipids* 37, 1007–1086.

Gibson, R.A., Neumann, M.A., Lien, E.L., Boyd, K.A. and Tu, W.C. (2012) Docosahexaenoic acid synthesis from alpha-linoleic acid is inhibited by diets high polyunsaturated fatty acids. *Prostaglandins, Leukotrienes and Essential Fatty Acids*. Available at: http://dx.doi.org/10.1016/j.plfa.2012.04.003.

Gissi, H.J. (2008) Effect of n-3 polyunsaturated fatty acids in patients with chronic heart failure, a randomised double blind placebo controlled trial. *Lancet* 372, 1223–1230.

González-Mañan, D., Tapia, G., Gormaz, J.G., Espesailles, A.D., Espinoza, A., Masson, L., Varela, P., Valenzuela, A. and Valenzuela, R. (2012) Bioconversion of α-linolenic acid to n-3 LCPUFA and expression of PPAR-alpha, acylcoenzyme A oxidase 1 and carnitine acyl transferase 1 are incremented after feeding rats with α-linolenic acid-rich oils. *Food and Function* DOI: 10.1039/c2fo30012e.

Goyens, P.L., Spilker, M.E., Zock, P.L., Katan, M.B. and Mensink, R.P. (2006) Conversion of α-linolenic acid in humans is influenced by the absolute amounts of α-linolenic acid and linoleic acid in the diet and not by the ratio. *American Journal of Clinical Nutrition* 84, 44–53.

Guil-Guerrero, J.L., Belabi, E.-H. and Rebolloso Fuentes, M.N. (2000) Eicosapentaenoic and arachidonic acids purification from red microalga *Porphyridium creuntum. Bioseparation* 9, 299–306.

Harris, W.S., Mozzaffarian, D., Rimm, E., Kriss-Etherton, P., Rudel, L.L., Appel, L.J., Engler, M.M., Engler, M.B. and Sacks, F. (2009) Omega-6 Fatty acids and risk for Cardiovascular Disease. *Circulation* 119, 902–907.

Hegsted, D., McGandy, R., Myers, M. and Stare, F. (1965) Quantitative effects of dietary fat on serum cholesterol in man. *American Journal of Clinical Nutrition* 17, 281–295.

Henry, J. (2009) Processing, manufacturing, uses and labelling of fats in the food supply. *Annals Nutrition Metabolism* 55, 273–300.

Hibbeln, J.R., Nieminen, L.R.G., Blasbalg, T.L., Riggs, J.A. and Lands, W.E.M. (2006) Healthy intakes of n-3 and n-6 fatty acids: estimations considering worldwide diversity. *American Journal of Clinical Nutrition* 83(Suppl.), 1483S–1493S.

Hu, F.B., Manson, J.E. and Willet, W.C. (2001) Types of dietary fat and risk of coronary heart disease a critical review. *Journal American Collaboration Nutrition* 20, 5–19.

Innis, S.M. (2000) The role of dietary n-6 and n-3 fatty acids in the developing brains. *Development Neuroscience* 22, 474–480.

Keys, A., Anderson, J. and Grande, F. (1965) Serum cholesterol response to changes in diet. IV. Particular saturated fatty acids in the diet. *Metabolism* 14, 776–787.

Kim, H.-W., Jagadeesh, S.R., Rapaport, S. and Igarachi, M.I. (2011) Regulation of rat brain polyunsaturated fatty acids (PUFA) metabolism during grades dietary n-3 deprivation. *Prostaglandins, Leukotrienes and Essential Fatty Acids* doi: 10.1016/j.plefa.2011.08.002.

Kraweczk, T. (2001) Fat in Dietary Guidelines Around the World. *Inform* 12, 132–140.

Kriss-Etherton, P., Fleming, J. and Harris, W. (2010) The Debate about n-6 Polyunsaturated Fatty Acid Recommendations for Cardiovascular Health. *Journal of the American Dietetic Association* 110, 201–204.

Kushi, L. and Giovanuce, E. (2002) Dietary fat and cancer. *American Journal of Medicine* 113(Suppl. 98), 635–705.

Lands, B. (2008) A critique of paradoxes in current advise on dietary lipids. Review. *Progress in Lipid Research* 47, 77–106.

Lands, W.E.M. (2003a) Diets could prevent many diseases. *Lipids* 18, 317–321.

Lands, W.E.M. (2003b) Primary prevention in cardiovascular disease: moving out of the shadows of the truth about death. *Nutrition Metabolism and Cardiovascular Disease* 13, 154–164.

Lands, W.E.M. (2005a) Dietary fat and health: the evidence and the politics of prevention; careful use dietary fats can improve life and prevent disease. *Annals of the New York Academy of Sciences* 1055, 179–192.

Lands, W.E.M. (2005b) *Fish, Omega-3 and Human Health*, 2nd edn. AOCS Press, Urbana, Illinois, pp. 3–220.

Martin, C.A., Almeida, V.V., Ruiz, M.R., Visentainer, J.E.L., Matshushita, M., Souza, N.E. and Visentainer, J.V. (2006) Acidos graxos poliinsaturados omega-3 e omega-6: importancia e ocurrencia em alimentos. *Revista de Nutrición,* Campinas, Brasil 19, 761–770.

Masson, L. and Mella, M.A. (1985) Materias Grasas de Consumo Habitual y Potencial en Chile. Composición en Ácidos Grasos. Universidad de Chile, Santiago, Chile. Editorial Universitaria, Inscripción N° 60.867, pp. 1–31.

Masson, L., Chamorro, H., Generini, G., Donoso, V., Pérez-Olea, J. and Mella, M. (1990) Fish oil intake in coronary artery disease patients, serum lipid profiles and progression of coronary heart disease. *Medical Science Research* 18, 905–907.

Masson, L., Camilo, C., González, K., Cáceres, A., Jorge, N. and Torija, M.E. (2008) New Sources of Oilseeds from Latin American Native Fruits. *Natural Product Communication* 3, 357–362.

Moreau, A. and Kamal-Eldin, A. (eds) (2009) Gourmet and Health Promoting Specialty Oils, Chapter 4, Hall III, C., Fitzpatrick, C. and Kamal-Eldin, A. Flax, Perilla, and Camelina Seed Oils: α-Linolenic Acid-Rich Oils, pp 151–183; Chapter 5, Callaway, J.C. and Pats, D.W. Heempseed Oil, pp 185–213, Chapter 19. Astiazarán, I. and Ansorema, D. Algal Oil, pp 491–513; Chapter 20, Pickova, J. Fish Oils, pp 515–526. AOCS Press, Urbana, Illinois.

Mozaffarian, D. (2005) Review. Does alpha-linolenic acid intakes reduce the risk of coronary heart disease? A review of the evidence. *Alternative Therapeutic Health Medicine* 11, 24–30.

Noaghiul, S. and Hibbein, J.R. (2003) Cross-national comparison of seafood consumption and rates of bipolar disorders. *American Journal of Psychiatry* 160, 2222–2227.

Paeratakul, S., Ferdinand, D.P., Champagne, C.M., Ryan, D.H. and Bray, G.A. (2003) Fast-Food consumption among US adults and children. Dietary and nutrient intake profile. *Journal of the American Dietetic Association* 1023, 1332–1338.

Paulsrud, J., Pensler, L., Whitten, C., Stewart, S. and Holman, R. (1972) Essential fatty acid deficiency in infants by fat-free intravenous feeding. *American Journal of Clinical Nutrition* 25, 897–904.

Pawlosky, R.J., Hibbein, J.R., Novotny, J.A. and Salem, N. Jr, (2001) Physiological compartmental analysis of α-linolenic acid metabolism in adult humans. *Journal of Lipid Research* 42, 1257–1265.

Peiretti, P.G. and Gai, F. (2009) Fatty acid and nutritive quality of chia (*Salvia hispanica* L.) seeds and plant during growth. *Animal Science and Technology* 148, 267–275.

Pischon, T., Hankinson, S.E., Hotamisligil, G.S., Rifai, N., Willet, W.C. and Rimm, E.B. (2003) Habitual Dietary Intake of n-3 and n-6 Fatty Acids in Relation to Inflammatory Markers Among US Men and Women. *Circulation* 108, 155–160.

Postle, A.D. (2012) Lipidomics. *Current Opinion in Clinical Nutrition Metabolism Care* 15, 127–133.

Ramsden, C.F., Hibbeln, J.R., Majchrzak, S.F. and Davis, J.M. (2010) n/6 Fatty acid-specific and mixed polyunsaturated dietary interventions have different effects on CHD risk: a meta-analysis of randomized controlled trials. *British Journal of Nutrition* 104, 1586–1600.

Rapoport, S.L., Ramadan, E. and Basselin, M. (2011) Docosahexaenoic acid (DHA) incorporation into the brain from plasma, as an *in vivo* biomarker of brain DHA metabolim and neurotransmission. *Prostaglandins and Other Lipids Mediators*, doi:10.1016/j.prostaglandins.2011.06.003.

Rincón-Cervera, M.A., Suárez-Medina, M.D. and Guil-Guerrero, J.L. (2009) Fatty acid composition of selected roes from some marine species. *European Journal of Lipid Research* 111, 920–925.

Rodríguez-Pérez, W. (2005) Estudio comparativo de la composición en ácidos grasos del aceite de semillas de plantas de la Amazonía colombiana. *Momentos de la Ciencia* 2, 75–81.

Ruiz-Rodriguez, A., Reglero, G. and Ibañez, E. (2010) Recent trends in the advanced analysis of bioactive fatty acids. *Journal of Pharmaceutical and Biomedical Analysis* 51, 305–326.

Salem, N., Jr, Wegher, B., Mena, P. and Uauy, R. (1996) Arachidonic and docosaheaenoic acids are biosynthesized from their 18-carbons precursors in human infants. *Proceedings of the National Academy of Sciences* 93, 49–54.

Sanders, T.A. (2009) Fat and fatty acid intake and metabolic effects in the human body. *Annals Nutrition Metabolism* 55, 162–172.

Sanders, T.A. and Reddy, S. (1992) The influence of a vegetarian diet on the fatty acid composition of human milk and the essential fatty acids status of the infant. *Journal of Pediatrics* 120 (4Pt 2), S71–S77.

Scollan, N., Hocquette, J.-F., Nuernberg, K., Dannenberger, D., Richardson, I. and Moloney, A. (2006) Innovation in beef production systems that enhance the nutritional and health value of beef lipids and their relationship with meat quality. *Meat Science* 74, 17–33.

Serhan, C.N., Hong, S., Gronest, K., Colgan, S.P., Devch, P.R., Manick, G. and Moussigne, R.L. (2002) Resolvins – a family of bioactive products of omega-3 fatty acids transformation circuits initiated by aspirin treatment that counter proinflammation signals. *Journal of Experimental Medicine* 196, 1025–1037.

Serhan, C.N., Ciang, N. and Van Dyke, T.E. (2008) Resolving inflammation dual anti-inflammatory and pro-resolution lipid. *Nature Reviews/Immunology* 8, 349–361.

Shaikh, S.R. (2012) Biophysical and biochemical mechanism by which dietary N-3 polyunsaturated fatty acids from fish oil disrupt membrane lipid rafts. *Journal of Nutritional Biochemistry* 23, 101–105.

Simopoulos, A.P. (2002) The importance of the ratio of omega-6/omega-3 essential fatty acids. *Biomedicine and Pharmakotherapy* 56, 365–379.

Simopoulos, A.P. (2004) Omega-3 Fatty acids and Antioxidants in edible Wild Plants. *Biological Research* 37, 263–277.

Simopoulos, A.P. (2006) Evolutionary aspects of diet, the omega-6/omega-3 ratio and genetic variation: nutritional implications for chronic diseases. *Biomedical Pharmacotherapy* 60, 502–507.

Simopoulos, A.P. (2008) The importance of the omega-6/omega-3 fatty acid ratio in cardiovascular disease and other chronic diseases. *Experimental Biology Medicine* 233, 674–688.

Sinclair, A.J., Attar-Bashi, N.M. and Li, D. (2000) What is the role of α-linolenic acid for mammals? *Lipids* 37, 1113–1123.

Sjövall, P., Lausmaa, J. and Jhoansson, B. (2004) Mass Spectrometric Imaging of Lipids in Brain Tissue. *Analytical Chemistry* 76, 4271–4278.

Speedy, A.W. (2003) Global production and consumption of animal source foods. *Journal of Nutrition* 133, 4048S–4053S.

Sprecher, H. (2000) Metabolism of highly unsaturated n-3 and n-6 fatty acids. *Biochimical et Biophysica Acta* 1486, 219–231.

Strawford, A., Antelo, F., Christiansen, M. and Hellerstein, M.K. (2004) Adipose tissue trygliceride turn over, de novo lipogenesis, and cell proliferation in humans measured with 2H2O. *American Journal Physiology Endrocrinology Metabolism* 286, E577–E588.

Strijbosch, R.A., Lee, S., Arsenault, D.A., Andersson, C., Gura, K.M., Bistrian, B.R. and Puder, M. (2008) Fish oil prevents essential fatty acids deficiency and enhances growth, clinical and biochemical implications. *Metabolism* 57, 698–707.

Tapiero, H., Ba, G.N., Couvreur, P. and Tew, K.D. (2002) Polyunsaturated Fatty Acids (PUFA) and Eicosanoids in Human Health and Pathologies. *Biomedicine and Pharmacotherapy* 56, 216–222.

Tu, W.C., Cook-Jhonson, R.J., James, M.J., Muhlhausler, B.S. and Gibson, R.A. (2012) Omega-3 long chain fatty acids synthesis is regulated more by substrate levels then gene expression. *Prostaglandins, Leukotrienes and Essential Fatty Acids* 83, 61–68.

Uauy, R., Mena, R. and Rojas, C. (2000a) Essential fatty acids in early life: structural and functional role. *Proceedings of the Nutrition Society* 59, 3–15.

Uauy, R., Mena, P., Wegher, B., Nieto, S. and Salem, N. Jr, (2000b) Long chain polyunsaturated fatty acids formation effect of gestational age and intrauterine growth. *Pediatric Research* 47, 127–135.

Valenzuela, R., Gormaz, J.G., Masson, L., Vizcarra, M., Cornejo, Z., Valenzuela, A. and Tapia, G. (2012) Evaluation of the hepatic bioconversion of α-linolenic acid (ALA) to eicosapentaenoic acid (EPA) and docosahexaenoic acid (DHA) in rats fed with oils drom chia (*Salvia hispanica*) or rosa mosqueta (*Rosa rubiginosa*). *Grasas y Aceites* 63, 61–69.

Villeneuve, P., Muderhwa, J., Graille, J. and Haas, M. (2000) Customizing lipases for biocatalysis. A survey of chemical, physical and molecular biological approaches. *Journal of Molecular Catalysis, B: Enzymatic* 9, 113–148.

Vuksan, V., Whithtam, D., Sievenpiper, J.L., Jenkins, A.L., Rocovick, A.L., Bazinet, R.P., Vidgen, E. and Hanna, A. (2007) Supplementation of Conventional Therapy With the Novel Grain Salba (*Salvia hispanica* L.) improves Major and Emerging Cardiovascular Risks Factors in Type 2 Diabetes. *Diabetes Care* 30, 2804–2809.

Wenk, M.R. (2005) The Emerging Field of Lipidomics. *Nature Reviews Drug Discovery* 4, 594–610.

WHO (2003) *Diet, Nutrition and the Prevention of Chronic Diseases*. Report of the WHO/FAO Consultation Technical Report, Series 916, WHO, Geneva.

Wood, J.D., Enser, M., Fisher, A.V., Nute, G.R., Sheard, P.R., Richardson, R.I., Hughes, S.I. and Whittington, F.M. (2008) Fat deposition, fatty acids composition and meat quality: a review. *Meat Science* 78, 343–358.

9 Glucosinolates: The Phytochemicals of Nutraceutical Importance

Dhan Prakash* **and Charu Gupta**
Amity Institute for Herbal Research and Studies, Amity University, Noida, India

9.1 Introduction

The edible plants within the *Brassica* genus (family *Brassicaceae*) contain an important health-promoting group of compounds known as glucosinolates (GLSs) (Fig. 9.1) and are β-D-thioglucoside-N-hydroxysulfates (Tsiafoulis *et al.*, 2003). They are a well-studied example of a structurally diverse class of defence compounds (Fahey *et al.*, 2001; Mithen, 2001). These are generally classified as alkyl, aliphatic, alkenyl, hydroxyalkenyl, aromatic, or indole (Bennett *et al.*, 2004) and are limited to the order of *Caparales*, which includes the agriculturally important oilseed rape, vegetable crops, such as broccoli and various cabbages, as well as *Arabidopsis thaliana* and *Barbarea vulgaris* (Brown *et al.*, 2003). GLSs are derived from amino acids and can thus be divided into three groups according to their amino acid precursor: aliphatic GLSs, derived from Ala, Leu, Ile, Val and Met; benzenic GLSs, derived from Phe or Tyr; and indolic GLSs, derived from Trp. Biosynthesis proceeds through three independent stages: (i) chain elongation of selected precursor amino acids (only Met and Phe); (ii) formation of the core glucosinolate structure; and (iii) secondary modifications of the amino acid side chain.

GLSs and their metabolites such as isothiocyanates (ITCs), show anticarcinogenic action and these phytochemicals, in concert with other constituents such as flavonoids, vitamins and minerals, could be the major efficacious agents (Barillari *et al.*, 2005; Hintze *et al.*, 2005). Foods having chemopreventive properties have attracted a lot of interest amongst common man. Most of the drugs contain the bioactive chemicals originally discovered in plant foods. Epidemiological evidences show that consumption of cruciferous vegetables can significantly reduce the risk of a number of tumours and cancers (Lampe and Peterson, 2002; Vallejo *et al.*, 2002, 2003; Jones *et al.*, 2006; Oerlemans *et al.*, 2006).

9.1.1 Occurrence

GLSs constitute a well-defined group of secondary plant metabolites in cruciferous plants. They undergo hydrolysis, catalysed by an endogenous plant enzyme, known as myrosinase, into a range of biologically active compounds (Bones and Rossiter, 2006; Cartea and Velasco, 2008). The unique properties of GLSs and ITCs were first reported at the

* E-mail: dprakash_in@yahoo.com

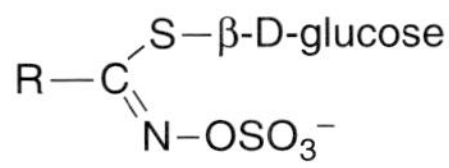

Fig. 9.1. General structure of glucosinolate.

beginning of the 17th century in an effort to understand the chemical origin of the sharp taste of mustard seeds. Earlier researches have highlighted the negative aspects of these compounds because of the prevalence of certain antinutritional or goitrogenic GLSs in the protein-rich defatted seed meals from widely grown oilseed crops and in some vegetables. There is, however, an opposite and positive side of this picture represented by the therapeutic and prophylactic properties of GLSs as nutritional or functional properties. To date, more than 120 GLSs have been characterized and their potent odour and taste suggests a role in herbivore and microbial defence. GLSs and their breakdown products are now known for their fungicidal, bactericidal, nematocidal and allelopathic properties. They are a major constituent of folk medicines and have recently attracted intense research interest because of their cancer chemoprotective attributes (Fahey *et al.*, 2001; Anilakumar *et al.*, 2006). The genes involved in the biosynthesis of all three phases of their biosynthesis, i.e. side-chain elongation of precursor amino acids, formation of the core glucosinolate structure and side-chain decoration, has been unveiled. Major breakthroughs include the ability to produce glucosinolates in *Nicotiana benthamiana*, the finding that specific GLSs play a key role in *Arabidopsis* innate immune response, and a better understanding of the link between primary sulfur metabolism and glucosinolate biosynthesis (Sønderby *et al.*, 2010).

GLSs are exclusively found in dicotyledonous plants, although closely related taxonomic groups contain only a small number of such compounds. Family *Brassicaceae* (syn. *Cruciferae*; including *Brassica* sp. and *Raphanus* sp.) genus *Brassica* alone contains more than 350 genera and 3000 species (Fahey *et al.*, 2001, 2002). Crucifers contain very high concentration of GLSs. Many commonly consumed vegetables, condiments, forages and oil-containing plants, such as cabbage, broccoli, cauliflower, collards, kale, mustard, Brussels sprouts and rapeseeds are good sources of GLSs (Song *et al.*, 2005; Oerlemans *et al.*, 2006). Broccoli derived from a species of wild cabbage, *Brassica oleracea*, is consumed widely in Europe (Margen, 2002). These vegetables are an excellent dietary source of phytochemicals including GLSs and its breakdown products, phenolics and other antioxidants such as vitamins C and K, as well as dietary essential minerals (Finley *et al.*, 2001; Jeffery and Araya, 2009). Hundreds of cruciferous species that have been investigated are able to synthesize GLSs. However, GLSs are by no means confined to crucifers; at least 500 species of non-cruciferous dicotyledonous angiosperms have been reported to contain one or more of the over 120 known GLSs. Some of the important sources of GLSs are *Arabis hirsuta, Barbarea praecox, B. vulgaris, Brassica campestris, B. juncea, B. napus, B. nigra, B. oleracea* var. *botrytis* subvar. *cymosa, Conringia orientalis, Isatis tinctoria, Lepidium sativum, Nasturtium officinalis, Reseda luteola, Reseda alba, Sibara virginica* and *Tropaeolum majus* (Fahey *et al.*, 2001, 2002). Most GLS-containing genera are clustered within the *Brassicaceae, Capparaceae* and *Caricaceae*, these including the largest number of GLS-containing species (Table 9.1).

9.1.2 Localization of glucosinolates in plant tissues

It has been found in a recent study on *A. thaliana* that glucosinolates are accumulated differentially in specific cells of reproductive organs. Using matrix-assisted laser desorption/ionization (MALDI) mass spectrometry imaging (MSI), distribution patterns of three selected compounds, 4-methyl-sulfinyl-butyl (glucoraphanin), indol-3-ylmethyl (glucobrassicin) and 4-benzoyloxybutyl glucosinolates, were mapped in the tissues of whole flower buds, sepals and siliques. The results showed that tissue localization patterns of aliphatic glucosinolate glucoraphanin and 4-benzoyloxybutyl glucosinolate were similar, but indole glucosinolate glucobrassicin had different localization, indicating a possible difference in function (Sarsby *et al.*, 2012).

Table 9.1. Important plant families of GLS-containing angiosperms (Fahey *et al.*, 2001).

Family	Chemical class	Glucosinolates
Brassicaceae	Sulfur in side chain, olefins, alcohols, ketones, aromatic, benzoates, indole, aliphatic straight and branch chain, multiple glycosylates	Sinigrin, glucobrassicanapin, glucopangulin, glucoalyssin, glucoerucin
Capparaceae	Sulfur-containing side chain, alcohols, ketones, aromatic, indole olefins, aliphatic straight and branch chain	Glucocappasalin, glucoiberin, gluconapin
Caricaceae	Aromatic	Glucotropaeolin
Limnanthaceae	Aliphatic alcohols, aromatic	Glucolimnanthin
Moringaceae	Olefins, aliphatic alcohols, aromatic, multiple glycosylates	Glucosisymbrin
Phytolaccaceae	Olefins, aliphatic alcohols, aromatic	Glucolepigramin
Resedaceae	Aliphatic alcohols, aromatic, indole, multiple glycosylates	Gluconasturtiin
Salvadoraceae	Olefins, aromatic	Glucoputranjivin
Tovariaceae	Olefins	Neoglucobrassicin
Tropaeolaceae	Aliphatic straight chain, olefins, aliphatic alcohols, aromatic	Glucoaubrietin

9.1.3 Structure

GLSs are β-thioglucoside N-hydroxysulfates (also known as (Z)-(or *cis*)-N-hydroximinosulfate esters or S-glucopyranosyl thiohydroximates), with a side chain (R) and a sulfur-linked β-D-glucopyranose moiety. A thioglycosylated sulfated oxime is an important structural feature of all known GLSs, which are mainly distinguished by variations in the amino acid derived carbon skeleton known as the 'side chain' (Mithen *et al.*, 2000; Rungapamestry *et al.*, 2007). GLSs share a similar basic structure consisting of a D-thioglucose group, a sulfonated oxime group and a side chain derived from methionine, phenylalanine, tryptophan or branched-chain amino acids.

Maximum GLSs contain either straight or branched carbon chains. Many of these compounds also contain double bonds (olefins), hydroxyl or carbonyl groups, or sulfur linkages in various oxidation states. The largest single group (one-third of all GLSs) contains a sulfur atom in various states of oxidation (e.g. methyl-thio-alkyl- or methyl-sulfonyl-alkyl-). Another small group of benzyl GLSs has an additional sugar moiety like rhamnose or arabinose, in glycosidic linkage to the aromatic ring. The presence of these sugars is of unknown significance, although it is intriguing that they are present in two families of plants (*Moringaceae* and *Resedaceae*) containing certain genera that are widely exploited for their pharmacological properties. Additionally, a number of sinapoyl and cinnamoyl salts and esters of some of the common GLSs are substituted on the thioglucoside moiety (Fahey *et al.*, 2001, 2002; West *et al.*, 2004).

9.1.4 Classification

GLSs have been classified according to their structure as aliphatic, aromatic, α-methyl-thio-alkyl and heterocyclic, e.g. indole (Fahey *et al.*, 2001; Cartea and Velasco, 2008). There are seven major classes of GLSs:

1. Methyl-sulfonyl-alkyl GLSs (glucoiberin, glucoraphanin and glycoalyssin).
2. Olefenic GLSs (sinigrin, gluconapin and progoitrin).
3. Aromatic GLSs (gluconasturtiin).
4. Ketonic GLSs (glucocappasalin, glucopangulin).
5. Alcoholic GLSs (gluconapoleiferin, progoitrin, epiprogoitrin).
6. ω-Hydroxyalkyl (benzoates) GLSs (glucomalcomiin, glucobenzosisymbrin).
7. Heterocyclic (indole) GLSs (glucobrassicin, neoglucobrassicin).

9.2 Biosynthesis of Glucosinolates

GLSs are sulfur-rich secondary metabolites characteristic of the Brassicales order with important biological and economic roles in plant defence and human nutrition. Application of systems biology tools continues to identify genes involved in the biosynthesis of GLSs. Recent progress includes genes in all three phases of the pathway, i.e. side-chain elongation of precursor amino acids, formation of the core GLS structure and side-chain decoration (Sønderby *et al.*, 2010). Biosynthesis of GLSs proceeds in three stages: (i) side-chain elongation of amino acids; (ii) development of the core structure; and (iii) secondary side-chain modifications. Gene identification in *Arabidopsis* followed by *in vitro* and *in vivo* characterization of the gene products have confirmed the tripartite biosynthetic concept derived from early biochemical studies (Mikkelsen *et al.*, 2002; Wittstock and Halkier, 2002). The core pathway, common to all GLSs, has received the most study in *Arabidopsis*. Side-chain elongation and modification strongly influence the bioactivities of GLS breakdown products. The evolution and ecological relevance of GLS variation has been reviewed (Kliebenstein *et al.*, 2005).

Synthesis of primary glucosinolates is accomplished in five steps and begins with the oxidation of precursor amino acids to aldoximes by side chain-specific cytochrome P450 mono-oxygenases (cytochromes P450) of the CYP79 family. The initial oxidation is not necessarily a committed step because the tryptophan-derived aldoxime is also an intermediate in the synthesis of indole-3-acetic acid (IAA) and camalexin (Glawischnig *et al.*, 2004; Hansen and Halkier, 2005). Region-specific post-aldoxime enzymes are less specific for the side chain because they transform non-endogenous and even artificial aldoximes into GLSs. The aldoximes are further oxidized by cytochromes P450 of the CYP83 family to aci-nitro compounds or nitrile oxides, strong electrophiles that spontaneously react with thiols to form S-alkylthiohydroximate conjugates (Hansen *et al.*, 2001). Cysteine is the likely thiol donor *in vivo*; however, it is not clear whether this conjugation is enzyme-mediated (Mikkelsen *et al.*, 2002). S-alkylthiohydroximate conjugates are cleaved by a C-S lyase into thiohydroximates, pyruvate and ammonia. This indicates that only one C-S lyase acts in GLS synthesis, and that this enzyme lacks side-chain specificity (Mikkelsen *et al.*, 2004).

Precursor amino acid elongation is analogous to the valine-to-leucine conversion and requires five reactions: an initial and final transamination, acetyl-CoA condensation, isomerization and oxidative decarboxylation (Mikkelsen *et al.*, 2002; Wittstock and Halkier, 2002). Labelling studies have confirmed this pathway (Graser *et al.*, 2001) and methylthioalkylmalate (MAM) synthases, which catalyse the condensation reaction (Textor *et al.*, 2004). Predicted plastid targeting signals for MAM gene products, purification of MAM synthase activities from enriched chloroplast preparations, and catalytic properties reminiscent of stromal enzyme regulation by light (basic pH optimum, dependence on ATP and divalent metal ions) strongly suggest that methionine side-chain elongation occurs in the chloroplast (Textor *et al.*, 2004). The chain-elongated α-keto acid can be transaminated and enter the core pathway, or it can pass through additional elongation cycles that insert up to nine methylene units. Given that the core pathway is proposed to be cytosolic (Chen *et al.*, 2003), chain-elongated α-amino (or α-keto) acids are likely to be exported from the chloroplast. Three partially redundant MAM genes control the variation in side-chain length of methionine-derived GLSs (Kroymann *et al.*, 2003). Secondary modification of the side chain is generally considered to be the final stage in GLS synthesis; however, desulfoglucosinolates could be the true substrates in some cases (Graser *et al.*, 2001). Side-chain decorations entail various kinds of oxidations, eliminations, alkylations and esterifications (Mikkelsen *et al.*, 2002; Wittstock and Halkier, 2002).

Methionine-derived GLSs are extensively modified; in *Arabidopsis*, this structural variety is generated by four polymorphic genetic loci. The substantial natural variation of aliphatic glucosinolates in *Arabidopsis* has expedited identification of two α-ketoglutarate-dependent dioxygenases, encoded by the tightly linked and duplicated *AOP2* and *AOP3* genes, which control production of alkenyl and hydroxyalkyl GLSs, respectively. The AOP enzymes and their orthologues in *Brassica oleracea* act after the methyl-thio to methyl-sulfinyl side-chain oxidation. Enzymes responsible for sulfur oxidation as well as for the methoxylations of indolyl glucosinolates remain to be identified, which is likely to be facilitated by comparative QTL mapping (Kliebenstein *et al.*, 2001a, b; Gao *et al.*, 2004).

9.3 Properties of Glucosinolates

9.3.1 Glucosinolates as phagostimulants

Aphids are phloem sap feeders and aphids such as *Brevicoryne brassicae* are specialists, feeding only on plants that contain glucosinolates. GLSs are, therefore, crucial feeding stimulants (phagostimulants) and this was elegantly proved by inducing feeding on broad bean (*Vicia faba*, not a host for the cabbage aphid) with the cut stem dipped in a solution containing 2-propenyl glucosinolate. The study strongly suggested that *B. brassicae* has a mechanism for sequestering GLSs, while in contrast *Myzus persicae* accumulates little GLS (which is instead found in the excreted honeydew). The actual location of GLSs in the aphid is not known, although circulation in the haemolymph is a distinct possibility, much as cyanogenic glycosides do in the larvae of *Zygaena trifolii* (Bridges *et al.*, 2002).

9.3.2 Hydrolysed products

Hydrolysis of GLSs is catalysed by an endogenous plant enzyme myrosinase (thioglucohydrolase; E.C. 3.2.1.147). Although GLSs are not protective themselves, they are converted by the co-existing myrosinases to bitter isothiocyanates (ITC), which defend the plant against predators (Fahey *et al.*, 2012). A wide range of biologically active breakdown products such as nitriles, ITCs, thiocyanates, epithionitriles and vinyl oxazolidinethiones are produced. Some compounds, for example ITCs, indoles, thiocyanates or nitriles, showed anticarcinogenic activity by inducing phase II biotransformation enzyme activity (Rungapamestry *et al.*, 2006).

Early studies reported that myrosinase is localized in the cytoplasm of specialized plant cells, myrosin cells. GLSs and myrosinase are segregated in intact plants (Brandt *et al.*, 2004; Sarikamiş*et al.*, 2009). Autolysis or tissue damage during freezing and thawing, chopping or chewing brings myrosinase in contact with GLSs and hydrolysis occurs (Rungapamestry *et al.*, 2007). The products of hydrolysis have important roles in the plant defence system against insects, fungi and microorganism infections. Similarly, in animals consuming plants, GLSs are not bioactive until they have been hydrolysed to an associated ITC (Rouzaud *et al.*, 2003) by myrosinase enzyme. The latter is released by disruption of the plant cell through harvesting, processing, or mastication (Finley, 2005). Myrosinase activity results in the release of the glucose moiety leaving behind an unstable intermediate, which spontaneously rearranges to produce several products (Mithen *et al.*, 2000; Oerlemans *et al.*, 2006). The type of product formed depends on several factors, such as pH, substrate or availability of ferrous ions (Kristal and Lampe, 2002; Lund, 2003; Guerrera, 2005; Shapiro *et al.*, 2006).

Several ITCs produced in hydrolysis and through rearrangement of GLSs are nutritionally important products (Song and Thornalley, 2007).The decreased risk of cancer linked to a diet rich in *Brassica* vegetables is widely associated to ITCs absorbed following ingestion of GLSs. ITC reacts with free amino and sulfhydryl groups of various proteins. These products are also responsible for the characteristic flavour and odour of *Brassica* vegetables (Das *et al.*, 2000) and for the biting taste of important condiments such as horseradish and mustard. Sinigrin and progoitrin, important GLSs, are related to bitterness in Brussels sprouts (Traka and Mithen, 2009), while in cooked cauliflower sinigrin and neoglucobrassicin are responsible for the bitter taste. The characteristic odour and taste of radish is due to the formation of 4-methyl-thio-3-butenyl ITC-derived GLSs (Engel *et al.*, 2002; Anilakumar *et al.*, 2006; Volden *et al.*, 2008).

The breakdown products of indolylmethyl GLSs consist of indole-3-acetonitrile, indole-3-carbinol (I-3-C) and 3,3′-diindoylmethane. Progoitrin, a major component of cabbage, cauliflower, Brussels sprouts and kale have antithyroidal properties because of their two reactive metabolites progoitrin ITC (2-hydroxy-3-butenyl) and goitrin (5-vinyloxazolidine-2-thione). The latter is produced from the former in cyclization reaction (Anilakumar *et al.*, 2006; Vandermeiren *et al.*, 2009). Sulforaphane (1-isothiocyanato-C (methyl-sulfinyl)-butane) has been identified in broccoli as a product of enzymatic or acid hydrolysis of the corresponding ω-(methyl-sulfinyl)-alkyl-GLS (glucoraphanin) (Rungapamestry *et al.*, 2007). Sulforaphane reduces the incidence of a number of forms of tumour in various experimental models and cell cultures. The chemoprotective effect of sulforaphane was thought to be due solely to its ability to behave as a mono-functional inducer of phase II enzymes, which are known to represent the most important group of detoxication enzymes of the human organism. However, sulforaphane has also been shown to inhibit the CYP2EI isoenzyme of the cytochrome P450, thus emerging as an inhibitor of phase I enzymes. Natural bioactives, GLSs breakdown products in broccoli such as I-3-C, benzyl ITC and phentyl ITC, may also be responsible for selective induction of apoptosis in cancer cells (Kristal and Lampe, 2002; Jackson and Singletary, 2004; Finley, 2005; Bialecki *et al.*, 2010).

9.3.3 Myrosinase

Myrosinase has been purified and characterized from several sources, including white mustard (*Sinapis alba*), cress (*Lepidium sativum*), yellow mustard (*Brassica juncea*), rapeseed (*Brassica napus*) and wasabi (*Wasabia japonica*). They are activated to various degrees by ascorbic acid and in some instances the enzyme is almost inactive in its absence. It has been suggested that ascorbate provides a nucleophilic catalytic group and activation is not dependent on the redox reactivity of ascorbate. Early work had shown that ascorbate creates an allosteric effect on the activity of the enzyme. Subsequently it was shown that ascorbate acts as a catalytic base.

Evidence strongly suggests that upon ingestion by humans, β-thioglucosidase activity of gut microflora is largely responsible for converting ingested GLSs to ITCs. Similar observations have also been made in numerous animal studies. After hydrolytic cleavage of the β-glucosyl moiety, the sulfate moiety is released non-enzymatically to form the thiohydroxamate-O-sulfonate from both aliphatic and aromatic GLSs. This unstable intermediate then rearranges to form ITCs, or other breakdown products (e.g. thiocyanates, nitriles, epithionitriles, oxazolidine-2-thiones) in a manner that depends upon the GLS substrate as well as the reaction conditions such as pH, or the presence of Fe^{2+} or epithiospecifier protein (Burmeister *et al.*, 2000; Volden *et al.*, 2008; Traka and Mithen, 2009).

9.3.4 Ascorbigens (ABG)

Some indole products of GLS are claimed to demonstrate breast cancer-preventing actions, due to their affinity and ability to bind with oestrogen receptors. Such an activity has been displayed by 3,3′-diindolylmethane, I-3-C and indolo(3,2-*b*) carbazole, which is formed from ascorbigen (ABG) (Fig. 9.2a) or I-3-C in a strongly acidic environment upon the activity of gastric juice (Horn *et al.*, 2002; Traka and Mithen, 2009). ABG is a natural derivative of L-ascorbic acid (AA) and was identified as a biotransformation product of the alkaloid glucobrassicin. ABG can be isolated from some fresh, non-fresh or sour cruciferous vegetable tissues (cabbage, kohlrabi, savoy cabbage, etc.). Their biological evaluation showed that the most active substance is 1′-methyl-ascorbigen (MeABG) (Fig. 9.2b) that inhibits tumour growth in animals, protects animals from some bacterial and viral infections and also has an immunomodulating activity. MeABG has a pronounced apoptotic effect in which formaldehyde from the methyl group of MeABG plays a crucial role (Moldrup *et al.*,

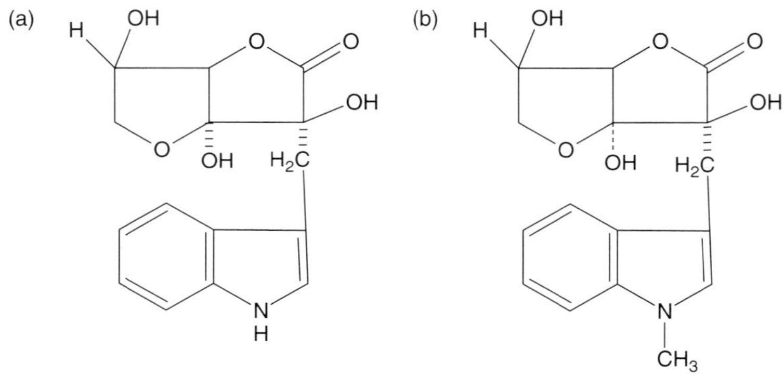

Fig. 9.2. Chemical structures of (a) ascorbigen (ABG) and (b) 1′-methyl-ascorbigen (MeABG).

2011). Ascorbigen is able to induce phase I and II enzymes that are centrally involved in the detoxification of xenobiotics. Cosmeceuticals containing ABG as an active principle are becoming increasingly popular, although the underlying cellular and molecular mechanisms regarding its potential anti-ageing and ultraviolet-protective properties have not been fully established (Wagner and Rimbach, 2009).

9.4 Biological Activity of Glucosinolates

GLSs are a highly diverse and variable group of phytochemicals. Studies show that they cause an increase in the activities of biotransformation enzymes in various tissues (Anilakumar *et al.*, 2006; Halkier and Gershenzon, 2006; Kos *et al.*, 2011). The antioxidant enzymes such as glutathione peroxidases (GSH-Px), glutathione reductase (GSSGR), glutathione S-transferase (GST) and superoxide dismutase (SOD) play an important role in cellular oxidative stress. It was noted that I-3-C at normal dietary levels does not induce the oxidative enzymes. However, in mice fed on semi-purified diets containing I-3-C, there was a significant increase in both hepatic and intestinal GST. I-3-C was found to reduce GSSGR and induce GSH-Px and SOD in rat liver. Glucoraphanin also induced hepatic quinone reductase (QR) and GST in mice (Guerrera, 2005; Anilakumar *et al.*, 2006).

9.4.1 Anticarcinogenic activities

In recent years, cancer prevention by natural products has received considerable attention. The potential protective role of *Brassica* vegetables and bioactive phytochemicals of these vegetables, such as flavonoids (e.g. quercetin), minerals (e.g. selenium) and vitamins (e.g. vitamin C) are well established. ITCs and I-3-C have been extensively studied and have shown chemoprotective activities during initiation and promotion phases of cancer development (Cartea and Velasco, 2008; Jeffery and Araya, 2009). Results clearly point towards a positive correlation between cancer prevention in many target organs and consumption of *Brassica* vegetables or their bioactive phytochemicals. The epidemiological literature also supports the hypothesis that high intakes of *Brassica* vegetables reduce prostate, lung and gastrointestinal tract cancer risk. There are clear indications that they block tumour initiation by modulating the activities of Phase I and Phase II biotransformation enzymes and suppress tumours by apoptosis (Moldrup *et al.*, 2011). *In vitro* and *in vivo* studies have reported that ITCs affect many steps of cancer development, including modulation of phases I and II detoxification enzymes. They function as a direct or indirect antioxidant by phase II enzyme induction thus modulating cell signalling, induction of apoptosis, control of the cell cycle and reduction of *Helicobacter* infections.

Mechanism of action

Dietary glucosinolates have been reported to block formation of endogenous or exogenous carcinogens for preventing initiation of carcinogenesis (Vig *et al.*, 2009). It is well known that living organisms are continuously exposed to a number of naturally occurring chemicals. The ability of the carcinogens to exert their effects depends largely on the interaction between activating and deactivating enzymes. Any imbalance will result in a change in the biological effect. Glucosinolates and their hydrolytic products modulate the activity of xenobiotic metabolizing phase I and II enzymes. Phase I enzymes generally increase the reactivity of the lipophilic compounds. On the other hand, phase II enzymes increase the water solubility and facilitate the removal of metabolites from the body. For the protection of cells against DNA damage by carcinogens and reactive oxygen species, inhibition of phase I and induction of phase II enzymes are required.

The genes for the phase II enzyme contain a specific sequence of DNA called antioxidant response element (ARE). Activities of phase II enzymes have been reported to be enhanced by glucosinolates and their hydrolytic products (Holst and Williamson, 2004). Isothiocyanates are an important group of breakdown products of glucosinolates and act at a number of points in the tumour development by blocking the metabolism of carcinogenic compounds through biotransformation. They generally enhanced the activity of phase II enzymes and inhibited phase I enzymes (Tawfiq *et al.*, 1995; Fahey *et al.*, 1997), thereby reducing the carcinogenic activity and enhancing the detoxification and clearance of carcinogens. Further, they serve as suppressors during the promotion phase of neo-plastic process. Induction of apoptosis and action of signal transduction pathways within the cell activities of glucosinolates has also been reported (Smith *et al.*, 2004).

Benzyl-p-hydroxybenzyl- and 2-hydroxybut-3-enyl glucosinolates have been reported to induce mammalian phase 2 enzymes of detoxification (Tawfiq *et al.*, 1995; Fahey *et al.*, 1997). Sulforaphane (SFN), the enzymatic degradation product of glucosinolate glucoraphenin, activated gene expression, thereby helping to clear carcinogenic substances from the body. SFN also increased levels of mammalian phase 2 enzymes through antioxidant response element (ARE)-mediated transcriptional activation (Hwang and Jeffery, 2005; Khor *et al.*, 2006). SFN supported a healthy immune system by significantly enhancing the production of chemicals involved in immune response (Thejass and Kuttan, 2006). In a study in which animals were genetically bred to develop intestinal polyps, a condition that led to tumour formation, the group of animals that were fed SFN had higher rates of apoptosis (cell suicide) and smaller tumours growing more slowly than animals not receiving SFN (Wang *et al.*, 2004).

Watercress and broccoli are reported to be rich sources of phenethyl isothiocyanate (PEITC), which may block the cytochrome P450-mediated metabolic activation of the common nitrosoamine to its potent carcinogenic forms (Palaniswamy *et al.*, 2003). Extracts of watercress and broccoli suppressed metalloproteinase-9, an enzyme closely associated with invasive potential of breast cancer (Conaway *et al.*, 1996). It also suppressed production of pro-inflammatory compounds such as nitric oxide (NO) and prostaglandins (Ribnicky *et al.*, 2001). PEITC also inhibited induction of lung and oesophagus cancer in both rat and mouse tumours (Stoner and Morse, 1996).

Indole-3-carbinol (I-3-C) is produced from indole-3-glucosinolates such as glucobrassicin through hydrolysis. Under acidic conditions I-3-C and elemental sulfur are formed. Anticarcinogenic, antioxidant and anti-atherogenic activities of I-3-C have been reported (Jongen, 1996). Further, I-3-C modulates the activities of both phase I and II enzymes. It suppressed cancer growth and induced programmed cell deaths in tumours of the breast, prostate, leukaemia, cervix and colon because of its ability to favourably influence the human body's balance of oestrogens. I-3-C also inhibited cancer cell growth by interfering in the production of proteins involved in abnormal cellular reproduction and by promoting the production of tumour suppressor proteins (Aggarwal and Ichikawa, 2005). I-3-C has also been reported to prevent cancer by interfering with angiogenesis, process of formation of new blood vessels that tumours require for their survival and spread (Wu *et al.*, 2005). 3-3′-diindolyl methane (DIM), a condensation product of I-3-C, enhanced beneficial effects of I-3-C by influencing the expression of genes involved in carcinogenesis, cell survival and physiological behaviour (Pappa *et al.*, 2007; Pledgie-Tracy *et al.*, 2007).

The most characterized GLS compounds are sulforaphane, phenethyl ITC, allyl ITC and I-3-C, but many other ITCs that are present in lower quantities may also contribute to the anticarcinogenic properties of *Brassicaceae* (Song and Thornalley, 2007). Various cited examples of dietary anticancer bioactives from broccoli include antiproliferative effects of sulforaphane in human breast cancer (Jackson and Singletary, 2004; Brandi *et al.*, 2005), reduced risk of cancer via decreased damage to DNA (Gill *et al.*, 2004; Jeffery and Araya, 2009), effects on the regulation of intestinal cell growth and death in colon cancer (Parnaud *et al.*, 2004), as well as the cancer-protective effect of high-selenium broccoli (Shapiro *et al.*, 2006) or the exertion of a protective effect in prostatic tumours (Giovannucci *et al.*, 2003; Canene-Adams *et al.*, 2005). For example, sulforaphane-induced apoptosis in prostate cancer cells is initiated by reactive oxygen species generation and the fact that both intrinsic and extrinsic caspase cascades contribute to the cell death caused by this highly promising cancer chemopreventive agent (Singh *et al.*, 2005). Additional effects of bioactives like ITCs from broccoli on bladder carcinoma cells (Munday and Munday, 2002; Tang and Zhang, 2004), on antioxidant capacity and on cellular oxidative stress, as well as cholesterol lowering effects (Suido *et al.*, 2003) and protective effects on cardiovascular disease (Sesso *et al.*, 2003) and *Helicobacter pylori* infections (Galan *et al.*, 2004), support the fact that the dose level of bioactives may be effective through human consumption of *Brassica* vegetables. So this could contribute to the lower incidence of different types of cancer and diseases in individuals who regularly consume such vegetables. Unfortunately, the biological activity of these molecules is compromised by the removal of the sulfate. After desulfation, they can no longer serve as substrates for myrosinase and thus their cognate ITCs are not available for bioassay or for direct measurement by cyclo-condensation-key tools in the study of the pharmacokinetics, pharmacodynamics and bioactivity of these compounds.

Selenium (Se) is a nutritionally essential element and its deficiency results in disease conditions in humans and domestic livestock (Raskin *et al.*, 2002). There are evidences that Se intake offers protection against cancer (Combs *et al.*, 2001). Se-methylated amino acids such as Se-methyl-selenocysteine (SeMSC) are metabolized primarily in the excretory pathway, and data suggest that methyl-selenol generated in this pathway is the metabolite, which is most responsible for preventing cancer (Cartea and Velasco, 2008). Broccoli accumulates Se in methylated forms and many other *Brassicaceae* species also accumulate Se (Finley, 2005). It has been reported that I-3-C has an inhibitory effect on cell growth in human cervical and endometrial cancer cells (Chinni *et al.*, 2001; Anilakumar *et al.*, 2006). It was shown that the ITC metabolite of sulforane was a major inducer of quinone reductase (QR) and phenylethyl isothiocynate (PEIT), a hydrolysed product of gluconasturtiin, was effective against nitrosamine-induced raise in oesophageal cancer (Traka and Mithen, 2009).

Little is known about the direct effect of broccoli sprouts on human health, even though *in vitro* and *in vivo* data provided evidence that supports the belief that young cruciferous sprouts with their high concentrations of phytochemicals may be a potent source of protective chemicals against cancer (Gill *et al.*, 2004). Recently, a phase I study of multiple biomarkers for metabolism and oxidative stress after 1-week intake of broccoli sprouts was carried out and it revealed that only 1 week of broccoli sprouts intake improved cholesterol metabolism and decreased oxidative stress markers (Murashima *et al.*, 2004). Broccoli sprouts are a rich source of GLSs and ITCs that induce phase II detoxication enzymes, boost antioxidant status and protect animals against chemically induced cancer. The ITCs are about six times more bioavailable than GLSs, which must first be hydrolysed. Thorough chewing of fresh sprouts exposes the GLSs to plant myrosinase and significantly increases dithiocarbamate excretion.

The anticarcinogenic properties of cruciferous vegetables have been attributed to I-3-C content while the protective effect has been attributed to induction of enzymes such as cytochrome P-450. A high urinary excretion of ITCs from Brussels sprouts conferred a low risk of lung cancer in a Cohort study of Chinese men (London *et al.*, 2000). The effect of broccoli extract on oxidative stress in HepG2 cells using the dichlorofluorescein-diacetate assay (Kurilich *et al.*, 2003) is reported. This study confirms the association between broccoli extracts and enhanced antioxidant activity while providing additional evidence for protection against reactive oxygen species at the cellular level (Anilakumar *et al.*, 2006).

9.4.2 Disease prevention by plant glucosinolates

Plant-based diets that are rich in cruciferous vegetables are found to be effective in preventing cancer and other chronic diseases. Crucifers contain a very high concentration of glucosinolates. Although they are not protective by themselves, glucosinolates are converted by coexisting myrosinases to bitter isothiocyanates (ITC), which defend plants against predators. Coincidentally, ITC also induce mammalian genes that regulate defences against oxidative stress, inflammation, and DNA-damaging electrophiles. Consequently, the efficiency of conversion of GLS to ITC may be critical in controlling the health-promoting benefits of crucifers. If myrosinase is heat-inactivated by cooking, the gastrointestinal microflora converts GLS to ITC, a process abolished by enteric antibiotics and bowel cleansing (Fahey *et al.*, 2012).

9.4.3 Antinutritional activities

It has been reported that GLSs have been condemned due to their goitrogenic and growth retardation activities. GLS breakdown products (oxazolidine-2-thiones) found in several oil meals may induce morphological and histological abnormalities of internal organs (Brandt *et al.*, 2004; Halkier and Gershenzon, 2006) as exemplified in increased thyroid weight in pigs and poultry, as well as depressed growth, goiters, poor egg production and liver damage. Goitrogenic activity has been associated with 5-vinyl oxozolidine-2-thione (goitrin) and thiocyanate ions. Goitrin shows its effect by interfering with thyroid hormone synthesis. In contrast, thiocyanate ion, derived from glucobrassicin, competes with iodine for uptake by the thyroid gland. Another possible hazard from indoles is their ability to react with nitriles to form carcinogenic N-nitroso compounds. Indole acetronitrile (IAN) can react with nitrite *in vitro* to form compounds that have been found to be mutagenic. It is likely that the extraction process would have destroyed the ascorbic acid, a key factor to regulate their bioactivity.

ITCs are the most toxic among the hydrolysis products, because they even affect herbivores (Agrawal and Kurashige, 2003; Kos *et al.*, 2011). Nitriles and thiocyanates have a lesser toxicity to insects (Lambrix *et al.*, 2001; Husebye *et al.*, 2005), whereas hardly anything is known about the biological effects of GLS-derived epithionitriles and oxazolidine-2-thiones on insect herbivores (Wittstock *et al.*, 2003; Moldrup *et al.*, 2011). It was also reported that growth retardation, liver lesions and necrosis as well as thyroid hypertrophy or hyperplasia appeared to occur when rabbits consumed diet containing 2–5 mg g^{-1} of GLSs. It is reported that I-3-C acts as scavenger of free radicals and reactive electrophiles and stabilizes biological membranes against fluidity changes. However, its antinutritional effects cannot be ruled out as indicated by its potential to enhance carcinogenic promotion in mouse skin, producing hepatotoxicity and neurological impairment (Lund, 2003; Halkier and Gershenzon, 2006). There are various evidences available from the studies carried out in countries like Egypt and Japan about the antinutritional factors in glucosinolates, commonly referred to as goitrogens. Intact glucosinolates are apparently free of toxicity, but on hydrolysis by an endogenous enzyme, myrosinase (thioglucoside glucohydrolase,

E.C.3.2.3.1), present in the seed and unheated meal, yield undesirable and potentially toxic products. Some of these products are goitrogenic; others are potentially hepatotoxic, whilst the majority are volatile and strongly pungent and responsible for the 'bite' of mustard, radish and horseradish (Hill, 1991). Other symptoms of the ingestion of large amounts of glucosinolates in animals and poultry include hyper-thyroidism, reduced feed intake and performance, enlarged thyroid gland and reduced levels of circulating thyroid hormones (Darroch and Bell, 1991).

9.5 Conclusions

The era of structural gene discovery in glucosinolate research, greatly aided by a combination of molecular, genetic and genomic approaches in *Arabidopsis*, has passed its peak. Early biochemical models of glucosinolate synthesis, which provided the first guidance in this area of research, have largely been confirmed. Current research increasingly focuses on glucosinolate transport and turnover, on regulatory mechanisms of glucosinolate biosynthesis, and on the feasibility of customizing glucosinolate profiles by molecular breeding and transgenic approaches. Furthermore, virtually nothing is known about the transport of glucosinolates from their production site to the proper storage site. Regulation of flux through the pathway might be affected by post-translational regulation of the enzymes either directly by e.g. phosphorylation or indirectly through a shift in redox potential. The regulatory network that controls glucosinolate accumulation includes primary sulfur and amino acid metabolism as well as biotic and abiotic signalling cascades. This emphasizes the importance of changing our thinking from a linear biosynthetic pathway to more complex integrated networks. A thorough understanding of glucosinolate pathway regulation will not only require the study of pathway gene expression in response to internal and external factors and their corresponding signalling networks, but also have to address 'neoclassic' questions of enzyme biochemistry, such as subcellular localization of enzymes, intracellular trafficking and channelling of intermediates, metabolon organization and flux control, regulation of enzyme activity by effectors and covalent modification, or protein structure–function relationships as recently reported for myrosinase. One of the next goals in glucosinolate research would be to understand the channelling of intermediates to the final product, which may be enabled by the presence of a biosynthetic multi-enzyme complex, also termed 'metabolon'. Understanding the dynamics of the glucosinolate network will not only advance our basic knowledge about secondary metabolites but also facilitate future efforts in metabolic engineering. Thus in the long term, by the usage of metabolic engineering of customized glucosinolate profiles, plant protection can be enhanced and functional foods with a nutritional and cancer-prevention strategy could be designed.

References

Aggarwal, B.B. and Ichikawa, H. (2005) Molecular targets and anti cancer potential of indole-3-carbinol and its derivative. *Cell Cycle* 4, 1201–1215.

Agrawal, A.A. and Kurashige, N.S. (2003) A role for isothiocyanates in plant resistance against the specialist herbivore *Pieris rapae*. *Journal of Chemical Ecology* 29, 1403–1415.

Anilakumar, K.R., Farhath, K. and Bawa, A.S. (2006) Dietary role of glucosinolate derivatives: a review. *Journal of Food Science and Technology* 43, 8–17.

Barillari, J., Cervellati, R., Paolini, M., Tatibouet, A., Rollin, P. and Iori, R. (2005) Isolation of 4-methylthio-3-butenyl glucosinolate from *Raphanus sativus* sprouts (Kaiware Daikon) and its redox properties. *Journal of Agricultural Food and Chemistry* 5, 9890–9896.

Bennett, R.N., Mellon, F.A. and Kroon, P.A. (2004) Screening crucifer seeds as sources of specific intact glucosinolates using ion-pair high-performance liquid chromatography negative ion electrospray mass spectrometry. *Journal of Agricultural and Food Chemistry* 52, 428–438.

Bialecki, J.B., Ruzicka, J., Weisbecker, C.S., Haribal, M. and Attygalle, A.B. (2010) Collision-induced dissociation mass spectra of glucosinolate anions. *Journal of Mass Spectrometry* 45, 272–283.

Bones, A.M. and Rossiter, J.T. (2006) The enzymic and chemically induced decomposition of glucosinolates. *Phytochemistry* 67, 1053–1067.

Brandi, G., Schiavano, G.F., Zaffaroni, N., De Marco, C., Paiardini, M., Cervasi, B. and Magnani, M. (2005) Mechanisms of action and anti-proliferative properties of *Brassica oleracea* juice in human breast cancer cell lines. *Journal of Nutrition* 135, 1503–1509.

Brandt, K., Christensen, L.P., Hansen-Møller, J., Hansen, S.L., Haraldsdottir, J., Jespersen, L., Purup, S., Kharazmi, A., Barkholt, V., Frøkiær, H. and Kobæk-Larsen, M. (2004) Health promoting compounds in vegetables and fruits: a systematic approach for identifying plant components with impact on human health. *Trends in Food Science and Technology* 15, 384–393.

Bridges, M., Jones, A.M.E., Bones, A.M., Hodgson, C., Cole, R., Bartlet, E., Wallsgrove, R., Karapapa, V.K., Watts, N. and Rossiter, J.T. (2002) Spatial organization of the glucosinolate–myrosinase system in brassica specialist aphids is similar to that of the host plant. *Proceedings of the Royal Society B: Biological Sciences* 269, 187–191.

Brown, P.D., Tokuhisa, J.G., Reichelt, M. and Gershenzon, J. (2003) Variation of glucosinolate accumulation among different organs and developmental stages of *Arabidopsis thaliana. Phytochemistry* 62, 471–481.

Burmeister, W.P., Cottaz, S., Rollin, P., Vasella, A. and Henrissat, B. (2000) High resolution X-ray crystallography shows that ascorbate is a cofactor for myrosinase and substitutes for the function of the catalytic base. *Journal of Biological Chemistry* 275, 39385–39393.

Canene-Adams, K., Campbell, J.K., Zaripheh, S., Jeffery, E.H. and Erdman, J.W., Jr (2005) The tomato as a functional food. *Journal of Nutrition* 135, 1226–1230.

Cartea, M.E. and Velasco, P. (2008) Glucosinolates in Brassica foods: Bioavailability in food and significance for human health. *Phytochemistry Reviews* 7, 213–229.

Chen, S., Glawischnig, E., Jorgensen, K., Naur, P., Jorgensen, B., Olsen, C.E., Hansen, C.H., Rasmussen, H., Pickett, J.A. and Halkier, B.A. (2003) CYP79F1 and CYP79F2 have distinct functions in the biosynthesis of aliphatic glucosinolates in Arabidopsis. *Plant Journal* 33, 923–937.

Chinni, S.R., Li, Y., Upadhyay, S., Koppolu, P.K. and Sarkar, F.H. (2001) Indole-3-carbinol induced cell growth inhibition, G cell cycle arrest apoptosis in prostate cancer cells. *Oncogene* 20, 2927–2936.

Combs, G.F., Clark, L.C. and Turnbull, B.W. (2001) An analysis of cancer prevention by selenium. *BioFactors* 14, 153–159.

Conaway, C.C., Jiao, D. and Chung, F.L. (1996) Inhibition of rat liver cytochrome p 450 isozymes by isothiocyanates and their conjugates: a structure activity relationship study. *Carcinogenesis* 17, 2423–2427.

Darroch, C.S. and Bell, J.M. (1991) Potential goitrogenic and toxic effects of an indole glucosinolate extract injected into the developing chick embryo. *Canadian Journal of Animal Science* 71, 481–487.

Das, S., Tyagi, A.K. and Kaur, H. (2000) Cancer modulation by glucosinolates: a review. *Current Science* 79, 1665–1671.

Engel, E., Baty, C., Le Corre, D., Souchon, I. and Martin, N. (2002) Flavor-active compounds potentially implicated in cooked cauliflower acceptance. *Journal of Agricultural and Food Chemistry* 50, 6459–6467.

Fahey, J.W., Zhang, Y.S. and Talalay, P. (1997) Broccoli sprouts: an exceptionally rich source of inducers of enzymes that protects against chemical carcinogens. *Proceedings of the National Academy of Sciences, USA* 94, 10367–10372.

Fahey, J.W., Zalcmann, A.T. and Talalay, P. (2001) The chemical diversity and distribution of glucosinolates and isothiocyanates among plants. *Phytochemistry* 56, 5–51.

Fahey, J.W., Haristoy, X., Dolan, P.M., Kensler, T.W., Scholtus, I., Stephenson, K.K., Talalay, P. and Lozniewski A. (2002) Sulforaphane inhibits extracellular, intracellular, and antibiotic-resistant strains of *Helicobacter pylori* and prevents benzo[a]pyrene-induced stomach tumors. *Proceedings of the National Academy of Sciences USA* 99, 7610–7615.

Fahey, J.W., Wehage, S.L., Holtzclaw, W.D., Kensler, T.W., Egner, P.A., Shapiro, T.A. and Talalay, P. (2012) Protection of humans by plant glucosinolates: efficiency of conversion of glucosinolates to isothiocyanates by the gastrointestinal microflora. *Cancer Prevention Research* 5, 603–611.

Finley, J.W. (2005) Proposed criteria for assessing the efficacy of cancer reduction by plant foods enriched in carotenoids, glucosinolates, polyphenols and seleno-compounds. *Annals of Botany* 95, 1075–1096.

Finley, J.W., Ip, C., Lisk, D.J., Davis, C.D., Hintze, K.J. and Whanger, P.D. (2001) Cancer-protective properties of high-selenium broccoli. *Journal of Agricultural Food and Chemistry* 49, 2679–2683.

Galan, M.V., Kishan, A.A. and Silverman, A.L. (2004) Oral broccoli sprouts for the treatment of *Helicobacter pylori* infection: a preliminary report. *Digestive Diseases and Sciences* 49, 1088–1090.

Gao, M., Li, G., Yang, B., McCombie, W.R. and Quiros, C.F. (2004) Comparative analysis of a Brassica BAC clone containing several major aliphatic glucosinolate genes with its corresponding *Arabidopsis* sequence. *Genome* 47, 666–679.

Gill, C.I.R., Haldar, S., Porter, S., Matthews, S., Sullivan, S., Coulter, J., McGlynn, H. and Rowland, I. (2004) The effect of cruciferous and leguminous sprouts on geno-toxicity, *in vitro* and *in vivo*. *Cancer Epidemiology, Biomarkers and Prevention* 13, 1199–1205.

Giovannucci, E., Rimm, E.B., Liu, Y., Stampfer, M.J. and Willett, W.C. (2003) A prospective study of cruciferous vegetables and prostate cancer. *Cancer Epidemiology, Biomarkers and Prevention* 12, 1403–1409.

Glawischnig, E., Hansen, B.G., Olsen, C.E. and Halkier, B.A. (2004) Camalexin is synthesized from indole-3-acetaldoxime, a key branching point between primary and secondary metabolism in Arabidopsis. *Proceedings of National Academy of Sciences USA* 101, 8245–8250.

Graser, G., Oldham, N.J., Brown, P.D., Temp, U. and Gershenzon, J. (2001) The biosynthesis of benzoic acid glucosinolate esters in *Arabidopsis thaliana*. *Phytochemistry* 57, 23–32.

Guerrera, P.M. (2005) Traditional phytotherapy in Central Italy. *Fitoterapia* 76, 1–25.

Halkier, B.A. and Gershenzon, J. (2006) Biology and biochemistry of glucosinolates. *Annual Review of Plant Biology* 57, 303–333.

Hansen, B.G. and Halkier, B.A. (2005) New insight into the biosynthesis and regulation of indole compounds in *Arabidopsis thaliana*. *Planta* 221, 603–606.

Hansen, C.H., Du, L., Naur, P., Olsen, C.E., Axelsen, K.B., Hick, A.J., Pickett, J.A. and Halkier, B.A. (2001) CYP83B1 is the oxime-metabolizing enzyme in the glucosinolate pathway in *Arabidopsis*. *Journal of Biological Chemistry* 276, 24790–24796.

Hill, R. (1991) Rapeseed meal in the diets of ruminants, a review. *Nutrition Abstracts and Reviews B* 61, 139–155.

Hintze, K.J., Wald, K. and Finley, J.W. (2005) Phytochemicals in broccoli transcriptionally induces thioredoxin reductase. *Journal of Agricultural and Food Chemistry* 53, 5535–5540.

Holst, B. and Williamson, G. (2004) A critical review of the bioavailability of glucosinolates and related compounds. *Natural Product Reports* 21, 425–447.

Horn, T.L., Reichert, M.A., Bliss, R.L. and Malejka-Giganti, D. (2002) Modulations of P450 mRNA in liver and mammary gland and P450 activities and metabolism of estrogen in liver by treatment of rats with indole-3-carbinol. *Biochemical Pharmacology* 64, 393–404.

Husebye, H., Arzt, S., Burmeister, W.P., Haertel, F.V., Brandt, A., Rossiter, J.T. and Bones, A.M. (2005) Crystal structure & resolution of an insect myrosinase from *Brevicoryne brassicae* shows its close relationship to β-glucosidases. *Insect Biochemistry and Molecular Biology* 35, 1311–1320.

Hwang, E.S. and Jeffery, E.H. (2005) Induction of quinine reductase by sulforaphane N-acetyl cystein conjugate and in murine hepatoma cells. *Journal of Medicinal Food* 8, 198–203.

Jackson, S.J.T. and Singletary, K.W. (2004) Sulforaphane inhibits human MCF-7 mammary cancer mitotic progression and tubulin polymerization. *Journal of Nutrition* 134, 2229–2236.

Jeffery, E.H. and Araya, M. (2009) Physiological effects of broccoli consumption. *Phytochemistry Review* 8, 283–298.

Jones, R.B., Faragher, J.D. and Winkler, S.A. (2006) A review of the influence of postharvest treatments on quality and glucosinolate content in broccoli (*Brassica oleracea* var. *italica*) heads. *Postharvest Biology and Technology* 41, 1–8.

Jongen, W.M.F. (1996) Glucosinolates in brassica: occurrence and significance as cancer modulating agents. *Proceedings of the Nutrition Society* 55, 433–446.

Khor, T.O., Hu, R., Shen, G., Jeong, W.S., Hebbar, V., Chen, C., Xu, C., Nair, S., Reddy, B., Chada, K. and Kong, A.N.T. (2006) Pharmacogenomics of cancer chemopreventive isothiocyanate compound sulforaphane in the intestinal polyps of Apc Min/+ mice. *Biopharmaceutics and Drug Disposition* 27, 407–420.

Kliebenstein, D.J., Kroymann, J., Brown, P., Figuth, A., Pedersen, D., Gershenzon, J. and Mitchell-Olds, T. (2001a) Genetic control of natural variation in *Arabidopsis* glucosinolate accumulation. *Plant Physiology* 126, 811–825.

Kliebenstein, D.J., Lambrix, V., Reichelt, M., Gershenzon, J. and Mitchell-Olds, T. (2001b) Gene duplication in the diversification of secondary metabolism: tandem 2-oxoglutarate-dependent dioxygenases control glucosinolate biosynthesis in *Arabidopsis*. *Plant Cell* 13, 681–693.

Kliebenstein, D.J., Kroymann, J. and Mitchell-Olds, T. (2005) The glucosinolate-myrosinase system in an ecological and evolutionary context. *Current Opinion in Plant Biology* 8, 264–271.

Kos, M., Kabouw, P., Noordam, R., Hendriks, K., Vet, L.E.M., Van Loon, J.J.A. and Dicke, M. (2011) Prey-mediated effects of glucosinolates on aphid predators. *Ecological Entomology* 36, 377–388.

Kristal, A.R. and Lampe, J.W. (2002) Brassica vegetables and prostate cancer risk: a review of the epidemiological evidence. *Nutrition and Cancer* 42, 1–9.

Kroymann, J., Donnerhacke, S., Schnabelrauch, D. and Mitchell-Olds, T. (2003) Evolutionary dynamics of an *Arabidopsis* insect resistance quantitative trait locus. *Proceedings of National Academy of Sciences USA* 100, 14587–14592.

Kurilich, A.C., Jeffery, E.H., Juvk, J.A., Walig, M.A. and Klein, B.P. (2003) Broccoli extracts protect against reactive oxygen species in HepG2 cells. *Journal of Nutraceuticals, Functional and Medical Foods* 4, 5–16.

Lambrix, V., Reichelt, M., Mitchell-Olds, T., Kliebenstein, D.J. and Gershenzon, J. (2001) The Arabidopsis epithiospecifier protein promotes the hydrolysis of glucosinolates to nitriles and influences *Trichoplusia ni* herbivory. *Plant Cell* 13, 2793–2807.

Lampe, J.W. and Peterson, S. (2002) Brassica, biotransformation and cancer risk: genetic polymorphisms alter the preventive effects of cruciferous vegetables. *Journal of Nutrition* 132, 2291–2994.

London, S.J., Yaun, J.M., Chung, F.C., Gao, Y.T., Coetzee, G.A., Ross, R.K. and Yu, M.C. (2000) Isothiocyanates, glutathione S-transferase M_1 and T_1 polymorphism and lung cancer risk: a prospective study of men in Shanghai, China. *Lancet* 35, 724–729.

Lund, E.K. (2003) Non-nutritive bioactive constituents of plants: dietary sources and health benefits of glucosinolates. *International Journal of Vitamin and Nutrition Research* 53, 135–143.

Margen, S. (2002) Recommended intakes for vitamins and minerals. In: UC Berkely Wellness Letter (eds) *Wellness Foods A to Z: an indispensable guide for health conscious food lovers.* REBUS Inc., New York, pp. 88–89.

Mikkelsen, M.D., Petersen, B.L., Olsen, C.E. and Halkier, B.A. (2002) Biosynthesis and metabolic engineering of glucosinolates. *Amino Acids* 22, 279–295.

Mikkelsen, M.D., Naur, P. and Halkier, B.A. (2004) *Arabidopsis* mutants in the C-S lyase of glucosinolate biosynthesis establish a critical role for indole-3-acetaldoxime in auxin homeostasis. *Plant Journal* 37, 770–777.

Mithen, R. (2001) Glucosinolates – biochemistry, genetics and biological activity. *Plant Growth Regulation* 34, 91–103.

Mithen, R.F., Dekker, M., Verkerk, R., Rabot, S. and Johnson, I.T. (2000) The nutritional significance, biosynthesis and bioavailability of glucosinolates in human foods. *Journal of the Science of Food Agriculture* 80, 967–984.

Moldrup, M.E., Geu-Flores, F., Olsen, C.E. and Halkier, B.A. (2011) Modulation of sulfur metabolism enables efficient glucosinolate engineering. *BMC Biotechnology* 11, 12–20.

Munday, R. and Munday, C.M. (2002) Selective induction of phase II enzymes in the urinary bladder of rats by allyl isothiocyanate, a compound derived from *Brassica* vegetables. *Nutrition and Cancer* 44, 52–59.

Murashima, M., Watanabe, S., Zhuo, X.G., Uchara, M. and Kurashige, A. (2004) Phase I study of multiple biomarkers for metabolism and oxidative stress after one-week intake of broccoli sprouts. *BioFactors* 22, 271–275.

Oerlemans, K., Barrett, D.M., Bosch Suades, C., Verkerk, R. and Dekker, M. (2006) Thermal degradation of glucosinolates in red cabbage. *Food Chemistry* 95, 19–29.

Palaniswamy, U.R., McAvoy, R.J., Bible, B.B. and Stuart, J.D. (2003) Ontogenic variations of ascorbic acid and phenethyl isothiocyanate concentrations in watercress (*Nasturtium officianale* R.Br.) leaves. *Journal of Agricultural and Food Chemistry* 51, 5504–5509.

Pappa, G., Strathmann, J., Lowinges, M., Bartsch, H. and Gerhausewr, C. (2007) Quantitative combination effects between sulforaphane and 3,3′-diindoylmethane on proliferation of human colon cancer cells *in vitro*. *Carcinogenesis* 28, 1471–1477.

Parnaud, G., Li, P., Cassar, G., Rouimi, P., Tulliez, J., Combaret, L. and Gamet-Payrastre, L. (2004) Mechanism of sulforaphane-induced cell cycle arrest and apoptosis in human colon cancer cells. *Nutrition and Cancer* 48, 198–206.

Pledgie-Tracy, A., Sobolewski, M. and Davidson, N.E. (2007) Sulforaphane induces cell type-specific apoptosis in human breast cancer cell lines *in vitro*. *Molecular Cancer Therapy* 6, 1013–1021.

Raskin, I., Ribnicky, D.M., Komarnytsky, S., Ilic, N., Poulev, A., Borisjuk, N., Brinker, A., Moreno, D.A., Ripoll, C., Yakoby, N., O'Neal, J.M., Cornwell, T., Pastor, I. and Fridlender, B. (2002) Plants and human health in the twenty-first century. *Trends in Biotechnology* 20, 522–531.

Ribnicky, D.M., Poulev, A., Henry, E. and Raskin, J. (2001) Seed of *Barbarea verna* as a rich source of phenethyl isothiocyanate to provide natural protection from environmental and dietary toxins. *Journal of Nutraceuticals, Functional and Medical Foods* 3, 43–65.

Rouzaud, G., Rabot, S., Ratcliffe, B. and Duncan, A.J. (2003) Influence of plant and bacterial myrosinase activity on the metabolic fate of glucosinolates in gnotobiotic rats. *British Journal of Nutrition* 90, 395–404.

Rungapamestry, V., Duncan, A.J., Fuller, Z. and Ratcliffe, B. (2006) Changes in glucosinolate concentrations, myrosinase activity, and production of metabolites of glucosinolates in cabbage (*Brassica oleracea* var. *capitata*) cooked for different durations. *Journal of Agricultural and Food Chemistry* 54, 7628–7634.

Rungapamestry, V., Duncan, A.J., Fuller, Z. and Ratcliffe, B. (2007) Effect of cooking brassica vegetables on the subsequent hydrolysis and metabolic fate of glucosinolates. *Proceedings of the Nutrition Society* 66, 69–81.

Sarıkamış, G., Balkaya, A. and Yanmaz, R. (2009) Glucosinolates within a collection of white head cabbages (*Brassica oleracea* var. *capitata* sub. var. *alba*) from Turkey. *African Journal of Biotechnology* 8, 5046–5052.

Sarsby, J., Towers, M.W., Stain, C., Cramer, R. and Koroleva, O.A. (2012) Mass spectrometry imaging of glucosinolates in *Arabidopsis* flowers and siliques. *Phytochemistry* 77, 110–118.

Sesso, H.D., Gaziano, J.M., Liu, S. and Buring, J.E. (2003) Flavonoid intake and the risk of cardiovascular disease in women. *American Journal of Clinical Nutrition* 77, 1400–1408.

Shapiro, T.A., Fahey, J.W., Dinkova-Kostova, A.T., Holtzclaw, W.D., Stephenson, K.K., Wade, K.L., Ye, L. and Talalay, P. (2006) Safety, tolerance, and metabolism of broccoli sprout glucosinolates and isothiocyanates: a clinical phase I study. *Nutrition and Cancer* 55, 53–62.

Singh, S.V., Srivastava, S.K., Choi, S., Lew, K.L., Antosiewicz, J., Xiao, D., Zeng, Y., Watkins, S.C., Johnson, C.S., Trump, D.L., Lee, Y.J., Xiao, H. and Herman-Antosiewicz, A. (2005) Sulforaphane-induced cell death in human prostate cancer cells is initiated by reactive oxygen species. *Journal of Biological Chemistry* 280, 19911–19924.

Smith, T.K., Lund, E.K., Parker, M.L., Clarke, R.J. and Johnson, I.T. (2004) Allyl isothiocyanate causes mitotic block, loss of cell adhesion and disrupted cytoskeletal structure in HT 29 cells. *Carcinogenesis* 25, 1409–1415.

Sønderby, I.E., Geu-Flores, F. and Halkier, B.A. (2010) Biosynthesis of glucosinolates – gene discovery and beyond. *Trends in Plant Science* 15, 283–290.

Song, L. and Thornalley, P.J. (2007) Effect of storage, processing and cooking on glucosinolate content of *Brassica* vegetables. *Food and Chemical Toxicology* 45, 216–224.

Song, L., Morrison, J.J., Botting, N.P. and Thornalley, P.J. (2005) Analysis of glucosinolates, isothiocyanates and amine degradation products in vegetable extracts and blood plasma by LC–MS/MS. *Analytical Biochemistry* 347, 234–243.

Stoner, G.D. and Morse, M.A. (1996) Isothiocyanates as inhibitors of oesophageal cancer. *Advances in Experimental Medicine and Biology* 401, 13–23.

Suido, H., Takeuchi, A., Makino, T. and Tanaka T. (2003) Serum cholesterol lowering effects of a broccoli and cabbage mixture in rats. *Atherosclerosis Supplements* 4(2), 238.

Tang, L. and Zhang, Y. (2004) Dietary isothiocyanates inhibit the growth of human bladder carcinoma cells. *Journal of Nutrition* 134, 2004–2010.

Tawfiq, N., Heaney, R.K., Plumb, J.A., Fenewick, G.R., Musk, S.S.R. and Williamson, G. (1995) Dietary glucosinolates as blocking agents against carcinogenesis: glucosinolates breakdown products assessed by induction of quinine reductase activity in murinehepa1C1c 7 cells. *Carcinogenesis* 16, 1191–1194.

Textor, S., Bartram, S., Kroymann, J., Falk, K.L., Hick, A., Pickett, J.A. and Gershenzon, J. (2004) Biosynthesis of methionine-derived glucosinolates in *Arabidopsis thaliana*: recombinant expression and characterization of methylthioalkylmalate synthase, the condensing enzyme of the chain-elongation cycle. *Planta* 218, 1026–1035.

Thejass, P. and Kuttan, G. (2006) Augmentation of natural killer cell and antibody dependent cellular cytotoxicity in BALB/c mice by sulforphane, a naturally occurring isothiocyanate from broccoli 1L-2 and IFN-gamma. *Immunonopharmacology and Immunnotoxicology* 28, 443–457.

Traka, M. and Mithen, R. (2009) Glucosinolates, isothiocyanates and human health. *Phytochemistry Reviews* 8, 269–282.

Tsiafoulis, C.G., Prodromidis, M.I. and Karayannis, M.I. (2003) Development of a flow amperometric enzymatic method for the determination of total glucosinolates in real samples. *Analytical Chemistry* 75, 927–934.

Vallejo, F., Tomás-Barberán, F.A. and García-Viguera, C. (2002) Potential bioactive compounds in health promotion from broccoli cultivars grown in Spain. *Journal of the Science of Food and Agriculture* 82, 1293–1297.

Vallejo, F., Tomás-Barberán, F.A. and García-Viguera, C. (2003) Health promoting compounds in broccoli as influenced by refrigerated transport and retail sale period. *Journal of Agricultural Food and Chemistry* 51, 3029–3034.

Vandermeiren, K., De Bock, M., Horemans, N. and Gielen, B. (2009) Impact of tropospheric ozone on glucosinolate and vitamin C content of oilseed rape and broccoli. In: Sirko, A., De Kok, L.J., Haneklaus, S., Hawkesford, M.J., Rennenberg, H., Saito, K., Schnug, E. and Stulen, I. (eds) *Sulfur Metabolism in Plants*. Backhuys Publishers, Leiden, Margraf Publishers, Weikersheim, the Netherlands, pp. 245–252.

Vig, A.P., Rampal, G., Singh, T.S. and Arora, S. (2009) Bioprotective effects of glucosinolates – a review. *LWT-Food Science and Technology* 42, 1561–1572.

Volden, J., Wicklund, T., Verkerk, R. and Dekker, M. (2008) The kinetics of changes in glucosinolate concentrations during long term cooking of white cabbage (*Brassica oleracea* L. ssp. *capitata* f. *alba*). *Journal of Agricultural and Food Chemistry* 56, 2068–2073.

Wagner, A.E. and Rimbach, G. (2009) Ascorbigen: chemistry, occurrence, and biologic properties. *Clinics in Dermatology* 27, 217–224.

Wang, L.J., Giovannucci, E.L., Hunter, D., Neuberg, D., Su, L. and Christiani, D.C. (2004) Dietary intake of Cruciferous vegetables glutathione-S-transferase (GST) polymorphisms and lung cancer risk in a Caucasian population. *Cancer Causes and Control* 15, 977–985.

West, L.G., Meyer, K.A., Balch, B.A., Rossi, F.J., Schultz, M.R. and Haas, G.W. (2004) Glucoraphanin and 4-hydroxyglucobrassicin contents in seeds of 59 cultivars of broccoli, raab, kohlrabi, radish, cauliflower, brussels sprouts, kale, and cabbage. *Journal of Agricultural and Food Chemistry* 52, 916–926.

Wittstock, U. and Halkier, B.A. (2002) Glucosinolate research in the *Arabidopsis* era. *Trends in Plant Science* 7, 263–270.

Wittstock, U., Kliebenstein, D.J., Lambrix, V., Reichelt, M. and Gershenzon, J. (2003) Glucosinolate hydrolysis and its impact on generalist and specialist insect herbivores. In: Romeo, J.T. (ed.) *Recent Advances in Phytochemistry: Integrative Phytochemistry: from Ethnobotany to Molecular Ecology*. Pergamon, Oxford, UK, pp. 101–125.

Wu, H.T., Lin, S.H. and Chen, Y.H. (2005) Inhibition of cell proliferation and *in vitro* markers of angiogenesis by indole-3-carbinol, a major indole metabolite present in cruciferous vegetables. *Journal of Agricultural and Food Chemistry* 53, 5164–5169.

10 Role of Phytoestrogens as Nutraceuticals in Human Health

Dhan Prakash* **and Charu Gupta**
Amity Institute for Herbal Research and Studies, Amity University, Noida, India

10.1 Introduction

Phytoestrogens (PE) are non-steroidal oestrogen-like chemical compounds produced by plants and present in many natural dietary sources, such as soybeans, wheat, barley, maize, lucerne and oats. They structurally or functionally mimic mammalian oestrogens and therefore are considered to play an important role in the prevention of cancers, heart diseases, menopausal symptoms and osteoporosis (Setchell, 1998; Adlercreutz, 2002; Kronenberg and Fugh-Berman, 2002). Oestrogens influence the growth and functioning of female and male reproductive tissues, maintain the skeletal and central nervous system, provide cardioprotective effects and protect against colon cancer and ageing of skin (Gruber *et al.*, 2002; Ruggiero and Likis, 2002). Plants with oestrogen-like biological activity are being used in traditional systems of medicine and folklore, for example, the pomegranate is associated with fertility, the Thai vine *Pueraria mirifica* as rejuvenator and aphrodisiac and hops were used to lower libido by the German clergy in the middle ages. To the present time, several hundred plants have been found to exhibit oestrogenic activity due to the presence of phytochemicals called phytoestrogens. They are recommended for the prevention of disturbed hormone-related diseases (Price and Fenwick, 1985; Murkies *et al.*, 1998).

Phytoestrogens are substances that promote oestrogenic actions in mammals and structurally are similar to mammalian oestrogen 17β-oestradiol (E_2) (Price and Fenwick, 1985; Knight and Eden, 1996). The diverse biological activity of PE is due in part to their ability to act oestrogenically as oestrogen agonists and anti-oestrogenically as antagonists. They can mimic endogenous oestrogens and cause oestrogenic effects. As oestrogen antagonists, they may block or alter oestrogen receptors (ER) and prevent oestrogenic activity, causing anti-oestrogenic effects (Brzezinski and Debi, 1999). Mechanistically, PE have been shown to bind to two types of oestrogen receptors: oestrogen receptor α (ERα) and oestrogen receptor β (ERβ) (Kuiper *et al.*, 1996). The two receptors differ in their tissue distribution and affinity to ligands, yet there is some overlap. Phytoestrogens show a lower binding affinity than E_2 and some show a higher binding affinity for ERβ than for ERα, which may suggest different pathways for their actions and explain tissue-specific variability of phytoestrogenic action (Kuiper *et al.*, 1998; Setchell, 1998).

* E-mail: dprakash_in@yahoo.com

Both genomic and non-genomic mechanisms have been proposed to explain phytoestrogenic effects on human health (Anderson *et al.*, 1999). PE are able to interact with enzymes and receptors, and because of their stable structure and low molecular weight they can pass through cell membranes (Adlercreutz, 1998). These interactions allow them to bind to ERs, induce specific oestrogen-responsive gene products, stimulate ER-positive breast cancer cell growth (Kurzer and Xu, 1997), interfere with steroid hormone metabolism or action (Adlercreutz, 1998) and alter ER structure and affect transcription (Santti *et al.*, 1998). Some genomic mechanisms of action include oestrogenic and anti-oestrogenic effects on ERs, while other effects may not involve direct interaction with ERs (Messina and Loprinzi, 2001). Non-genomic effects that do not involve ERs include: induction of cancer cell differentiation, inhibition of tyrosine kinase and DNA topoisomerase activities, suppression of angiogenesis and antioxidant effects of PE (Kurzer and Xu, 1997). Other effects can take place at the cellular and molecular level and potentially influence the biosynthesis and metabolism of steroids and fatty acids, the serum steroid carrier proteins (sex steroid binding proteins and α-fetoprotein), and the intracellular and trans-membrane transfer of hormones to a membrane and to nuclear receptors (Benassayag *et al.*, 2002). The different activities and the bioavailability of PE vary depending on such factors as the form of administration, dosage, individual metabolism and the ingestion of other pharmacological substances (Kelly *et al.*, 1995; Wiseman, 1999). Target tissue, concentration dependency, number and type of ER and the presence or absence of endogenous oestrogens also influence the effect of PE (Glazier and Bowman, 2001).

There are several classes of PE: steroidal oestrogens, found in few plants and the more ubiquitous phenolic oestrogens, isoflavones, stilbenes, coumestans and lignans. Other classes of PE that have been reported include: anthraquinones (Matsuda *et al.*, 2001), chalcones (Rafi *et al.*, 2000), flavones (Milligan *et al.*, 1999), prenylflavonoids (Kitaoka *et al.*, 1998) and saponins (Chan *et al.*, 2002). Phytoestrogens have been categorized based on their chemical structures, which resemble E_2. Oestrogen receptors bind with steroidal as well as numerous non-steroidal compounds. An aromatic ring and a hydroxyl group are important for binding effectiveness and the remainder of the ER will accept hydrophobic groups (Anstead *et al.*, 1997).

The mechanisms through which the PE may influence sex hormone production, metabolism and biological activity could depend, at least in part, on their mixed oestrogen agonist/antagonist properties and binding to oestrogen receptors. Furthermore, these weakly oestrogenic molecules have been demonstrated to affect intracellular enzymes, protein synthesis, growth factor action, malignant cell proliferation, cell differentiation, cell adhesion, angiogenesis and apoptosis. Experimental studies in animals suggest that both lignans and isoflavonoids are among the dietary factors affording protection against atherosclerotic vascular disease and cancer (Clarkson and Anthony, 1998; Murkies *et al.*, 1998; Tham *et al.*, 1998).

Important features that enable chemicals to bind to an ER are the steric and hydrophobic properties of a compound, as well as the hydrogen bonding between the phenolic hydroxyl group and the ER binding site (Hu and Aizawa, 2003). Oestrogenic flavonoids are similar in structure to E_2. They are composed of a planar ring system that includes a *p*-hydroxy-substituted aromatic ring that is approximately 12 Å away from a second in-plane hydroxyl group (Hu and Aizawa, 2003). Two ring structures separated with two carbon atoms as well as spacing between hydrophobic and hydrogen bond interactions are also important in binding affinity to ERs (Brzozowski *et al.*, 1997). Other characteristics for ER-binding affinity of a chemical are the degree and size of branching of the alkyl group and its location on the phenolic ring and the distribution range of electron density on the A ring (Hu and Aizawa, 2003). The biological activity of individual PE varies and is often reported as less active than mammal or synthetic oestrogens (Knight and Eden, 1996; Tham *et al.*, 1998). Differences in oestrogenic activity of similarly classified chemicals may be due to the structural features or deviations in those structures. Some phytoestrogenic compounds may show different oestrogenicity due to the bioassay employed (Messina and Loprinzi, 2001) and others may not show oestrogenic activity in bioassays because only their metabolized derivatives are hormonally active (Miksicek, 1994).

10.2 The Naturally Occurring Groups of Phytoestrogens

The major PE groups are isoflavones, flavones, stilbenes, lignans and coumestans (Table 10.1). As studies continue to evaluate the biological effects of PE on human health, the complexity is more evident as oestrogenic and anti-oestrogenic effects are observed as well as a variety of mechanisms of action.

10.2.1 Isoflavones

Isoflavones (Fig. 10.1) are the most studied group of PE. These are found exclusively in the family *Fabaceae* (*Leguminosae*) (King and Young, 1999). Soybeans are a very rich source of isoflavones and contain approximately 2 g of isoflavones kg^{-1} fresh weight (Reinli and Block, 1996). The isoflavonoids encompass several structurally and biosynthetically related classes such as flavonols, anthocyanins, flavanones, coumestans and chalcones. Isoflavonoids differ structurally from other classes of flavonoids in having the phenyl ring attached at the 3- rather than at the 2-position of the heterocyclic ring. In addition, isoflavonoids differ on account of their greater structural variation and the greater frequency of isoprenoid substitution. Isoflavones have similar structure to oestrogen and have the capacity to exert both oestrogenic and anti-oestrogenic effects, they may block the effects of oestrogen in some tissues, e.g. the breast and womb lining, but act like an oestrogen in providing possible protection against bone loss and heart diseases. In this subclass, the most thoroughly investigated and interesting compounds with regard to oestrogenicity are genistein (Fig. 10.1a), daidzein (Fig. 10.1b), biochanin A (Fig. 10.1c) and formononetin (Fig. 10.1d) (Shutt and Braden, 1968).

The main isoflavones, genistein and daidzein, commonly exist as inactive glucosides. They are also derived from precursors, biochanin A and formononetin, which are converted to genistein and daidzein, respectively, after breakdown by intestinal glucosidases. Daidzein is further partially metabolized to O-desmethyl-angiolensin (O-DMA) (Fig. 10.2) and equol (Fig. 10.3). Because of their non-steroidal skeleton and different special structure, PE when bound to the oestrogen receptors (ER) were expected to act totally differently. They share structural features with oestrogen, in the sense that the presence of particular hydroxyl groups that can be positioned in a stereochemical alignment virtually identical to one of the oestrogen. They can exist as glucosides or as aglycones, the glucosides being readily hydrolysed in the

Table 10.1. Dietary phytoestrogens of human interest and their food sources.

Class	Phytoestrogens	Food sources
Isoflavones	Genistein, biochanin A, diadzein (with its metabolites: O-DMA and equol), formononetin, glycetin	Soy, groundnut, clover, sunflower seed, walnut
Flavones	Apigenin, chrysin, quercetogetin, luteolin, tricetin	Parsley, celery, citrus peels, capsicum, pepper
Stilbenes	Resveratrol	Grape, groundnut
Lignans	Secoisolariciresinol, matairesional, enterodiol, enterolactone	Soybean, groundnut, broccoli, cashew nut, kiwi, pomegranate, triticale straw, flaxseed, cereals
Coumestans	Coumestrol	Mung beans or soy sprouts, lucerne sprouts, clover

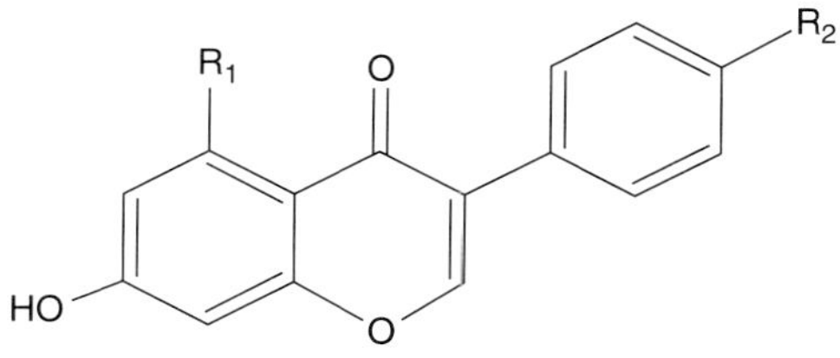

Fig. 10.1. Chemical structures of some isoflavones: R_1 = OH, R_2 = OH, genistein; R_1 = H, R_2 = OH, daidzein; R_1 = OH, R_2 = OMe, biochanin A; and R_1 = H, R_2 = OMe, formononetin.

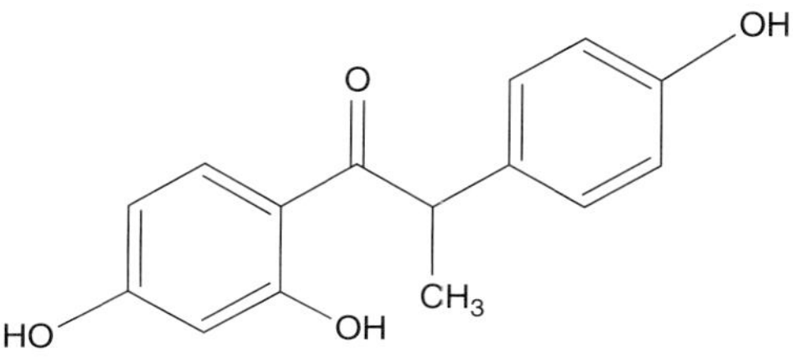

Fig. 10.2. O-Desmethyl-angiolensin (O-DMA).

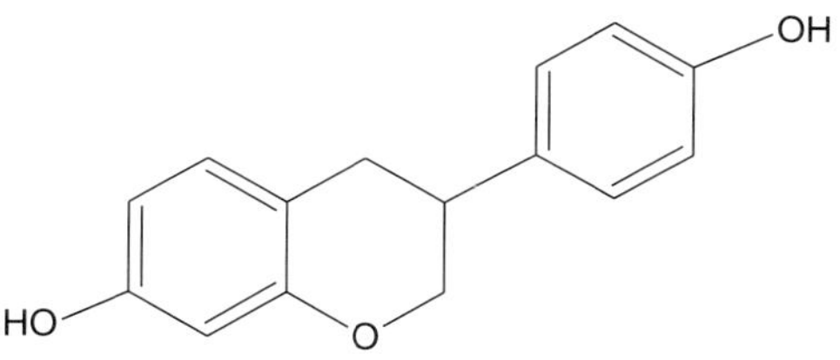

Fig. 10.3. Equol.

gut to their aglycones. The aglycones are easily transported across intestinal epithelial cells. Once ingested, the absorption of these compounds requires initial hydrolysis of the sugar moiety or demethylation, respectively, by gut/bacteria-released intestinal enzymes in the digestive tract. The metabolization of isoflavones to equol and O-desmethyl-angolensin (O-DMA) for daidzein and to 2-(4-hydroxyphenyl) propanoic acid and tri-hydroxybenzene (THB) for genistein by gut bacteria are subjects of large inter-individual variation, depending on gastrointestinal microflora and diet.

The oestrogen effect of isoflavones is much less powerful than the oestrogen hormones. This is why isoflavones and PE exercise a balancing effect when the level of oestrogens is low, such as during the menopause, and cause less menopause symptoms. Isoflavones can also reduce the effect of the oestrogen on cells and skin layers when the hormone levels are high, and then essentially reduce the risk of oestrogen-linked cancers. Some isoflavones are termed as antioxidants because of their ability to trap singlet oxygen (Heber *et al.*, 2008). Some isoflavones, in particular soy isoflavones, when studied in populations eating soy protein, have indicated that there is a lower incidence of breast cancer and other common cancers because of its role in influencing sex hormone metabolism and biological activity through intracellular enzymes, protein synthesis, growth factor actions, malignant cell proliferations, differentiation and angiogenesis (Heber *et al.*, 2008). Their main food sources (Table 10.2) are soy cheese, soy flour, soybean, tofu and legumes.

Genistein is one of the several known isoflavones. It is found in a number of plants including lupin, fava beans, soybeans and tofu (Kaufman *et al.*, 1997), *Flemingia vestita* (Rao and Reddy, 1991) and coffee (Alves *et al.*, 2010). Soybeans, a cholesterol-free, high-protein legume, contain the most genistein. Other legumes, such as chickpeas (garbanzo beans), contain small amounts of genistein. Genistein can be found in many food products containing soy such as soy-based infant formulas, tofu, soymilk, soy flour, textured soy protein, soy protein isolates and tempeh as well as over-the-counter dietary supplements. Genistein was first isolated in 1899 from the dyer's broom *Genista tinctoria*; hence, the chemical name derived from the generic name. Genistein is a PE that binds to oestrogen receptors and has both weak oestrogenic and weak anti-oestrogenic effects. *In vitro* studies have shown that the growth of both oestrogen receptor-positive breast cancer cells and oestrogen receptor-negative breast cancer cells is inhibited when high levels of genistein (>10 μM) are added to the culture medium; however, the growth of oestrogen receptor-positive breast cancer cells is actually stimulated when low and physiologically relevant concentrations of genistein are added (Messina *et al.*, 2006). The association between genistein and breast cancer risk

Table 10.2. Isoflavonoid contents of some commonly used foods.

Sources	Isoflavones ($\mu g\ g^{-1}$)	Sources	Isoflavones ($\mu g\ g^{-1}$)
Soy	610–2440	Kidney bean	0.1–4.1
Soy sprouts	250–530	Black gram	6.4–12.6
Soy protein	465–1993	Green gram	7.0
Soymilk	13–211	Red gram	2–5.6
Soy cheese	33–593	Beans	0.3–1.3
Tofu	79–635	Lentil	0.23–0.4
Miso	227–892	Barley	0.21
Soy sauce	12.7–23.0	Peas	0.4
Groundnut	3.5–8.4	Coconut	0.19
Tea	2.34	Currants	2.25

in vitro is complex and depends on both the concentration of genistein and the concentration of oestrogen. Genistein has consistently been shown to inhibit the development of oestrogen-sensitive mammary tumours when given to prepubertal rats (Trock *et al.*, 2006).

The isoflavones such as genistein found in soy should be an integral part of everyone's diet. They help to reduce cholesterol, prevent atherosclerosis, protect or slow prostate and breast cancer growth, prevent the kind of cell mutation that causes DNA damage, inhibit blood supply to already existing tumours, ease menopause and lower the risk of osteoporosis. Genistein is considered the natural analogue to the drug tamoxifen, which is an anti-oestrogen compound used to treat breast cancer. Genistein has also shown the ability to destroy certain cancer gene enzymes that can change a normal cell into a cancer cell, which simultaneously inhibit blood vessel growth to larger tumours. Genistein can diminish the possibilities of cellular mutations which can result in malignant tumours, especially in tissue which is oestrogen-sensitive. Genistein is the isoflavone that bumps oestrogen away from oestrogen receptor sites on cells and inhibits an enzyme called tyrosine kinase, which is involved in the formation of malignant tumours (Padilla-Banks, 2006; Trock *et al.*, 2006).

A closely related compound to the isoflavonoids is 8-prenyl-naringenin (Fig. 10.4), an isoflavanone, found in hops (*Humulus lupulus*), an ingredient used in beer. Populations in China, Japan, Taiwan and Korea are estimated to consume high quantities of isoflavones and women of these countries complain of fewer incidences of osteoporosis and related health problems, especially hot flushes, cardiovascular diseases, lower incidence of hormone-dependent breast and uterine cancer. Isoflavones have also been reported to inhibit angiogenesis, cell cycle progression, aromatase enzyme inhibition, stimulation of sex hormone binding globulin (SHBG) synthesis and digitalis-like activity.

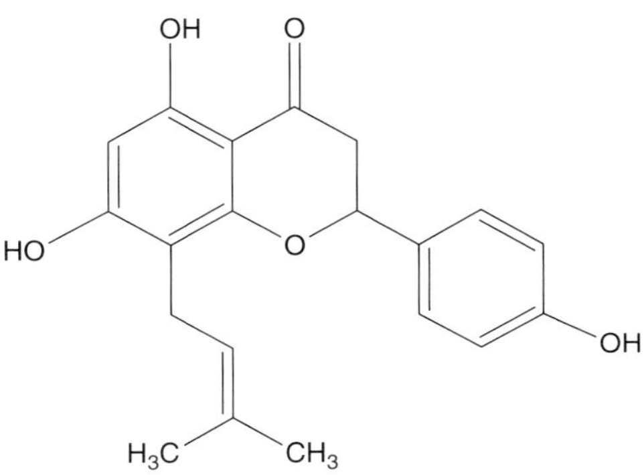

Fig. 10.4. 8-Prenyl-naringenin.

10.2.2 Flavones

The flavones are a group of naturally occurring chemical compounds widely distributed in plants. Natural flavones include apigenin, chrysin, quercetogetin, luteolin and tricetin. Their major food sources are parsley, celery, citrus peels, capsicum and pepper. Apigenin (4′,5,7-trihydroxyflavone) (Fig. 10.5) is commonly present in fruits and vegetables with proven anti-inflammatory and anticarcinogenic effects in various animal tumour model systems (Birt *et al.*, 1986; Liang *et al.*, 1999). It has been shown to suppress angiogenesis in melanoma and carcinoma of the breast, skin and colon (Caltagirone *et al.*, 2000; Liu *et al.*, 2005). The effects of apigenin seem to be primarily mediated through suppression of the expression of hypoxia-inducible factor 1-α, cyclooxygenase-2, nitric oxide synthase-2, vascular endothelial growth factor, and lipoxygenase (Liang *et al.*, 1999; Fang *et al.*, 2005). Apigenin has shown potential to inhibit growth in several human cancer cells, including breast, colon, skin, thyroid, leukaemia and prostate (Wang *et al.*, 1999, 2000; Knowles *et al.*, 2000; Yin *et al.*, 2001). These cell inhibitory effects are mediated via cell cycle arrest and induction of apoptosis. The molecular targets of apigenin-mediated cell growth inhibition and apoptosis are through activation of caspases, inhibition of fatty acid synthase, topoisomerase inhibition, nuclear factor-nB inhibition, and modulation in Bax and Bcl-2 ratio (Wang *et al.*, 1999; Brusselmans *et al.*, 2005). Apigenin is a potent inhibitor of several protein tyrosine kinases, including epidermal growth factor receptor and src tyrosine kinase. Apigenin has also been shown to inhibit activation of phosphatidylinositol 3-kinase, protein kinase B/Akt, mitogen-activated protein kinase/extracellular signal-regulated kinase 1/2, casein kinase-2, and other upstream kinases involved in the development and progression of cancer (Llorens *et al.*, 2004; Way *et al.*, 2004).

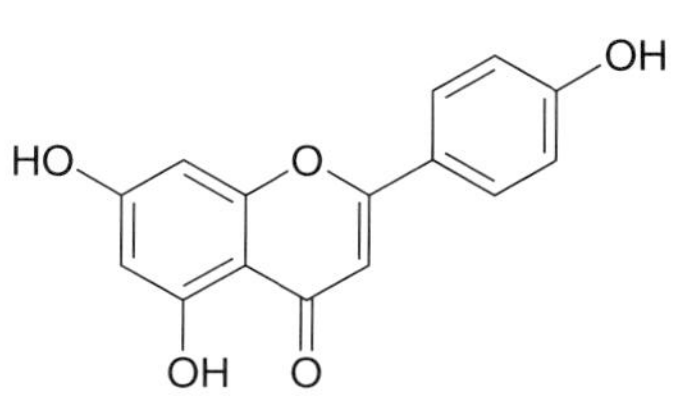

Fig. 10.5. Apigenin.

10.2.3 Stilbenes

In biochemical terms, stilbenes (Fig. 10.6) belong to the family of phenylpropanoids and share most of their biosynthesis pathway with chalcones (Sobolev *et al.*, 2006). An example of stilbene is resveratrol (Fig. 10.6a) found in grapes which has several health benefits (Jang *et al.*, 1997). It exists in two structural isomeric forms, *cis* and *trans*, with the *trans* form being more common and possessing greater biological activity. One of the richest sources of this is *Polygonum cuspidatum*, a weed that is used in traditional Chinese and Japanese medicines. Trees such as *Eucalyptus* and spruce have also been found to contain resveratrol (Rolfs and Kindle, 1984). The primary dietary sources in the human diet (Table 10. 3) are groundnuts, grapes and wine.

The potential health benefits of resveratrol depend upon its absorption, bioavailability and metabolism. Using the Caco-2 human intestinal cell model, Kaldas *et al.* (2003) demonstrated that resveratrol uptake remained linear, for 1 h, and transportation was non-directional. Metabolites identified in the Caco-2 cells were resveratrol sulfate and resveratrol glucuronide, with the former being

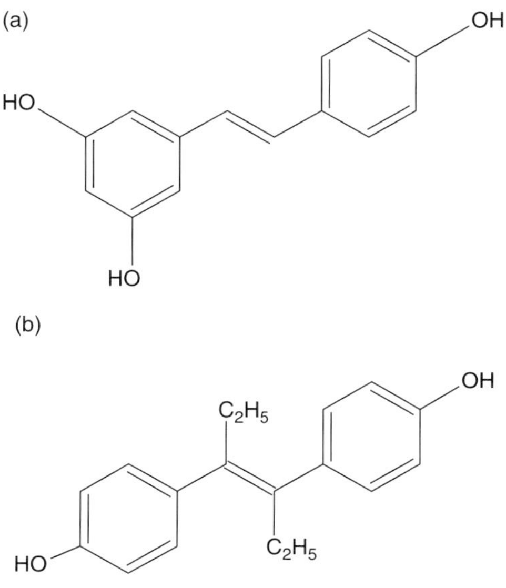

Fig. 10.6. Chemical structures of some stilbenes: (a) *trans*-resveratrol and (b) diethyl-stilbestrol.

Table 10.3. Resveratrol content of some foods.

Sources	Resveratrol (μg g^{-1})
Wine	0.316–15.348
Peanut butter	0.015–0.982
Groundnuts	0.003–0.0725
Green groundnuts	0.183–0.716
Polygonum cuspidatum	296–377
Green grapes	0.016–0.318
Black grapes	0.945–1.874
Raisins	0.0005–0.003
Grape juice – black	Traces–0.087
Grape juice – green	Traces–0.0015

predominant. It has exhibited antioxidant, cardioprotective, chemopreventive, anti-inflammatory and oestrogenic properties, as well as interaction with signal transduction pathways. It has been shown to inhibit oxidative-induced apoptosis in a variety of cell lines including Swiss 3T3 mouse fibroblasts, human peripheral blood mononuclear (PBM) and human retinal pigment epithelium (RPE) cells (Kutuk *et al.*, 2004; King *et al.*, 2005). Reduced oxidative stress in RPE cells by resveratrol may be associated with reduced incidence of age-related macular degeneration (AMD), a leading cause of blindness in the elderly. The antioxidant activity of resveratrol may also be associated with protection against the progression of atherosclerosis.

Red wine is one of the few dietary sources of resveratrol and it is believed that this compound is responsible, in part, for the positive cardiovascular effects associated with moderate wine consumption (Constant, 1997). The most accepted mechanism of cardioprotection by resveratrol is the inhibition of platelet aggregation (Bhat *et al.*, 2001). Excessive aggregation can lead to the development of cardiovascular disease. Pretreatment of platelets with resveratrol has been shown to inhibit lipopolysaccharide (LPS) and LPS + thrombin-stimulated platelet adhesion to collagen and fibrinogen in a non-dose-dependent manner (Olas *et al.*, 2002; Wang *et al.*, 2002).

The antiproliferative activity of resveratrol has been observed in a number of cancer cell lines and may be due, in part, to the induction of apoptosis (Ding and Adrian, 2002). Proliferation inhibition may also be caused by the arrest of the cell cycle (Castello and Tessitore, 2005). Piceatannol, a naturally occurring analogue of resveratrol, has been observed to inhibit the proliferation of cancer cell lines via apoptosis and cell cycle arrest (Wolter *et al.*, 2002; Larrosa *et al.*, 2004). Resveratrol has also been observed to decrease induced COX-2 activity by inhibiting the expression of the enzyme via signal transduction pathways. Resveratrol also inhibits the inflammatory actions of cytokines, such as tumour necrosis factor-α (TNF-α) and interleukin-1β (IL-1β) (Culpitt *et al.*, 2003).

The structural similarity of resveratrol to the synthetic oestrogen diethyl-stilbestrol (DES) (Fig. 10.6b) suggests that it may have oestrogenic activity, cardioprotection and prevention of oestrogen-dependent cancers. The oestrogenic activity of resveratrol may also help prevent bone loss in post-menopausal women. Resveratrol was shown to increase the proliferation of osteoblastic MC3T3-E1 cells and induce alkaline phosphatase activity, an enzyme believed to be involved in bone mineralization (Liu *et al.*, 2005).

Resveratrol has been shown to exert neuroprotective effects (de Ruvo *et al.*, 2000), as well as beneficial effects on the cardiovascular system (Pendurthi *et al.*, 1999). These effects are mostly attributed to its antioxidant properties. Resveratrol acts on the process of carcinogenesis by affecting the three phases: tumour initiation, promotion and progression phases, and suppresses the final steps of carcinogenesis, i.e. angiogenesis and metastasis. It is also able to activate apoptosis, arrest the cell cycle or inhibit kinase pathways. Most noticeable biological activities are antithrombogenic, anti-inflammatory, cardioprotective, neuroprotective, anti-ageing and cancer preventive and therapeutic activities. The *trans*-resveratrol can also act as a preventive agent against vascular diseases, cancers, viral infection or neurodegenerative processes. Resveratrol inhibits the proliferation and induces apoptotic cell death in multiple cancer cell types *in vitro* (Kim *et al.*, 2002; Pozo-Guisado *et al.*, 2002); moreover, in animal models of cancer, resveratrol has been shown to inhibit angiogenesis and delay tumour growth (Tseng *et al.*, 2004), impede carcinogenesis (Gusman *et al.*, 2001) and reduce experimental metastasis

(Busquets *et al.*, 2007). It is produced in huge amounts in grape skin in response to infection by *Botrytis cinerea*. This production of resveratrol blocks the proliferation of the pathogen, thereby acting as a natural antibiotic.

10.2.4 Lignans

The lignan (Fig. 10.7) family is a large group of naturally abundant molecules that can be found in a plethora of superior plants of which flaxseed is a particularly rich source (Table 10.4). Lignans, along with isoflavones and coumestans, comprise the three major classes of PE. When plant lignans are consumed, intestinal bacteria convert some into two mammalian lignans, enterolactone (Fig. 10.7a) and enterodiol (Fig. 10.7b). These compounds are absorbed from the digestive tract, circulate and are excreted in the urine (Beejmohun *et al.*, 2007; Bonzanini *et al.*, 2009; Hosseinian and Mazza, 2009; Attoumbre *et al.*, 2010).

Among lignans, secoisolariciresinol (SECO) (Fig. 10.7c) and matairesinol (Fig. 10.7d) are of particular interest. Secoisolariciresinol and matairesinol are two lignan dimers that are not oestrogenic by themselves, but readily convert to the mammalian lignans, enterodiol and enterolactone, respectively, which are oestrogenic. These are of great interest because of their oestrogenic, anti-oestrogenic, anticarcinogenic, antiviral, antifungal and antioxidant activities. Particularly abundant in flaxseed, these molecules can also be found, for example, in soybean, groundnut (Mazur and Adlercreutz, 1998), broccoli, cashew nut (Schwartz and Sontag, 2006), kiwi (Milder *et al.*, 2005) and pomegranate (Bonzanini *et al.*, 2009), triticale straw (Hosseinian and Mazza, 2009), greater burdock (Cai *et al.*, 2006) or *Forsythia intermedia* (Umezawa *et al.*, 1991), asparagus, whole grains and tea.

SECO, like other lignans, may occur in different natural forms. They can be free (aglycone) or attached to other molecules such as glucose or organic acids to form more complex structures. In flaxseed, SECO is mostly mono- or diglycosylated (SMG or SDG). SMG and SDG are linked by 3-hydroxy-3-methyl-glutaryl units (HMG), derived from

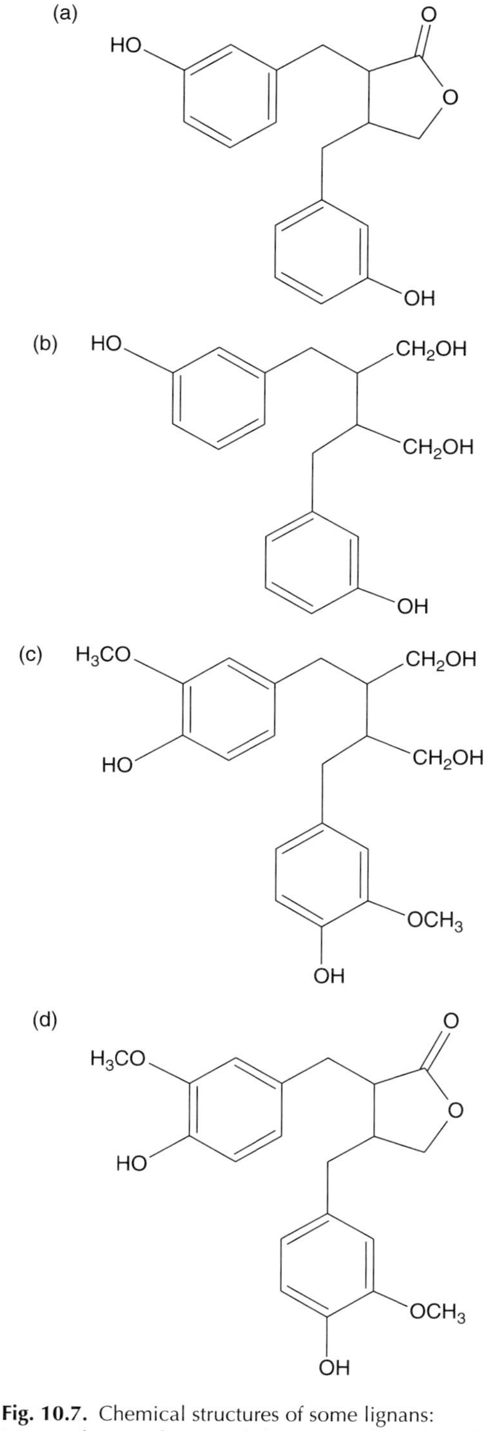

Fig. 10.7. Chemical structures of some lignans: (a) enterolactone, (b) enterodiol, (c) secoisolariciresinol and (d) matairesinol.

Table 10.4. Lignan content of some commonly used foods.

Sources	Lignans (μg g^{-1})	Sources	Lignans (μg g^{-1})
Soybean	9.6	Banana	2.8
Flaxseed	347.5–1140.3	Orange	2.1
Lentils	19.6	Apple	2.5
Cowpea	1.9	Mango	1.7
Green gram	1.7	Carrot	29.3
Walnut	1.7	Cabbage	18.6
Cashew	2.6	Cauliflower	16.2
Groundnut	1.8	Onion	10.3
Wheat	0.8–2.8	Garlic	10.5
Rice	0.2–6.0	Potato	4.1
Barley	1.3	Cucumber	5.9
Strawberry	7.7	Tomato	3.3

3-hydroxy-3-methyl-glutaric acid (HMGA), to form an ester-linked biopolymer (Ford *et al.*, 2001). SECO is known to have many physiological properties and health benefits. Indeed, SECO is converted into enterolignans (enterodiol (END) and enterolactones (ENL)) by the anaerobic intestinal microflora (Wang *et al.*, 2010). Three other components of flaxseed, matairesinol (MATA), lariciresinol (LARI) and pinoresinol (PINO), are also converted into enterolignans. These are mammalian oestrogen precursors, also called PE (Raffaelli *et al.*, 2002; Bartkiene *et al.*, 2011).

Due to the structural similarity of enterolignans with mammalian oestrogens, these compounds are potentially interesting for combating some hormone-dependent cancers (Apers *et al.*, 2003; Duncan *et al.*, 2003; Boccardo *et al.*, 2006; Albertazzi and Purdie, 2008). Some epidemiologic investigations have shown that the risk of breast, prostate and colon cancers is lower in countries or regions in which the diet is particularly rich in lignans. Lignans could influence ER-negative and ER-positive tumours by decreasing insulin-like growth factor-1 (IGF-1), epidermal growth factor receptor (EGFR), HER2 and the vascular endothelial growth factor (VEGF), which supports angiogenesis (Bergman *et al.*, 2007; Chen *et al.*, 2007). One trial that involved up to 15 g flaxseed day^{-1} suggests that flaxseed's impact on serum oestrogens may be greater among overweight and obese postmenopausal women than those of normal weight (Sturgeon *et al.*, 2008); these women generally have higher levels of circulating oestrogen. The range in response to lignans may also reflect polymorphisms in genes related to hormone metabolism (McCann *et al.*, 2007).

The protective role of lignans can be in part explained by the fact that they can bind to oestrogen receptors. They can stimulate the production of the sex hormone-binding globulin in the liver, which results in reducing the concentration of free hormones in the plasma. They interact with sex steroid binding protein and act as inhibitors of several steroid metabolizing enzymes such as aromatase and cholesterol 7α-hydrolase, these inhibitions being positive against breast and colon cancers, respectively. Their antioxidant activity is also one of the possible anticarcinogenic mechanisms of the compounds. For example, they can reduce the endogenous generation of oxidized DNA bases. SECO, SDG, END and ENL were demonstrated to have antioxidant activity (Hu *et al.*, 2007).

10.2.5 Coumestans

Coumestans (Fig. 10.8) are another important group of plant (family *Fabaceae*) phenols that show oestrogenic activity. The main coumestans with phytoestrogenic effects are coumestrol (Fig. 10.8a) and 4′-methoxy-coumestrol (Fig. 10.8b). Coumestrol was first isolated from ladino clover (*Trifolium repens* L.), strawberry clover (*Trifolium fragiferum* L.) and lucerne (*Medicago sativa* L.). Coumestrol and genistein have higher binding affinities to ER-β than

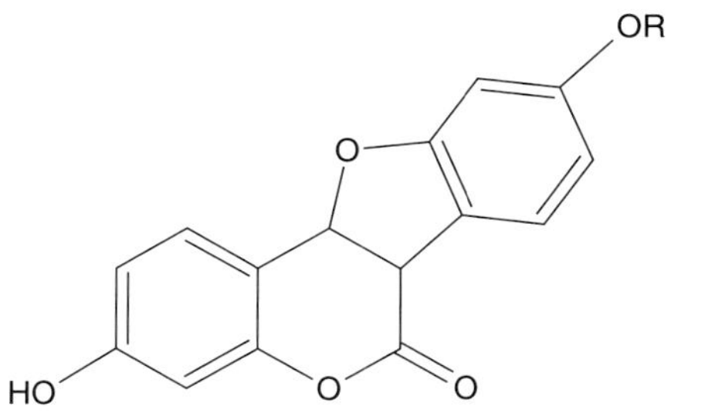

Fig. 10.8. Chemical structures of some coumestans: R = H, coumestrol; R = Me, 4′-methoxy-coumestrol.

the other PE compounds (Whitten and Naftolin, 1998). Under *in vitro* conditions, coumestrol has been reported to inhibit bone resorption and to stimulate bone mineralization (Tsutsumi, 1995). Coumestans are less common in the human diet than isoflavones (Ibarreta *et al.*, 2001) yet similar to isoflavones, in that they are also found in legumes, particularly sprouts of lucerne and mung bean (*Vigna radiata*) (Adams, 1995; Mazur and Adlercreutz, 1998) and they are especially high in clover (Franke *et al.*, 1995). Soy sprouts also show high levels of coumestrol (71.1μg g^{-1}) (Ibarreta *et al.*, 2001), however low levels have been reported in Brussels sprouts and spinach.

10.2.6 Terpenoids

Ikeda *et al.* (2000) surveyed oestrogenic and anti-oestrogenic activities of terpenoid (Fig. 10.9) phytochemicals found in the *Umbelliferae* family and revealed that three compounds, tschimgine (Fig. 10.9a), tschimganidine (Fig. 10.9b) and ferutinine (Fig. 10.9c), have agonistic and/or antagonistic activities for ER-α and ER-β. Ferutinine and tschimganidine are sesquiterpenoids and tschimgine is a monoterpenoid. The structures and biosynthesis of these terpenoids are distinct from the well-known PE such as isoflavones (Mazur and Adlercreutz, 2000). Ferutinine isolated from *Ferula jaeschkeana* was reported to increase uterine weight and prevent pregnancy when administered orally in rats. It may modulate oestrogen signalling similar to PE, specifically oestrogen receptor subtype selective PE, and may be useful as natural SERMs. It is an

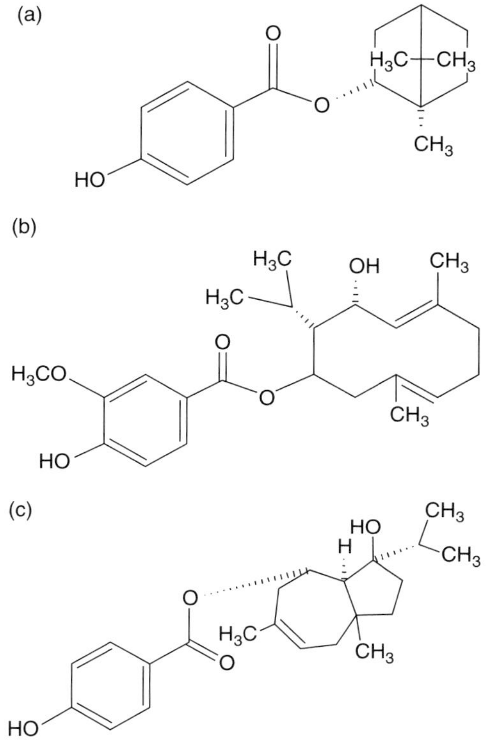

Fig. 10.9. Chemical structures of some oestrogenic terpenoids: (a) tschimgine, (b) tschimganidine and (c) ferutinine.

agonist for ER-α and an agonist/antagonist for ER-β, tschimgine is an agonist for both ER-α and ER-β and tschimganidine is an agonist for ER-α only. It was assumed that they affect the endocrine system similar to other man-made endocrine disrupters to exert their effects through oestrogen receptors, specifically ER-α and ER-β (Cherdshewasart *et al.*, 2009).

10.3 Human Health and Phytoestrogens

Researchers have proposed the hypothesis that lowered cardiovascular diseases, osteoporotic fractures and rates of breast cancer in Asian populations are related to a diet rich in soy or in other words PE (Adlercreutz, 1998, 2002). However, when evaluating this relationship, confounding factors such as lifestyle, diet, socio-cultural and morphological differences that distinguish Asian and Western populations must be considered in the analysis (This *et al.*, 2001). Several studies have discussed the potential effects of PE in treating breast cancer, prostate cancer, endometrial cancer and liver disease (Adlercreutz *et al.*, 2000b; Lei *et al.*, 2002). Additional research has shown that intestinal bacteria are seen as important in the metabolism of PE and have the ability to refine PE into compounds similar in structure to E_2 that protect against cancer (Xu *et al.*, 1995). Some of the proposed mechanisms by which PE may inhibit cancer cells are: inhibition of DNA topoisomerase, suppression of angiogenesis, induction of differentiation in cancer cell lines and induction of apoptosis (Glazier and Bowman, 2001). As studies continue to evaluate the biological effects of PE on human health, the complexity is more evident as oestrogenic and anti-oestrogenic effects are observed as well as a variety of mechanisms of action (Kronenberg and Hughes, 1999). More clinical trials are needed to assess the beneficial effects of PE on health (Naftolin and Stanbury, 2002).

10.3.1 Breast cancer

In the Western world, breast cancer is the most common cancer affecting women. Historically, the risk of breast cancer was much higher in American women than in Asian women prior to the influence of Western foods in Asian cultures (Bouker and Hilakivi-Clarke, 2000). Epidemiological studies of breast cancer and the dietary intake of soy and lignin have been reviewed (Peeters *et al.*, 2003), as well as the mechanisms of phytoestrogenic action in breast tissue (Adlercreutz, 2003). One task has been to find an oestrogen replacement therapy for women at risk for breast cancer or who have survived breast cancer. A diet rich in PE has been suggested as a preventive agent against breast cancer although there is conflicting evidence (Wagner *et al.*, 2001). Phytoestrogens act as weak oestrogens and exhibit oestrogenic activity in a low-oestrogen environment; therefore it has been postulated that they show anti-oestrogenic activity in a high-oestrogen environment (Messina and Loprinzi, 2001). This explanation suggests that prior to menopause when there is a high-oestrogen environment PE may protect against breast cancer and after menopause when there is a low oestrogen environment they may stimulate breast cancer (Anderson *et al.*, 1999). A commonly reported study found that dietary genistein may stimulate the growth of oestrogen-dependent tumours in humans with low oestrogen levels (Hsieh *et al.*, 1998). There are several isoforms of ER that may play a role in ER-β hetero-dimerization with ER-α resulting in decreased oestrogenic effects (Adlercreutz, 2002). Additional mechanisms proposed are: inhibition of tyrosine as well as other protein kinases, inhibition of angiogenesis, alteration of growth-factor activity and binding proteins (Adlercreutz, 2002).

10.3.2 Prostate cancer

Prostate cancer is one of most common cancers for men in the USA, however, little is known about its aetiology. *In vitro* studies using human prostate cancer cells have shown the inhibition of cell growth with high concentrations of PE (Adlercreutz *et al.*, 2000a). Rats consuming soy and rye bran had delayed growth of implanted prostate tumours (Landstrom *et al.*, 1998). Further testing with the same PE increased apoptosis of the tumours and reduced tumour growth in nude mice implanted with human prostate tumours. However, oestrogen has shown controversial effects, such as growth of prostate cancer and benign prostatic hyperplasia and, therefore, PE may have similar effects (Adlercreutz *et al.*, 2000a). Adlercreutz (2002) reviewed some of the more recent studies on prostate cancer and stated that findings support the hypothesis that soy consumption prevents prostate cancer, yet more studies are needed.

Several epidemiological studies suggest the beneficial use of PE in reducing prostate cancer (Severson *et al.*, 1989; Jacobsen *et al.*, 1998; Strom *et al.*, 1999). A human study of 83 cases and 107 controls used a dietary questionnaire to evaluate PE consumption for prostate cancer risk (Strom *et al.*, 1999). The results showed slightly protective effects on prostate cancer risk with greater consumption of PE. Severson *et al.* (1989) showed that increased tofu consumption was associated with a decreased risk of prostate cancer in men of Japanese ancestry living in Hawaii. Another study reported that Adventist men who consumed soy milk daily were at lower risk for prostate cancer (Jacobsen *et al.*, 1998). A herbal mixture including liquorice and ginseng as well as six other herbs has shown oestrogenic activity and was effective in two cases of hormone-refractory prostate cancer (de la Taille *et al.*, 2000). Other studies have evaluated alternative therapies, such as soy, black cohosh, vitamin E and red clover for their potential use in alleviating hot flushes for prostate cancer patients (Moyad, 2002). A randomized cross-over study on soy food consumption and serum prostate specific antigen (PSA) in men with hyperlipidaemia showed lowered LDL, no significant effects on serum PSA, and a reduced calculated risk for coronary heart disease (Jenkins *et al.*, 2003).

10.3.3 Cardiovascular disease

The leading cause of death in women in industrialized nations is coronary heart disease (CHD). In menopause the risk of CHD greatly increases, which may be due to the loss of oestrogen (Wroblewski and Cooke, 2000). Lipid profiles, vascular reactivity, cellular proliferation and thrombosis are factors that affect CHD and on which PE have shown beneficial effects (Anderson *et al.*, 1999). Mechanisms suggested explaining the prevention of cardiovascular disease and the reduction of atherosclerosis are: improvement of plasma lipid concentrations, reduction of thrombus formation such as inhibition of platelet action, improvement of systemic arterial compliance and antioxidant activity (van der Schouw *et al.*, 2000). Studies suggest that isoflavones as antioxidants may affect atherogenesis by reducing the oxidation of LDL (Ruiz-Larrea *et al.*, 1997). Kurzer and Xu (1997) reported that soy isoflavones act as antioxidants by directly or indirectly enhancing the activities of catalase, superoxide dismutase, glutathione peroxidase and glutathione reductase enzymes. Hwang *et al.* (2001) reported that extracts of soy, lucerne and acerola cherry (*Malpighia glabra* L., *Malpighiaceae*) may synergistically interact to prevent LDL oxidation.

10.3.4 Osteoporosis/bone health

Osteoporosis is often associated with menopause. The evidence supports ERT in the prevention of osteoporosis in post-menopausal women and therefore PE have been evaluated for their effects on bone mineral density. Researchers hypothesize that a diet rich in isoflavones has a protective effect on bone (Tham *et al.*, 1998). Ipriflavone, a synthetic isoflavone derivative (7-isopropoxy-isoflavone), has been used extensively in animal and human studies to evaluate bone health and PE with beneficial results (Scheiber and Rebar, 1999). Vincent and Fitzpatrick (2000) observed that genistein has a biphasic effect, lower doses improved bone mineral as opposed to high doses, on bone mineral density in ovariectomized rats. Van der Shouw *et al.* (2000) reviewed three studies in bone mineral density with PE consumption that were conducted with post-menopausal women. Two of the studies showed an increase in bone mineral density and the third study, a 10-year follow up study conducted in the Netherlands, reported a loss of bone associated with higher urinary equol and enterolactone excretion. Kurzer and Xu (1997) reviewed several other studies that include possible mechanisms of action to explain the beneficial effect of PE on bone loss. These mechanisms include preventing urinary calcium loss, beneficial effects on osteoblasts, and influences on the secretion of calcitonin which suppresses bone resorption.

10.3.5 Menopausal symptoms

The symptoms associated with menopause may cause many women to seek medical care. Hormone replacement therapy (HRT) has proven effective in the reduction of hot flushes, yet it is still controversial if HRT may be associated with increased risks of breast and endometrial cancers. Initial findings from the WHI randomized controlled trial in which women received a daily dose of conjugated equine oestrogen (0.625 mg) and medroxyprogesterone acetate (2.5 mg) have shown an increased risks to benefits ratio (Rossouw *et al.*, 2002). The investigators detected increased risks for invasive breast cancer and CHD with the consumption of the combined hormone preparation after 5.2 years of average follow up. Due to controversial evidence on HRTs, alternative therapies have been sought, such as PE and SERMs. Some botanicals used in Western countries for menopausal syndromes are black cohosh, dong quai, ginseng, red clover, hops (*Humulus lupulus* L., *Cannabaceae*), oil of evening of primrose (*Oenothera biennis* L., *Onagraceae*) and chasteberry (*Vitex agnus-castus* L., *Verbenaceae*) (Kronenberg and Fugh-Berman, 2002). Several reviews have discussed studies conducted on PE and menopausal symptoms and still much contradictory evidence exists as to the benefits of PE (Glazier and Bowman, 2001; Merritt, 2001; Kronenberg and Fugh-Berman, 2002).

10.3.6 Cognition

Cognition and memory functioning have been reported to decrease around menopause, and therefore studies have investigated the association of ERT and cognition, as well as PE and cognition (Vincent and Fitzpatrick, 2000). However, limited studies are available on the effects of PE on cognitive functioning. The mechanisms are not understood, but it has been suggested that PE act as oestrogen agonists and may increase spine density and synapse formation in the hippocampus of adults. In addition, PE may interact with the transcription of neurotrophin genes (File *et al.*, 2003).

10.4 Herbal Sources of Phytoestrogens

There is a higher concentration of PE in leguminous plants even though they are also found in grains, vegetables and fruits. The most common PE found in legumes are isoflavonoids (Dewick, 1993). Of the monocotyledons, species of *Iris* (*Iridaceae*) are a major source of isoflavonoids. Of the gymnosperms, the genera *Juniperus* and *Podocarpus* have been reported to produce isoflavonoids (Dewick, 1993). Cucurbitales contains *Cucurbitaceae* (pumpkin), Rosales contains *Cannabaceae* (hops) and Fabales contains *Fabaceae* (soy, liquorice, red clover) and Malpighiales (flax). Other plant species that have shown oestrogenic activity are found in different orders located at basal positions from the initial group discussed: Myrtales (evening primrose), Apiales (dong quai, ginseng) and Lamiales (chasteberry) (Daly *et al.*, 2001).

10.4.1 Soybean

Soybean (*Glycine max, Fabaceae*) is a good source of genistein, its glycosides and daidzein, which possess oestrogenic activity. A detailed discussion of soy for women who have survived breast cancer has been presented (Messina and Loprinzi, 2001). It is unclear whether soy protein with trace amounts of isoflavones, PE-intact soy protein, or a combination of both causes the beneficial cholesterol effects seen in animal studies (van der Schouw *et al.*, 2000). Consumption of soy protein has shown a decrease in lipid peroxidation compared with the case in consumption in post-menopausal cynomolgus monkeys and lowered atherosclerosis in rabbits (Wagner *et al.*, 1997; van der Schouw *et al.*, 2000). Isoflavone-intact soy protein has lowered LDL and raised high density lipids (HDL) cholesterol suggesting that the active components are found in the extractable protein portion (Clarkson and Anthony, 1998).

10.4.2 Black cohosh

Black cohosh (*Actaea racemosa, Cimicifuga racemosa*, family *Ranunculaceae*) grows in eastern North America, from southern Maine to Georgia. It is also known as baneberry, black snakeroot and rattle weed. Native Americans used the roots and rhizomes for a variety of indications such as stimulation of menstrual flow, dysmenorrhea, suppression of cough, treatment of diarrhoea, childbirth and rheumatism (Foster and Tyler, 1999). It contains a number of compounds with potential bioactivity including triterpene, glycosides, resin, salicylates, isoferulic acid, sterols and alkaloids. It contain formononectin, biochanin A, genistein 4′-methyl ether, flavonoids, such as kaempferol, that are thought to account for the reported reduction in hot flushes and menopausal symptoms (McCoy and Kelly, 1996; Kronenberg and Fugh-Berman, 2002). Black cohosh does not appear to alter the hormonal pattern associated with menopause, low oestrogen accompanied by elevated luteinizing hormone (LH) and follicle-stimulating hormone (FSH). It is possible that it affects the pathway downstream from the oestrogen (Blumenthal, 2004). Clinical trials on effects of *Actaea racemosa* extracts on menopausal symptoms have yielded excellent efficacy against classic menopausal complaints and osteoprotective properties, and extracts were deemed safe even when the dosage was increased threefold. Furthermore, several studies suggest that its extracts might help control psychic problems typically found during menopausal transition (Viereck *et al.*, 2005). The study by Liske *et al.* (2002) suggested that *C. racemosa* extract is associated with improvement in menopause symptoms without evidence of any significant side effects.

10.4.3 Flax

Flax (*Linum usitatissimum*, family *Linaceae*) seeds contain dietary lignan secoisolariciresinol diglucoside (SDG), present in high concentrations, and its metabolites enterolactone and enterodiol are thought to decrease the risk of hormone-dependent cancers, cardiovascular disease and other ‘welfare’ diseases. Flaxseed also contains other biologically active phenolic compounds, such as phenolic acids. The SDG (Bambagiotti-Alberti *et al.*, 1994a) is often referred to as secoisolariciresinol or SECO, the aglycone of SDG (Mazur *et al.*, 1996); it exists in two isomeric forms in flaxseed (Bambagiotti-Alberti *et al.*, 1994b). Smaller quantities of matairesinol, isolariciresinol, lariciresinol, demethoxy-secoisolariciresinol and pinoresinol have also been identified in flaxseed (Meagher *et al.*, 1999; Sicilia *et al.*, 2003). Flaxseed oil contains polyunsaturated fatty acids such as α-linolenic acid, which may lower cholesterol and have antioxidant effects for health. The lignan content in flaxseed differ between varieties but is also dependent on growing location and year (Thompson *et al.*, 1997). The SDG levels remain unchanged during the manufacture of breads and cookies that contained flaxseed (Muir and Westcott, 2000). Flaxseed is considered one of the richest sources of lignan PE (Thompson *et al.*, 1991). A study of 145 women with climacteric complaints showed a reduction in menopausal symptoms (including hot flush and vaginal dryness) with the consumption of a diet rich in PE (Brzezinski *et al.*, 1997).

Role of flaxseed in obesity and diabetes

In hyperlipidaemic subjects, ingestion of whole flaxseed lowers serum cholesterol and postprandial glucose (Cunnane *et al.*, 1995). This effect may be due to the presence of n-3-α-linolenic acid in flaxseed oil. It also improved insulin sensitivity, increased HDL cholesterol and decreased LDL oxidation (Nestel *et al.*, 1997). However, Jenkins *et al.* (1999) reported that in non-obese, non-diabetic, hypercholesterolaemic subjects, diets supplemented with partially defatted flaxseed lowered total and LDL cholesterol but had no effect on serum HDL cholesterol, possibly a result of the fibre present in defatted flaxseed. Because partially defatted flaxseed is low in α-linolenic acid, the hypocholesterolaemic effect may be due to other ingredients in flaxseed. Prasad *et al.* (1998) further showed that secoisolariciresinol diglucoside, a lignan present in flaxseed, also lowers serum total cholesterol and LDL cholesterol and reduces hypercholesterolaemic atherosclerosis in rabbits.

Like soy isoflavones, lignans have antioxidant activity (Xue *et al.*, 1992). However, whole flaxseed had no significant effect on markers of lipid peroxidation in humans (Cunnane *et al.*, 1995), but partially defatted flaxseed lowered serum protein thiol groups, indicating increased oxidation (Jenkins *et al.*,

1999). Secoisolariciresinol diglucoside (Prasad, 1997), the lignan present in flaxseed, and its mammalian metabolites secoisolariciresinol, enterodiol and enterolactone have been shown to have antioxidant activity. The antioxidant activity of secoisolariciresinol and enterodiol is higher than that of vitamin E or the parent glucoside present in flaxseed (Prasad, 2000).

10.4.4 Red clover

Red clover (*Trifolium pratense*, family *Fabaceae*) is a legume rich in isoflavonoid PE including genistein, daidzein, formononetin and biochanin-A, phytochemicals that are now recognized for supporting critical hormone levels without having any negative side effects (Mazur and Adlercreutz, 1998). Controlled clinical trials show that PE from red clover help to maintain proper bone density in menopausal women, as well as relieving hot flushes and night sweats. Red clover has been a valued medicine since ancient times and was particularly valued for treating respiratory problems, colds, flu and infections in the 19th century. Red clover is recognized as a detoxification herb or 'blood cleanser'. The *in vivo* oestrogenic and anti-oestrogenic effects of red clover extract have been studied in the uterus, vaginal cells and mammary glands of ovariectomized Sprague-Dawley rats (Burdette *et al.*, 2002). Red clover is considered to have an advantage compared to other plants containing PE as it is the only plant having four of the most important isoflavones (biokain A, daidzein, formononetin and genistein) characterized by pro-oestrogenic activity, by rapid intestinal absorption and increased receptor affinity (Beck *et al.*, 2005). Red clover has been reviewed for menopausal symptoms (Kronenberg and Fugh-Berman, 2002), showing reduced hot flush count (van der Weijer and Barentsen, 2002) and was also found to have a favourable metabolic impact on serum lipids in post-menopausal women (Terzic *et al.*, 2012). In another recent study, the effect of red clover supplementation on menopausal women's quality of life showed no difference with the placebo (Ehsanpour *et al.*, 2012). Additional studies are needed to further elucidate the benefits or adverse effects of red clover.

10.4.5 Hops

Hops (*Humulus lupulus*, family *Cannabaceae*) medicinally has been valued as a sedative, for inflammation and as a tonic (Foster and Tyler, 1999). Menstrual disturbances were frequently observed in women hop pickers and their oestrogenic activity was associated with the plant (Verzele, 1986). The female flowers of hops are considered oestrogenic. The most potent PE in hops is 8-prenyl-naringenin (8-PN), which is found in beer in low quantities (Milligan *et al.*, 1999). Other studies have tested the presence of hops in dietary supplements used for breast enhancement (Coldham and Sauer, 2001). *In vitro* and animal data suggest that 8-PN might exhibit several biological activities (Diel *et al.*, 2004; Effenberger *et al.*, 2005; Humpel *et al.*, 2005), and hop-containing dietary supplements are marketed to reduce menopausal complaints and used for breast enhancement (Coldham and Sauer, 2001).

An important factor influencing the bioavailability and activity of PE is their metabolic fate upon ingestion. In general, after reaching the colon, flavonoids are partially degraded, depending on their structure, thereby leading to lower bioavailability (Simons *et al.*, 2005). However, microbial transformation in the colon may also increase the biological activity of the ingested compounds, a process that has been described for different PE (Rowland *et al.*, 1999). For hops, isoxanthohumol (IX) is the prevailing prenyl-flavonoid in beer and is 10–30 times more abundant as 8-PN (Stevens *et al.*, 1999). Schaefer *et al.* (2005) noted the activation of IX into the PE 8-PN inside the human body. The intestinal microbial community might be responsible for this production of 8-PN after IX consumption (Possemiers *et al.*, 2005). Microbial O-demethylation of IX in the human intestine could readily increase intestinal prenyl-flavonoids (8-PN) concentrations ten-fold (Possemiers *et al.*, 2005), leading to the uptake of active oestrogen doses after moderate beer consumption that might fall within the range of biological activities (Prestwood *et al.*, 2003).

10.4.6 Dong quai

Dong quai (*Angelica sinensis*, family *Apiaceae*) has been referred to as the 'female ginseng' and is used for a variety of conditions such as a blood tonic and decongestant for body organs (Hardy, 2000). Its roots are used for women as a tonic often in combination with other herbs. Other women's conditions treated with dong quai are dysmenorrhea, irregular menstruation, constipation, anaemia and abdominal pain. *In vitro* it has acted as a growth inhibitor with breast cancer cell lines (Zava *et al.*, 1998; Dixon-Shanies and Shaikh, 1999) but has also been observed to stimulate the growth of MCF-7 cells (Amato *et al.*, 2002). Dong quai has shown weak binding affinity for ER-α and ER-β and weak stimulation of progesterone receptor (*PR*) expression (Liu *et al.*, 2001).

10.4.7 Liquorice

The root of liquorice (*Glycyrrhiza glabra*, family *Fabaceae*) has been consumed for thousands of years in China for its health benefits and detoxification effects as well as its use as a flavouring and sweetening agent (Wang and Nixon, 2001). Medicinally it has been used as a demulcent, expectorant and has been shown to have antioxidant and antimicrobial activity. The main components of liquorice are glycyrrhizin (glycyrrhizinic acid), which is sweeter than sugar, and glycyrrhetinic acid. Both have been clinically used in the treatment of hyperlipidaemia, allergic inflammation, atopic dermatitis and atherosclerosis (Tamir *et al.*, 2001). The PE in liquorice have a mild oestrogenic effect, making the herb potentially useful in easing certain symptoms of premenstrual syndrome, such as irritability, bloating and breast tenderness. Although the glycyrrhizin in liquorice actually inhibits the effect of the body's own oestrogens, the mild oestrogenic effect produced by liquorice's PE manages to override this inhibiting action. Another compound in liquorice, licochalcone-A, has shown oestrogenic activity with ERs and induced apoptosis in MCF-7 and HL-60 cell lines (Rafi *et al.*, 2000).

10.4.8 Other herbal sources

Other botanical sources with oestrogenic effects and potential health benefits are evening primrose oil (*Oenothera biennis*, *Onagraceae*), chasteberry (*Vitex agnus-castus*, *Verbenaceae*), lucerne (*Medicago sativa*, *Fabaceae*) and ginseng (*Panax* spp. and *Eleutherococcus senticosus*, *Araliaceae*) (Glazier and Bowman, 2001; Liu *et al.*, 2001; Amato *et al.*, 2002; Kronenberg and Fugh-Berman, 2002). Coffee (*Coffea arabica* L., *Rubiaceae*) has also shown weak oestrogenic activity (Kitts, 1987). Evening primrose is a common herb in North America that has been used medicinally by the Native Americans. Coumestrol was the first PE identified in lucerne and is considered one of the richest food sources for this PE (Kurzer and Xu, 1997). Ginseng has a long history of medicinal use in Asia as a tonic and stimulant. There are several different plants that are referred to as ginseng and all are in the *Araliaceae* family: *Panax ginseng* C.A. Mey. (Chinese or Korean ginseng), *P. quinquefolium* L. (American ginseng) and *Eleutherococcus senticosus* (Siberian ginseng).

10.5 Adverse Effects of Phytoestrogens

Some concerns have been discussed about the risks associated with PE, such as increased plasma concentration of isoflavones in babies that ingest soymilk, the ability of non-hormonal secondary plant metabolites to modify sex steroid metabolism, and the effects of PE on the thyroid (Ibarreta *et al.*, 2001). In addition, the genetic toxicity potential of PE has been reviewed (Kulling *et al.*, 2002). It has been reported that sheep consuming large amounts of clover which has with high amounts of PE showed infertility and reproductive disorders (Adams, 1995). Cheetahs in captivity also had reduced fertility rates when consuming a feline diet composed of a soybean product, which was reversed when it was removed from the diet (Setchell *et al.*, 1987). Toxicities associated with herbal medicines that include PE have also been presented in the literature (Sheehan, 1998).

As potential endocrine disrupters, PE may act as anti-oestrogens and harm the reproductive health of males (Santti *et al.*, 1998). Reduced sperm quality, undescended testes and urogenital tract abnormalities were increased in the sons of mothers taking diethyl-stilbestrol (DES) compared with those who did not take the miscarriage preventative drug (Sheehan, 1998). Studies in cultured human lymphoblastoid cells reported that coumestrol was mutagenic and clastogenic (Domon *et al.*, 2001). Other concerns related to PE are their effect on thyroxine, insulin and glucagon (Ohno *et al.*, 1993). Hypothyroid cases were associated with infants fed soybean diets (Fort *et al.*, 1990). In the USA, regulation is limited on standardization, preparation and extraction methods of the PE products being sold and marketed as nutritional supplements. As nutritional supplements these products are not supported by clinical trials and, therefore, should be administered and taken with this in mind (This *et al.*, 2001).

10.6 Conclusions

Diets rich in plant-derived products may supply a variety of phytoestrogens capable of producing a range of pharmacological effects in the human body. As people live longer, women are spending more of their lives in menopause, affected by a variety of oestrogen-related conditions such as osteoporosis, cognitive and cardiovascular disease, increased risk of breast cancer and other symptoms that decrease the overall quality of life. Epidemiological evidence and experimental data from animal studies are highly suggestive of the beneficial effects of PE on human health, but the clinical data supportive of such effects are either not available, or are awaiting design and execution of appropriate prospective large-scale clinical studies. Due to the functional and structural differences of PE, their biological activities are also highly variable and there may be other effects that have not yet been studied.

References

Adams, N.R. (1995) Detection of the effects of phytoestrogens on sheep and cattle. *Journal of Animal Science* 73, 1509–1515.

Adlercreutz, H. (1998) Evolution, nutrition, intestinal microflora, and prevention of cancer: a hypothesis. *Proceedings of the Society for Experimental Biology and Medicine* 217, 241–246.

Adlercreutz, H. (2002) Phytoestrogens and cancer. *Lancet Oncology* 3, 364–373.

Adlercreutz, H. (2003) Phytoestrogens and breast cancer. *Journal of Steroid Biochemistry and Molecular Biology* 1803, 1–6.

Adlercreutz, H., Mazur, W., Bartels, P.V., Elomaa, V., Watanabe, S., Wahala, K., Landstrom, M., Lundin, E., Bergh, A., Damber, J.E., Aman, P., Widmark, A., Johansson, A., Zhang, J.X. and Hallmans, G. (2000a) Phytoestrogens and prostate disease. *Journal of Nutrition* 130, 658S–659S.

Adlercreutz, H., Mazur, W., Stumpf, K., Kikkinen, P., Hulten, K. and Hallmans, G. (2000b) Food containing phytoestrogens and breast cancer. *BioFactors* 12, 89–93.

Albertazzi, P. and Purdie, D.W. (2008) The nature and utility of the phytoestrogens: a review of the evidence. *Maturitas* 61, 214–226.

Alves, R.C., Almeida, I.M.C., Casal, S. and Oliveira, M.B.P.P. (2010) Isoflavones in coffee: influence of species, roast degree, and brewing method. *Journal of Agricultural and Food Chemistry* 58, 3002–3007.

Amato, P., Christophe, S. and Mellon, P.L. (2002) Estrogenic activity of herbs commonly used as remedies for menopausal symptoms. *Menopause* 9, 145–150.

Anderson, J.J.B., Anthony, M., Messina, M. and Garner, S.C. (1999) Effects of phyto-oestrogens on tissues. *Nutrition Research Reviews* 12, 75–116.

Anstead, G.M., Carlson, K.E. and Katzenellenbogen, J.A. (1997) The estradiol pharmacophore: ligand structure-estrogen receptor binding affinity relationships and a model for the receptor binding site. *Steroids* 62, 268–303.

Apers, S., Vlietinck, A. and Pieters, L. (2003) Lignans and neolignans as lead compounds. *Photochemistry Reviews* 2, 201–217.

Attoumbre, J., Bienaime, C., Dubois, F., Fliniaux, M.A., Chabbert, B. and Baltora-Rosset, S. (2010) Development of antibodies against secoisolanciresinol – Application to the immunolocalization of lignans in *Linum usitatissimum* seeds. *Phytochemistry* 71, 1979–1987.

Bambagiotti-Alberti, M., Coran, S.A., Ghiara, C., Giannelini, V. and Raffaelli, A. (1994a) Revealing the mammalian lignan precursor secoisolariciresinol diglucoside in flax seed by ionspray mass spectrometry, *Rapid Communications in Mass Spectrometry* 8, 595–598.

Bambagiotti-Alberti, M., Coran, S.A., Ghiara, C., Moneti, G. and Raffaelli, A. (1994b) Investigation of mammalian lignan precursors in flax seed: first evidence of secoisolariciresinol diglucoside in two isomeric forms by liquid chromatography/mass spectrometry. *Rapid Communications in Mass Spectrometry* 8, 929–932.

Bartkiene, E., Juodeikiene, G., Basinskiene, L., Liukkonen, K.-H., Adlercreutz, H. and Kluge, H. (2011) Enterolignans enterolactone and enterodiol formation from their precursors by the action of intestinal microflora and their relationship with non-starch polysaccharides in various berries and vegetables. *Food Science and Technology* 44, 48–53.

Beck, V., Rohr, U. and Jungbauer, A. (2005) Phytoestrogens derived from red clover: an alternative to estrogen replacement therapy? *Journal of Steroid Biochemistry and Molecular Biology* 94, 499–518.

Beejmohun, V., Fliniaux, O., Hano, C., Pilard, S., Grand, E., Lesur, D., Cailleu, D., Lamblin, F., Laine, E., Kovensky, J., Fliniaux, M.A. and Mesnard, F. (2007) Coniferin dimerisation in lignan biosynthesis in flax cells. *Phytochemistry* 68, 2744–2752.

Benassayag, C., Perrot-Applanat, M. and Ferre, F. (2002) Phytoestrogens as modulators of steroid action in target cells. *Journal of Chromatography B – Analytical, Technological and Biomedical Life Science* 777, 233–248.

Bergman, J.M., Thompson, L.U. and Dabrosin, C. (2007) Flaxseed and its lignans inhibit estradiol-induced growth, angiogenesis, and secretion of vascular endothelial growth factor in human breast cancer xenografts *in vivo*. *Clinical Cancer Research* 13, 1061–1067.

Bhat, K.P.L., Kosmeder, J.W. II and Pezzuto, J.M. (2001) Biological effects of resveratrol. *Antioxidants and Redox Signaling* 3, 1041–1064.

Birt, D.F., Walker, B., Tibbels, M.G. and Bresnick, E. (1986) Anti-mutagenesis and antipromotion by apigenin, robinetin and indole-3-carbinol. *Carcinogenesis* 7, 959–963.

Blumenthal, M. (2004) The use of black cohosh to treat symptoms of menopause. *Sexuality, Reproduction and Menopause* 2, 27–34.

Boccardo, F., Puntoni, M., Guglielmini, P. and Rubagotti, A. (2006) Enterolactone as a risk factor for breast cancer: a review of the published evidence. *Clinica Chimica Acta* 365, 58–67.

Bonzanini, F., Bruni, R., Palla, G., Serlataite, N. and Caligiani, A. (2009) Identification and distribution of lignans in *Punica granatum* L. fruit endocarp, pulp, seeds, wood knots and commercial juices by GC-MS. *Food Chemistry* 117, 745–749.

Bouker, K.B. and Hilakivi-Clarke, L. (2000) Genistein: does it prevent or promote breast cancer? *Environmental Health Perspectives* 108, 701–708.

Brusselmans, K., Vrolix, R., Verhoeven, G. and Swinnen, J.V. (2005) Induction of cancer cell apoptosis by flavonoids is associated with their ability to inhibit fatty acid synthase activity. *Journal of Biological Chemistry* 280, 5636–5645.

Brzezinski, A. and Debi, A. (1999) Phytoestrogens: the 'natural' selective estrogen receptor modulators? *European Journal of Obstetrics, Gynecology and Reproductive Biology* 85, 47–51.

Brzezinski, A., Adlercreutz, H., Shaoul, R., Rosier, A., Shmueli, A., Tanos, V. and Schenker, J.G. (1997) Short-term effects of phytoestrogen-rich diet on postmenopausal women. *Menopause* 4, 89–94.

Brzozowski, A.M., Pike, A.C., Dauter, Z., Hubbard, R.E., Bonn, T., Engstrom, O., Ohman, L., Greene, G.L., Gustafsson, J.A. and Carlquist, M. (1997) Molecular basis of agonism and antagonism in the oestrogen receptor. *Nature* 389, 753–758.

Burdette, J.E., Liu, J., Lantvit, D., Lim, E., Booth, N., Bhat, K.P.L., Hedayat, S., van Breemen, R.B., Constantinou, A.I., Pezzuto, J.M., Farnsworth, N.R. and Bolton, J.L. (2002) *Trifolium pratense* (red clover) exhibits estrogenic effects in vivo in ovariectomized Sprague-Dawley rats. *Journal of Nutrition* 132, 27–30.

Busquets, S., Ametller, E., Fuster, G., Olivan, M., Raab, V., Argilés, J.M. and López-Soriano, F.J. (2007) Resveratrol, a natural diphenol, reduces metastatic growth in an experimental cancer model. *Cancer Letters* 245, 144–148.

Cai, Y.Z., Mei, S., Jie, X., Luo, Q. and Corke, H. (2006) Structure-radical scavenging activity relationships of phenolic compounds from traditional Chinese medicinal plants. *Life Science* 78, 2872–2888.

Caltagirone, S., Rossi, C., Poggi, A., Ranelletti, F.O., Natali, P.G., Brunetti, M., Aiello, F.B. and Piantelli, M. (2000) Flavonoids apigenin and quercetin inhibit melanoma growth and metastatic potential. *International Journal of Cancer* 87, 595–600.

Castello, L. and Tessitore, L. (2005) Resveratrol inhibits cell cycle progression in U937 cells. *Oncology Reports* 13, 133–137.

Chan, R.Y., Chen, W.F., Dong, A., Guo, D. and Wong, M.S. (2002) Estrogen like activity of ginsenoside Rg1 derived from *Panax notoginseng*. *Journal of Clinical Endocrinology and Metabolism* 87, 3691–3695.

Chen, J., Power, K.A., Mann, J., Cheng, A. and Thompson, L.U. (2007) Dietary flaxseed interaction with tamoxifen induced tumor regression in athymic mice with MCF-7 xenografts by down regulating the expression of estrogen related gene products and signal transduction pathways. *Nutrition and Cancer* 58, 162–170.

Cherdshewasart, W., Sutjit, W., Pulcharoen, K. and Chulasiri, M. (2009) The mutagenic and antimutagenic effects of the traditional phytoestrogen-rich herbs, *Pueraria mirifica* and *Pueraria lobata*. *Brazilian Journal of Medical and Biological Research* 42, 816–823.

Clarkson, T.B. and Anthony, M.S. (1998) Phytoestrogens and coronary heart disease. *Baillieres Clinical Endocrinology and Metabolism* 12, 589–604.

Coldham, N.G. and Sauer, M.J. (2001) Identification, quantitation and biological activity of phytoestrogens in a dietary supplement for breast enhancement. *Food and Chemical Toxicology* 39, 1211–1224.

Constant, J. (1997) Alcohol, ischemic heart disease, and the French paradox. *Coronary Artery Disease* 8, 645–649.

Culpitt, S.V., Rogers, D.F., Fenwick, P.S., Shah, P., Matos, C.D., Russell, R.E.K., Barnes, P.J. and Donnelly, L.E. (2003) Inhibition by red wine extract, resveratrol, of cytokine release by alveolar macrophages in COPD. *Thorax* 58, 942–946.

Cunnane, S.C., Hamadeh, M.J., Liede, A.C., Thompson, L.U., Wolever, T.M. and Jenkins, D.J. (1995) Nutritional attributes of traditional flaxseed in healthy young adults. *American Journal of Clinical Nutrition* 61, 62–68.

Daly, D.C., Cameron, K.M. and Stevenson, D.W. (2001) Plant systematic in the age of genomics. *Plant Physiology* 127, 1328–1333.

de la Taille, A., Hayek, O.R., Burchardt, M., Burchardt, T. and Katz, A.E. (2000) Role of herbal compounds (PC-SPES) in hormone refractory prostate cancer: two case reports. *Journal of Alternative and Complementary Medicine* 6, 449–451.

de Ruvo, C., Amodio, R., Algeri, S., Martelli, N., Intilangelo, A., D'Ancona, G.M. and Esposito, E. (2000) Nutritional antioxidants as anti-degenerative agents. *International Journal of Developmental Neuroscience* 18, 359–366.

Dewick, P.M. (1993) Isoflavonoids. In: Harborne, J.B. (ed.) *The Flavonoids: Advances in Research Since 1986*. Chapman and Hall, London, pp. 117–238.

Diel, P., Thomae, R.B., Caldarelli, A., Zierau, O., Kolba, S., Schmidt, S., Schwab, P., Metz, P. and Vollmer, G. (2004) Regulation of gene expression by 8-prenylnaringenin in uterus and liver of Wistar rats. *Planta Medica* 70, 39–44.

Ding, X.Z. and Adrian, T.E. (2002) Resveratrol inhibits proliferation and induces apoptosis in human pancreatic cancer cells. *Pancreas* 25, 71–76.

Dixon-Shanies, D. and Shaikh, N. (1999) Growth inhibition of human breast cancer cells by herbs and phytoestrogens. *Oncology Reports* 6, 1383–1387.

Domon, O.E., McGarrity, L.J., Bishop, M., Yoshioka, M., Chen, J.J. and Morris, S.M. (2001) Evaluation of the genotoxicity of the phytoestrogen, coumestrol, in AHH-1 TK+/– human lymphoblastoid cells. *Mutation Research* 474, 129–137.

Duncan, A.M., Phipps, W.R. and Kurzer, M.S. (2003) Phyto-oestrogens. *Best Practice and Research: Clinical Endocrinology and Metabolism* 17, 253–271.

Effenberger, K.E., Johnsen, S.A., Monroe, D.G., Spelsberg, T.C. and Westendorf, J.J. (2005) Regulation of osteoblastic phenotype and gene expression by hop-derived phytoestrogens. *Journal of Steroid Biochemistry and Molecular Biology* 96, 387–399.

Ehsanpour, S., Salehi, K., Zolfaghari, B. and Bakhtiari, S. (2012) the effect of red clover on quality of life in post-menopausal women. *Iranian Journal of Nursing and Midwifery Research* 17, 34–40.

Fang, J., Xia, C., Cao, Z., Zheng, J.Z., Reed, E. and Jiang, B.H. (2005) Apigenin inhibits VEGF and HIF-1 expression via PI3K/AKT/p70S6K1 and HDM2/p53 pathways. *The Federation of American Societies for Experimental Biology Journal* 19, 342–353.

File, S.E., Hartley, D.E., Alom, N. and Rattray, M. (2003) Soya phytoestrogens change cortical and hippocampal expression of BDNF mRNA in male rats. *Neuroscience Letters* 338, 135–138.

Ford, J.D., Huang, K.S., Wang, H.B., Davin, L.B. and Lewis, N.G. (2001) Biosynthetic pathway to the cancer chemopreventive secoisolariciresinol diglucoside-hydroxymethyl glutaryl ester-linked lignan oligomers in flax (*Linum usitatissimum*) seed. *Journal of Natural Products* 64, 1388–1397.

Fort, P., Moses, N., Fasano, M., Goldberg, T. and Lifshitz, F. (1990) Breast and soy-formula feedings in early infancy and the prevalence of autoimmune thyroid disease in children. *Journal of the American College of Nutrition* 9, 164–167.

Foster, S. and Tyler, V.E. (1999) Tyler's Honest Herbal (ed.) *A Sensible Guide to the Use of Herbs and Related Remedies*. Haworth Herbal Press, New York.

Franke, A.A., Custer, L.J., Cerna, C.M. and Narala, K. (1995) Rapid HPLC analysis of dietary phytoestrogens from legumes and from human urine. *Proceedings of Society for Experimental Biology and Medicine* 208, 18–26.

Glazier, M.G. and Bowman, M.A. (2001) A review of the evidence for the use of phytoestrogens as a replacement for traditional estrogen replacement therapy. *Archives of Internal Medicine* 161, 1161–1172.

Gruber, C.J., Tschugguel, W., Schneeberger, C. and Huber, J.C. (2002) Production and actions of estrogens. *New England Journal of Medicine* 346, 340–352.

Gusman, J., Malonne, H. and Atassi, G. (2001) A reappraisal of the potential chemo-preventive and chemotherapeutic properties of resveratrol. *Carcinogenesis* 22, 1111–1117.

Hardy, M.L. (2000) Herbs of special interest to women. *Journal of American Pharmacists Association* 40, 234–242.

Heber, D., Berdanier, C.D., Dwyer, J.T. and Feldman, E.B. (eds) (2008) *Plant Foods and Phytochemicals in Human Health*. CRC Press, Boca Raton, Florida, pp. 176–181.

Hosseinian, F.S. and Mazza, G. (2009) Triticale bran and straw: potential new sources of phenolic acids, proanthocyanidins, and lignans. *Journal of Functional Foods* 1, 57–64.

Hsieh, C.Y., Santell, R.C., Haslam, S.Z. and Helferich, W.G. (1998) Estrogenic effects of genistein on the growth of estrogen receptor positive human breast cancer (MCF-7) cells *in vitro* and *in vivo*. *Cancer Research* 58, 3833–3838.

Hu, C., Yuan, Y.V. and Kitts, D.D. (2007) Antioxidant activities of the flaxseed lignan secoisolariciresinol diglucoside, its aglycone secoisolariciresinol and the mammalian lignans enterodiol and enterolactone *in vitro*. *Food and Chemical Toxicology* 45, 2219–2227.

Hu, J.Y. and Aizawa, T. (2003) Quantitative structure–activity relationships for estrogen receptor binding affinity of phenolic chemicals. *Water Research* 37, 1213–1222.

Humpel, M., Isaksson, P., Schaefer, O., Kaufmanna, U., Ciana, P., Maggic, A. and Schleuning, W.D. (2005) Tissue specificity of 8-prenylnaringenin: protection from ovariectomy induced bone loss with minimal trophic effects on the uterus. *Journal of Steroid Biochemistry and Molecular Biology* 97, 299–305.

Hwang, J., Hodis, H.N. and Sevanian, A. (2001) Soy and alfalfa phytoestrogen extracts become potent low-density lipoprotein antioxidants in the presence of acerola cherry extract. *Journal of Agricultural and Food Chemistry* 49, 308–314.

Ibarreta, D., Daxenberger, A. and Meyer, H.H. (2001) Possible health impact of phytoestrogens and xenoestrogens in food. *Acta Pathologica, Microbiologica et Immunologica* 109, 161–184.

Ikeda, K., Arao, Y., Otsuka, H., Nomoto, S., Horiguchi, H., Kato, S. and Kayama, F. (2000) Terpenoids found in the *Umbelliferae* family act as agonists/antagonists for ER(alpha) and ER(beta): differential transcription activity between ferutinine-liganded ER(alpha) and ER(beta). *Biochemical and Biophysical Research Communications* 291, 354–360.

Jacobsen, B.K., Knutsen, S.F. and Fraser, G.E. (1998) Does high soy milk intake reduce prostate cancer incidence? The Adventist Health Study (United States). *Cancer Causes and Control* 9, 553–557.

Jang, M., Cai, L., Udeani, G.O., Slowing, K.V., Thomas, C.F., Beecher, C.W.W., Fong, H.H.S., Farnsworth, N.R., Kinghorn, A.D., Mehta, R.G., Moon, R.C. and Pezzuto, J.M. (1997) Cancer chemopreventive activity of resveratrol, a natural product derived from grapes. *Science* 275, 218–220.

Jenkins, D.J., Kendall, C.W., Vidgen, E., Agarwal, S., Rao, A.V., Rosenberg, R.S., Diamandis, E.P., Novokmet, R., Mehling, C.C., Perera, T., Griffin, L.C. and Cunnane, S.C. (1999) Health aspects of partially defatted flaxseed, including effects on serum lipids, oxidative measures, and *ex vivo* androgen and progestin activity: a controlled crossover trial. *American Journal of Clinical Nutrition* 69, 395–402.

Jenkins, D.J., Kendall, C.W., D'Costa, M.A., Jackson, C.J., Vidgen, E., Singer, W., Silverman, J.A., Koumbridis, G., Honey, J., Rao, A.V., Fleshner, N. and Klotz, L. (2003) Soy consumption and phytoestrogens: effect on serum prostate specific antigen when blood lipids and oxidized low density lipoprotein are reduced in hyperlipidemic men. *Journal of Urology* 169, 507–511.

Kaldas, M.I., Walle, U.K. and Walle, T. (2003) Resveratrol transport and metabolism by human intestinal CaCo-2 cells. *Journal of Pharmacy and Pharmacology* 55, 307–312.

Kaufman, P.B., Duke, J.A., Brielmann, H., Boik, J. and Hoyt, J.E. (1997) A comparative survey of leguminous plants as sources of the isoflavones, genistein and daidzein: implications for human nutrition and health. *Journal of Alternative and Complementary Medicine* 3, 7–12.

Kelly, G.E., Joannou, G.E., Reeder, A.Y., Nelson, C. and Waring, M.A. (1995) The variable metabolic response to dietary isoflavones in humans. *Proceedings of Society for Experimental Biology and Medicine* 208, 40–43.

Kim, Y.M., Yun, J., Lee, C.K., Lee, H., Min, K.R. and Kim, Y. (2002) Oxyresveratrol and hydroxystilbene compounds. Inhibitory effect on tyrosinase and mechanism of action. *Journal of Biological Chemistry* 277, 16340–16344.

King, A. and Young, G. (1999) Characteristics and occurrence of phenolic phytochemicals. *Journal of the American Dietetic Association* 99, 213–218.

King, R.E., Kent, K.D. and Bomser, J.A. (2005) Resveratrol reduces oxidation and proliferation of human retinal pigment epithelial cells via extracellular signal-regulated kinase inhibition. *Chemico Biological Interactions* 151, 143–149.

Kitaoka, M., Kadokawa, H., Sugano, M., Ichikawa, K., Taki, M., Takaishi, S., Iijima, Y., Tsutsumi, S., Boriboon, M. and Akiyama, T. (1998) Prenylflavonoids: a new class of non-steroidal phytoestrogen (Part 1). Isolation of 8-isopentenylnaringenin and an initial study on its structure–activity relationship. *Planta Medica* 64, 511–515.

Kitts, D.D. (1987) Studies on the estrogenic activity of a coffee extract. *Journal of Toxicology and Environmental Health Sciences* 20, 37–49.

Knight, D.C. and Eden, J.A. (1996) A review of the clinical effects of phytoestrogens. *Obstetrics and Gynecology* 87, 897–904.

Knowles, L.M., Zigrossi, D.A., Tauber, R.A., Hightower, C. and Milner, J.A. (2000) Flavonoids suppress androgen-independent human prostate tumor proliferation. *Nutrition and Cancer* 38, 116–122.

Kronenberg, F. and Fugh-Berman, A. (2002) Complementary and alternative medicine for menopausal symptoms: a review of randomized, controlled trials. *Annals of Internal Medicine* 137, 805–813.

Kronenberg, F. and Hughes, C. (1999) Exogenous and endogenous estrogens: an appreciation of biological complexity. *Menopause* 6, 4–6.

Kuiper, G.G., Enmark, E., Pelto-Huikko, M., Nilsson, S. and Gustafsson, J.A. (1996) Cloning of a novel estrogen receptor expressed in rat prostate and ovary. *Proceedings of National Academy of Sciences USA* 93, 5925–5930.

Kuiper, G.G., Lemmen, J.G., Carlsson, B., Corton, J.C., Safe, S.H., van der Saag, P.T., van der Burg, B. and Gustafsson, J.A. (1998) Interaction of estrogenic chemicals and phytoestrogens with estrogen receptor beta. *Endocrinology* 139, 4252–4263.

Kulling, S.E., Lehmann, L. and Metzler, M. (2002) Oxidative metabolism and genotoxic potential of major isoflavone phytoestrogens. *Journal of Chromatography B – Analytical Technology and Biomedical Life Sciences* 777, 211–218.

Kurzer, M.S. and Xu, X. (1997) Dietary phytoestrogens. *Annual Review of Nutrition* 17, 353–381.

Kutuk, O., Adli, M., Poli, G. and Basaga, H. (2004) Resveratrol protects against 4-HNE induced oxidative stress and apoptosis in Swiss 3T3 fibroblasts. *BioFactors* 20, 1–10.

Landstrom, M., Zhang, J.X., Hallmans, G., Aman, P., Bergh, A., Damber, J.E., Mazur, W., Wahala, K. and Adlercreutz, H. (1998) Inhibitory effects of soy and rye diets on the development of Dunning R3327 prostate adenocarcinoma in rats. *Prostate* 36, 151–161.

Larrosa, M., Tomas-Barberan, F.A. and Espín, J.C. (2004) The grape and wine polyphenol piceatannol is a potent inducer of apoptosis in human SK-Mel-28 melanoma cells. *European Journal of Nutrition* 43, 275–284.

Lei, B., Roncaglia, V., Vigano, R., Cremonini, C., de Maria, N., del Buono, M.G., Manenti, F. and Villa, E. (2002) Phytoestrogens and liver disease. *Molecular and Cellular Endocrinology* 193, 81–84.

Liang, Y.C., Huang, Y.T., Tsai, S.H., Lin-Shiau, S.Y., Chen, C.F. and Lin, J.K. (1999) Suppression of inducible cyclooxygenase and inducible nitric oxide synthase by apigenin and related flavonoids in mouse macrophages. *Carcinogenesis* 20, 1945–1952.

Liske, E., Hanggi, W., Henneicke-von, Z.H.H., Boblitz, N., Wustenberg, P. and Rahlfs, V.W. (2002) *Journal of Women's Health & Gender-Based Medicine* 11, 163–174.

Liu, J., Burdette, J.E., Xu, H., Gu, C., van Breemen, R.B., Bhat, K.P., Booth, N., Constantinou, A.I., Pezzuto, J.M., Fong, H.H., Farnsworth, N.R. and Bolton, J.L. (2001) Evaluation of estrogenic activity of plant extracts for the potential treatment of menopausal symptoms. *Journal of Agricultural and Food Chemistry* 49, 2472–2479.

Liu, L.Z., Fang, J., Zhou, Q., Hu, X., Shi, X. and Jiang, B.H. (2005) Apigenin inhibits expression of vascular endothelial growth factor and angiogenesis in human lung cancer cells: implication of chemoprevention of lung cancer. *Molecular Pharmacology* 68, 635–643.

Llorens, F., Miro, F.A., Casanas, A., Roher, N., Garcia, L., Plana, M., Gomez, N. and Itarte, E. (2004) Unbalanced activation of ERK1/2 and MEK1/2 in apigenin-induced HeLa cell death. *Experimental Cell Research* 299, 15–26.

Matsuda, H., Shimoda, H., Morikawa, T. and Yoshikawa, M. (2001) Phytoestrogens from the roots of *Polygonum cuspidatum* (Polygonaceae): structure-requirement of hydroxyanthraquinones for estrogenic activity. *Bioorganic and Medicinal Chemistry Letters* 11, 1839–1842.

Mazur, W. and Adlercreutz, H. (1998) Natural and anthropogenic environmental estrogens: the scientific basis for risk assessment. Naturally occurring oestrogens in food. *Pure and Applied Chemistry* 9, 1759–1776.

Mazur, W. and Adlercreutz, H. (2000) Overview of naturally occurring endocrine-active substances in the human diet in relation to human health. *Nutrition,* 16, 654–687.

Mazur, W., Fotsis, T., Wahala, K., Ojala, S., Salakka, A. and Adlercreutz, H. (1996) Isotope dilution gas chromatographic-mass spectrometric method for the determination of isoflavonoids, coumestrol, and lignans in food samples. *Analytical Biochemistry* 233, 169–180.

McCann, S.E., Wactawski-Wende, J., Kufel, K., Olson, J., Ovando, B., Kadlubar, S.N., Davis, W., Carter, L., Muti, P., Shields, P.G. and Freudenheim, J.L. (2007) Changes in 2-hydroxyestrone and 16alpha-hydroxyestrone metabolism with flaxseed consumption: modification by COMT and CYP1B1 genotype. *Cancer Epidemiology, Biomarkers and Prevention* 16, 256–262.

McCoy, J. and Kelly, W. (1996) Survey of *Cimicifuga racemosa* for phytoestrogenic flavonoids. In: *212th American Chemical Society National Meeting,* Orlando, Florida.

Meagher, L.P., Beecher, G.R., Flanagan, V.P. and Li, B.W. (1999) Isolation and characterization of the lignans, isolariciresinol and pinoresinol, in flaxseed meal. *Journal of Agricultural and Food Chemistry* 47, 3173–3180.

Merritt, J.C. (2001) Therapeutic options: hormone replacement therapy – soy therapy. *Journal of National Medical Association* 93, 288–292.

Messina, M., McCaskill-Stevens, W. and Lampe, J.W. (2006). Addressing the soy and breast cancer relationship: review, commentary, and workshop proceedings. *Journal of National Cancer Institute* 98, 1275–1284.

Messina, M.J. and Loprinzi, C.L. (2001) Soy for breast cancer survivors: a critical review of the literature. *Journal of Nutrition* 131, 3095S–3108S.

Miksicek, R.J. (1994) Interaction of naturally occurring nonsteroidal estrogens with expressed recombinant human estrogen receptor. *Journal of Steroid Biochemistry and Molecular Biology* 49, 153–160.

Milder, I.E.J., Arts, I.C.W., van de Putte, B., Venema, D.P. and Hollman, P.C.H. (2005) Lignan contents of Dutch plant foods: a database including lariciresinol, pinoresinol, secoisolariciresinol and matairesinol. *British Journal of Nutrition* 93, 393–402.

Milligan, S.R., Kalita, J.C., Heyerick, A., Rong, H., de Cooman, L. and de Keukeleire, D. (1999) Identification of a potent phytoestrogens in hops (*Humulus lupulus* L.) and beer. *Journal of Clinical Endocrinology and Metabolism* 84, 2249–2252.

Moyad, M.A. (2002) Complementary/alternative therapies for reducing hot flashes in prostate cancer patients: reevaluating the existing indirect data from studies of breast cancer and postmenopausal women. *Urology* 59, 20S–33S.

Muir, A.D. and Westcott, N.D. (2000) Quantitation of the lignin secoisolariciresinol diglucoside in baked goods containing flax seed or flax meal. *Journal of Agricultural Food Chemistry* 48, 4048–4052.

Murkies, A.L., Wilcox, G. and Davis, S.R. (1998) Phytoestrogens. *Journal of Clinical Endocrinology and Metabolism* 83, 297–303.

Naftolin, F. and Stanbury, M.G. (2002) Phytoestrogens: are they really estrogen mimics? *Fertility and Sterility* 77, 15–17.

Nestel, P.J., Pomeroy, S.E., Sasahara, T., Yamashita, T., Liang, Y.L., Dart, A.M., Jennings, G.L., Abbey, M. and Cameron, J.D. (1997) Arterial compliance in obese subjects is improved with dietary plant n-3 fatty acid from flaxseed oil despite increased LDL oxidizability. *Arteriosclerosis, Thrombosis and Vascular Biology* 17, 1163–1170.

Ohno, T., Kato, N., Ishii, C., Shimizu, M., Ito, Y., Tomono, S. and Kawazu, S. (1993) Genistein augments cyclic adenosine 3′5′-monophosphate (cAMP) accumulation and insulin release in min6 cells. *Endocrine Research* 19, 273–285.

Olas, B., Wachowicz, B., Saluk-Juszczak, J. and Zielinski, T. (2002) Effect of resveratrol, a natural polyphenolic compound, on platelet activation induced by endotoxin or thrombin. *Thrombosis Research* 107, 141–145.

Padilla-Banks, E. (2006) Neonatal exposure to the phytoestrogen genistein alters mammary gland growth and developmental programming of hormone receptor levels. *Endocrinology* 147, 4871–4882.

Peeters, P.H.M., Keinan-Boker, L., van der Schouw, Y.T. and Grobbee, D.E. (2003) Phytoestrogens and breast cancer risk. *Breast Cancer Research and Treatment* 77, 171–183.

Pendurthi, U.R., Williams, J.T. and Rao, L.V. (1999) Resveratrol, a polyphenolic compound found in wine, inhibits tissue factor expression in vascular cells: a possible mechanism for the cardiovascular benefits associated with moderate consumption of wine. *Arteriosclerosis, Thrombosis and Vascular Biology* 19, 419–426.

Possemiers, S., Heyerick, A., Robbens, V., de Keukeleire, D. and Verstraete, W. (2005) Activation of proestrogens from hops (*Humulus lupulus* L.) by intestinal microbiota; conversion of isoxanthohumol into 8-prenylnaringenin. *Journal of Agricultural and Food Chemistry* 53, 6281–6288.

Pozo-Guisado, E., Alvarez-Barrientos, A., Mulero-Navarro, S., Santiago-Josefat, B. and Fernandez-Salguero, P.M. (2002) The anti-proliferative activity of resveratrol results in apoptosis in MCF-7 but not in MDAMB-231 human breast cancer cells: cell-specific alteration of the cell cycle. *Biochemical Pharmacology* 64, 1375–1386.

Prasad, K. (1997) Hydroxyl radical-scavenging property of secoisolariciresinol diglucoside (SDG) isolated from flax-seed. *Molecular and Cellular Biochemistry* 168, 117–123.

Prasad, K. (2000) Antioxidant activity of secoisolariciresinol diglucoside derived metabolites, secoisolariciresinol, enterodiol, and enterolactone. *International Journal of Angiology* 9, 220–225.

Prasad, K., Mantha, S.V., Muir, A.D. and Westcott, N.D. (1998) Reduction of hypercholesterolemic atherosclerosis by CDC-flaxseed with very low alphalinolenic acid. *Atherosclerosis* 136, 367–375.

Prestwood, K.M., Kenny, A.M., Kleppinger, A. and Kulldorff, M. (2003) Ultralow-dose micronized 17 beta-estradiol and bone density and bone metabolism in older women – a randomized controlled trial. *Journal of the American Medical Association* 290, 1042–1048.

Price, K.R. and Fenwick, G.R. (1985) Naturally occurring oestrogens in foods –a review. *Food Additives and Contaminants* 2, 73–106.

Raffaelli, B., Hoikkala, A., Leppala, E. and Wahala, K. (2002) Enterolignans. *Journal of Chromatography B*, 777, 29–43.

Rafi, M.M., Rosen, R.T., Vassil, A., Ho, C.T., Zhang, H., Ghai, G., Lambert, G. and diPaola, R.S. (2000) Modulation of bcl-2 and cytotoxicity by licochalcone-A, a novel estrogenic flavonoid. *Anticancer Research* 20, 2653–2658.

Rao, H.S.P. and Reddy, K.S. (1991) Isofavones from *Flemingia vestita*. *Fitoterapia* 63, 458.

Reinli, K. and Block, G. (1996) Phytoestrogen content of foods. *Nutrition and Cancer* 26, 123–148.

Rolfs, C.H. and Kindl, H. (1984) Stilbene synthase and chalcone synthase. Two different constitutive enzymes in cultured cells of *Picea exelsa*. *Plant Physiology* 75, 489–492.

Rossouw, J.E., Anderson, G.L., Prentice, R.L., LaCroix, A.Z., Kooperberg, C., Stefanick, M.L., Jackson, R.D., Beresford, S.A., Howard, B.V., Johnson, K.C., Kotchen, J.M. and Ockene, J. (2002) Risks and benefits of estrogen plus progestin in healthy postmenopausal women: principle results from the Women's Health Initiative randomized controlled trial. *Journal of the American Medical Association* 288, 321–333.

Rowland, I., Wiseman, H., Sanders, T., Adlercreutz, H. and Bowey, E. (1999) Metabolism of oestrogens and phytoestrogens: role of the gut microflora. *Biochemical Society Transactions* 27, 304–308.

Ruggiero, R.J. and Likis, F.E. (2002) Estrogen: physiology, pharmacology, and formulations for replacement therapy. *Journal of Midwifery and Women's Health* 47, 130–138.

Ruiz-Larrea, M.B., Mohan, A.R., Paganga, G., Miller, N.J., Bolwell, G.P. and Rice-Evans, C.A. (1997) Antioxidant activity of phytoestrogenic isoflavones. *Free Radical Research* 26, 63–70.

Santti, R., Makela, S., Strauss, L., Korkman, J. and Kostian, M.-L. (1998) Phytoestrogens: potential endocrine disruptors in males. *Toxicology and Industrial Health* 14, 223–237.

Schaefer, O., Bohlmann, R., Schleuning, W.D., Schulze-Forster, K. and Humpel, M. (2005) Development of a radioimmunoassay for the quantitative determination of 8-prenylnaringenin in biological matrices. *Journal of Agricultural and Food Chemistry* 53, 2881–2889.

Scheiber, M.D. and Rebar, R.W. (1999) Isoflavones and postmenopausal bone health: a viable alternative to estrogen therapy. *Menopause* 6, 233–241.

Schwartz, H. and Sontag, G. (2006) Determination of secoisolariciresinol, lariciresinol and isolariciresinol in plant foods by high performance liquid chromatography coupled with coulometric electrode array detection. *Journal of Chromatography B* 838, 78–85.

Setchell, K.D. (1998) Phytoestrogens: the biochemistry, physiology, and implications for human health of soy iso-flavones. *American Journal of Clinical Nutrition* 68, 1333S–1346S.

Setchell, K.D., Gosselin, S.J., Welsh, M.B., Johnston, J.O., Balistreri, W.F., Kramer, L.W., Dresser, B.L. and Tarr, M.J. (1987) Dietary estrogens – a probable cause of infertility and liver disease in captive cheetahs. *Gastroenterology* 93, 225–233.

Severson, R.K., Nomura, A.M., Grove, J.S. and Stemmermann, G.N. (1989) A prospective study of demographics and prostate cancer among men of Japanese ancestry in Hawaii. *Cancer Research* 49, 1857–1860.

Sheehan, D.M. (1998) Herbal medicines, phytoestrogens and toxicity: risk: benefit considerations. *Proceedings of Society for Experimental Biology and Medicine* 217, 379–385.

Shutt, D.A. and Braden, A.W.H. (1968) The significance of equol in relation to the oestrogenic responses in sheep ingesting clover with a high formononetin content. *Australian Journal of Agricultural Research* 19, 545–553.

Sicilia, T., Niemeyer, H.B., Honig, D.M. and Metzler, M. (2003) Identification and stereochemical characterization of lignans in flaxseed and pumpkin seeds. *Journal of Agricultural and Food Chemistry* 51, 1181–1188.

Simons, A.L., Renouf, M., Hendrich, S. and Murphy, P.A. (2005) Human gut microbial degradation of flavonoids: structure-function relationships. *Journal of Agricultural and Food Chemistry* 53, 4258–4263.

Sobolev, V.S., Horn, B.W., Potter, T.L., Deyrup, S.T. and Gloer, J.B. (2006) Production of stilbenoids and phenolic acids by the peanut plant at early stages of growth. *Journal of Agricultural and Food Chemistry* 54, 3505–3511.

Stevens, J.F., Taylor, A.W. and Deinzer, M.L. (1999) Quantitative analysis of xanthohumol and related prenylflavonoids in hops and beer by liquid chromatography tandem mass spectrometry. *Journal of Chromatography A* 832, 97–107.

Strom, S.S., Yamamura, Y., Duphorne, C.M., Spitz, M.R., Babaian, R.J., Pillow, P.C. and Hursting, S.D. (1999) Phytoestrogen intake and prostate cancer: a case-control study using a new database. *Nutrition and Cancer* 33, 20–25.

Sturgeon, S.R., Heersink, J.L., Volpe, S.L., Bertone-Johnson, E.R., Puleo, E., Stanczyk, F.Z., Sabelawski, S., Wahala, K., Kurzer, M.S. and Bigelow, C. (2008) Effect of dietary flaxseed on serum levels of estrogens and androgens in postmenopausal women. *Nutrition and Cancer* 60, 612–618.

Tamir, S., Eizenberg, M., Somjen, D., Izrael, S. and Vaya, J. (2001) Estrogen-like activity of glabrene and other constituents isolated from licorice root. *Journal of Steroid Biochemistry and Molecular Biology* 78, 291–298.

Terzic, M., Micic, J., Dotlic, J., Maricic, S., Mihailovic, T. and Knezevic, N. (2012) Impact of phytoestrogens on serum lipids in post menopausal women. *Geburtsh Frauenheilk* 72, 527–531.

Tham, D.M., Gardner, C.D. and Haskell, W.L. (1998) Clinical Review 97: potential benefits of dietary phytoestrogens: a review of the clinical, epidemiological, and mechanistic evidence. *Journal of Clinical Endocrinology and Metabolism* 83, 2223–2235.

This, P., de la Rochefordiere, A., Clough, K., Fourquet, A. and Magdelenat, H. (2001) Phytoestrogens after breast cancer. *Endocrine-Related Cancer* 8, 129–134.

Thompson, L.U., Rickard, S.E., Cheung, F., Kenaschuk, E.O. and Obermeyer, W.R. (1997) Variability in anticancer lignan levels in flaxseed. *Nutrition and Cancer* 27, 26–30.

Thompson, L.U., Robb, P., Serraino, M. and Cheung, F. (1991) Mammalian lignan production from various foods. *Nutrition and Cancer* 16, 43–52.

Trock, B.J., Hilakivi-Clarke, L. and Clarke, R. (2006) Meta-analysis of soy intake and breast cancer risk. *Journal of National Cancer Institute* 98, 459–471.

Tseng, S.H., Lin, S.M., Chen, J.C., Su, Y.H., Huang, H.Y., Chen, C.K., Lin, P.Y. and Chen, Y. (2004) Resveratrol suppresses the angiogenesis and tumor growth of gliomas in rats. *Clinical Cancer Research* 10, 2190–2202.

Tsutsumi, N. (1995) Effect of coumestrol on bone metabolism in organ culture. *Biological and Pharmaceutical Bulletin* 18, 1012–1015.

Umezawa, T., Davin, L.B. and Lewis, N.G. (1991) Formation of lignans (-)-secoisolariciresinol and (-)-matairesinol with *Forsythia intermedia* cell-free extracts. *Journal of Biological Chem*istry 266, 10210–10217.

van der Schouw, Y.T., de Kleijn, M.J., Peeters, P.H. and Grobbee, D.E. (2000) Phyto-oestrogens and cardiovascular disease risk. *Nutrition, Metabolism and Cardiovascular Diseases* 10, 154–167.

van der Weijer, P.H.M. and Barentsen, R. (2002) Isoflavones from red clover (Promensil) significantly reduce menopausal hot flush symptoms compared with placebo. *Maturitas* 42, 187–193.

Verzele, M. (1986) Centenary review: 100 years of hop chemistry and its relevance to brewing. *Journal of the Institute of Brewing and Distilling* 92, 32–48.

Viereck, V., Emons, G. and Wuttke, W. (2005) Black cohosh: just another phytoestrogen? *Trends in Endocrinology & Metabolism* 16, 214–221.

Vincent, A. and Fitzpatrick, L.A. (2000) Soy isoflavones: are they useful in menopause? *Mayo Clinic Proceedings* 75, 1174–1184.

Wagner, J.D., Cefalu, W.T., Anthony, M.S., Litwak, K.N., Zhang, L. and Clarkson, T.B. (1997) Dietary soy protein and estrogen replacement therapy improve cardiovascular risk factors and decrease aortic cholesteryl ester content in ovariectomized cynomolgus monkeys. *Metabolism* 46, 698–705.

Wagner, J.D., Anthony, M.S. and Cline, J.M. (2001) Soy phytoestrogens: research on benefits and risks. *Clinical Obstetrics and Gynecology* 44, 843–852.

Wang, C.Z., Ma, X.Q., Yang, D.H., Guo, Z.R., Liu, G.R., Zhao, G.X., Tang, J., Zhang, Y.N., Ma, M., Cai, S.Q., Ku, B.S. and Liu, S.L. (2010) Production of enterodiol from defatted flaxseeds through biotransformation by human intestinal bacteria. *BMC Microbiology* 10, 115.

Wang, I.K., Lin-Shiau, S.Y. and Lin, J.K. (1999) Induction of apoptosis by apigenin and related flavonoids through cytochrome release and activation of caspase-9 and caspase-3 in leukaemia HL-60 cells. *European Journal of Cancer* 35, 1517–1525.

Wang, W., Heideman, L., Chung, C.S., Pelling, J.C., Koehler, K.J. and Birt, D.F. (2000) Cell cycle arrest at G2-M and growth inhibition by apigenin in human colon carcinoma cell lines. *Molecular Carcinogenesis* 28, 102–110.

Wang, Z., Huang, Y., Zou, J., Cao, K., Xu, Y. and Wu, J.M. (2002) Effects of red wine and wine polyphenol resveratrol on platelet aggregation *in vivo* and *in vitro*. *International Journal of Molecular Medicine* 9, 77–79.

Wang, Z.Y. and Nixon, D.W. (2001) Licorice and cancer. *Nutrition and Cancer* 39, 1–11.

Way, T.D., Kao, M.C. and Lin, J.K. (2004) Apigenin induces apoptosis through proteasomal degradation of HER2/neu in HER2/neu-overexpressing breast cancer cells via the phosphatidylinositol 3-kinase/Akt-dependent pathway. *Journal of Biological Chemistry* 279, 4479–4489.

Whitten, P.L. and Naftolin, F. (1998) Reproductive actions of phytoestrogens. *Baillieres Clinical Endocrinology and Metabolism* 12, 667–690.

Wiseman, H. (1999) The bioavailability of non-nutrient plant factors: dietary flavonoids and phyto-oestrogens. *Proceedings of the Nutrition Society* 58, 139–146.

Wolter, F., Clausnitzer, A., Akoglu, B. and Stein, J. (2002) Piceatannol, a natural analog of resveratrol, inhibits progression through the S-phase of the cell cycle in colorectal cancer cell lines. *Journal of Nutrition* 132, 298–302.

Wroblewski, L.L. and Cooke, J.P. (2000) Phytoestrogens and cardiovascular health. *Journal of the American College of Cardiology* 35, 1403–1410.

Xu, X., Harris, K.S., Wang, H.J., Murphy, P.A. and Hendrich, S. (1995) Bioavailability of soybean isoflavones depends upon gut microflora in women. *Journal of Nutrition* 125, 2307–2315.

Xue, J.Y., Liu, G.T., Wei, H.L. and Pan, Y. (1992) Antioxidant activity of two dibenzocyclooctene lignans on the aged and ischemic brain in rats. *Free Radical Biology and Medicine* 12, 127–135.

Yin, F., Giuliano, A.E., Law, R.E. and van Herle, A.J. (2001) Apigenin inhibits growth and induces G2-M arrest by modulating cyclin-CDK regulators and ERK MAP kinase activation in breast carcinoma cells. *Anticancer Research* 21, 413–420.

Zava, D.T., Dollbaum, C.M. and Blen, M. (1998) Estrogen and progestin bioactivity of foods, herbs, and spices. *Proceedings of Society for Experimental Biology and Medicine* 217, 369–378.

11 Phytosterols and their Healthy Effects

Lilia Masson*

Universidad de Chile, Santiago, Chile; Fundación CAPES, Universidad Federal de Rio de Janeiro, Instituto de Nutrición Josué de Castro, Rio de Janeiro, Brasil

11.1 Introduction

Phytosterols (PS) are related to cyclopentaphenanthren with four condensed rings of 28 or 29 carbons. They are classified according to the presence or absence of a methyl group on carbon 4, and a hydroxyl group on position 3. The principal active constituents present in plant seed oils are β-sitosterol, stigmasterol, Δ5-avenasterol, campesterol, Δ7-avenasterol and brassicasterol. They correspond to the 4-desmethylsterols; cholesterol is 4-desmethylsterol. All have a side chain with 9–10 carbons; β-sitosterol, stigmasterol, Δ5-avenasterol and Δ7-avenasterol have 29 carbons and an ethyl group in the side chain; campesterol and brassicasterol have 28 carbons and a methyl group in the side chain (Bloch, 1988). PS are unsaturated or saturated, in which case they are designated as phytostanols. Figure 11.1 shows the structure of the principal phytosterols and phytostanols, compared with cholesterol. The function of PS in the plant kingdom is similar to the function of cholesterol in animal cells, i.e. they are part of the cell membrane and stabilize the phospholipid bilayer (Hartman and Benveniste, 1987; Piironen *et al.*, 2000; Moreau *et al.*, 2002; Katan *et al.*, 2003; Dutta, 2004; Kritchevsky and Chen, 2005; Patel and Thompson, 2006).

This chapter throws an insight on principal natural sources of PS and their role in human health through their action as a functional component in foods by decreasing cholesterol absorption, and as a consequence to reduce total cholesterol and LDL-cholesterol plasma levels.

11.2 Principal Natural Sources

PS are natural components present in the unsaponifiable matter of fats and oils from vegetal origin. They are part of the cellular membrane of all plant organisms. They are produced by 'de novo synthesis' and generally are called vegetal sterols or more properly phytosterols, although frequently sterols and stanols are included together. A large number of different PS and related compounds present in plants have been described, however, β-sitosterol, stigmasterol, Δ5-avenasterol, campesterol, Δ7-avenasterol and brassicasterol are the more abundant (Plumb *et al.*, 2011). Their basic chemical structure is similar to cholesterol, with a difference in the side carbon chain.

* E-mail: masson_lilia@yahoo.es

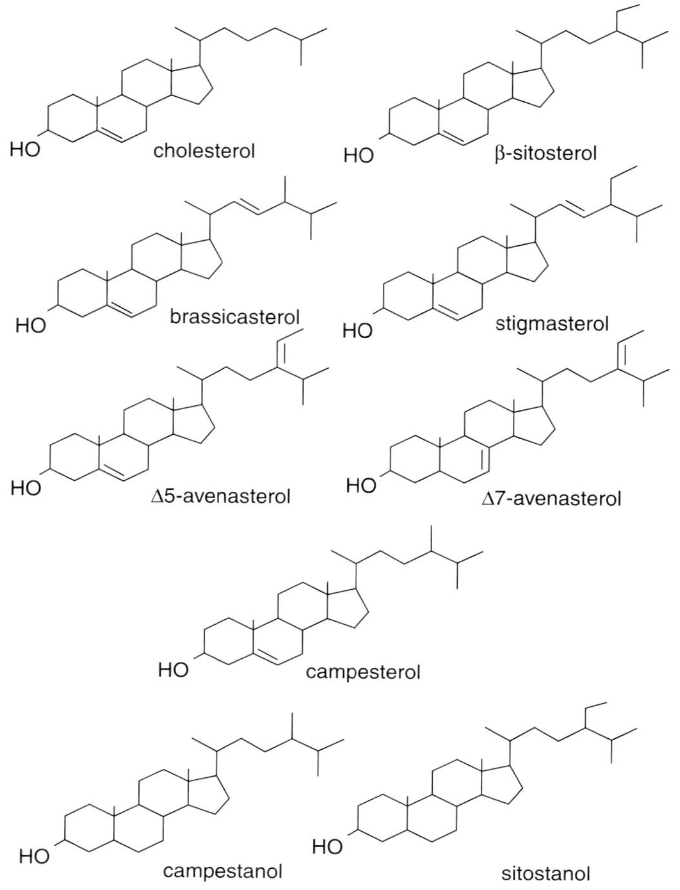

Fig. 11.1. Cholesterol, common phytosterols and phytostanols.

Naturally they are in free form, esterified with fatty acids or conjugated as stearyl glycosides; their hydrogenation produce stanols, which are less abundant in nature, but they can be produced by the hydrogenation of the double bond present in the second ring, forming sitostanol, campestanol, stigmastanol, etc. The PS are not synthesized by mammals and their presence in humans is exclusively provided by the diet (Grunwald, 1975). Plant oils are the principal sources of PS; normally, three are more abundant in the order β-sitosterol, which contributes around 80–90% of total PS, then campesterol, followed by stigmasterol. The other PS such as Δ5-avenasterol and Δ7-avenasterol only occur in minor fractions. Some exceptions occur in nature, e.g. brassicasterol is present in *Brassica* species (Weihrauch and Gadner, 1978).

In relation to their general distribution and occurrence in plants, and their oilseed world production, soybean oil is the one of the principal sources of PS in the common diet, followed by sunflower, canola rapeseed, maize, cotton and sesame oils (Moreau and Kamal-Eldin, 2009). Another source of PS is from groundnuts, but they have a lower consumption (Francisco and Resurrección, 2008).

Jiménes-Escrig *et al.* (2006) published the common sources of PS in the Spanish diet and listed the PS content of 54 foods. The estimated daily intake was 276 mg with β-sitosterol as usually the highest contribution. The highest values by group from major to minor were: sunflower oil, olive oil, then nuts with pistachio, sunflower seeds and almonds, followed by legumes. Among 24 vegetables, artichoke and cauliflower had the highest values and amongst 16 fruits, olives, white grapes and oranges were the highest. Cereals have been also considered as a good source of PS. Normén *et al.* (2002) studied 76 foods based on cereals that are consumed in Sweden and the Netherlands including flours, grains, germs, processed cereals, pasta, spreads, biscuits, cakes, etc., and confirmed that they are also a good dietary source of PS and phytostanols. Normén *et al.* (2007) published a PS database for fatty foods consumed in Sweden and the Netherlands; 87 items were included; the highest contribution was again by β-sitosterol. The maximum total PS content was 775 mg 100 g^{-1} in a fat spread; amongst oils, maize oil presented the highest value.

Nurmi *et al.* (2008) studied the PS content in wheat genotypes. Information of PS content in new natural alternatives such as special plant seed oils or fruit pulp has also been published. PS present in avocado oil is especially rich in sitosterol (Woolf *et al.*, 2009); another source is maize fibre oil (Moreau *et al.*, 2009). The sterols characterization in refined borage was published by Wretensjo and Kalberg (2002). The unsaponifiable composition of five Amazonian palm kernel oils was given by Bereau *et al.* (2003). PS content in the oil extracted from four seeds of native Latin American fruits, cherimola (*Annona cherimola*), prickly pear (*Opuntia ficus-indica*), mountain papaw (*Carica pubescens*) and palm coconut (*Jubaea chilensis* (Molina) Baillon) has been described by Masson *et al.* (2008a, b).

11.3 Industrial Sources of Phytosterols

There are two principal raw materials of PS that are available for use as ingredients in functional foods, cosmetics and pharmaceutical forms. Rich sources of PS are soybean oil with around 1.5% of unsaponifiable matter content, which is a convenient raw material and has the world's highest production, and from tall oil, a by-product obtained in Kraft cellulose procedure with high unsaponifiable matter of around 12%. Different industrial processes protected under patent were developed to recover PS, principally sitosterol and campesterol, from deodorized soybean oil distillates and from tall oil (Fernandes and Cabral, 2007). Supercritical CO_2 has been assayed for this purpose (Eisenmenger and Dunford, 2008). During the industrial refining process of the plant oils, alterations can be produced in PS (Ferrari *et al.*, 1997).

11.4 Nutritional and Health Benefits of Phytosterols

From the nutritional and health point of view, it has been proved that the ingestion of PS in higher amounts than the usual contribution by diet intake decreases cholesterol and LDL-cholesterol concentrations in plasma. Its absorption is inhibited in the intestine at the micelle structure level by reduction of the cholesterol solubility (Ellegárd *et al.*, 2007). Nguyen (1999) reported the effect of plant stanol esters for lowering cholesterol. Cholesterol absorption presents a great variation among individuals of between 30 and 80%, with a mean estimated close to 50% (Miettinen and Kesaniemi, 1989). On the other hand, PS absorption is very low, with a variation of around 0.1–5% depending on the specific PS structure. By increasing the amount of PS intake, the inhibition of cholesterol absorption also increases. It was shown that around 2–3 g of PS in a normal or low fat diet produced a reduction of total plasma cholesterol of around 7% and in the order of 10% of LDL-cholesterol reduction (Kozlowska-Wojciechowska *et al.*, 2003). This potential of PS as a dietary tool for decreasing plasma and LDL-cholesterol has been explained by Ling and Jones (1995) and Jones *et al.* (1997).

An evaluation of clinical trials in patients with hypercholesterolaemia and PS balance was done by Lees *et al.* (1997). Jones *et al.* (1998) assayed PS from tall oil source in patients with different cholesterol levels and obtained good results. Neil and Huxley (2002) studied the efficacy and therapeutic potential of PS, confirming that a daily intake of around 2–2.5 g of sterol or stanol esters can lower LDL-cholesterol level by 10–15%. Andersson *et al.* (2004) used a database of PS content of around 350 foods to relate PS intake of the habitual diet with serum cholesterol level. The results indicated an inverse relationship between PS intake and total serum cholesterol and LDL-cholesterol. The daily PS intake for the UK population was around 300 mg day^{-1} for both sexes and ranged between 100 and 700 mg day^{-1}. These findings suggested that natural PS in diet have a positive effect (Ellégard *et al.*, 2007). Ostlund Jr (2007) described different types of PS, their mechanism of action, PS supplements, PS and their natural sources. Francisco and Resurrección (2008) also worked on groundnuts as natural sources of PS. The data published for an occidental diet showed variation between 200 and 400 mg day^{-1} (Kozlowska-Wojciechowska *et al.*, 2003). For Asian populations and vegetarian people, the values reported are higher, ranging between 345 and 499 mg day^{-1} (Francisco and Resurrección, 2008). Chan *et al.* (2006) studied other aspects of plant sterol physiology and its possible relation on coronary heart disease. Studies have shown that PS is not absorbed and they are present in small amounts in diet. For decreasing the absorption of cholesterol, high amounts of PS are required. Studies on the possible increase of PS plasma concentration related with cardiovascular disease were also evaluated. A possible positive effect of PS in cancer has been studied by Awad and Fink (2000).

11.5 Phytosterols as Functional Ingredient to Decrease Plasma Cholesterol and LDL-Cholesterol Levels

Research has shown that the amounts of PS or phytostanols required for reducing plasma cholesterol and LDL-cholesterol in humans are high, ranging between 2 and 3 g (Kozlowska-Wojciechowska *et al.*, 2003). Industrial procedures were developed to produce concentrated PS of high purity and safety for human consumption (Thompson and Grundy, 2005); e.g. margarine with 1.8 g day^{-1} of phytostanol ester is a good example of a functional food that was developed after overcoming several physiological and technological difficulties. The main steps were to understand the mechanisms for lowering plasma cholesterol by PS and phytostanols, to select the best source of PS, to synthesize phytostanol esters that could be dispersed in a fat margarine phase, to develop the product, and to test dose and frequency. Related to safety, these functional foods are considered Generally Recognized as Safe (GRAS) by the USA and the European Union (Thomson and Grundy, 2005).

Another study published by Salo *et al.* (2005) is related to the production of stanol esters, their physical properties and chemical behaviour of the newly synthesized compounds. If margarine is the target product then a more rigid structure at a certain temperature range is necessary, but it must be spreadable in a wide range of temperatures. Different assays were done including oxidative stability; likewise, food applications were developed for 12 different food products.

Another development was to deliver PS or phytostanols ester as a nutraceutical. Earnest *et al.* (2007) assayed the effect of PS esters in capsules in adults suffering from hypercholesterolaemia. The daily dose was 2.6 g of PS ester divided into four capsules of 0.650 g each containing 88% PS ester: sitosterol 40–58%, campesterol 20–28% and stigmasterol 14–23%. Total time treatment for sample and placebo was 12 weeks. The results showed a moderate LDL-cholesterol reduction compared with placebo by using the developed functional foods.

A number of other studies have confirmed the role of PS and phytostanol in reducing plasma total cholesterol and LDL-cholesterol in humans with the consumption of different functional foods enriched with these compounds. Some other work done by Linchtenstein and Deckelbaum (2001) studied the stanol/sterol ester-content in foods and blood cholesterol levels. Christiansen *et al.* (2001) assayed yoghurt enriched in PS in patients with moderate hypertriglyceridaemia and obtained good results. Volpe *et al.* (2001) studied the effect of yoghurt enriched with PS on serum lipids in patients with moderated hypercholesterolaemia with positive results. De Graaf *et al.* (2002) used

PS from tall oil in a chocolate matrix and a significant decrease in plasma cholesterol and LDL-cholesterol was obtained. Nestel *et al.* (2001) obtained a positive effect on lowering cholesterol plasma using PS ester and free stanol in margarine, butter and low-fat foods.

Mensink *et al.* (2002) evaluated the effect of stanol esters incorporated in low-fat yoghurt on serum lipids and lipoproteins. After 1 week of treatment, a positive response was obtained and reduction in plasma tocopherol was observed along with a decrease in carotene. Ground beef added to soybean PS also showed a positive response (Matvienko *et al.*, 2002). Jones *et al.* (2003) assayed PS in low- and non-fat beverages as part of a controlled diet to test the response in this low-fat medium but reduction in plasma lipid levels was not observed. Kozlowska-Wojciechowsk *et al.* (2003) evaluated the results using margarine enriched with PS on blood lipids, platelet function and fibrinogen level in young men and obtained promising results. St-Onge *et al.* (2003) studied the consumption of a functional oil rich in PS and medium chain triglyceride oil, which improved plasma lipid profiles in men. Clifton *et al.* (2004) measured the relative effects of the intake equivalent to 1.6 g of PS as ester, using four low-fat foods, milk, yoghurt, bread and cereal, to test the influence of the matrix in the lowering cholesterol effects in plasma. The results showed that the four food matrix enriched with PS gave positive response, and milk gave the best response. Carotene pigments were measured and some small changes were observed considering the normal variability. Davaraj *et al.* (2004) assayed orange juice fortified with PS; the response was positive in mildly hypercholesterolaemic healthy individuals. Simojoki *et al.* (2005) used margarine enriched with stanol ester among persons with and without cardiovascular disease for the adoption of functional food in Finland. The results were positive, but a decrease in plasma carotene pigments was observed. According to researchers, the beneficial or adverse effects on health in long term use of a phytostanol ester in enriched margarine have to be studied further.

11.6 Health Benefits and Safety

Although chemical component(s) are present in nature, it does not mean that it is safe for human consumption, especially in larger amounts than those normally found in the common daily diet and specially when human physiology has restrictions to their absorption. This is exactly the situation for PS: they have a very similar cholesterol chemical structure, but they do not have an active transport mechanism for absorption at the intestine level, as for cholesterol (De Jong *et al.*, 2003).

Piironen *et al.* (2000) studied biosynthesis, biological function and human nutrition. The biological role of PS in plant cell membranes, mean daily intake and its serum cholesterol lowering effect was investigated along with chemical and physical properties, chemical structures and biosynthesis. Fats, oils and cereals as the principal sources along with their daily intake values were included. Changes in ripening, postharvest processing and PS lost during oil refining, bleaching and deodorization are commented on. Methods of extraction and analysis are described. It finishes with information about PS bioavailability and physiological effects.

Moreau *et al.* (2002) described the structural diversity, occurrence and metabolism, quantification methods, healthy effects, action mechanisms, clinical studies, lowering efficacy on LDL-cholesterol, anticancer properties, effects on absorption of fat soluble vitamins and antioxidants of PS. There are evidences that PS and phytostanol esters reduce absorption of α-tocopherol, β-carotene and lycopene. They can also protect LDL-cholesterol oxidation. The researchers also observed that PS esters can increase PS absorption appearing as oxidized compounds, which can induce atherosclerotic damage, and they are cytotoxic. PS are considered GRAS substances.

Katan *et al.* (2003) reported randomized double-blind trials comparing foods with and without added plant stanols and sterols. According to the age, the percentage of LDL-cholesterol decreased between 8 and 16%. The effect of formulation, intake frequency, subgroups responses, effects related with diet, drugs to reduce cholesterol and potential effect for cardiovascular risk reduction was also discussed. No adverse effect had yet been reported. Toxicological studies in animals, absorption of PS and phytostanols, relation with possible atherosclerosis promotion, reduction of plasma fat-soluble vitamins and risk involved were also presented.

11.7 Conclusion

Phytosterols and synthesized phytostanols have been used in humans as functional ingredients to decrease serum cholesterol and LDL-cholesterol. The numerous publications related to this interesting subject using different food matrixes have showed positive effects, but it is necessary to incorporate some fat in the formulations of these functional foods to improve the positive effects. As the persons must have a high intake dose of PS to be effective, it is important to continue the evaluation if some decrease of cholesterol and LDL-cholesterol is obtained with a moderated or high direct dietary intake of PS. It is important to create the awareness amongst the new generation about the importance of PS content in foods by introducing new databases that can permit a better evaluation of PS in natural and processed foods for local populations. The principal concern related to a decrease of the absorption of fat-soluble antioxidants needs more evaluation; the potential risks to induce phytosterolaemia seems not to be so critical, but the research must continue. Phytosterols and phytostanols have been catalogued as GRAS by International Food Agencies.

Nature is so selective and wise in its decisions, that the door is not opened for all.

References

Andersson, S.W., Skinner, J., Ellegård, L., Welch, A.A., Binham, S., Mulligan, A., Andersson, H. and Khaw, K.-T. (2004) Intake of dietary sterols is inversely related to serum cholesterol concentration in men and women in the EPIC Norfolk population: a cross-sectional study. *European Journal of Clinical Nutrition* 58, 1378–1385.

Awad, A. and Fink, C. (2000) Phytosterols as anticancer dietary components: evidence and mechanism of action. *Journal of Nutrition* 130, 2127–2130.

Bereau, D., Benjelloun-Mlayah, B., Banoub, J. and Bravo, R. (2003) FA and Unsaponifiable Composition of Five Amazonian Palm Kernel Oils. *Journal of the American Oil Chemists' Society* 80, 49–53.

Bloch, K. (1988) Sterol structure and function. *Journal American Oil Chemists' Society* 65, 1763–1766.

Chan, Y.-M., Varady, K., Lin, Y., Trautwein, E., Mensink, R., Plat, J. and Jones, P. (2006) Plasma concentration of plant sterol: Physiology and relationship with coronary heart disease. *Nutrition Reviews* 64, 385–402.

Clifton, P.M., Noakes, M., Sullivan, D., Erichsen, N., Ross, D., Annison, G., Fassoulakis, A., Cehun, M. and Nestel, P. (2004) Cholesterol-lowering effects of plasma sterol esters differ in milk, yoghurt, bread and cereal. *European Journal of Clinical Nutrition* 58, 503–509.

Christiansen, L.I., Lahteenmaki, P.L.A., Mannelin, M.R., Seppanen-Laakso, T.E., Hiltunen, R.V.K. and Yliruusi, Y.K. (2001) Cholesterol-Lowering Effect of Spread Enriched with Microcrystalline Plant Sterols in Hypercholesterolemic Subjects. *European Journal of Nutrition* 40, 66–73.

Davaraj, S., Jialal, I. and Vega-López, S. (2004) Plant-Sterol-Fortified Orange Juice effectively Lowers Cholesterol Levels in Mildly Hypercholesterolemic Healthy Individuals. *Arterioesclerosis Thrombosis Vascular Biology* 23, e25–e28.

De Graaf, J., Nolting, P.R.W.D., van Dam, M., Belsey, E.M., Kastelein, J.J.P., Pritchard, P.H. and Stalenoef, A.F.H. (2002) Consumption of Tall Oil-Derivated Phytosterols in a Chocolate Matrix Significantly Decreases Plasma Total and Low-density Lipoprotein-Cholesterol Level. *British Journal of Nutrition* 88, 479–488.

De Jong, A., Plat, J. and Mensink, R. (2003) Metabolic Effects of Plant Sterol and Stanol (Review). *Journal of Nutritional Biochemistry* 14, 362–369.

Dutta, P.C. (ed.) (2004) *Phytosterols as Functional Foods Components and Nutracceuticals.* Marcel Dekker, New York.

Earnest, C.P., Mikus, C.R., Lemieux, I., Arsenault, B.J. and Church, T.S. (2007) Examination of encapsulated phytosterol ester supplementation on lipid indices associated with cardiovascular disease. *Nutrition* 23, 625–633.

Eisenmenger, M. and Dunford, N.T. (2008) Bioactive Components of Commercial and Supercritical Carbon Dioxide processed Wheat Germ Oil. *Journal of the American Oil Chemists' Society* 85, 55–61.

Ellegård, L., Andersson, S., Normén, L. and Andersson, H. (2007) Dietary plant sterol and cholesterol metabolism. *Nutrition Reviews* 65, 39–45.

Fernades, P. and Cabral, J.M.S. (2007) Phytosterols: applications and recovery methods. *Bioresourse Technology* 98, 2335–2350.

Ferrari, R.Ap., Esteves, W., Mukherjee, K. and Schulte, E. (1997) Alterations of sterols and steryl Esters in Vegetable Oils during Industrial Refining. *Journal of Agriculture and Food Chemistry* 48, 4753–4757.

Firestone, D. (2006) *Physical and Chemical Characteristics of Oils, Fats, and Waxes*, 2nd edn. AOCS Press, USA.

Francisco, M. and Resurrección, A. (2008) Functional Components in Peanuts. *Critical Reviews in Food Science and Nutrition* 48, 715–746.

Hartman, M.-A. and Benveniste, P. (1987) Plant membrane sterols: isolation, identification and biosynthesis. *Methods in Enzymology* 148, 632–650.

Grunwald, C. (1975) Plant Sterols. *Annual Review in Plant Physiology* 26, 209–236.

Jiménez-Escrig, A., Santos-Hidalgo, A.B. and Saura-Calixto, F. (2006) Common Sources and estimated Intake of Plant Sterols in the Spanish Diet. *Journal of Agriculture and Food Chemistry* 54, 3402–3471.

Jones, P.J.H., MacDougall, D.E., Ntanios, F. and Vanstone, C.A. (1997) Dietary phytosterols as cholesterol-lowering agents in humans. *Canadian Journal of Physiology and Pharmacology* 75, 217–227.

Jones, P.J.H., MacDougall, D.E., Feng, J.Y. and Parsons, W. (1998) Short-term administration of tall oil phytosterols improves plama lipid profiles in subjects with different cholesterol levels. *Metabolism* 47, 751–756.

Jones, P.J.H., Vanstone, C.A., Raeini-Sarjaz, M. and St-Onge, M.P. (2003) Phytosterols in low-and non fat beverages as part of a controlled diet fail to lower plasma lipid levels. *Journal of Lipid Research* 44, 1713–1719.

Katan, M.B., Grundy, S.M., Jones, P., Law, M., Miettinen, T. and Paoletti, R. (2003) Efficacy and Safety of Plant Stanols and Sterols In the Management of Blood Cholesterol levels. *Mayo Clinical Proceedings* 78, 965–978.

Kozlowska-Wojciechowska, M., Jastrzębska, M., Naruszewicz, M. and Foltyñska, A. (2003) Impact of Margarine Enriched With Plant Sterol on Blood Lipids, Platelet Function, and Fibrinogen Level in Young Men. *Metabolism* 52, 1373–1378.

Kritchevsky, D. and Chen, S.C. (2005) Phytosterols – health benefits and potential concerns: a review. *Nutrition Research* 25, 413–428.

Lees, A.M., Mok, H.Y., Lees, R.S., McCluskey, M.A. and Grundy, S.M. (1997) Plant sterols as cholesterol-lowering agents: clinical trials in patients with hypercholesterolemia and studies of sterol balance. *Atherosclerosis* 28, 325–338.

Linchtenstein, A. and Deckelbaum, R. (2001) Stanol/Sterol ester-containing foods and blood cholesterol levels. *Circulation* 103, 1177–1179.

Ling, W.H. and Jones, P.J.H. (1995) Minireview dietary phytosterols: a review of metabolism, benefits and side effects. *Life Science* 57, 195–206.

Masson, L., Camilo, C., Gonzalez, K., Cáceres, A., Jorge, N. and Torija, M.E. (2008a) New Sources of Oliseeds from Latin American Native Fruits. *Natural Products Communications* 3, 357–362.

Masson, L., Camilo, C.Y. and Torija, M. (2008b) Caracterización del aceite de coquito de palma chilena (*Jubaea chilensis*). *Grasas y Aceites* 59, 33–38.

Matvienko, O.A., Lewis, D.S., Swanson, M., Arndt, B., Rainwater, D.L., Steward, J. and Alekel, D.L. (2002) A Single Dose of Soybean Phytosterols in Ground Beef Decreases Serum Total Cholesterol and LDL Cholesterol in Young, Mildly Hypercholescholeremic Men. *American Journal of Clinical Nutrition* 76, 57–64.

Mensink, R.P., Ebbing, S., Lindhout, M., Plat, J. and van Heugten, M.M.A. (2002) Effect of plant stanol esters supplied in low-fat yoghurt on serum lipids and lipoproteins, non-cholesterol sterols and fat soluble antioxidant concentrations. *Atherosclerosis* 160, 205–213.

Miettinen, T.A. and Kesaniemi, Y.A. (1989) Cholesterol absorption: regulation of cholesterol synthesis and elimination and within-population variations of serum cholesterol levels. *American Journal of Clinical Nutrition* 49, 629–635.

Moreau, R.A. and Kamal-Eldin, A. (2009) *Gourmet and Health Promoting Specialty Oils*. AOCS Press, Urbana Illinois, pp. 1–587.

Moreau, R.A., Whitaker, B.D. and Hicks, K.B. (2002) Phytosterols, phytostanols, and their conjugates in foods: structural diversity, quantitative analysis, and health-promoting uses. *Progress in Lipid Research* 41, 457–500.

Moreau, R.A., Singh, V., Powell, M.J. and Hicks, K.B. (2009) Corn Kernel Oil and Corn Fiber Oil. In: Moreau, R.A. and Kamal-Eldin, A. (eds) *Gourmet and Health Promoting Specialty Oils*. AOCS Press, Urbana, Illinois, pp. 409–431.

Neil, H.A.W. and Huxley, R.R. (2002) Efficacy and Therapeutic Potential of Plant Sterols. *Atherosclerosis Supplements* 3, 11–15.

Nestel, P.M., Cehum, M., Pomeroy, S., Abbey, M. and Weldon, G. (2001) Cholesterol-Lowering Effects of Plant Sterol Ester and Non-Esterified Stanols in Margarine, Butter and Low-Fat Foods. *European Journal of Clinical Nutrition* 55, 1084–1090.

Nguyen, T. (1999) The cholesterol-lowering action of plant stanol esters. *Journal of Nutrition* 129, 2109–2112.

Normén, L., Bryngelsson, S., Jhonsson, M., Evheden, P., Ellegard, L., Brants, H., Andersson, H. and Dutta, P. (2002) The Phytosterol Content of Some Cereal Foods Commonly Consumed in Sweden and in the Netherlands. *Journal of Food Composition and Analysis* 15, 693–704.

Normén, L., Ellegard, L., Brants, H., Dutta, P. and Andersson, H. (2007) A phytosterol database: Fatty foods consumed in Sweden and the Netherlands. *Journal of Food Composition and Analysis* 20, 193–201.

Nurmi, T., Nystrom, L., Edelmann, M., Lampi, A.-M. and Piironen, V. (2008) Phytosterols in Wheat Genotypes in the HEALTHGRAIN Diversity Screen. *Journal of Agriculture and Food Chemistry* 56, 9710–9715.

Ostlund Jr, R.E. (2007) Phytosterols, Cholesterol Absorption and Healthy Diets. *Lipids* 42, 41–45.

Patel, M.D. and Thompson, P.D. (2006) Phytosterols and vascular disease. *Atherosclerosis* 186, 12–19.

Piironen, V., Lindsay, D.G., Miettinen, A., Toivo, J. and Lampi, A.M. (2000) Plant sterols: biosynthesis, biological function and their importance to human nutrition. *Journal of the Science of Food Agriculture* 80, 939–966.

Plumb, J.A., Rhodes, M.J.C., Lampi, A.M., Buchgraber, M. and Kroon, P.A. (2011) Phytosterols in plant foods: Exploring contents, data distribution and aggregated values using an online bioactive database. *Journal of Food Composition and Analysis* 24, 1024–1031.

Salo, P., Hopia, A., Ekblom, J., Lahtinen, R. and Paivi, L. (2005) Plant Stanol Ester as a Cholesterol-Lowering Ingredient of Benecol Foods. In: Akoh, C.C. and Ming Lai, O. (eds) *Healthful Lipids*. AOCS Press, USA, pp. 699–730.

Simojoki, M., Luoto, R., Uutela, A., Rita, H., Boice Jr, J.D., McLaughlin, J.K. and Puska, P. (2005) Use of plant stanol ester margarine among persons with and without cardiovascular disease: Early phases of the adoption of a functional food in Finland. *Nutrition Journal* 4, 20 doi:10.1186/1475-2891-4-20.

St-Onge, M.-P., Lamarche, B., Mauger, J.-F. and Jones, P. (2003) Consumption of a Functional Oil Rich in Phytosterols and Medium-Chain Triglyceride Oil Improves Plasma Lipid Profiles in Men. *The Journal of Nutrition* 133, 1815–1820.

Thompson, G.R. and Grundy, S.M. (2005) History and Development of Plant Sterol and Stanol Esters for Cholesterol-Lowering Purposes. *The American Journal of Cardiology* 96, 4D–9D.

Volpe, R., Niittynen, L., Korpela, R., Sirtori, C., Bucci, A., Fraone, N. and Pazzucconi, F. (2001) Effect of Yoghurt Enriched with Plant Sterols on Serum Lipids in Patients with Moderated Hypercholesterolaemia. *British Journal of Nutrition* 86, 233–239.

Weihrauch, J.L. and Gadner, J.M. (1978) Sterol content of food plants origin. *Journal American Dietetic Association* 73, 39–47.

Woolf, A., Wong, M., Eyres, L., McGhie, T., Lund, C., Olsson, S., Wang, Y., Bulley, C., Wang, M., Friel, E. and Requejo-Jackman, C. (2009) Avocado Oil. In: Moreau, R.A. and Kamal-Eldin, A. (eds) *Gourmet and Health Promoting Specialty Oils*. AOCS Press, Urbana, Illinois, pp. 73–125.

Wretensjo, I. and Karlberg, B. (2002) Characterization of sterols in Refined Borage Oil by GC-MS. *Journal of the American Oil Chemists' Society* 79, 1069–1074.

12 Carotenoids: Chemistry and Health Benefits

Dhan Prakash* and Charu Gupta
Amity Institute for Herbal Research and Studies, Amity University, Noida, India

12.1 Introduction

The term 'nutraceutical' combines 'nutrition' and 'pharmaceutical' to signify that food extracts can be used as preventive drugs. The nutraceuticals are foods or bioactive ingredients in foods that protect or promote health and occupy a place at the intersection of food and pharmaceuticals. Their main ingredients are the phytochemicals, which play a key role in the efficacy of nutraceuticals as health protective and disease preventive agents. These phytochemicals include terpenes, phytosterols, indoles, phenols and thiols etc. Identification is based on the protective functions, physical and chemical characteristics of the molecules, e.g. carotenoids, the precursor of vitamin A, have preventive action against many eye diseases and cancer. Isoprenoids are active against free radicals, while, omega-3 and omega-6 fatty acids support cardiovascular health. Thus, in humans and animals carotenoids play an important role in protection against photo-oxidative processes (Demming-Adams and Adams, 2002) by acting as oxygen and peroxyl radical scavengers. Their synergistic action with other antioxidants makes them even more potent compounds.

Carotenoids, the basic source of yellow, orange and red plant pigments, are widely distributed in nature. Carotenoids are ubiquitous phytochemicals playing a role in a range of special processes. Thus, they are regarded as essential compounds for life mainly due to their various roles. Carotenoids are present in all living organisms, from bacteria and algae to higher plants, in both non-photosynthetic and photosynthetic tissues and are present in most commonly consumed vegetables and fruits.

The group of carotenoids consists of more than 700 phytochemicals, which constitute photosynthetic membranes and produce colours in plants and animals. Out of these, only about 24 commonly occur in human foodstuffs. The most-studied carotenoids are α-carotene (Fig. 12.1), β-carotene (Fig. 12.2), lycopene (Fig. 12.3), lutein (Fig. 12.4) and zeaxanthin (Fig. 12.5). The principal carotenoids of foods are β-carotene, β-cryptoxanthin (Fig. 12.6), lycopene and lutein. Carotenoid pigments, which are abundant in many fruits and vegetables, have been studied for their diverse roles in phytochemistry and phytomedicine (Dutta *et al.*, 2005).

Carotenoids are mainly C_{40} isoprenoids, consisting of eight isoprene units. There are

* E-mail: dprakash_in@yahoo.com

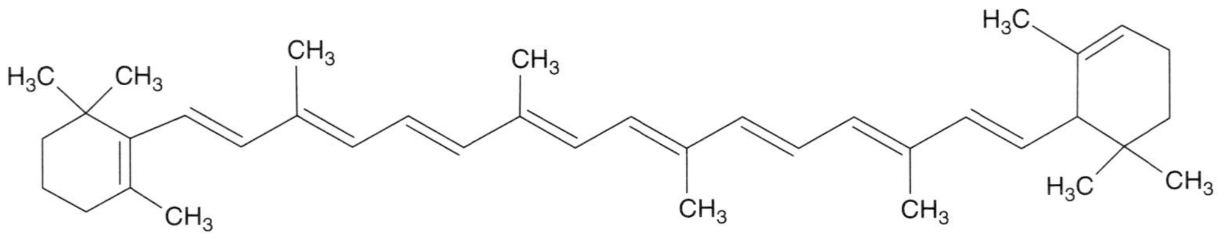

Fig. 12.1. Alpha-carotene.

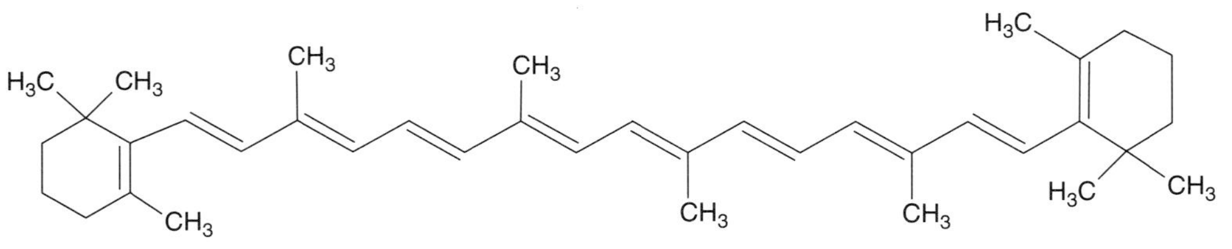

Fig. 12.2. Beta-carotene.

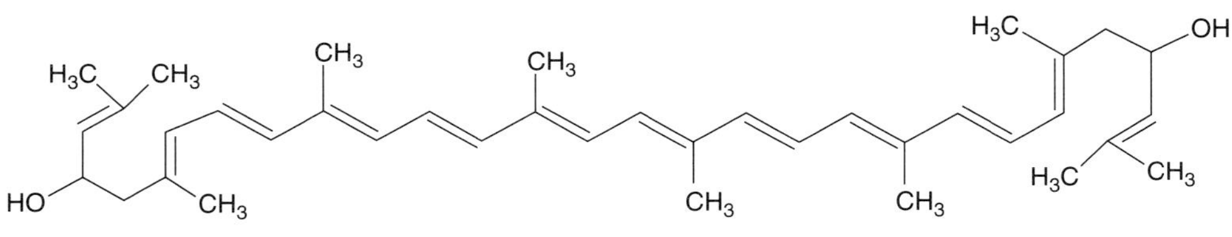

Fig. 12.3. Lycopene.

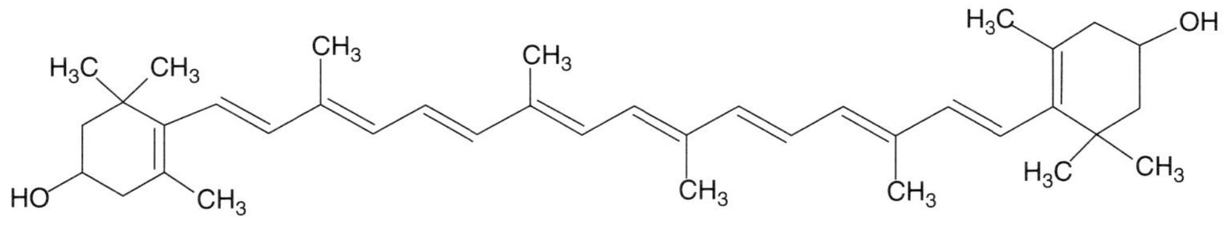

Fig. 12.4. Lutein.

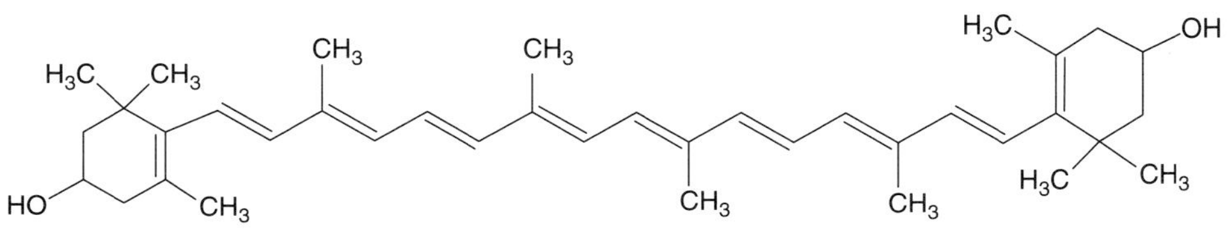

Fig. 12.5. Zeaxanthin.

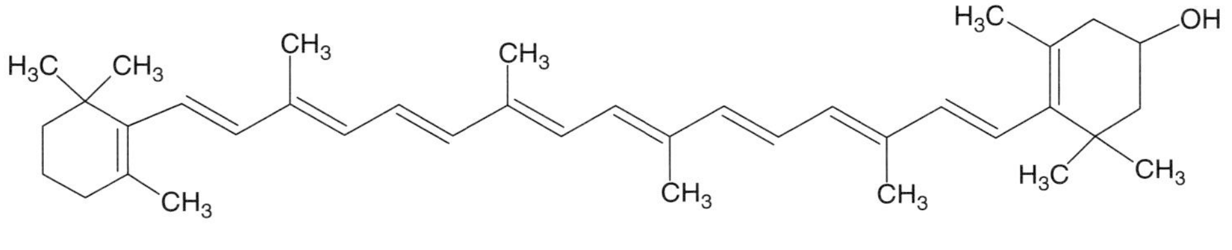

Fig. 12.6. Beta-cryptoxanthin.

two main groups of carotenoids: (i) carotenes (β-carotene, lycopene) contain only hydrogen and carbon and may be cyclic or linear; (ii) oxycarotenoids (xanthophylls, lutein) contain hydrogen, carbon and oxygen in the form of hydroxy, epoxy, or oxy groups. The polyene chain in carotenoids contains up to 15 conjugated double bonds, which are responsible for their characteristic absorption spectra and specific photochemical properties (Britton, 1995; Olson and Krinsky, 1995; Britton *et al.*, 1998). They might be responsible for quenching singlet oxygen and for intercepting deleterious free radicals and reactive oxygen species (ROS). These properties make them a part in diverse antioxidant defence systems. Most carotenoids are found in linear or all-*trans* configuration. Exposure to light or heat may facilitate the *trans* to *cis* isomerization of one or more double bonds (Krinsky, 1989). The physico-chemical properties and the biological activities of carotenoids are intimately related to their structures. Carotenoids with known structure can be conclusively identified by the combined use of chromatographic behaviour, UV-visible absorption spectra and specific group chemical reactions to confirm the type, location and number of functional groups in xanthophylls (Azevedo-Meleiro and Rodriguez-Amaya, 2004). The following structural properties of carotenoids (Papas, 1999) might be responsible for their antioxidant functions: (i) a multiplicity of closely spaced energy levels between the excited state and ground state, such that they can dissipate excited state energy via small collisional exchanges with the solvent; (ii) minimal tendency for their excited-state to sensitize other molecules; (iii) resonance states in the excited state by allowing delocalization and stabilization of the excited state; and (iv) multiple potential sites on the carotenoids for attack by active oxygen. Each double bond in their polyene chain can exist in two configurations, *trans* or *cis* geometrical isomers. In nature they occur predominantly or entirely in all-*trans* form. The presence of a *cis* double bond creates greater steric hindrance between nearby hydrogen atoms and/or methyl groups, so that *cis* isomers are generally less stable thermodynamically than that of *trans* forms (Dutta *et al.*, 2005).

Carotenoids are the precursors for biosynthesis of plant growth regulators and protect photosynthetic apparatus by quenching harmful ROS (Sarry *et al.*, 1994; Papas, 1999). Among the carotenes, only α- (Agarwal and Rao, 1998), β- and ε-carotene (Fig. 12.7) (Auldridge *et al.*, 2006) possess vitamin A activity, out of which β-carotene is the most active. Alpha-carotene possesses 50–54% and ε-carotene 42–50% of the antioxidant activity as compared to β-carotene. These carotenes along with γ- (Fig. 12.8) (Britton *et al.*, 1998), δ- (Fig. 12.9) and ζ-carotene (Fig. 12.10) (Krinsky, 1989), lycopene (di Mascio *et al.*, 1989) and lutein (García-Limones *et al.*, 2008), which do not convert to vitamin A, seem to offer

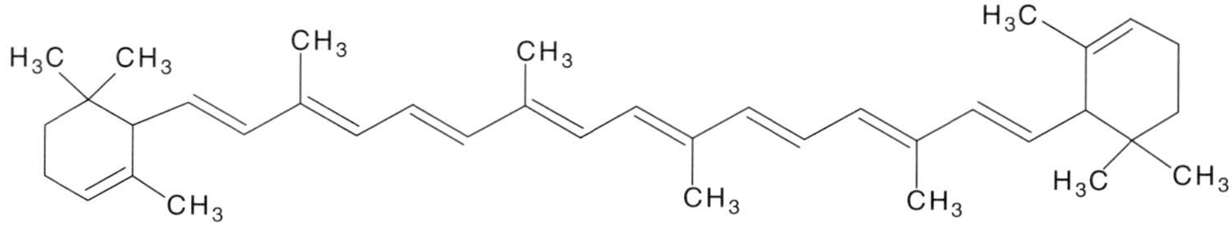

Fig. 12.7. Epsilon-carotene.

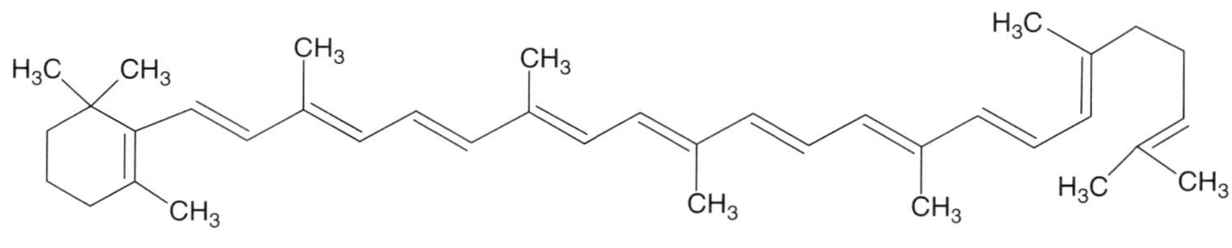

Fig. 12.8. Gamma-carotene.

protection against lung, colorectal, breast, uterine and prostate cancers. They are tissue specific in protection but overall protective effects are greater when all carotenes are taken together.

Dietary carotenoids are obtained from a number of fruits and vegetables, such as green leafy vegetables, spinach, carrots, peaches, apricots and sweet potatoes. Human diet supplemented with carotenoids is beneficial in reducing chronic conditions related to coronary heart diseases (CHD), certain cancers and macular degeneration (Sarry *et al.*, 1994; Mayne, 1996; Woodl *et al.*, 1997; van het Hof *et al.*, 1999, 2002). Carotenoids accumulate in light exposed tissues, such as skin, and as such have gained increased value in the cosmetic industries as suitable compounds for photoprotection due to their scavenging action on ROS and anti-inflammatory properties (Stahl and Sies, 2007). Photo-oxidative damage affects cellular lipids, proteins and DNA and is involved in the patho-biochemistry of erythema formation, premature ageing of the skin, development of photodermatoses and skin cancer. Evidence shows that β-carotene, lutein and perhaps even lycopene can prevent UV-induced erythema formation and contribute to lifelong protection against exposure to harmful effects of sunlight (Stahl and Sies, 2007).

The nutraceutical industry synthetically manufactures five major carotenoids on an industrial scale (e.g. lycopene, β-carotene, canthaxanthin, zeaxanthin and astaxanthin) for use in a range of food products and cosmetics, such as vitamin supplements and health products and as feed additives for poultry, livestock, fish and crustaceans (del Campo *et al.*, 2007; Jackson *et al.*, 2008). One of the most commercially valuable pigments, astaxanthin (Fig. 12.11), is primarily synthesized by marine microorganisms, such as the green alga *Haematococcus pluvialis* and accumulates

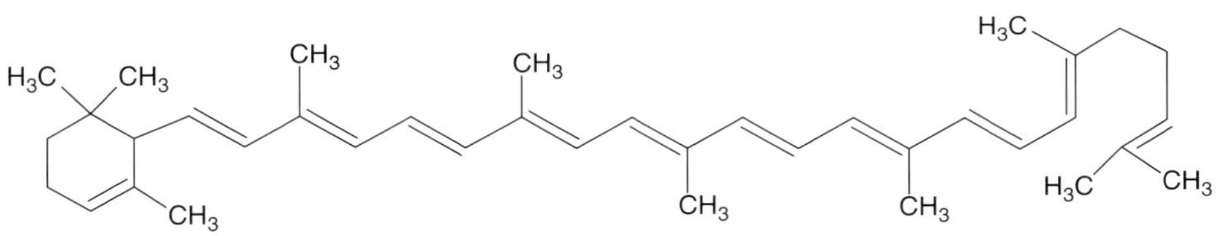

Fig. 12.9. Delta-carotene.

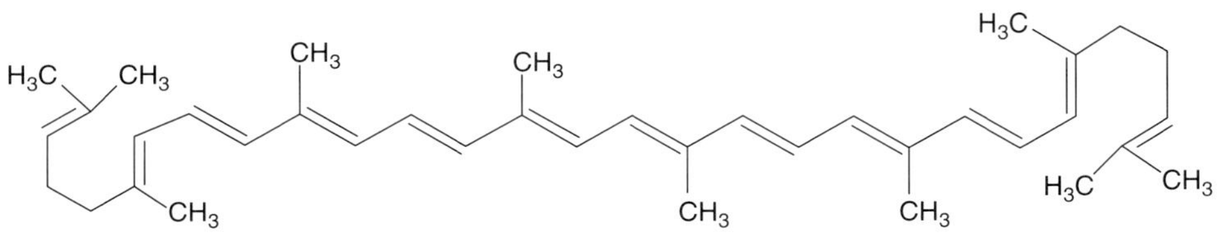

Fig. 12.10. Zeta-carotene.

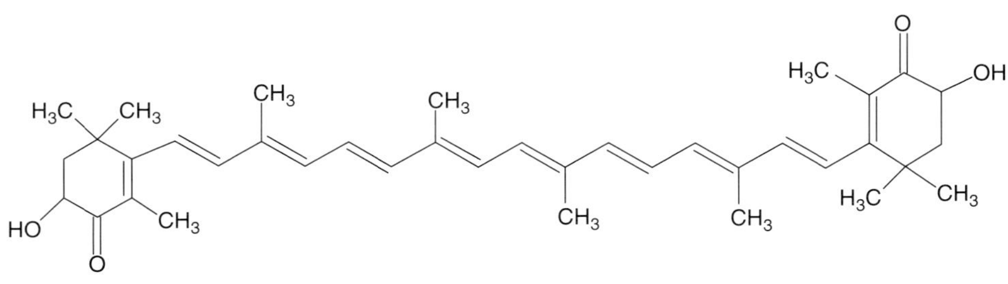

Fig. 12.11. Astaxanthin.

in fish such as salmon, thus colouring their flesh red. Astaxanthin has been implicated as a potential therapeutic agent treating cardiovascular disease and prostate cancer (Fassett and Coombes, 2011).

12.2 Accumulation, Storage and Insights from Biofortification

The storage of carotenoids requires a lipophilic environment, usually within the membranes of plastid organelles, which behave as a sink for their accumulation. Carotenoids are usually synthesized *de novo* in differentiated plastids of roots, flowers, fruits and seeds, accumulated mostly in chloroplasts, chromoplasts, and amyloplasts (starch-storing plastids), leucoplasts (colourless plastids), etioplasts (dark-brown precursors of the chloroplast) and elaioplasts (lipid-storing plastids) (Cazzonelli and Pogson, 2010; Pogson and Albrecht, 2011).

There are links between changes in carotenoid composition and plastid biogenesis, morphology and protein translocation; in particular, it is noteworthy that carotenoids (e.g. lutein) are necessary for the differentiation of an etioplast into a chloroplast. The regulation of carotenoids targeting, storage and sequestration within various plastid types is a process to modulate a sink for carotenoid accumulation (Lu *et al.*, 2006; Cuttriss *et al.*, 2007). A naturally occurring mutation in the *Brassica oleracea* orange-curd (*or*) gene changes a normally white cauliflower curd into an orange *Or* mutant, which accumulates high levels of β-carotene (Lu and Li, 2008). Carotenoids accumulate in lipoprotein structures within the chromoplast (Bartley and Scolnik, 1995; Vishnevetsky *et al.*, 1999), which might allow for additional carotenoid biosynthesis. Therefore chromoplasts serve as a metabolic sink to control carotenoid accumulation in plants and reveal the importance of plastid differentiation in controlling their accumulation in plants (Lu *et al.*, 2006; Li and van Eck, 2007).

Xanthophylls mostly found in green vegetables and yellow fruits are stored in the retina of the eye. They are expected to protect vitamin A, E and other carotenoids from oxidation and skin from adverse effects of sunlight. Lutein and zeaxanthin (van het Hof *et al.*, 1999) protect photoreceptor cells from light-generated oxygen radicals. They are more effective than α-, β-carotenes and lycopene in chemopreventive activity. Lycopene exerts greater antioxidant activity compared to β-carotene and it has also been reported to protect cholesterol against oxidative damage (di Mascio *et al.*, 1989; Mayne, 1996; Woodl *et al.*, 1997; Papas, 1999; van het Hof *et al.*, 1999). It does not convert to vitamin A but may provide important health benefits such as protection against cancer by quenching the destructive potential of singlet oxygen. β-cryptoxanthin (Sarry *et al.*, 1994) occurs in oranges, mango, papaya, cantaloupe, peaches, prunes and squash etc. It exhibits provitamin A activity. Astaxanthin (Woodl *et al.*, 1997) and capsanthin (Fig. 12.12) (Krinsky, 1993) are other naturally occurring xanthophylls with potent antioxidant properties. Their antioxidant effects enable these compounds to play a vital role in protecting organisms against the damage of photoradiation during photosynthesis. They function as antioxidants by protecting lipid peroxidation, blood and other

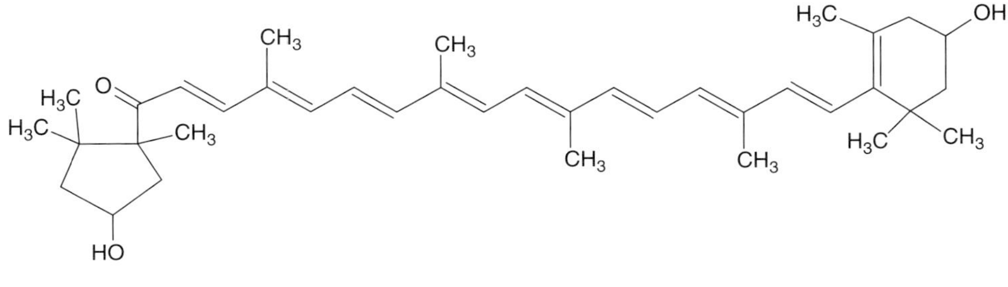

Fig. 12.12. Capsanthin.

body fluids from free ROS including singlet oxygen, hydroxyl, peroxide and superoxide radicals (Krinsky, 1989; Sarry *et al.*, 1994; Papas, 1999). Thus the all-*trans* isomer of β-carotene is the major source of retinol (Fig. 12.13) (di Mascio *et al.*, 1989) due to its high provitamin A activity. One molecule of it can theoretically provide two molecules of *trans*-retinaldehyde or retinal (Fig. 12.13) (Mayne, 1996). It quenches singlet oxygen, induces gap junction communication and inhibits lipid peroxidation. Its high serum levels are correlated with low incidences of cancer. It is similar to β-carotene in biological activity, but quenches reactive oxygen more effectively (di Mascio *et al.*, 1989; Krinsky, 1993; Krinsky *et al.*, 2003; Rissanen *et al.*, 2003).

12.3 Turnover and Degradation of Carotenoids

Carotenoids are relatively stable compounds that accumulate in diverse types of tissues (photosynthetic and non-photosynthetic). Recently, it was demonstrated by $^{14}CO_2$ uptake experiments that their turnover appears to be much greater than expected (Beisel *et al.*, 2010). Given the continued synthesis in mature leaves, the active degradation of carotenoids by CCD (carotenoid cleavage dioxygenases) and NCED (9-*cis*-epoxycarotenoid dioxygenase) enzymatic turnover has become an exciting area of discovery (Bouvier *et al.*, 2005; Lewinsohn *et al.*, 2005; Walter *et al.*, 2010). Members of these gene families are involved in the biosynthesis of the phyto-hormone ABA (NCEDs), which controls abiotic stress signalling pathways and strigolactone (CCDs) which controls shoot growth and root–mycorrhizal symbiosis.

The *CCD* gene family (*CCD1, 4, 7* and *8*) play essential roles in synthesis of colour, apo-carotenoids flavour, aroma volatiles and phytohormones such as strigolactone. The active degradation of the xanthophylls by CCD activity can reduce lutein content in strawberries as well as cause changes in the pigmentation in chrysanthemums from white to yellow (Ohmiya *et al.*, 2006; García-Limones *et al.*, 2008). In maturing *Arabidopsis* seeds a loss of function of *CCD1* activity leads to higher carotenoids level and may have a role in synthesis of apo-carotenoids flavour and aroma volatiles (Auldridge *et al.*, 2006). Similarly, in tomato (*Lycopersicon esculentum*) *LeCCD1* activity contributes to the formation of the flavour volatiles β-ionone, pseudoionone and geranylacetone (Simkin *et al.*, 2004). Two of the genes that affect shoot branching encoding *CCD7* and *CCD8* can sequentially cleave β-carotene to form the C18 compound, i.e. 13-apo-carotenone (Schwartz *et al.*, 2004). *CCD7* appears to be a biosynthetic cross point, controlling both strigolactone and AM-induced C13 cyclohexenone and C14 mycorradicin apo-carotenoids. Finally, there are examples of cross talk where inhibition of ABA biosynthesis reduces *CCD7* and *CCD8* transcript abundance as well as strigolactone levels (Lopez-Raez *et al.*, 2010).

12.4 Properties of Carotenoids

Carotenoids perform several functions; they are involved in light harvesting, but also contribute to stabilize the structure and aid in the

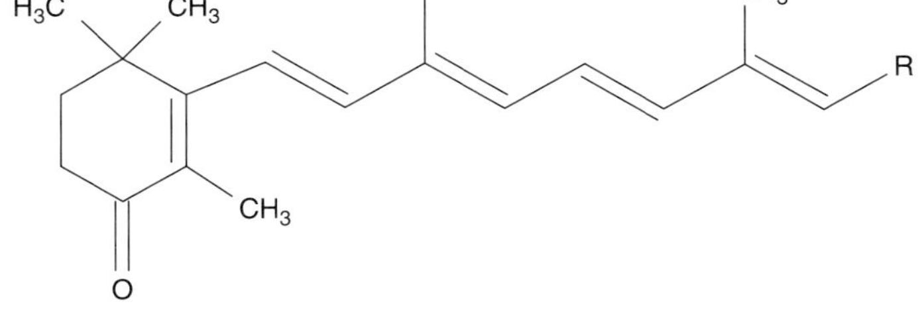

Fig. 12.13. R=CH_2OH, retinol; R=CHO, retinal; and R=COOH, retinoic acid.

function of photosynthetic complexes, besides quenching chlorophyll triplet states, scavenging free radicals and dissipating excess energy (Guedes *et al.*, 2011). The intrinsic antioxidant activity of carotenoids constitutes the basis for their protective action against oxidative stress; however, not all biological activities claimed for carotenoids relate to their ability to inactivate free radicals and ROS. The lycopene molecule represents the basic structure of the carotenoids. It is composed of 40 carbon atoms arranged in a long structure with alternating single and double bonds. The distinctive pattern of alternating single and double bonds in the polyene backbone of carotenoids allows them to absorb excess energy from other molecules, while the nature of the specific end groups on carotenoids may influence their polarity. The former may account for their antioxidant properties and the latter may explain the differences of ionic interaction with biological membranes (Krinsky, 1993; Britton, 1995; Krinsky *et al.*, 2003). Carotenoids can undergo many reactions with a wide variety of chemical reagents; some of them might be similar to chemicals found in biological systems. The process that draws much attention is the oxidation of carotenoids, during which they may function as biological antioxidants. Their ability to react with radical species has served as the basis for the determination of various lipoxygenase activities. They inhibit lipid peroxidation, enhance eye health and immune systems and reduce the risk of certain cancers. The basis for the assay involves an oxidative interruption of the conjugated double-bond system, which is invariably accompanied by a loss of the visible absorption, or a 'bleaching' of the carotenoids. Carotenoids are heat stable up to 50°C and degrade rapidly above this temperature. On heating, the naturally occurring *trans*-double-bond configuration rearranges to *cis*- configuration. This causes a slight shift in the absorbance maxima. Common unit operations of food processing, blanching, retorting and freezing etc., in general have either minor or no effect on their degradation. Frozen and heat-sterilized foods, with few exceptions, exhibit excellent stability. They are also stable in foods over a pH range of 2 to 7 (Stahl *et al.*, 1997; Clinton, 1998; Cooper *et al.*, 1999; Papas, 1999; van den Berg, 1999; Anonymous, 2001; Holick *et al.*, 2002; Krinsky *et al.*, 2003).

12.5 Carotenoids as Antioxidants

Free radicals can damage the body's DNA, RNA, enzymes, carbohydrates, proteins, lipids and cell membranes and thus weaken the natural defences. DNA damage can cause cancer while damage in arteries may cause hardening and increase the risk of heart attack, several other diseases, premature ageing and death (Mayne, 1996; Papas, 1999; Rissanen *et al.*, 2001). Halliwell and Gutteridge (1995) defined an antioxidant as 'any substance when present at low concentrations compared to those of an oxidizable substrate, significantly delays or prevents oxidation'. Antioxidants, such as carotenoids, polyphenols, vitamins C and E are known to have synergistic interactions through their recycling mechanisms (Papas, 1999; Anonymous, 2000; Flood *et al.*, 2002). Antioxidants help to control free radicals by quenching, by donating electrons to molecules before they damage other cells. Antioxidants may have additional activities, such as reducing the energy of a free radical or stopping it from forming by interrupting an oxidizing chain reaction. They may also trap free radicals and lipid peroxides, delaying the onset of lipid peroxidation, stopping production of further free radicals and inhibiting the damaging effects of certain enzymes that can degrade connective tissues. The mechanisms of reactions between carotenoids and radical species may involve radical addition, hydrogen abstraction and electron transfer, but its precise mechanisms remain unclear (Liebler, 1993; Agarwal and Rao, 1998; Papas, 1999).

A large number of the evidences support the hypothesis that lipid oxidation or oxidative stress may be the underlying mechanism in chronic diseases and that β-carotene acts as an antioxidant *in vivo* (Liebler, 1993; Krinsky, 1998). Carotenoids have been considered as antioxidants, rather than pro-oxidants, based on *in vitro* experimental evidence. The antioxidant and/or pro-oxidant properties of carotenoids are affected by their concentration, oxygen partial pressure and the nature of the environment. Such actions of carotenoids on lipid oxidation have been of interest in food lipids as well as in biological membrane lipids. Since the major constituents of biological membranes are lipid and proteins, oxidation can damage membrane lipids. In foods, extensive investigation has been focused on lipid oxidation, which is of importance as it leads to rancidity. In all, β-carotene has been the prototype for examining the antioxidant action of carotenoids in different *in vitro* lipid models (Jung and Min, 1991; Levy *et al.*, 1995; Young and Lowe, 2001).

Research has been focused upon manipulation of carotenoid content and composition in crop plants through biotechnological techniques to improve their nutritional value for human consumption. The genes and cDNAs encoding enzymes of the pathway have been identified, sequenced and characterized. Thus, the enormous progress in cloning carotenogenic genes has opened up the possibility of genetic manipulation of the carotenoid biosynthetic pathway in plants (Bertram, 1999; Naik *et al.*, 2003).

12.6 Light Absorption and Energy Transfer

The best-known property of carotenoids is their ability to absorb light. The ultraviolet and visible spectrum is the first diagnostic tool for the identification of carotenoids. The evidence of this is all through nature, in the various colours of plants and animals. The process of absorbing light involves the formation of singlet state carotenoids (^{1}carotenoids) and subsequently transfer of this excitation energy to a photoresponsive pigment, chlorophyll, to initiate the process of photosynthesis. This type of process can effectively extend the wavelength of light available to an organism for photosynthesis, since the bulk of carotenoids present in photosynthesizing systems absorb at wavelengths different from those of the photoactive chlorophylls or bacterio-chlorophylls (Bartley and Scolnik, 1995; Papas, 1999). In addition to absorbing light directly, they can also be excited by an energy transfer reaction to form the triplet state species (^{3}carotenoids). The important reactions are the transfer of energy from a suitable triplet sensitizer or excitation energy from singlet oxygen ($^1O_2$) to carotenoids. In each case, ^{3}carotenoid species are formed.

$$\text{Carotenoids} + h\nu \rightarrow {}^3\text{Carotenoids}$$

$$^3\text{Carotenoids} + \text{Chlorophyll} \rightarrow \text{Carotenoids} + {}^3\text{Chlorophyll}$$

12.6.1 Photosynthesis

Life on Earth is based on the energy of solar radiation, which is captured by higher plants, algae and photosynthetic bacteria. These organisms contain photosynthetic pigments such as chlorophylls, phycobilins and carotenoids, which absorb light in a wide range of wavelengths, covering the whole visible region and extending even to the near infrared region (Paavo and Tuomo, 2001). Finally, the energy is stored in the form of carbohydrates and other hydrogen-containing organic compounds.

12.6.2 Singlet oxygen scavenging

Carotenoids have a wonderful ability to interact with and neutralize oxidants, chemically reactive oxygen species known as singlet oxygen and free radicals. Natural astaxanthin has the greatest ability to serve in this antioxidant function, which is why it is the world's strongest natural antioxidant. But many other carotenoids also have an antioxidant effect (Paavo and Tuomo, 2001). Scavenging of free radicals may be by obtaining its missing electron by removing an electron from another molecule or to add itself to another to pair its single electron, forming an adduct. In either case, the electron-rich character of carotenoids make them attractive to radicals, thus sparing other cell components (lipids, proteins, DNA) from radical damage. For example, the nitrogen dioxide radical (NO_2^-) may obtain an electron from a carotenoid, resulting in the formation of a cationic (positively charged) radical. The observed order of reactivity with

NO_2^- was (lycopene = zeaxanthin) > lutein > (astaxanthin = crypothaxanthin).

Carotenoids 'quench' singlet oxygen primarily by a physical mechanism, in which the excess energy of singlet oxygen is transferred to the carotenoid's electron-rich structure. The carotenoid is excited by this added energy into a triplet state (^{3}Car*), and then relaxes into its ground state (^{1}Car) by losing the extra energy as heat. During this physical mechanism the carotenoid structure remains unchanged to protect against further singlet oxygen. The conjugated system of alternating double and single bond in which the electrons are effectively delocalized over the entire length of polyene chain is responsible for the physical and chemical properties of carotenoids and is primarily responsible for the excellent ability of carotenoids to physically quench singlet oxygen without degradation.

$$^1O_2^* + {}^1\text{Car} \rightarrow {}^3O_2 + {}^3\text{Car}^*$$

$$^3\text{Car}^* \rightarrow {}^1\text{Car} + \text{Heat}$$

12.7 Sources of Carotenoids

Because plants are able to synthesize carotenoids *de novo*, they are widely distributed in plant-derived foods and the composition is enriched by the presence of small amounts of biosynthetic precursors and derivatives of the major carotenoids. In general the level of carotenoids is directly proportional to the intensity of colour. Egg yolks, dairy products, fruits, vegetables, legumes, grains and seeds are the major food sources. In green leafy vegetables, β-carotene is predominant while in the orange-coloured fruits and vegetables such as carrots, apricots, mangoes, yams, winter-squash, other carotenoids typically predominate. Yellow vegetables have higher concentrations of xanthophylls with a low provitamin A activity, but some of these compounds, such as lutein, may have significant health benefits. The red and purple vegetables and fruits such as tomatoes, red cabbage, berries and plums contain a large portion of non-vitamin A active carotenoids. Tomato and watermelon are major sources of lycopene (del Campo *et al.*, 2007).

Carotenoids constitute a class of terpenoid pigments, derived from a 40-carbon polyene chain, which can be envisaged as their molecular backbone – indeed, it provides carotenoids with distinctive molecular structures, and the associated chemical properties including light-absorption features that are essential for photosynthesis and, in general, for life in the presence of oxygen (del Campo *et al.*, 2007). The aforementioned backbone may be complemented by cyclic groups (rings) and oxygen-containing functional groups. Hence, hydrocarbon carotenoids are denoted as carotenes as a whole, but oxygenated derivatives are known specifically as xanthophylls; with oxygen being present as –OH groups (e.g. lutein), as oxi-groups (e.g. cantaxanthin) or as a combination of both (e.g. astaxanthin) (del Campo *et al.*, 2007). All xanthophylls synthesized by higher plants (e.g. violaxanthin, antheraxanthin, zeaxanthin, neoxanthin and lutein) can also be synthesized by green microalgae; however, these possess additional xanthophylls (e.g. loroxanthin, astaxanthin and canthaxanthin). Diatoxanthin, diadinoxanthin and fucoxanthin can also be produced by brown algae or diatoms (Eonseon *et al.*, 2003).

The large number of existing species of microalgae constitutes a unique reservoir of biodiversity, which supports potential commercial exploitation of many novel products besides vitamins, pigments and polyunsaturated fatty acids. The key factor for their eventual economic feasibility is the possibility of operating large photobioreactors, able to handle biomass and metabolites to sufficiently high levels (Sanchez *et al.*, 2008). A distinction is usually made between primary and secondary carotenoids: primary carotenoids (i.e. xanthophylls) are structural and functional components of the cellular photosynthetic apparatus, so they are essential for survival (Eonseon *et al.*, 2003; Guedes *et al.*, 2011), whereas secondary ones encompass those produced by microalgae to large levels, but only after exposure to specific environmental stimuli (via carotenogenesis).

Xanthophylls are relatively hydrophobic molecules, so they are typically associated with membranes and/or involved in non-covalent binding to specific proteins; they are usually localized in the thylakoid membrane, whereas secondary carotenoids are found in lipid vesicles, in either the plastid stroma or the cytosol. Most xanthophylls in cyanobacteria and oxygenic photosynthetic bacteria are associated with chlorophyll-binding polypeptides of the photosynthetic apparatus, however, in most green microalgae, carotenes and xanthophylls are synthesized within plastids, and accumulate therein only. Conversely, secondary xanthophylls in some green microalgae (e.g. astaxanthin in *Haematococcus* sp.) accumulate in the cytoplasm; this realization raises the possibility of an extra-plastidic site of carotenoid biosynthesis in that genus. Alternatively, xanthophylls synthesized in the chloroplast may be exported, and consequently accumulate in the cytoplasm (Tardy and Havaux, 1996; Rabbani *et al.*, 1998; Eonseon *et al.*, 2003) so they may be found in essentially all cellular compartments.

12.8 Health Benefits and Concerns about Carotenoids

Carotenoids are generally regarded as safe, based primarily on studies with β-carotene. Increased consumption of carotenoids may cause the skin to turn orange or yellow, known as carotenodermia. This occurrence is completely benign and is unrelated to jaundice that can result from liver disease or other causes. Vitamin A plays an important role in vision, bone growth, reproduction, cell division and differentiation. However, unlike β-carotene, high doses have a negative impact on bone health and increase levels of retinoic acid (Fig. 12.13) (Rissanen *et al.*, 2003). It can also affect the ability of vitamin D to maintain normal calcium levels in the body, resulting in weakened bone structure. Hypervitaminosis can result in dry, itchy skin, headache, fatigue, vomiting, liver damage, loss of hair and appetite. It maintains the surface linings of eye and respiratory, urinary and intestinal tracts. When those linings break down, bacteria can enter the body and cause infection. The immune system helps to prevent or fight off infections by making white blood cells that destroy harmful bacteria and viruses. It may help lymphocytes function more effectively in fighting against infections, prevent bacteria and viruses from entering the integrity of skin and mucous membranes (Reaven *et al.*, 1994; Gaziano *et al.*, 1995).

Vitamin A deficiency is still a major public health problem in the developing world. It is most often associated with protein/calorie malnutrition and affects over 120 million children worldwide. In countries where immunization programmes are not effectively monitored, its deficiency is common, leading to the death of millions of children each year from complications of childhood blindness, infectious diseases such as measles, xerophthalmia and pneumonia etc. As a result of the worldwide significance of vitamin A deficiency in children, the WHO and the UNICEF recommend its administration for all children diagnosed with measles in communities where its deficiency is a serious problem and death from measles is greater than 1% (West *et al.*, 1989).

Lycopene, α- and β-carotenes help to prevent heart disease by inhibiting the formation of harmful LDL cholesterol (Reaven *et al.*, 1994). Cataracts are caused, at least in part, by long-term free radical damage to eyes. Carotenoids may delay that risk by scavenging free radicals. Mixed carotenoid supplement along with vitamins C and E reduces the risk of developing diabetes and fights against Alzheimer's disease by protecting nerve cells in the brain from deterioration. They may protect sperm from damage by free radicals, so can be used as treatment for male infertility. β-carotene may also protect against chromosome abnormalities and/or their damage. It suppresses the activity of a gene involved in inflammation and reddening of skin, which is a marker for oxidative stress. Lower intakes of carotenoids in the body are associated with a higher risk of colorectal cancer. It has been found that high supplemental intakes of lutein, zeaxanthin, cryptoxanthin, α- and β-carotene etc. reduced the risk of breast, cervical and lung cancer. Lycopene appears to be particularly effective against cancers of the prostate, digestive tract and lungs and may also protect the body against the effects of chemotherapy or radiation. They protect against sun damage because of their effect on the immune system, scavenger role towards oxidative substances and shield-like influence on the skin (West *et al.*, 1989; Stahl *et al.*, 1992; Reaven *et al.*, 1994; Gaziano *et al.*, 1995; Khachik *et al.*, 1997; Palozza, 1998; Cooper *et al.*, 1999; Papas, 1999; Holick *et al.*, 2002).

Most oxidation reactions in foods are deleterious, e.g. degradation of vitamins, pigments and lipids, with consequent loss of nutritional value and development of off-flavours (Halliwell and Gutteridge, 2007; Guedes *et al.*, 2011). Antioxidants, which are adventitious in, or deliberately added to foods, can inhibit oxidation or slow down initiation by free alkyl radicals, as well as interrupt propagation of such free radical chains. On the other hand, carotenoids are particularly strong dyes, even at levels of parts per million. Specifically, canthaxanthin, astaxanthin and lutein from *Chlorella* have been in regular use as pigments, and have accordingly been included as ingredients of feed for salmonid fish and trout, as well as poultry, to enhance the reddish colour of fish or the yellowish colour of egg yolk (Guerin *et al.*, 2003; Cysewski and Lorenz, 2004; Plaza *et al.*, 2009).

The dietary β-carotene is a protective agent against cancer, therefore carotenoids have received wide research interest as potential antioxidants in both *in vitro* studies and animal models (Krinsky, 1998; Krinsky *et al.*, 2003). In the early 1980s, scientific knowledge of the antioxidant action of β-carotene was mainly based on observational studies, which widely reported that a higher consumption of carotenoid-rich vegetables and fruits was associated with a lower risk of cancer and cardiovascular disease (Krinsky, 1998; Rapola *et al.*, 1998). Contrary to observational studies, the major intervention trials on β-carotene supplementation have reported a lack of protection against cancer and cardiovascular disease. In Finland, the α-tocopherol, β-carotene (ATBC) cancer study reported that supplemental β-carotene may in fact have harmful effects, whereas dietary β-carotene had an adverse protective effect among smokers (Anonymous, 1994; Albanes *et al.*, 1997). In addition, the supplementation of β-carotene did not protect healthy men or smokers and workers exposed to asbestos. On the other hand, the supplementation with the combination of β-carotene, vitamin E and selenium may inhibit cancer development. The Linxian trial observed a significant reduction in total mortality, due mostly to a lowered risk of cancer, among general adult population receiving the combination of β-carotene, vitamin E and selenium (Hennekens *et al.*, 1996; Omenn *et al.*, 1996; Albanes *et al.*, 1997; Krinsky, 1998; Rapola *et al.*, 1998; Krinsky *et al.*, 2003). Mayne (1996) and Omaye *et al.* (1997) stated that more evidence is needed to understand many of the associations between carotenoids and the observed effects in risks of disease. Mayne (1996) stated that it is difficult to interpret whether the apparent benefits are due to antioxidant vitamins, nutrients, dietary habits or other non-dietary lifestyle characteristics. Much more experimentation is needed to establish the important dietary antioxidants and their optimal intake. In human beings, carotenoids can serve several important functions (Blot, 1993; Krinsky, 1998). The most widely studied and well-understood nutritional role for carotenoids is their provitamin A activity. It has been shown that pro- and non-provitamin A carotenoids are capable of inhibiting the growth of transformed fibroblasts (Omaye *et al.*, 1997; Papas, 1999). There is increasing evidence that growth arrest is due to the stimulation of gap-junctional communication between transformed and surrounding normal cells. These findings suggest that carotenoids or their derivatives play a role in intercellular signals involved in growth control. Inhibitory effects of β-carotene and lycopene on cell proliferation have also been described for several human cancer cell lines (Bartley and Scolnik, 1995; Bertram and Bortkiewicz, 1995; Sharoni and Levi, 1996; Halliwell, 1997; Osborne *et al.*, 1997).

As potent biological antioxidants, carotenoids are able to absorb the excitation energy of singlet oxygen radicals into their complex ringed chain, thus promoting energy dissipation, while protecting tissues from chemical damage. They can also delay propagation of such chain reactions as those initiated by degradation of polyunsaturated fatty acids, which are known to dramatically contribute to the decay of lipid membranes, thus seriously hampering cell integrity

(Guerin *et al.*, 2003). Carotenoids have also the ability to stimulate the immune system, thus being potentially involved in more than 60 life-threatening diseases, including various form of cancer, coronary heart diseases, premature ageing and arthritis (Mojaat *et al.*, 2008); this is specifically the case for canthaxanthin and astaxanthin, and other non-provitamin A carotenoids.

Although it is tempting to formulate a cocktail of carotenoids with the desire to provide a full spectrum of health benefits, it is advised to take into consideration the specific health benefit to be targeted. For example, it might not be beneficial to include lycopene for an eye health formula because it has not been demonstrated to improve vision. The significance of pro-oxidant actions of carotenoids in biological systems remains to be investigated. To achieve a better understanding of carotenoids and dietary antioxidants, further investigations should concentrate on the interactions between carotenoids and tocopherols as well as other complex interactions between naturally occurring components of foods (Gaziano *et al.*, 1995; Cooper *et al.*, 1999; Holick *et al.*, 2002).

Carotenoids are ubiquitous and are essential in nature, clearly serving numerous functions during animal and plant development. At present, there is growing research interest in the interactions between carotenoids and a variety of free radicals involved in oxidative stresses. Knowledge on the antioxidant/pro-oxidant properties of carotenoids at the gene level would be of interest. On the other hand, the current *in vitro* and *in vivo* evidence of pro-oxidant activity of carotenoids has awakened a new area of interest in this field. Future research prospects should deal with the complexities of diet, genetics and environment in the disease process (Papas, 1999).

References

Agarwal, S. and Rao, A.V. (1998) Tomato lycopene and low density lipoprotein oxidation: a human dietary intervention study. *Lipids* 33, 981–984.

Albanes, D., Virtamo, J., Taylor, P.R., Rautalahti, M., Pietinen, P. and Heinonen, O.P. (1997) Effects of supplemental β-carotene, cigarette smoking and alcohol consumption on serum carotenoids in the α-Tocopherol, β-Carotene cancer prevention study. *American Journal of Clinical Nutrition* 66, 366–372.

Anonymous (2001) A randomized, placebo-controlled, clinical trial of high-dose supplementation with vitamins C and E, β-carotene and zinc for age-related macular degeneration and vision loss: Age Related Eye Disease Study Research Group report no. 8. *Archives of Ophthalmology* 119, 1417–1436.

Anonymous (1994) The effect of vitamin E and β-carotene on the incidence of lung cancer and other cancers in male smokers. The α-Tocopherol, β-Carotene Cancer Prevention Study Group. *New England Journal of Medicine* 330, 1029–1035.

Anonymous (2000) β-carotene and other carotenoids. Dietary reference intakes for vitamin C, vitamin E, selenium, and carotenoids. In: Food and Nutrition Board. *National Academy Press,* Washington DC, USA.

Auldridge, M.E., Block, A., Vogel, J.T., Dabney-Smith, C., Mila, I., Bouzayen, M., Magallanes-Lundback, M., DellaPenna, D., McCarty, D.R. and Klee, H.J. (2006) Characterization of three members of the *Arabidopsis* carotenoid cleavage dioxygenase family demonstrates the divergent roles of this multifunctional enzyme family. *The Plant Journal* 45, 982–993.

Azevedo-Meleiro, C.H. and Rodriguez-Amaya, D.B. (2004) Confirmation of the identity of the carotenoids of tropical fruits by HPLC-DAD and HPLC-MS. *Journal of Food Composition and Analysis* 17, 385–396.

Bartley, G.E. and Scolnik, P.A. (1995) Plant carotenoids: pigments for photoprotection, visual attraction and human health. *The Plant Cell* 7, 1027–1038.

Beisel, K.G., Jahnke, S., Hofmann, D., Koppchen, S., Schurr, U. and Matsubara, S. (2010) Continuous turnover of carotenes and chlorophyll a in mature leaves of *Arabidopsis* revealed by $^{14}CO_2$ pulse-chase labelling. *Plant Physiology* 152, 2188–2199.

Bertram, J.S. (1999) Carotenoids and gene regulation. *Nutrition Reviews* 57, 182–191.

Bertram, J.S. and Bortkiewicz, H. (1995) Dietary carotenoids inhibit neoplastic transformation and modulate gene expression in mouse and human cells. *American Journal of Clinical Nutrition* 62, 1322S–1326S.

Blot, W.J. (1993) Nutrition intervention trials in Linxian, China: supplementation with specific vitamin/mineral combinations, cancer incidence, and disease-specific mortality in the general population. *Journal of National Cancer Institute* 85, 1483–92.

Bouvier, F., Isner, J.C., Dogbo, O. and Camara, B. (2005) Oxidative tailoring of carotenoids: a prospect towards novel functions in plants. *Trends in Plant Science* 10, 187–194.

Britton, G. (1995) Structure and properties of carotenoids in relation to function. *FASEB Journal* 15, 1551–1558.

Britton, G. Liaaen-Jemsen, S. and Pfander, H. (1998) *Carotenoids, Biosynthesis and Metabolism*. Birkauser verlag. Basel, Switzerland.

Cazzonelli, C.I. and Pogson, B.J. (2010) Source to sink: regulation of carotenoid biosynthesis in plants. *Trends in Plant Science* 15, 266–274.

Clinton, S.K. (1998) Lycopene: chemistry, biology, and implications for human health and disease. *Nutrition Reviews* 56, 35–51.

Cooper, D.A., Eldridge, A.L. and Peters, J.C. (1999) Dietary carotenoids and lung cancer: a review of recent research. *Nutrition Reviews* 57, 133–145.

Cuttriss, A.J., Chubb, A., Alawady, A., Grimm, B. and Pogson, B. (2007) Regulation of lutein biosynthesis and prolamellar body formation in *Arabidopsis*. *Functional Plant Biology* 34, 663–672.

Cysewski, G.R. and Lorenz, R.T. (2004) Industrial production of microalgal cell-mass and secondary products-species of high potential: *Haematococcus*. In: Richmond, A. (ed.) *Handbook of Microalgal Culture, Biotechnology and Applied Phycology*, Blackwell Science, Oxford, UK, pp. 281–288.

del Campo, A.J., García-González, M. and Guerrero, M.G. (2007) Outdoor cultivation of microalgae for carotenoid production: current state and perspectives. *Applied Microbiology and Biotechnology* 74, 1163–1174.

Demming-Adams, B. and Adams III, W.W. (2002) Antioxidants in photosynthesis and human nutrition. *Science* 298, 2149–2153.

di Mascio, P., Kaiser, S. and Sies, H. (1989) Lycopene as the most efficient biological carotenoid singlet oxygen quencher. *Archives of Biochemistry and Biophysics* 274, 532–538.

Dutta, D., Chaudhuri, U.R. and Chakraborty, R. (2005) Structure, health benefits, antioxidant property and processing and storage of carotenoids. *African Journal of Biotechnology* 4, 1510–1520.

Eonseon, J., Polle, J.E.W., Lee. H.K., Hyund, S.M. and Chang, M. (2003) Xanthophylls in microalgae: from biosynthesis to biotechnological mass production and application. *Microbial Biotechnology* 13, 165–174.

Fassett, R.G. and Coombes, J.S. (2011) Astaxanthin: a potential therapeutic agent in cardiovascular disease. *Marine Drugs* 9, 447–465.

Flood, V., Smith, W., Wang, J.J., Manzi, F., Webb, K. and Mitchell, P. (2002) Dietary antioxidant intake and incidence of early age-related maculopathy: the Blue Mountains Eye Study. *Ophthalmology* 109, 2272–2278.

García-Limones, C., Schnabele, K., Blanco-Portales, R., Luz, B.M., Caballero, J.L., Schwab, W. and Munoz-Blanco, J. (2008) Functional characterization of FaCCD1: a carotenoid cleavage dioxygenase from strawberry involved in lutein degradation during fruit ripening. *Journal of Agricultural and Food Chemistry* 56, 9277–9285.

Gaziano, J.M., Manson, J.E., Branch, L.G., Colditz, G.A., Willett, W.C. and Buring, J.E. (1995) A prospective study of consumption of carotenoids in fruits and vegetables and decreased cardiovascular mortality in the elderly. *Annals of Epidemiology* 5, 255–260.

Guedes, A.C., Amaro, H.M. and Malcata, F.X. (2011) Microalgae as Sources of Carotenoids. *Marine Drugs* 9, 625–644.

Guerin, M., Huntley, M.E. and Olaizola, M. (2003) *Haematococcus* astaxanthin: applications for human health and nutrition. *Trends in Biotechnology* 21, 210–215.

Halliwell, B. (1997) Antioxidants and human disease- a general introduction. *Nutrition Review* 55, 544–549.

Halliwell, B. and Gutteridge, J. (1995) *Free radicals in biology and medicine*. Clarendon Press, Oxford, UK.

Halliwell, B. and Gutteridge, J.M.C. (2007) *Free Radicals in Biology and Medicine*. Fourth Edition, Oxford University Press, New York, USA.

Hennekens, C.H., Buring, J.E. and Manson, J.E. (1996) Lack of effect of long-term supplementation with β carotene on the incidence of malignant neoplasms and cardiovascular disease. *New England Journal of Medicine* 334, 1145–1149.

Holick, C.N., Michaud, D.S. and Stolzenberg-Solomon, R. (2002) Dietary carotenoids, serum β-carotene, and retinol and risk of lung cancer in the α-tocopherol, β-carotene cohort study. *American Journal of Epidemiology* 156, 536–547.

Jackson, H., Braun, C.L. and Ernst, H. (2008) The chemistry of novel xanthophyll carotenoids. *American Journal of Cardiology* 101, S50–S57.

Jung, M.Y. and Min, D.B. (1991) Effects of Quenching Mechanisms of Carotenoids on the Photosensitized Oxidation of Soybean Oil. *Journal of American Oil Chemists' Society* 68, 653–658.

Khachik, F., Spangler, C.J. and Cecil, J. (1997) Identification, quantification and relative concentrations of carotenoids and their metabolites in human milk and serum. *Analytical Biochemistry* 69, 1873–1881.

Krinsky, N.I. (1989) Antioxidant functions of carotenoids. *Free Radical Biology and Medicine* 7, 617–635.

Krinsky, N.I. (1993) Actions of carotenoids in biological systems. *Annual Review of Nutrition* 13, 561–587.
Krinsky, N.I. (1998) The antioxidant and biological properties of the carotenoids. *Annals of the New York Academy of Sciences* 854, 443–447.
Krinsky, N.I., Landrum, J.T. and Bone, R.A. (2003) Biologic mechanisms of the protective role of lutein and zeaxanthin in the eye. *Annual Review of Nutrition* 23, 171–201.
Levy, Y., Ben-Amotz, A. and Aviram, M. (1995) Effect of dietary supplementation of β carotene to human on its binding to plasma LDL and on the lipoprotein susceptibility to undergo modification: comparison of the synthetic all-trans isomer with the natural algae β carotene. *Journal of Nutritional and Environmental Medicine* 5, 13–22.
Lewinsohn, E., Sitrit, Y., Bar, E., Azulay, Y., Ibdah, M., Meir, A., Yosef, E., Zamir, D. and Tadmor, Y. (2005) Not just colours- carotenoid degradation as a link between pigmentation and aroma in tomato and watermelon fruit. *Trends in Food Science & Technology* 16, 407–415.
Li, L. and van Eck, J. (2007) Metabolic engineering of carotenoid accumulation by creating a metabolic sink. *Transgenic Research* 16, 581–585.
Liebler, D.C. (1993) Antioxidant Reactions of Carotenoids. *Annals of the New York Academy of Sciences* 691, 20–31.
Lopez-Raez, J.A., Kohlen, W., Charnikhova, T., Mulder, P., Undas, A.K., Sergeant, M.J., Verstappen, F., Bugg, T.D., Thompson, A.J., Ruyter-Spira, C. and Bouwmeester, H. (2010) Does abscisic acid affect strigolactone biosynthesis? *New Phytologist* 187, 343–354.
Lu, S. and Li, L. (2008) Carotenoid metabolism: biosynthesis, regulation and beyond. *Journal of Integrative Plant Biology* 50, 778–785.
Lu, S., van Eck, J., Zhou, X., Lopez, A.B., O'Halloran, D.M., Cosman, K.M., Conlin, B.J., Paolillo, D.J., Garvin, D.F., Vrebalov, J., Kochian, L.V., Kupper, H., Earle, E.D., Cao, J. and Li, L. (2006) The cauliflower *Or* gene encodes a DnaJ cysteine-rich domain-containing protein that mediates high levels of β-carotene accumulation. *The Plant Cell* 18, 3594–3605.
Mayne, S.T. (1996) β-carotene, carotenoids and disease prevention in humans. *FASEB Journal* 10, 690–701.
Mojaat, M., Pruvost, J., Foucault, A. and Legrand, J. (2008) Effect of organic carbon sources and Fe^{2+} ions on growth and β-carotene accumulation by *Dunaliella salina*. *Biochemical Engineering Journal* 39, 177–184.
Naik, P.S., Chanemougasoundharam, A., Khurana, S.M.P. and Kalloo, G. (2003) Genetic manipulation of carotenoid pathway in higher plants. *Current Science* 85, 1423–1430.
Ohmiya, A., Kishimoto, S., Aida, R., Yoshioka, S. and Sumitomo, K. (2006) Carotenoid cleavage dioxygenase (CmCCD4a) contributes to white colour formation in *chrysanthemum* petals. *Plant Physiology* 142, 1193–1201.
Olson, J.A. and Krinsky, N.I. (1995) Introduction: the colorful, fascinating world of the carotenoids: important physiologic modulators. *FASEB* 9, 1547–1550.
Omaye, S.T., Krinsky, N.I., Kagan, V.E., Mayne, S.T., Liebler, D.C. and Billack, W.R. (1997) β-carotene: Friend or foe? *Fundamental and Applied Toxicology* 40, 163–174.
Omenn, G.S., Goodman, G.E., Thornquist, M.D., Balmes, J., Cullen, M.R., Glass, A., Keogh, J.P., Meyskens, F.L., Valanis, B., Williams, J.H., Barnhart, S. and Hammar, S. (1996) Effects of a combination of β-carotene and vitamin A on lung cancer and cardiovascular disease. *New England Journal of Medicine* 334, 1150–1155.
Osborne, M., Boyle, P. and Lipkin, M. (1997) Cancer prevention. *Lancet* 349, 27–30.
Paavo, H.H. and Tuomo, S.L. (2001) The functions of chlorophylls in photosynthesis. *Physiology and Maintenance*- Vol V. University of Helsinki, Finland.
Palozza, P. (1998) Pro-oxidant actions of carotenoids in biologic systems. *Nutrition Review* 56, 257–265.
Papas, A.M. (1999) *Antioxidant status, Diet, Nutrition and Health*. CRC Press, Washington DC.
Plaza, M., Herrero, M., Cifuentes, A. and Ibáñez, E. (2009) Innovative natural functional ingredients from microalgae. *Journal of Agricultural and Food Chemistry* 57, 7159–7170.
Pogson, B.J. and Albrecht, V. (2011) Genetic dissection of chloroplast biogenesis and development: an overview. *Plant Physiology* 155, 1545–1551.
Rabbani, S., Beyer, P., von Lintig, J., Hugueney, P. and Kleinig, H. (1998) Induced β-carotene synthesis driven by triacylglycerol deposition in the unicellular alga *Dunaliella bardawil*. *Plant Physiology* 116, 1239–1248.
Rapola, J., Virtamo, J., Ripatti, S., Haukka, J., Huttunen, J., Albanes, D., Taylor, P. and Heinonen O. (1998) Effects of α tocopherol and β carotene supplements on symptoms, progression, and prognosis of angina pectoris. *Heart* 79, 454–458.

Reaven, P.D., Ferguson, E., Naveab, M. and Powell, F.L. (1994) Susceptibility of human LDL to oxidative modification. Effects of variations in β-carotene concentration and oxygen tension. *Arteriosclerosis and Thrombosis* 14, 1162–1169.

Rissanen, T.H., Voutilainen, S. and Nyyssonen, K. (2001) Low serum lycopene concentration is associated with an excess incidence of acute coronary events and stroke: the Kuopio ischaemic heart disease risk factor study. *British Journal of Nutrition* 85, 749–754.

Rissanen, T.H., Voutilainen, S., Nyyssonen, K., Salonen, R., Kaplan, G.A. and Salonen, J.T. (2003) Serum lycopene concentrations and carotid atherosclerosis: the Kuopio ischaemic heart disease risk factor study. *American Journal of Clinical Nutrition* 77, 133–138.

Sanchez, J.F., Fernandez, J.M., Acien, F.G., Rueda, A., Perez-Parra, J. and Molina, E. (2008) Influence of culture conditions on the productivity and lutein content of the new strain *Scenedesmus almeriensis*. *Process Biochemistry* 43, 398–405.

Sarry, J.E., Montillet, J.L., Sauvaire, Y. and Havaux, M. (1994) The protective function of the xanthophyll cycle in photosynthesis. *FEBS Letters* 353, 147–150.

Schwartz, S.H., Qin, X. and Loewen, M.C. (2004) The biochemical characterization of two carotenoid cleavage enzymes from *Arabidopsis* indicates that a carotenoid-derived compound inhibits lateral branching. *Journal of Biological Chemistry* 279, 46940–46945.

Sharoni, Y. and Levi, J. (1996) In: Kumpulainen, J.T. and Salonen, J. (ed.) *Natural Antioxidants and Food Quality in Atherosclerosis and Cancer Prevention*. Royal Society of Chemistry, Cambridge, UK.

Simkin, A.J., Schwartz, S.H., Auldridge, M., Taylor, M.G. and Klee, H.J. (2004) The tomato carotenoid cleavage dioxygenase 1 genes contribute to the formation of the flavor volatiles β-ionone, pseudoionone and geranylacetone. *The Plant Journal* 40, 882–892.

Stahl, W. and Sies, H. (2007) Carotenoids and flavonoids contribute to nutritional protection against skin damage from sunlight. *Molecular Biotechnology* 37, 26–30.

Stahl, W., Shwarz, W., Sundquist, A.R. and Sies, H. (1992) *Cis-trans* isomers of lycopene and β-carotene in human serum and tissues. *Archives of Biochemistry and Biophysics* 294, 173–177.

Stahl, W., Nicolai, S. and Briviba, K. (1997) Biological activities of natural and synthetic carotenoids: induction of gap junctional communication and singlet oxygen quenching. *Carcinogenesis* 18, 89–92.

Tardy, F. and Havaux, M. (1996) Photosynthesis, chlorophyll fluorescence, light-harvesting system and photoinhibition resistance of a zeaxanthin-accumulating mutant of *Arabidopsis thaliana*. *Journal of Photochemistry and Photobiology A: Chemistry* 34, 87–94.

van den Berg, H. (1999) Carotenoid interactions. *Nutrition Reviews* 57, 1–10.

van het Hof, K.H., Brouwer, I.A., West, C.E., Haddeman, E., Steegers-Theunissen, R.P., van Dusseldorp, M., Weststrate, J.A., Eskes, T.K. and Hautvast, J.G. (1999) Bioavailability of lutein from vegetables is 5 times higher than that of β-carotene. *American Journal of Clinical Nutrition* 70, 261–268.

van het Hof, K.H., West, C.E., Weststrate, J.A. and Hautvast, J.G. (2002) Dietary factors that affect the bioavailability of carotenoids. *Nutrition Research* 130, 503–506.

Vishnevetsky, M., Ovadis, M. and Vainstein, A. (1999) Carotenoid sequestration in plants: the role of carotenoid-associated proteins. *Trends in Plant Science* 4, 232–235.

Walter, M.H., Floss, D.S. and Strack, D. (2010) Apocarotenoids: hormones, mycorrhizal metabolites and aroma volatiles. *Planta* 232, 1–17.

West Jr, K.P., Howard, G.R. and Sommer, A. (1989) Vitamin A and Infection: Public Health Implications. *Annual Review of Nutrition* 9, 63–86.

Woodl, A.A., Britton, G. and Jackson, M.G. (1997) Careteniods and protection of phospholipids in solution or in liposomes against oxidation by peroxyl radicals: Relationship between carotenoid structure and protective ability. *Biochimica et Biophysica Acta* 1336, 575–586.

Young, A.J. and Lowe, G.M. (2001) Antioxidant and prooxidant properties of carotenoids. *Archives of Biochemistry and Biophysics* 385, 20–27.

13 Phenolic Acids as Natural Antioxidants

Lilia Masson*
Universidad de Chile, Santiago, Chile; Fundación CAPES, Universidad Federal de Rio de Janeiro, Instituto de Nutrición Josué de Castro, Rio de Janeiro, Brasil

13.1 Introduction

The word antioxidant is related with a function or activity that can have a chemical structure to interfere in any natural oxidative process that happens *in vivo* or *in vitro* and to delay the process until the compound that presents this activity is consumed as part of its protective activity (Choe and Min, 2005). On the other hand, oxidative processes are part of natural living organisms generating chemical compounds that are highly reactive and necessary in some physiological events, and in other events can produce damage to tissues structures. They must be neutralized by the antioxidants present in the medium. Normally compounds with antioxidant activity are present in very small amounts related to the oxidative substrate (Becker *et al.*, 2004).

From a human nutrition perspective, two biological systems need antioxidant protection: food and human tissues. As a consequence, two different fields for their presence, activity and application are distinguished: *in vitro* for food and *in vivo* for humans. In both systems, adequate chemical or biological markers must be selected for evaluating the antioxidant activity. It is easier to test the protection of an antioxidant in a food matrix than in a live organism due to natural difficulties associated with the intake dose, biotransformation, absorption, metabolism, excretion, biomarker used, etc. (Kenner and Lapidot, 2001). Many organic compounds can present antioxidant activity. In general, the safer chemical structure associated with an antioxidant activity in foods and in humans is the phenol ring. The more labile substrates to suffer oxidative alteration are lipids due to their chemical structure (Frankel *et al.*, 1993).

The oxidation rate is related with the fatty acid structure, saturated fatty acids being more stable. The presence of one double bond in the carbon chain is enough to increase the oxidative process. Two or more double bonds enhance the susceptibility according to an exponential rate. Polyunsaturated fatty acids present in any matrix, *in vitro* or *in vivo*, have the highest susceptibility for suffering oxidative alteration and if the matrix is not naturally protected, adequate and safe synthetic or natural antioxidants will be used (Pokorny, 1991). Normally, unsaturated fatty acids are oxidized by the mechanism of free radical formation followed by a propagation step, starting a chain reaction self-catalysed with

* E-mail: masson_lilia@yahoo.es

production of more free radicals. Normally phenolic antioxidants act in this step as a chain-breaking step, donating its hydrogen as electron donors, acting as free radicals scavengers to stop the chain reaction, and they are oxidized to stable structures according to this chemical reaction (Sthal, 2000).

Phenolic acids are described as natural antioxidants belonging to an abundant and wide group of chemical structures where a phenol ring is always present. They are constituents of vegetal structure, principally in fruits, seeds, leaves and flowers, where they exert different functions such as antioxidants, antimicrobials, contribution to colour, flavour, etc. (Kuhnau, 1976). Some have a very complex structure that is highly polymerized; others are simpler as in the case of phenolic acids. Phenolic acids can be part of more complex compounds that are divided into different groups according to some common phenolic structures. Phenolic acids are sub-grouped into a number of common simple phenolic molecules with a carboxylic group and with phenolic OH. For this reason they are named hydroxy acids, and they can also have a methoxy group associated with the phenolic ring (Hermann, 1989). Interest in these phenolic acids has increased from the biological point of view related to their possible protective antioxidant activity for preventing or delaying some diseases related to the oxidative lipid tissue processes, such as atherosclerosis caused by LDL-cholesterol oxidation. In a more general way, they can act at cellular level as a free radical scavenging agent, protecting the tissue from these aggressive biological compounds generated in the human organism by lipid oxidation (Sroka and Cisowski, 2003).

This chapter deals with the structure, presence in plant kingdom and foods, aspects related with their antioxidant, physiological functions, applications and bioavailability of phenolic compounds.

13.2 Structure and Antioxidant activity

Phenolic acids belong to a large family of plant compounds quite common in the plant kingdom designated as phenolic compounds with a more complex structure, but maintaining a phenolic ring in their molecules (Bravo, 1998). The interest in these phenolic molecules arises from the biological point of view because they are considered as the important source of natural antioxidants for humans (Del Rio *et al.*, 2010). From phenolic acids, two chemical structures corresponding to benzoic and cinnamic acids (hydroxybenzoic and hydroxycinnamic acid) are the parent molecules which originated from two families: the benzoic acid (hydroxybenzoic) family with seven carbons; p-hydroxybenzoic, protocatechuic, vanillic, syringic and gallic acids belong to this family. The other family formed by cinnamic acid (p-hydroxycinnamic) with nine carbons and two carbon chains with one double bond is linked to the carboxylic group. This family is constituted principally by p-cumaric, ferulic, caffeic and sinapic acids; when caffeic acid is associated with quinic acid, chlorogenic acid is formed (Cuvelier *et al.*, 1992). A relationship between antioxidant capacity and chemical structures of these phenolic acids exists (Fig. 13.1). The parent acids have an important difference: p-hydroxycinnamic has a group -CH=CH-COOH, which replaces the -COOH in benzoic acid, contributing to a better phenolic ring stabilization; the presence of phenolic OH groups in the 'ortho' or 'para' position, or both, improves the antioxidant activity of the respective phenolic acid. In addition, the presence of one or two -OCH_3 groups, favour the antioxidant behaviour of the respective phenolic acid (Rice-Evans *et al.*, 1996). According to these chemical structures, in general, the cinnamic acid family must have a better antioxidant activity than the benzoic family, and following with this criteria in the benzoic family, 3,4-dihydroxybenzoic and syringic acid should present the highest antioxidant activity. On the other hand, in the cinnamic family, caffeic and sinapic acid should be the most potent. Comparing the two families, caffeic acid should have the best antioxidant capacity (Natella *et al.*, 1999).

These phenolic acids are widely spread in the plant kingdom; they are part of our daily diet through vegetables, fruits and grains. The interest in human health is related with their chemical structure corresponding

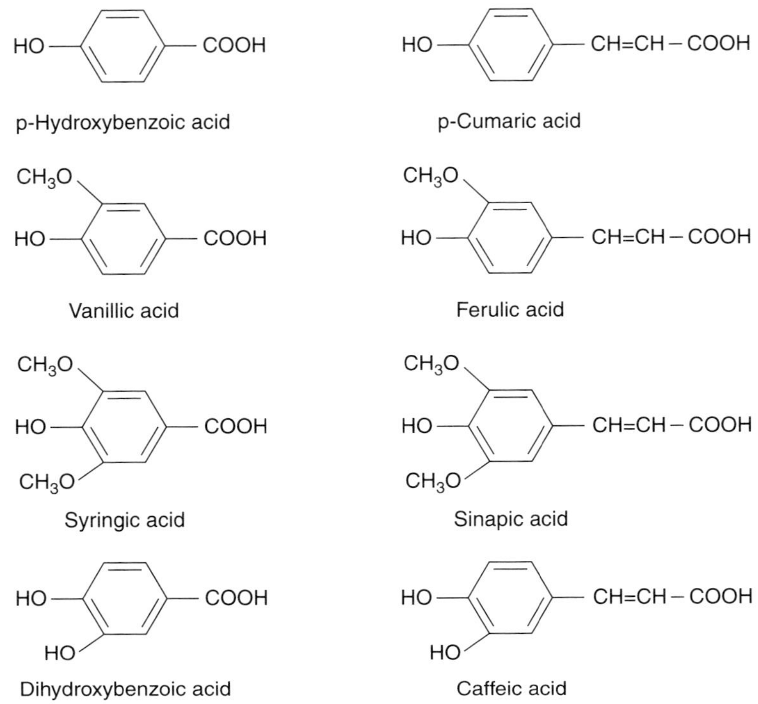

Fig. 13.1. Structure of some phenolic acids and their methoxy derivatives.

to primary antioxidants and the biological possibility that they can have some protective effects to prevent or delay some non-transmissible diseases such as cardiovascular disease and cancer, which affects many people in the world population (Cheng *et al.*, 2007).

13.3 Occurrence in the Plant Kingdom

It has been described that phenolic acids are produced in plants as a defensive response to any aggressive environmental situation. In addition, different factors can influence the concentration of phenolic acids in a fruit or vegetable, such as climate, variety, culture conditions and state of maturity (Asami *et al.*, 2003). As an example, phenolic acids decrease during ripening; this situation means that the phenolic acid content in determined species can present variations among the same species. Different studies have been conducted to modify phenol content or profile (Clifford, 1999; Parr and Boldwell, 2000). Storage conditions, domestic and industrial processing can also change the phenolic content of fruits or vegetables such as apple, pear, grape juice and potato (Spanos and Wrolstad, 1992; Friedman, 1997). The two families of phenolic acids are present in fruits, vegetables and grains such as cereals. The hydroxybenzoic acid family is less abundant compared with the hydroxycinnamic acid family. The changes produced in natural and industrialized fruits and vegetables and the consumers' preference influence the phenolic acid intake for caffeic acid and has been estimated around 500–1000 mg day^{-1} (Clifford, 2000). Onions, some red fruits such as strawberries, raspberries, blackberries and black radish contain members of the hydroxybenzoic acids family; gallic acid being one of the most important and it is also present in leaves. This family is present in more complex phenolic molecules such as gallotanins in mango fruit and in ellagitanins (Clifford and Scalbert, 2000). As a consequence of this direct contribution to the daily human diet, the studies of the hydroxybenzoic acid family is less relevant. The other family of hydroxycinnamic acids is more abundant in nature, principally in the external part of the foods, in fruits bonded with glucose, or esterified with three acids: quinic, shikimic or tartaric (Naczj and Shahidi, 2006). Phenolic acids principally from the hydroxycinnamic family are present in berries. They are considered as the principal dietary source and they can constitute around 70–85% of the total phenolic compounds; more than 20 phenolic acids can be present in berries. Their content can change due to many circumstances such as cultivars, ripening and pre- and postharvest management (Zadenowski *et al.*, 2005). Beans are also a source of phenolic acids (Lutria and Pastor-Corrales, 2006).

Caffeic and ferulic acids are the principal representatives of the hydroxycinnamic family in plant foods. Caffeic acid can be in a free form or a combined form. It is the principal phenolic acid in fruits. One of the important esterified structures of caffeic acid is with quinic acid corresponding to chlorogenic acid. It is found in a high amount in coffee and berries such as strawberries, raspberries and blackberries (Farah and Donangelo, 2006). Humans do not have the esterase to liberate caffeic acid from quinic acid, but it can pass to the free form by the intestinal microflora or by processing such as with a low or high temperature and fermentation (Plumb *et al.*, 1999). The second important hydroxycinnamic acid is ferulic acid; it can be present free or conjugated associated with dietary fibre and linked with an ester bond to hemicelluloses. Cereals are the principal dietary sources, especially wheat germ (Kroon *el al.*, 1996, 1997; Kroon and Williamson, 1999). Distribution is not uniform; the external layer and bran are the richest parts. Therefore it is recommended to increase the consumption of integral grains (Nystrom *et al.*, 2005). Wheat flour is refined, which removes outer bran and external layers thereby decreasing ferulic acid content. Maize grains have the highest content among cereals; normally ferulic acid is bonded to hemicelluloses. Detailed ferulic acid content in grains, peel fruits and vegetables such as potato, commercial foods and beverages and other hydroxycinnamic acids have been studied (Hermann, 1989; Clifford, 1999, 2000; Scalbert and Williamson, 2000; Manach *et al.*, 2004; Mattila *et al.*, 2006; Mattila and Hellstrom, 2007; Zhao and Moghadassian, 2008).

13.4 Occurrence in Some Foods and Beverages

Potatoes have been widely studied due to their importance as one of the more relevant crops in the world. In addition to their traditional nutritive value, potatoes represent a good source of phytochemicals with healthy properties as phenolics compounds: flavonoids, anthocyanins and carotenoids pigments (Malmberg and Theander, 1985). Among phenolic acids, chlorogenic acid is the principal with the highest percentage, and skin of potato is a better source of phenolic acids than pulp (Lewis *et al.*, 1999). Lukascewicz *et al.* (2004) reported that purple red skin potatoes contain twice the phenolic acid concentration as to white skin tubers; purple or red pulp cultivars contain three or four times the phenolic acid concentration compared with white potato pulp. The predominating phenolic acid is chlorogenic acid followed by protocatechuic, vanillic and p-cumaric acid. Peel, an industrial waste from potato processing, can be used as 'added value' to other food products. Transgenic approaches have demonstrated that it is possible to increase phenolic acids in potato cultivars. Andre *et al.* (2007) have reported high variations, up to 11-fold in total phenolic content in native Andean potato, signalling that genotype has a high influence in this content.

Other contributions by potato are the phytochemicals present in skin and pulp including flavonoids, folate and carotenoids. Factors influencing the phytochemical content and stability are genotypes, agronomic factors, postharvest storage, cooking, processing and co-pigmentation. The antioxidant activity of the compounds depends on their chemical structure. Human health benefits along with methods for determining these bioactive components were also discussed. Potato is not only starch, it also contains many phytochemicals that function as health promoters; they should be an important part of the daily human diet (Ezekiel *et al.*, 2011).

Flowers can also be a good source of phenolic acids, such as *Hemerocallis*. This flower, known as Chinese day lily, is part of the traditional Chinese food products. The chlorogenic acids in this flower were analysed by LC-MS. Three chlorogenic acids, three p-coumaroilquinic acids and two feruloilquinic acids were also identified. The extract is especially rich in the 3 and 4-acyl chlorogenic acids, making it a good source of these phenolic acids (Clifford *et al.*, 2006).

Total and individual phenolic acids were studied in 34 commercial beers in China. They protected beer from oxidation during storage, contributing to flavour. The individual phenolic acids were determined by HPLC: gallic, protocatechuic, vanillic, caffeic, syringic, p-cumaric and ferulic acids were quantified; gallic and ferulic acids were the principal. A great variation in the values among beers was obtained. The results will permit the improvement of beer flavour, stability and protecting endogenous antioxidants in the raw materials and beers, through selectively raising certain varieties with improved phenolic acid contents (Zhao *et al.*, 2010).

Coffee, an important beverage consumed all around the world, is a good source of chlorogenic acid. An analysis of 17 green coffee seeds from Brazilian Arabica cultivars and progenies, 4 Arabica and 13 hybrids was obtained after wet and semi-dry postharvest procedures. Chlorogenic acid was determined by HPLC and structure by HPLC-MS. The results indicated an increase of chlorogenic acid in the seeds obtained by the wet procedure. Sensory analysis was done to test if the difference in chlorogenic acid had any influence on sensory parameters (Duarte *et al.*, 2010).

Palm oil is an important commodity and during processing water-soluble waste contains phenolic antioxidants. Phenolic compounds present in Malaysian fruit oil palm (*Elaeis guineensis*) in soluble-free, insoluble-bound and esterified forms were studied. Eight phenolic acids were determined: ferulic, p-hydroxybenzoic and p-cumaric were the most important; as minor acids gallic, protocatechuic, vanillic, caffeic, syringic were determined by HPLC/MS/MS; the antioxidant capacities of the fractions were also determined. The results indicated that the principal phenolic content obtained from palm fruit oil is in the insoluble-bound form (Neo *et al.*, 2010).

Cereals are other good sources of phenolic acids. The profile was determined in the flour from two Canadian wheat varieties 'Western Red Spring' and 'Western Amber Durum' at different sprout degrees by UPLC method. Seven phenolic acids in both wheat varieties were determined: 4-hydroxybenzoic, vanillic, caffeic, syringic, p-cumaric, ferulic and sinapic acids. Syringic and ferulic acids were the principal acids in both wheat varieties. During germination, syringic acid increased, and it was concluded that sprouted wheat had better nutritional properties, which could improve the nutritive value in food products (Van Hung *et al.*, 2011).

Phenolic acids together with small amounts of proteins in white wines can produce turbidity or precipitate in bottled white wines affecting its commercial value. A Sauvignon white wine produced experimentally and bottled was studied. Phenolic compounds were analysed by GLC/MS and HPLC/ESI-TOF. Eight acids were quantified: tyrosol, vanillic, protocatechuic, syringic, *trans*-p-cumaric, gallic, ferulic and trans-caffeic acids. Differences in the concentration before and after the protein precipitate were found: gallic and shikimic acids decreased and syringic and p-cumaric increased. Phenolic compounds can be involved in turbid appearance in white wines (Esteruelas *et al.*, 2011).

13.5 Bioavailability

Phenolic acids are widely distributed in the plant kingdom because they are considered secondary metabolites of higher molecular weight phenolic compounds. Chemically they can be free or esterified or in the form of glycoside. They are recognized as potent antioxidants *in vitro* and *in vivo*, through different mechanisms (Silva *et al.*, 2000; Nystrom *et al.*, 2005; Amorati *et al.*, 2006). Their antioxidant activity has been evaluated in LDL oxidative protection (Laranjinha *et al.*, 1994; Cheng *et al.*, 2007; Srinivasan *et al.*, 2007; Wu *et al.*, 2007; Maurya and Devasagayam, 2010). They can also present pro-oxidant properties (Inoue *et al.*, 1994). As in nature they are found in different forms, free or bound, this affects their bioavailability and their distribution is not uniform among foods. There is a great dependence between the dietary habits of the population and their possible biological functions (Kylli *et al.*, 2008).

Manach *et al.* (2004, 2005) studied the bioavailability of polyphenols in humans, including general information on the two families of phenolic acids. Work on ferulic acid was published by Zhao and Moghadasian (2008). These studies showed valuable information about occurrence, dietary intake and pharmacokinetic properties of these natural phenol acids. With regard to absorption in nature, the occurrence of hydroxycinnamic acids, as caffeic and ferulic acid principally as esters with sugars, organic acids and lipids has some effects on biological properties such as intestinal absorption (Kroon *et al.*, 1996, 1997). It has been described that humans do not have specific esterases in their tissues to hydrolyse this ester union. Chlorogenic acid, the ester formed by caffeic acid and quinic acid, is quite abundant in plant foods and seeds. As caffeic and ferulic ester derivatives are practically not absorbed in the small intestine, the possibility for absorption and to be metabolized is with the participation of enzymatic systems such as xylanases and esterases of the intestinal microflora (Plumb *et al.*, 1999). The metabolism of phenolic acids by rat microflora was studied by Scheline (1968). Nardini *et al.* (2002) studied the absorption of phenolic acids in humans after coffee ingestion. Caffeic acid and other related metabolites were detected in plasma, confirming that the chlorogenic absorption occurred mainly in the colon after hydrolysis by the microflora. Hydroxybenzoic acids are well absorbed and gallic acid has been determined in plasma. Nardini *et al.* (2009) studied the absorption of hydroxycinnamic acids present in white wine in humans. The bioavailability was good: caffeic, ferulic and p-cumaric acids were absorbed as tartaric acid esters, and metabolized as conjugated glucoronides or sulfates. Zhao *et al.* (2003a, b, 2004) have confirmed that the bonded form of ferulic acid is mandatorily related to its absorption, as the free form is rapidly absorbed in the plasma and excreted in the urine. Zhang *et al.* (2010) determined chlorogenic and cinnamic acids in plasma and studied their pharmacological application using liquid chromatography tandem MS assay.

13.6 Biological Importance and Applications

Phenolic acids are natural antioxidants and act as trapping reactive oxygen species (ROS), avoiding oxidation of cellular structures (Blair, 2001) and food matrixes (Choe and Min, 2005). Singh *et al.* (2008) published detailed information of oxidative stress, uptake and metabolism in the brain, mechanism of action and dietary intake of these compounds.

The biological effect of the phenolic acids in humans are directly related with their bioavailability, which can present different levels related principally with the chemical bonded structure in nature and after the absorption with their metabolic and excretion routes (Manach *et al.*, 2004, 2005). In general, free phenolic acids as more simple chemical structures are absorbed in the intestine and rapidly metabolized as conjugated derivatives but, for bonded structures, the absorption occurs mainly in the colon after hydrolysis by the intestinal microflora (Nardini *et al.*, 2009).

Studies on neurodegenerative disorders and natural antioxidant activity were carried out by Sultana (2011). This is a new research area with the possibility to control Alzheimer's disease oxidative stress as a neurodegenerative disease by searching the protective effect of a natural product such as ferulic acid ethyl ester, improving its capacity to cross cell membranes that are rich in lipids, such as in the brain. This ferulic ethyl ester maintains the antioxidant activity for scavenging biologically generated free radicals and represents a potential nutraceutical product.

A novel application of cinnamic acid derivatives is in treatment of different diseases such as cancer. Qian *et al.* (2010) developed a complete procedure for the synthesis of cinnamic acid metronidazole ester derivatives. The corresponding assays for inhibitory and cell proliferation were performed; one of the synthesized compounds presented a great inhibitory activity in tumour growth, which could be a new potential agent against cancer.

Ota *et al.* (2011) studied the interactions of three cinnamic acids: p-cumaric, caffeic and ferulic, using liposomes as a membrane lipid model and their styrene derivatives for testing their comparative effect in membrane structural properties. p-Cumaric acid, which is less polar than the others, presented the highest destabilization effect on membrane lipid structure, followed by caffeic and ferulic.

Maurya and Devasagayam (2010) studied the antioxidant and possible pro-oxidant activity of ferulic and caffeic acids due to their beneficial effect for protecting different tissues from excessive generated free radicals. Ferulic acid showed a better antioxidant activity than caffeic acid in front of nitric oxide and ABTS. Both presented a pro-oxidative behaviour in Fenton reaction above a limit concentration. The study suggested that both phenolic acids are good antioxidants at low concentrations.

Ferulic acid belonging to the cinnamic acid family, and is one of the more common phenolic acids present in the plant kingdom in seeds, leaves, fruits and vegetables. It exhibits antioxidant activity by means of electron donation due to its chemical structure. Srinivasan *et al.* (2007) reviewed its bioavailability and potential antioxidant activity in different physiological situations as protective agent in inflammatory process such as atherosclerosis, cancer, brain and pulmonary alterations, and as skin protector, concluding that ferulic acid can have different protective biological applications.

Caffeic acid is the third phenolic acid of the cinnamic acid family. It is a potent primary antioxidant scavenging free radicals. Important research in cardiovascular disease has been carried out. Kumaran and Prince (2010) studied its antioxidant effect in infarct Wistar male rats. Three doses were used to determine dose-response in isoproterenol (ISO-induced myocardial-infarcted rats). The highest dose was selected for the assay; different biological markers were determined. Heart histopathology studies were also done. In conclusion, the authors considered that caffeic acid had an effective and safe response as antioxidant in the experimental animal model.

Physiological effects of caffeic acids, related with inflammatory process and damage in Parkinson's disease, was studied in rats by Tsai *et al.* (2011). Three concentrations were used to observe inflammatory injury, measuring neurothophic factors. Results indicated a positive effect to elevate caffeic acid in the brain, alleviated inflammatory damage and dopamine loss. It was suggested that caffeic acid is a neuroprotective agent in the case of Parkinson's disease; it is recommended to study its effects at higher doses and for a longer period.

Another recent research line associated with caffeic and cinnamic acids is their hypoglycaemic effect in mice promoting glucose utilization (Huang and Shen, 2012). This activity can have an effect on type-2 diabetes treatments. Mouse FL83B cells treated with tumour necrosis factor-x (TNF-x) to induce insulin resistance, were used to test the hypoglycaemic effect of caffeic and cinnamic acids; glycogen and phosphoenol pyruvate carboxylase were determined. The results indicated that both phenolic acids were effective to improve glucose utilization in these induced insulin resistant mouse hepatocytes.

Other biological activity of caffeic acid and its amides is related with its antimicrobial potency (Fu *et al.*, 2010). A total of 23 amides were synthesized and tested against *Bacillus subtilis, Escherichia coli, Pseudomonas fluorescens, Staphylococcus aureus* and some fungi such as *Aspergillus niger, Candida albicans* and *Trichophyton rubrum*. Five of these amides were effective and those with caffeic anilide group had the better inhibitory activities. In the same area, Zhao, M. *et al.* (2010) studied the antimicrobial activity of chlorogenic acid extracted from tobacco leaves, a good natural source of this derived phenolic acid. To improve its stability, the authors prepared an inclusion complex with β-cyclodextrin; free and complexed chlorogenic acid were tested for their antimicrobial activity against *E. coli, B. subtilis* and *S. aureus.* The results showed no difference between chlorogenic acid forms, but the authors considered that the inclusion complex form is more convenient for its application.

One of the principal activities of phenolic acids, their antioxidant power, was investigated using chlorogenic acid and its metabolite caffeic acid (Sato *et al.*, 2011). *In vitro* assays in Caco-2 cells were used; caffeic acid showed stronger antioxidant activity than chlorogenic acid. Considering that caffeic acid is a metabolite of chlorogenic acid produced at intestine level, the authors commented that the chlorogenic antioxidant activity could be related to the liberated caffeic acid.

Aytekin *et al.* (2011) evaluated the antioxidant activities of different molecular weight synthetic chitosan-caffeic acid derivatives with the purpose of increasing chitosan antioxidant properties. The procedures for measuring antioxidant activity were based on reactions by radical-scavenging activity and reducing power. The results showed that caffeic acid was the principal component with antioxidant activity in the developed synthetic products.

Other biological activity investigated in phenolic acids is related with their possible positive anticancer action. El-Refaei and El-Naa (2010) studied the anti-tumour effect of caffeic acid phenethyl ester against animal carcinogenesis. The procedure consisted of implanting tumour Ehrlich carcinoma cells *in vivo* into a Swiss mice strain and to administer a dose of the caffeic acid phenethyl ester. The results showed a significant decrease of tumour and an improvement in the survival of the animal; this may be related with an apoptotic effect.

A study comparing the antioxidant activity of the whole hydroxycinnamic phenolic acids family including chlorogenic acid was carried out by Cheng *et al.* (2007). Caffeic, chlorogenic, sinapic, ferulic and p-cumaric were evaluated in an *in vitro* peroxidation human LDL model using AAPH or cupric ion. The kinetic analysis for APPH showed this order: caffeic similar chlorogenic > sinapic > ferulic > p-cumaric; for cupric ion: caffeic similar chlorogenic > sinapic similar ferulic similar p-cumaric; these different activities are related with the presence of ortho-dihydroxyl or 4-hydroxy-3 methoxyl groups in the respective molecules of the phenolic acids tested.

Related to the other phenolic acid family of p-hydroxybenzoic acid, vanillic acid was studied relating to its protective effect in isoproterenol-induced cardiotoxic rats. It was assayed at two dose levels; the highest dose had better results with regard to protective antioxidant effect. The mechanisms involved were free radical scavenging, antioxidant and anti-inflammatory properties (Prince *et al.*, 2011).

13.7 Applications of Phenolic Acids in Food

Applications of the antioxidant properties of phenolic acids were tested in stripped sunflower triacylglycerols. Syringic, 3,4-dihydroxybenzoic, sinapic and caffeic acids were studied at different concentrations at 22 and 90°C. According to the results, the effectiveness increased in this order: syringic > 3,4-dihydroxibenzoic > sinapic > caffeic acids. Sinapic and caffeic acids showed a higher activity at 90°C (Marinova and Yanishlieva, 2003). In the second finding, caffeic and chlorogenic acids were tested in stripped sunflower triacylglycerols at different concentrations; the fatty acid profile was determined by GLC. The samples with the phenolic antioxidants were heated at 100°C. The results indicated that at the same concentration both phenolic acids had the same protective behaviour, but at higher concentrations, caffeic acid presented a better protection against oxidative processes (Marinova *et al.*, 2009).

Other applications of phenolic acids have been done in fish products. Caffeic acid was tested in minced fish muscle with and without addition of wheat dietary fibre, which showed a pro-oxidant effect in the fish. The addition of caffeic acid during the storage period improved the stability. The mechanism could be the regeneration of endogenous α-tocopherol in a redox cycle involving endogenous ascorbic acid. These results were considered promising by the authors, for using caffeic acid as a natural antioxidant in new restructured minced fish products (Sanchez-Alonso *et al.*, 2011).

Recent studies on caffeic acid activity in fish lipids matrices were done by Medina *et al.* (2012). The different aspects related to caffeic acid in marine lipids as effects of lipid oxidation were analysed. Special discussion was included related to caffeic acid activity in different fish lipid systems including liposomes, emulsions and minced fish muscle. In conclusion, caffeic acid was an effective antioxidant in fish minces stored at cold temperatures in liposomes and bulk fish oil.

Two procedures for phenolic acid extraction have been reported for ferulic and chlorogenic acids. Salleh *et al.* (2011) used waste material from paddy straw rice in Malaysia optimizing the ferulic acid extraction according to the developed model. The other procedure was applied for chlorogenic acid extraction from green coffee beans by microwave-assisted extraction (Upadhyay *et al.*, 2012). According to researchers, these procedures constitute a potential alternative to conventional solvent extraction. The yields were higher than those obtained by the conventional extraction method. The extracts showed a high radical-scavenging activity; they proposed that these processes can be projected for industrial application.

13.8 Conclusions

Phenolic acids constitute a group of natural chemical compounds that by their structure have antioxidant activity. They are organized in two families, widely distributed in the plant kingdom. Humans are suffering from many physiological disorders that can have a direct link with oxidative stress and through the use of these natural and simple molecules, such conditions can be alleviated. They are already present in fruits, grains and vegetables, and maybe now we are discovering the secret of their presence in our life.

Simple things can have high impact in our life

References

Amorati, R., Pedulli, G.F., Cabrini, L. and Zamborin, L. (2006) Solvent and pH effect on the antioxidant activity of caffeic and other phenolic acids. *Journal of Agriculture and Food Chemistry* 54, 2932–2937.

Andre, C.M., Oufir, M., Guigmard, C., Hoffman, L., Haussman, J.F.F., Evers, D.S. and Larondelle, Y. (2007) Antioxidant profiling of native Andean potato tuber (*Solanum tuberosum* L.) reveals cultivars with high levels of β-carotene, α tocopherol, chlorogenic acid, and petanin. *Journal of Agriculture and Food Chemistry* 55, 10839–10849.

Asami, D.K., Hong, Y.J., Barret, D.M. and Mitchel, A.E. (2003) Comparison of the total phenolic and ascorbic acid content of freeze-dried and air-dried marionberry, strawberry, and corn growing conventional, organic, and sustainable agricultural practices. *Journal of Agriculture and Food Chemistry* 51, 1237–1241.

Aytekin, A.O., Morimura, S. and Kida, K. (2011) Synthesis of chitosan-caffeic acid derivatives and evaluation of their antioxidants activities. *Journal of Bioscience and Bioengineering* 111, 212–216.

Becker, E.M., Nissen, E.R. and Skbisted, L.H. (2004) Antioxidant Evaluation Protocols: Food Quality or Health Effects. *European Food Research* 221, 382–386.

Blair, I.A. (2001) Lipid Hydroperoxide–Mediated DNA Damage. *Experimental Gerontology* 36, 1473–1481.

Bravo, L. (1998) Polyphenols: Chemistry. Dietary Sources, Metabolism, and Nutritional Significance. *Nutrition Reviews* 56, 317–333.

Cheng, J.-C., Dai, F., Zhou, B., Yang, L. and Liu, Z.L. (2007) Antioxidant activity of hydroxycinnamic acid derivatives in human low density lipoprotein: Mechanism and structure-activity relationship. *Food Chemistry* 104, 132–139.

Choe, E. and Min, D.B. (2005) Chemistry and Reactions of Reactive Oxygen Species in Foods. *Journal of Food Science* 70, 142–145.

Clifford, M.N. (1999) Chlorogenic acids and other cinnamates – nature occurrence and dietary burden. *Journal of the Science of Food and Agriculture* 79, 362–372.

Clifford, M.N. (2000) Chlorogenic acids and other cinnamates – nature, occurrence, dietary burden, absorption and metabolism. *Journal of the Science of Food and Agriculture* 80, 1033–1043.

Clifford, M.N. and Scalbert, A. (2000) Ellagitanins occurrence in foods, bioavailability and cancer prevention. *Journal of the Science of Food and Agriculture* 80, 1118–1125.

Clifford M.N., Wu, W. and Kuhnert, N. (2006) The chlorogenic acid in *Hemerocallis*. *Food Chemistry* 95, 574–578.

Cuvelier, M.E., Richard, H. and Berset, C. (1992) Comparison of the antioxidative activity of some acid-phenolic: structure-activity relationship. *Bioscience Biotechnology and Biochemistry* 56, 324–325.

Del Rio, D., Costa, L.G., Lean, M.E.J. and Crozier, A. (2010) Polyphenols and Health; What compounds are involved? *Nutrition, Metabolism and Cardiovascular Diseases* 29, 1–6.

Duarte, G., Pereira, A.A. and Farah, A. (2010) Chlorogenic acids and other relevant compounds in Brazilian coffees processed by semi-dry and wet post-harvesting methods. *Food Chemistry* 118, 851–855.

El-Rafaei, M. and El-Naa, M. (2010) Inhibitory effect of caffeic acid phenethyl ester on mice bearing tumor involving agiostatic and apoptotic activities. *Chemical-Biological Interactions* 186, 152–156.

Esteruelas, M., Kontoudakis, N., Gil, M. and Fort, M.F. (2011) Phenolic compounds present in natural haze protein of Sauvignon white wine. *Food Research International* 44, 77–83.

Ezekiel, R., Narpinder, S., Shagun, S. and Kaur, A. (2011) Beneficial phytochemicals in potato – a review. *Food Research International* doi:10.1016/j.foodres.2011.04.025.

Farah, A. and Donangelo, C.M. (2006) Phenolic compounds in coffee. *Brazilian Journal of Plant Physiology* 18, 23–36.

Frankel, E.N., Kanner, J., German, J.B., Parks, E. and Kinsella, J.E. (1993) Inhibition of oxidation of human low-density lipoprotein by phenolic substances in red wine. *Lancet* 341, 454–457.

Friedman, M. (1997) Chemistry, biochemistry, and dietary role of potato polyphenols. A review. *Journal of Agriculture and Food Chemistry* 45, 1523–1540.

Fu, J., Cheng, K,. Zhang, Z., Fang, R. and Zhu, H. (2010) Synthesis, structure and structure-activity relationship analysis of caffeic acid amides as potential antimicrobials. *European Journal of Medicinal Chemistry* 45, 2638–2643.

Hermann, K. (1989) Occurrence and content of hydroxycinnamic and hydroxybenzoic acid components in foods. *Critical Reviews in Food Science and Nutrition* 28, 315–347.

Huang, D.-W. and Shen, S.-C. (2012) Caffeic acid and cinnamic acid ameliorate glucose metabolism via modulating glycogenesis and gluconeogenesis in insulin-resistant mouse hepatocytes. *Journal of Functional Foods* doi: 10.1016/j.ff.2012.01.005.

Inoue, M., Suzuki, R., Koide, T., Sakagushi, N., Ogihara, Y. and Yabu, Y. (1994) Antioxidant gallic acid induces apoptosis in HL60RG cells. *Biochemistry Biophysics Research Communications* 204, 898–904.

Kenner J. and Lapidot, T. (2001) The Stomach as a Bioreactor: Dietary Lipid Peroxidation in the Gastric Fluid and the Effect of Plant Derived Antioxidants. *Free Radical Bio Medicine* 31, 1388–1395.

Kroon, P.A. and Williamson, G. (1999) Hydrocynnamates in plant and foods: current and future perspectives. *Journal of the Science of Food and Agriculture* 79, 355–361.

Kroon, P.A., Faulds, C.B., Ryden, P. and Williamson, G. (1996) Solubilization of ferulic acid from plant cell wall materials in a model human gut system. *Biochemistry Society Transmission* 24, 8384–8384.

Kroon, P.A., Faulds, C.B., Ryden, P., Robertson, J.A. and Williamson, G. (1997) Release of covalently bound ferulic acid from fiber in the human colon. *Journal of Agriculture and Food Chemistry* 45, 661–667.

Kuhnau, J. (1976) The flavonoids. A class of semi-essential food components: their role in human nutrition. *World Review Nutrition Diet* 24, 117–191.

Kumaran, K.S. and Prince, P.S.M. (2010) Protective effect of caffeic acid on cardiac markers and lipid peroxide metabolism in cardiotoxic rats: an in vivo and in vitro study. *Metabolism Clinical and Experimental* 59, 1172–1180.

Kylli, P., Nousiainen, P., Sipila, J., Tenkanen, M. and Heinonen, M. (2008) Antioxidant potential of hydroxicinannamic acid glycoside esters. *Journal of Agriculture and Food Chemistry* 56, 4797–4805.

Laranjinha, J.A.N., Almeida, L.M. and Madeira, V.M.C. (1994) Reactivity of dietary phenolic acids with peroxyl radicals: antioxidant activity upon low density lipoprotein peroxidation. *Biochemistry and Pharmacology* 48, 487–494.

Lewis, C.E., Walker, J.L.R. and Lancaster, J.E. (1999) Changes in anthocyanin, flavonoid and phenolic acid concentrations during development and storage of coloured potato (*Solanum tuberosum* L.) tubers. *Journal of the Science of Food and Agriculture* 79, 311–316.

Lukascewicz. M., Matyniak-Kata, L., Skada, J., Fecka, L., Cisowski, W. and Szopa, J. (2004) Antioxidant capacity manipulation in transgenic potato tubers by changes in phenolic compounds content. *Journal of Agriculture and Food Chemistry* 52, 1526–1533.

Lutria, D.L. and Pastor-Corrales, M.A. (2006) Phenolic acids content of 15 dry edible bean (*Phaseolus vulgaris* L.) varieties. *Journal of Food Composition and Analysis* 19, 205–211.

Malmberg, A.C. and Theander, O. (1985) Determination of chlorogenic acid in potato tubers. *Journal of Agriculture and Food Chemistry* 33, 549–551.

Manach, C., Scalbert, A., Morand, C., Rémésy, C. and Jiménez, L. (2004) Polyphenols: food sources and bioavailability. *American Journal of Clinical Nutrition* 79, 727–742.

Manach, C., Williamson, G., Morand, C., Scalbert, A. and Rémésy, C. (2005) Bioavailability and bioefficiency of polyphenols in humans. I. Review of 97 bioavailability studies. *American Journal of Clinical Nutrition* 81(Suppl.), 230S–240S.

Marinova, E.M. and Yanishlieva, N.V. (2003) Antioxidant activity and mechanism of action of some phenolic acids at ambient and high temperatures. *Food Chemistry* 81, 180–197.

Marinova, E.M., Tonevaa, A. and Yanishlieva, N. (2009) Comparison of the antioxidative properties of caffeic and chlorogenic acids. *Food Chemistry* 114, 1498–1502.

Mattila, P. and Hellstrom, J. (2007) Phenolic acids in potatoes, vegetables, and some of their products. *Journal of Food Composition and Analysis* 20, 152–160.

Mattila, P., Hellstrom, J. and Torronen, R. (2006) Phenolic acids in berries, fruits and beverages. *Journal of Agriculture and Food Chemistry* 54, 7193–7199.

Maurya, D.K. and Devasagayam, T.P.A. (2010) Antioxidant and prooxidant nature of hydroxycinnamic acid derivatives ferulic and caffeic acids. *Food and Chemical Toxicology* 48, 3369–3373.

Medina, I., Undeland, I., Larsson, K., Storro, I., Rustad, T., Jacobsen, C., Kristinovà, V. and Gallardo, J.M. (2012) Activity of caffeic acid in different fish lipids matrices: a review. *Food Chemistry* 131, 730–740.

Naczj, M. and Shahidi, F. (2006) Phenolics in cereals, fruits and vegetables. Occurrence, extraction and analysis. *Journal of Pharmaceutical and Biomedical Analysis* 41, 1523–1542.

Nardini, M., Cirillo, F., Natella, F. and Scaccini, C. (2002) Absorption of phenolic acids in humans after coffee consumption. *Agriculture and Food Chemistry* 50, 5735–5741.

Nardini, M., Forte, M., Vrhovsek, U., Mattivi, F., Viola, R. and Scaccini, C. (2009) White wine phenolics are absorbed and extensively metabolized in humans. *Journal of Agriculture and Food Chemistry* 57, 2711–2718.

Natella, F., Nardini, M., Di Felice, M. and Scaccini, C. (1999) Benzoic and cinnamic acids derivatives as antioxidant: structure activity relation. *Journal of Agricultural and Food Chemistry* 47, 1453–1459.

Neo, Y.-P., Ariffin, A. and Tan, Y.A. (2010) Phenolic acid analysis and antioxidant activity assessment of oil palm (*E. guineensis*) fruits extracts. *Food Chemistry* 122, 353–359.

Nystrom, L., Makinen, M., Lampi, A.M. and Piironen, V. (2005) Antioxidant activity of steryl ferulate extracts from rye and wheat bran. *Journal of Agricultural and Food Chemistry* 53, 2503–2510.

Ota, A., Abramovic, H., Abram, V. and Poklar Ulrih, N. (2011) Interactions of *p*-cumaric, caffeic and ferulic acids and their styrenes with model lipid membranes. *Food Chemistry* 125, 1256–1261.

Parr, A.J. and Boldwell, G.P. (2000) Phenols in the plant and in men. The potential for possible nutritional enhancement of the diet by modifying the phenol content or profile. *Journal of Agriculture and Food Chemistry* 80, 985–1012.

Plumb, G.W., Garcia-Conesa, M.T., Kroon, P.A., Rhodes, M., Ridley, S. and Williamson, G. (1999) Metabolism of chlorogenic acid by human plasma, liver and intestine and gut microflora. *Journal of Food Science and Agriculture* 79, 390–392.

Pokorny, J. (1991) Natural antioxidants for food use. *Trends in Food Science and Technology* 2, 223–227.

Prince, P.S.M., Rajakumar, S. and Dhanasekar, K. (2011) Protective effects of vainillic acid on electrocardiogram, lipid peroxidation, antioxidants, proinflammatory markers and histopathology in isoprotenol induced cardiotoxic rats. *European Journal of Pharmacology* 668, 233–240.

Qian, J., Zhang, H.-J., Zhang, H., Xu, C., Zhao, J. and Zhu, H.-L. (2010) Synthesis, molecular modeling and biological evaluation of cinnamic acid metronidazole ester derivatives as novel anticancer agent. *Bioorganic and Medicinal Chemistry* 18, 4991–4996.

Rice-Evans, C.A., Miller, N.J. and Paganga, G. (1996) Structure-antioxidant activity relationships of flavonoids and phenolic acids. *Free Radical Bio Medicine* 20, 933–956.

Salleh, N.H.M., Daud, M.Z.M. and Arbain, D. (2011) Optimization of alkaline hydrolysis of paddy straw for ferulic acid extraction. *Industrial Crops and Products* 14, 1635–1640.

Sanchez-Alonso, I., Careche, M., Moreno, P., Gonzalez, M.J. and Medina, I. (2011) Testing caffeic acid as a natural antioxidant in functional fish-fibre restructured products. *LWT-Food Science and Technology* 44, 1149–1155.

Sato, Y., Itagaki, S., Kurokawa, T., Ogura, J. and Kobayashi, M. (2011) *In vitro* and *in vivo* antioxidant properties of chlorogenic acid and caffeic acid. *International Journal of Pharmaceutics* 403, 136–138.

Scalbert, A. and Williamson, G. (2000) Dietary Intake and Bioavailability of Polyphenols. *Journal of Nutrition* 130, 2073S–2085S.

Scheline, R.R. (1968) Metabolism of phenolic acids by the rat intestinal microflora. *Acta Pharmacologica et Toxicologica* 26, 189–205.

Silva, F.A., Borges, F., Guimares, C., Lima, J.L., Matos, C. and Reis, S. (2000) Phenolic acids and derivatives: studies on the relationship among structure, radical scavenging activity, and physicochemical parameters. *Journal of Agriculture and Food Chemistry* 48, 2122–2126.

Singh, M., Arsenault, M., Sanderson, T., Murthy, V. and Ramassamy, C. (2008) Challenges for research on polyphenols from foods in Alzheimer's disease: Bioavailability, metabolism, and cellular and molecular mechanisms. *Journal of Agriculture and Food Chemistry* 56, 4855–4873.

Spanos, G.A. and Wrolstad, R.E. (1992) Phenolics of apple, pear, and white grape juice and their changes with processing and storage – a review. *Journal of Agriculture and Food Chemistry* 40, 1478–1487.

Srinivasan, M., Sudheer, A.R. and Menon, V.P. (2007) Ferulic Acid: therapeutic potential through its antioxidant property. *Journal of Clinical Biochemistry Nutrition* 40, 92–100.

Sroka, Z. and Cisowski, W. (2003) Hydrogen peroxide scavenging, antioxidant and anti-radical activity of some phenolic acids. *Food and Chemical Toxicology* 41, 753–758.

Sthal, W. (2000) Lipid oxidation and antioxidants. *Current Opinion in Clinical Nutrition Metabolic Care* 3, 121–126.

Sultana, R. (2011) Ferulic acid ethyl ester as a potential therapy in neurodegenerative disorders. *Biochimica and Biophysica Acta* doi:10.1016/j.bbadis.2011.10.015.

Tsai, S., Chao, C. and Yin, M. (2011) Preventive and therapeutic effect of caffeic acid against inflammatory injury in striatum of MPTP-treated mice. *European Journal of Pharmacology* 670, 441–447.

Upadhyay, R., Rammalaksmi, K. and Rao, L.J.M. (2012) Microwave-assisted extraction of chlorogenic acids from green coffee beans. *Food Chemistry* 130, 184–188.

Van Hung., P., Hatcher, D.W. and Barker, W. (2011) Phenolic acid composition of sprouted wheats by ultra-performance liquid chromatography (UPLC) and their antioxidants activities. *Food Chemistry* 126, 1896–1901.

Wu, W.M., Lu, L., Long, Y., Wang, T., Liu, L., Chen, Q. and Wang, R. (2007) Free radical scavenging and antioxidative activities of caffeic and phenyl ester (CAPE) and its related compounds in solution and membranes: a structure-activity insight. *Food Chemistry* 105, 107–115.

Zadenowski, R., Naczk, M. and Nesterowicz, J. (2005) Phenolic acids profiles in some small berries. *Journal of Agriculture and Food Chemistry* 53, 2118–2124.

Zhang, J., Chen, M., Ju, W., Liu, S., Xu, M., Ch, J. and Wu, T. (2010) Liquid chromatograph tandem mass spectrometry assay for the simultaneous determination of chlorogenic acid and cinnamic acid in plasma and its application to a pharmacokinetic study. *Journal of Pharmaceutical and Biomedical Analysis* 51, 685–680.

Zhao, H., Chen, W., Lu, J. and Zhao, M. (2010) Phenolic profiles and antioxidant activities of commercial beers. *Food Chemistry* 119, 1150–1158.

Zhao, M., Wang, H., Yang, B. and Tao, H. (2010) Identification of cyclodextrin inclusion complex of chlorogenic acid and its antimicrobial activity. *Food Chemistry* 120, 1138–1142.

Zhao, Z. and Moghadasian, M.H. (2008) Chemistry, natural sources, dietary intake and pharmacokinetics properties of ferulic acid: a review. *Food Chemistry* 109, 691–702.

Zhao, Z., Egashira, Y. and Sanoda, H. (2003a) Digestion and absorption of ferulic acid sugars esters in rat gastrointestinal tract. *Journal of Agriculture and Food Chemistry* 51, 5534–5539.

Zhao, Z., Egashira, Y. and Sanoda, H. (2003b) Ferulic acid sugar ester are recovered in rat plasma and urine mainly as the sulfoglucuronide of folic acid. *Journal of Nutrition* 133, 1355–1361.

Zhao, Z., Egashira, Y. and Sanoda, H. (2004) Ferulic acid is quickly absorbed from rat stomach as the free form and then conjugated mainly in liver. *Journal of Nutrition* 134, 3083–3088.

14 Role of Antioxidant Polyphenols in Nutraceuticals and Human Health

Dhan Prakash* **and Charu Gupta**
Amity Institute for Herbal Research and Studies, Amity University, Noida, India

14.1 Introduction

Antioxidants are known to defuse free radicals leading to limited risk of oxidative stress and associated disorders. Phytochemicals with antioxidant capacity naturally present in food are of great interest due to their beneficial effects on human health as they offer protection against oxidative deterioration (Scalbert and Williamson, 2000; Tiwari, 2001). Epidemiological and animal studies suggest that the regular consumption of fruits, vegetables and whole grains reduces the risk of chronic diseases associated with oxidative damage (Kris-etherton *et al.*, 2002; Nichenametla *et al.*, 2006). Carotenoids, tocopherols, ascorbates, lipoic acids and polyphenols are strong natural antioxidants with free-radical scavenging activity. Synthetic antioxidants such as butylated hydroxy anisole (BHA) and butylated hydroxy toluene (BHT) play a useful role in food and pharmaceutical industries (Escarpa and Gonzalez, 2001; Kondratyuk and Pezzuto, 2004; Scalbert *et al.*, 2005).

Polyphenols are a large family of naturally occurring plant products that are widely distributed in plant foods, including fruits, vegetables, nuts, seeds, flowers and bark. A wide variety of dietary plants including grains, berries, legumes, tea, grapes, olive oil, cocoa, walnuts, groundnuts, spices, fruits, vegetables etc. contain polyphenols (Bravo, 1998). Polyphenols, with approximately 8000 structural variants, are characterized by the presence of aromatic rings bearing one or more hydroxyl moieties, which have proven pivotal roles in mediating their properties (Leiro *et al.*, 2004). Although the knowledge of absorption, bioavailability and metabolism of polyphenols is not entirely known, it appears that some polyphenols are bioactive and are absorbed in their native or modified form by the microflora of the intestine. The active components of dietary phytochemicals (e.g. curcumin, resveratrol, capsaicin, catechins, vitamins and β-carotene) are believed to suppress the inflammatory processes, moderate cell signalling pathways, proliferation, apoptosis, redox balance and most often appear to be protective against cancer, neurodegenerative disorders and cardiovascular diseases among others (Aggarwal and Shishodia, 2006; Rahman *et al.*, 2006).

Polyphenols are known for their unique property of activation at multiple levels, through the modulation of MAPK, Akt and NF-κB signalling pathways, inhibiting the

* E-mail: dprakash_in@yahoo.com

production of inflammatory cytokines and chemokines, suppressing the activity of COX and iNOS and decreasing the production of free radicals (Chang *et al.*, 2003). Several phytochemicals, including genistein (Li and Sarkar, 2002), curcuminoids (Aggarwal and Shishodia, 2006) and catechins (Tang *et al.*, 2003), are known to suppress the activation of Akt, thus inhibiting cancer cell growth. Some phenols such as resveratrol, curcumin and green tea catechins have been shown to suppress COX-2, giving the benefit of decreasing the production of reactive oxygen species (Gerhauser *et al.*, 2003; Babu and Liu, 2008).

Furthermore, several polyphenols suppress lipid peroxidation to maintain the cellular status of antioxidant enzymes like superoxide dismutase, catalase and glutathione peroxidase (Labinskyy *et al.*, 2006). Due to the NF-κB suppressing effect of polyphenols, some of them (e.g. curcumin, resveratrol, quercetin and green tea polyphenols) have been shown to decrease the expression of chemokines and cytokines (Hidaka *et al.*, 2002; Kowalski *et al.*, 2005). Polyphenols present in healthy foods or drinks are readily metabolized to phenolic acids and aldehydes by the microflora of the intestine, raising the possibility that these metabolites are responsible for their anti-inflammatory properties (Rios *et al.*, 2003). A wide variety of polyphenols, most of which are dietary supplements, have been reported to possess substantial skin photoprotective effects (Nichols and Katiyar, 2010).

In the recent years, there has been much awareness about functional foods and nutraceuticals fortified with natural polyphenols and their health benefits like their potent antioxidant activity, anticarcinogenic, anti-inflammatory, antineurodegenerative, antidiabetic, antiviral, skin photoprotective, anti-allergic, antiplatelet, anti-ageing, cytoprotective and DNA-protective properties.

14.2 Natural Polyphenols and their Potential Health Benefits

Polyphenols comprise several distinct groups based on their chemical structures: flavonoid polyphenols like epigallocatechin-3-gallate (EGCG) from green tea and quercetin from apples, and non-flavonoid polyphenols such as curcumin from tumeric and resveratrol from grapes. Flavonoids are polyphenolic compounds that are ubiquitous in nature and are categorized according to their chemical structures into distinct groups, i.e. flavonols, flavones, flavanones, flavanols, isoflavones, and anthocyanins. Over 4000 flavonoids have been identified, many of which occur in widely occurring fruits, vegetables and beverages (tea, coffee, beer, wine and fruit drinks). The flavonoids have been of considerable interest because of their potential health effects. The compounds based on C6–C3 backbone represent another important class, the non-flavonoids. These compounds are directly linked to lignin (polymer phenyl propanoid) biosynthesis in vascular plants. The most important examples are curcumin and resveratrol. Phenolic acids are non-flavonoid polyphenolic compounds, which can be further divided into two main types, benzoic acid and cinnamic acid derivatives based on C1–C6 and C3–C6 backbones. While fruits and vegetables contain many free phenolic acids, in grains and seeds, particularly in the bran or hull, phenolic acids are often in the bound form (Adom and Liu, 2002; Kim *et al.*, 2006).

14.2.1 Flavonoids

Flavonoids are polyphenolic compounds present in berries, legumes, tea, grapes, olive oil, cocoa, walnuts, groundnuts, spices, fruits and vegetables. In particular, green vegetables, onion, apple, berries and tea are rich sources of flavonoids. Flavonoids have received the most attention and have been studied extensively, since they have many curative effects such as antibacterial, antioxidant, antiviral, analgesic activities, etc. (Burda and Oleszek, 2001). Flavonoids form a group of many different compounds of which more than 5000 have been currently characterized. Flavonoids can be classified into several distinct subclasses (Table 14.1) based on their chemical structures and exert various health promoting effects in the human body for disease prevention.

Table 14.1. Dietary sources of polyphenols.

Classes/subclasses	Polyphenols	Sources
Anthocyanidins, Flavonoids	Cyanidin 3-glycosides Delphinidin, malvidin, pelargonidin	Blackberries, black currant, black grape, blueberries, cherries, cranberry, plums, pomegranate, raspberry, red wine, strawberries
Anthoxanthins Flavonols	Fisetin, isorhamnetin, kaempferol, myricetin, quercetin	Apples, apricots, beans, berries, black currant, broccoli, buckwheat, celery, cherries, cherry tomatoes, chives, cocoa, grapes, kale, lettuce, onions, peppers, plums, red wine, spinach, sweet potato, tea
Flavanones	Hesperetin, naringenin, eriodictyol	Citrus fruits and their juices, grapes, tangerine juice
Flavones	Apigenin, luteolin	Celery, fresh parsley, olives, oregano, peppers, rosemary
Flavanols (Flavan-3-ols)	Epicatechin and their gallates, morin, procyanidins, prodelphinidins, catechin	Apples, apricots, berries, cherries, chocolate, grapes, peaches, pears, plums, raisins, red wine, tea
Isoflavones (Flavans)	Daidzein, equol, genistein	Grape seeds/skin, soy cheese and sauces, soy products, soybean
Flavonoid glycoside	Hesperidin, naringin, rutin	Grapefruit, lemon, orange juice, orange, tangerine juice
Phenolic acids	Caffeic acid, chlorogenic acid, ferulic acid, *p*-cumaric acid	Apple, apple juice, blueberry, cider, cranberry, grapefruit, lemon, lettuce, coffee beans, orange, peach, pear, cherry, potato, spinach, tea
Hydroxybenzoic acids	Ellagic and gallic acids	Grape juice, pomegranate juice, raspberry juice, longan seed, strawberry
Trihydroxy-stilbenes	Resveratrol	Grapes, groundnuts, red wine
Tannins	Catechin, epicatechin polymers, ellagitannins, proanthocyanidins, tannic acids	Apple juice, blackberry, chick pea, cocoa, coffee, grape seeds and skin, lentils, olive, peach, peas, plum, pomegranate, raspberries, red wine, strawberries, tea, walnuts
Diferuloylmethane	Curcumin	Turmeric

Flavonols

Flavonols (Fig. 14.1) are a class of flavonoids that have the 3-hydroxyflavone backbone. Their diversity stems from the different positions of the phenolic OH groups. Good sources of flavonols are onion, curly kale, leek, cherry, tomato, broccoli, apple, green and black tea, black grapes and blueberries. The common compounds of this group are kaempferol, quercetin, quercetrin, rutin and myricetin. Quercetin, a dietary polyphenol, is beneficial in cancer prevention and therapy because different concentrations of quercetin counter the transformation and growth processes of cancer (Watjen *et al.*, 2005). Malignant tumours result from uncontrolled cell growth due to mutations. Mutations are a result of DNA damage, which is commonly incurred through exposure to reactive oxygen species (ROS). Quercetin is able to donate electrons to ROS (Awad *et al.*, 2000) and thereby reduces their ability to damage cellular DNA. Quercetin can trigger apoptotic cascades by multiple mechanisms and via both the mitochondrial and death-domain pathways in various cell lines (Siegelin *et al.*, 2009). It occurs in food as glycoside and its dietary sources are red onions, fennel, cherries, blueberries, black and green tea, capers, red grapes, red wine and apples (de Boer *et al.*, 2005). Quercetin shows anti-inflammatory action by its direct antioxidant activity and inhibition of inflammatory mediators and enzymes, such as lipoxygenase. It also has the ability to prevent the oxidation of low-density lipoproteins (LDL) by scavenging free radicals and chelating transition metal ions. As a result, quercetin may aid in the prevention of certain diseases, such as atherosclerosis, and chronic inflammation. In addition, quercetin has anticarcinogenic and anti-inflammatory properties. These actions have been attributed to quercetin's potential for reducing the occurrence of cardiovascular, metabolic and neurodegenerative diseases and cancers (Murota and Terao, 2003; Davis *et al.*, 2009; Davis, 2010).

Flavones

Flavones (Fig. 14.2) are a class of flavonoids based on the backbone of 2-phenylchromen-4-one (2-phenyl-1-benzopyran-4-one). The flavones are a group of naturally occurring chemical compounds widely distributed in

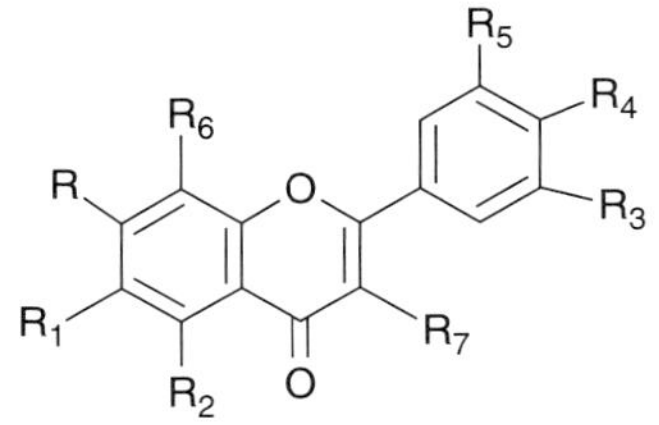

	R	R_1	R_2	R_3	R_4	R_5	R_6	R_7
Kaempferol	OH	H	OH	H	OH	H	H	OH
Quercetin	OH	H	OH	H	OH	OH	H	OH
Quercetrin	OH	H	OH	H	OH	OH	H	O-Glucose
Rutin	OH	H	OH	H	OH	OH	H	O-Rutinose
Myricetin	OH	H	OH	OH	OH	OH	H	OH

Fig. 14.1. Structure of flavonols.

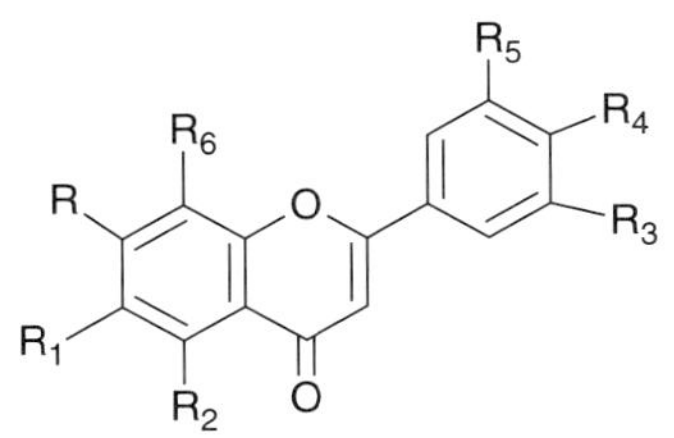

	R	R_1	R_2	R_3	R_4	R_5	R_6
Apigenin	OH	H	OH	H	OH	H	H
Vitexin	OH	H	OH	H	OH	H	O-Glucose
Isovitexin	OH	O-Glucose	OH	H	OH	H	H
Luteolin	OH	H	OH	H	OH	OH	H

Fig. 14.2. Structure of flavones.

the plant world. Natural flavones include apigenin, vitexin, isovitexin and luteolin. Their major food sources are parsley, celery, capsicum and pepper. Apigenin (4′,5,7-trihydroxyflavone) is commonly present in fruits and vegetables with proven anti-inflammatory and anticarcinogenic effects in various animal models. It has been shown to suppress angiogenesis in melanoma and carcinoma of the breast, skin and colon (Caltagirone *et al.*, 2000; Liu, L.Z. *et al.*, 2005).

Flavanones

Flavanones (Fig. 14.3) are a particular type of flavonoids, polyphenolic compounds that act as pigments giving colour to plants. Flavanones, such as naringenin and hesperidin have a more restricted distribution than other flavonoid compounds and are specific to citrus fruits. Citrus flavonoids have anti-inflammatory, anticarcinogenic and antitumour activities. Naringenin (naringenin-7-rhamnoglucoside) is the predominant flavanone in grapefruit (*Citrus paradisi*) and is responsible for the bitterness of grapefruit juices. A cohort study found that the intake of hesperidin and naringenin reduces the risk of chronic diseases such as cerebro-vascular disease and asthma (Benavente-Garcia *et al.*, 1997; Montanari *et al.*, 1998; Knekt *et al.*, 2002). Hesperidin and naringenin directly inhibit TNF-α-stimulated FFA secretion and may be useful for developing treatments to ameliorate FFA-induced insulin resistance (Yoshida *et al.*, 2010).

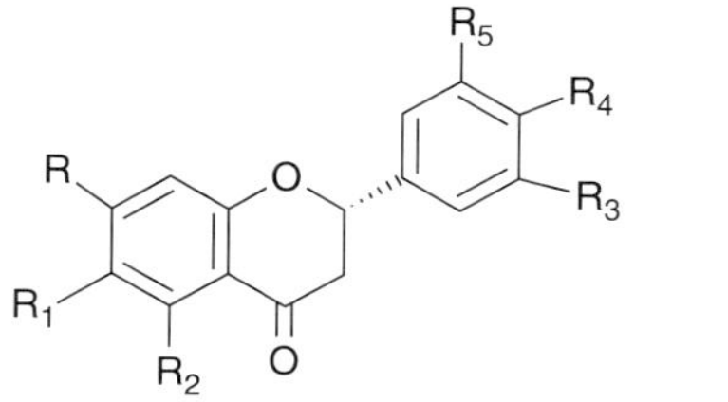

	R	R_1	R_2	R_3	R_4	R_5
Naringenin	OH	H	OH	H	OH	H
Hesperidin	OH	H	OH	H	OCH_3	H

Fig. 14.3. Structure of flavanones.

Flavanols (catechins)

These flavanols (Fig. 14.4) include the catechins and the catechin gallates. The major compounds are catechin, epicatechin, catechin gallate and epicatechin gallate. Their primary food sources are chocolate, beans, apricot, cherry, grapes, peach, red wine, cider, green tea, black tea, blackberry, and cocoa beverages. The flavanols have biological effects, including antimutagenicity, antitumourigenesis, free radical scavenging etc. The green tea catechins make up approximately 60–80% weight of tea polyphenols. (-)-Epigallocatechin-gallate (EGCG) is the most abundant of the four major catechins, which also include (-)-epicatechin gallate (ECG), (-)-epigallocatechin (EGC) and (-)-epicatechin (EC). EGCG is also the most active component of green tea leaves and has antimutagenic, antitumour, anti-inflammatory and free-radical scavenging activities. It also inhibits lipid peroxidation and induces of apoptosis of malignant cells by regulating various signal pathways (Azam *et al.*, 2004; He *et al.*, 2009; Khalatbary *et al.*, 2010). Numerous studies support the antioxidant potential of polyphenols as free radical scavengers and disease-preventative characteristics, such as antimicrobial, anti-inflammatory, anticancer, cardiovascular and the capacity to modulate blood glucose. Quercetin and EGCG are flavonoids that exhibit all these properties (Cheplick *et al.*, 2010; Xiao *et al.*, 2010).

Isoflavones

Isoflavones (Fig. 14.5) are secondary vegetable substances, which can act as oestrogens in the body and have protective functions. They exercise a balancing effect when the level of oestrogens is low, such as during the menopause, and cause less menopause systems. Isoflavones can also reduce the effect of oestrogen on cells and skin layers when the hormone levels are high, and then essentially reduce the risk of oestrogen-linked cancers. Some isoflavones are termed as antioxidants, in particular soy isoflavones; when studied in populations eating soy protein, there are indications that there is a lower incidence of breast

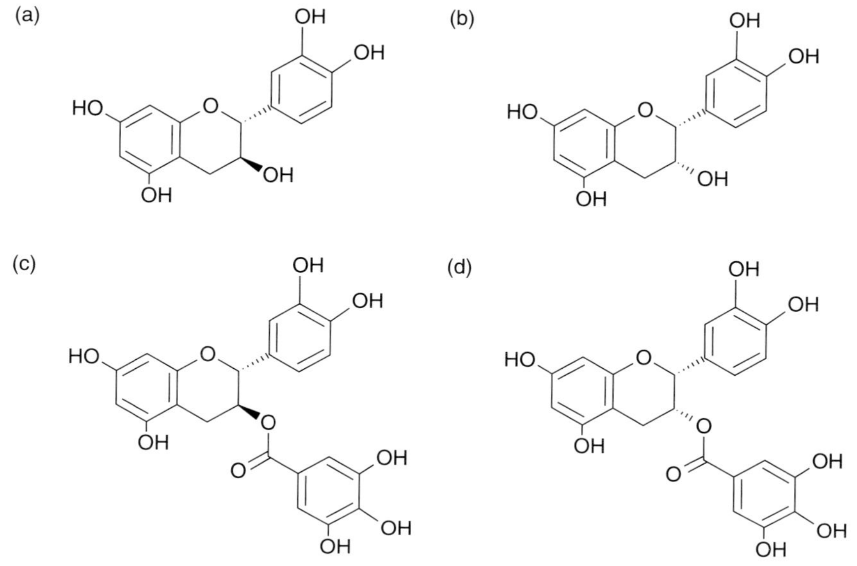

Fig. 14.4. Structure of flavanols (flavan-3-ols). a, catechin; b, epicatechin; c, catechin gallate and d, epicatechin gallate.

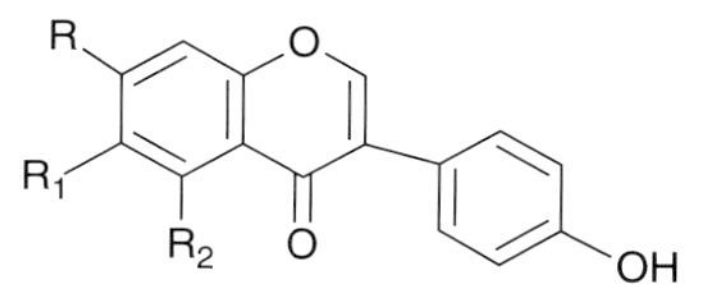

	R	R_1	R_2
Daidzein	OH	H	OH
Genistein	OH	H	H

Fig. 14.5. Structure of isoflavones.

cancer and other common cancers because of its role in influencing sex hormone metabolism and biological activity through intracellular enzymes, protein synthesis, growth factor actions, malignant cell proliferations, differentiation and angiogenesis (Heber *et al.*, 2008). Isoflavones are produced almost exclusively by the members of the *Fabaceae* (*Leguminosae*) family. Their main food sources are soy cheese, soy flour, soybean and tofu. Daidzein and genistein (Fig. 14.5) are among several known isoflavones. They should be an integral part of everyone's diet as they help to reduce cholesterol, prevent atherosclerosis, protect or slow prostate and breast cancer growth, prevent the kind of cell mutation that causes DNA damage, inhibit blood supply to already existing tumours, ease menopause and lower the risk of osteoporosis (Trock *et al.*, 2006).

Anthocyanins

Anthocyanins (Fig. 14.6) are the largest group of water-soluble pigments in the plant kingdom. They have been demonstrated to have potential health benefits and disease prevention properties and are known as potential antioxidants (Zafra-Stone *et al.*, 2007). Consumption of anthocyanin-enriched foods is associated with a reduced risk of several diseases such as atherosclerosis (Xia *et al.*, 2006), dyslipidaemia (Qin *et al.*, 2009) and diabetes (Ghosh and Konishi, 2007). Anthocyanins may appear red, purple, or blue depending on the pH. They are synthesized via the phenylpropanoid pathway; they are odourless and nearly flavourless, contributing to taste as a moderately astringent sensation. Anthocyanins occur in all tissues of higher plants, including leaves, stems, roots, flowers and fruits.

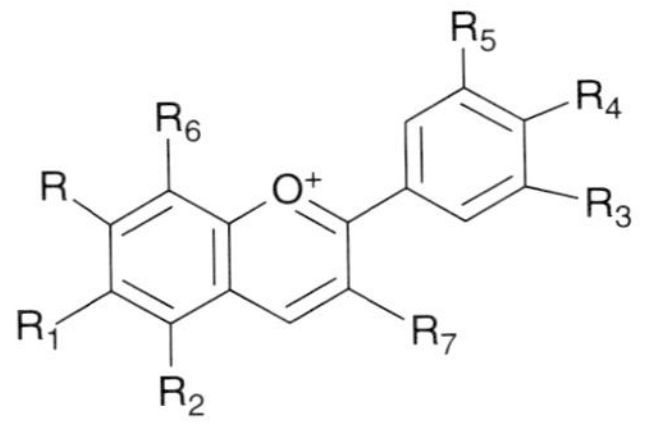

	R	R_1	R_2	R_3	R_4	R_5	R_6	R_7
Pelargonidin	OH	H	OH	H	OH	H	H	OH
Cyanidin	OH	H	OH	OH	OH	H	H	OH
Delphinidin	OH	H	OH	OH	OH	OH	H	OH

Fig. 14.6. Structure of anthocyanins.

Anthocyanins are derivatives of anthocyanidins, which include pendant sugars. Plants rich in anthocyanins are *Vaccinium* species, such as blueberry, cranberry, blackberry, cherry and red cabbage. The highest recorded amount appears to be especially in the seedcoat of black soybean (Choung *et al.*, 2001). Anthocyanins are considered secondary metabolites as a food additive. They are approved for use as a food additive in the European Union, Australia and New Zealand. The main anthocyanin compounds (Fig. 14.6) are pelargonidin, cyanidin and delphinidin. Cyanidin and its glycosides are naturally dietary pigments, which have been found with promising potential benefits to humans, especially in the prevention and treatment of diabetes mellitus (Ghosh and Konishi, 2007; Akkarachiyasit *et al.*, 2010).

14.3 Non-flavonoids

14.3.1 Curcuminoids

Curcuminoids are natural polyphenols and a mixture of curcumin, demethoxycurcumin and bisdemethoxycurcumin. Curcuminoids produce a pronounced yellow colour and are beneficial for health. Curcumin (Fig. 14.7), chemically known as diferuloylmethane, is a polyphenol present in the rhizomes of *Curcuma longa* and is one of its major components, being responsible for its various biological actions. *In vitro*, it exhibits antiparasitic, antispasmodic, anti-inflammatory, anticarcinogenic and gastrointestinal, antifungal, antiviral, antiprotozoal and nematocidal properties (Cui *et al.*, 2007). *In vivo*, it has shown antiparasitic and anti-inflammatory activity (Araujo and Leon, 2001; Pérez-Arriaga *et al.*, 2006). Curcumin has also been extensively studied as a potential drug for the treatment of lung fibrosis. The mechanism of blocking fibrosis by curcumin is related to decreasing collagen accumulation in the lungs (Smith *et al.*, 2010), which is attributed to its anti-inflammatory and antioxidant activities (Lee *et al.*, 2010; Zhang *et al.*, 2011). Ample evidence exists to support curcumin's use in cancer prevention through its antiproliferative and anticarcinogenic properties or as an adjunct in overall cancer treatment (Hasima and Aggarwal, 2012). Curcumin is generally regarded as safe in a clinical trial of cancer patients and marketed as a dietary supplement (Cheng *et al.*, 2001). Curcumin has been shown to possess apoptotic activity against human colon cancer cells (Agarwal *et al.*, 2003), stomach and skin tumours (Azuine and Bhide, 1992), breast cancer cells (Ramachandran *et al.*, 2002) and prostate cancer cells (Dorai *et al.*, 2001). Curcumin also has the ability to suppress UV irradiation-induced DNA mutagenesis and induction of cellular SOS functions (Wilken *et al.*, 2011).

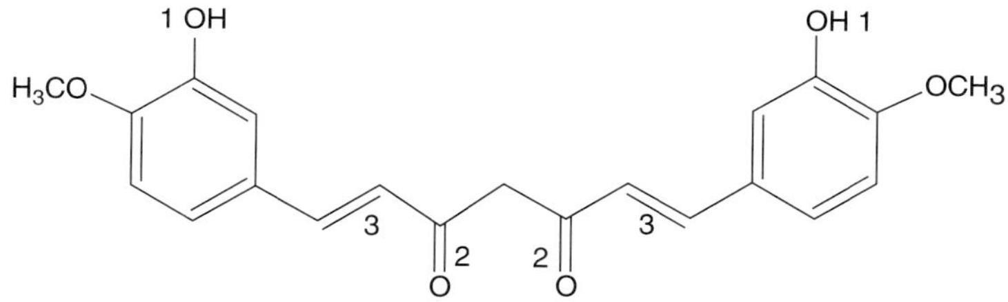

Fig. 14.7. Structure of curcumin: 1, phenolics groups; 2, keto groups; and 3, double bonds.

Riva *et al.* (2008) characterized the action of curcumin on HIV-1 persistently-infected CD4+ T-cells as a model for HIV cell reservoirs and found that curcumin interferes with viral production. It is able to inhibit the genotoxic and histochemical changes induced in the experimental animals by various chemical agents as it reduced the percentages of micronucleated polychromatic erythrocytes in bone marrow cells of mice and inhibited chromosomal aberrations, micronuclei formation and sister chromatid exchanges (SCEs) incidences in mouse bone marrow cells induced by benzo(a)pyrene (Shukla *et al.*, 2003) and lead acetate (Ramadan *et al.*, 2012). Studies carried out both *in vitro* and *in vivo* indicated that curcumin possesses a moderate antimalarial activity (Mimche, 2010). Konatham *et al.* (2010) reported the potential beneficial effects of curcumin against diabetes. In their study, curcumin has been shown to reduce hyperlipidaemia, delay the development of cataract, ameliorate renal lesions, and reduce the cross-linking of collagen in a streptozotocin-treated diabetic animal model. Curcumin has also been shown to lower blood glucose levels in type-2 diabetic KK-Ay mice (Nishiyama *et al.*, 2005). The anti-inflammatory and antioxidant properties of turmeric also have been proposed to lessen insulin resistance and prevent type-2 diabetes in a mice model by dampening the inflammatory response caused by obesity. It was also found that dietary curcumin could increase the expression of adiponectin, which in turn improves insulin sensitivity in insulin-resistant animal models (Weisberg *et al.*, 2008). Asai and Miyasawa (2001) reported that the dietary curcuminoids prevent high-fat diet-induced lipid accumulation in rat liver and epididymal adipose tissue. Wu *et al.* (2008) found that curcumin significantly reduced the plasma and hepatic cholesterol and triglyceride levels in rats. Manjunatha and Srinivasan (2007) have reported lowering of serum and liver cholesterol levels in induced hypercholesterolaemic rats. Curcumin showed reduction of lipid levels in peritoneal macrophages in LDL receptor knockout mice fed with a high fat diet (Zingg *et al.*, 2012).

14.3.2 Stilbenoids

Stilbenoids are secondary products of heartwood formation in trees that can act as phytoalexins. In chemical terms, they are hydroxylated derivatives of stilbene. In biochemical terms, they belong to the family of phenylpropanoids and share most of their biosynthesis pathway with chalcones (Sobolev *et al.*, 2006). An example of a stilbenoid is resveratrol (Fig. 14.8), which is found in grapes and which has been suggested to have many health benefits (Jang *et al.*, 1997). It exists in two structural isomeric forms, *cis* and *trans*, with the *trans* form being more common and possessing greater biological activity. One of the richest sources of this compound is *Polygonum cuspidatum*, a weed that is used in traditional Chinese and Japanese medicines. Trees such as *Eucalyptus* and spruce have also been found to contain resveratrol (Rolfs and Kindl, 1984). The primary dietary sources in the human diet are groundnuts, grapes and wine.

The potential health benefits of resveratrol depend upon its absorption, bioavailability and metabolism (Kaldas *et al.*, 2003). It has

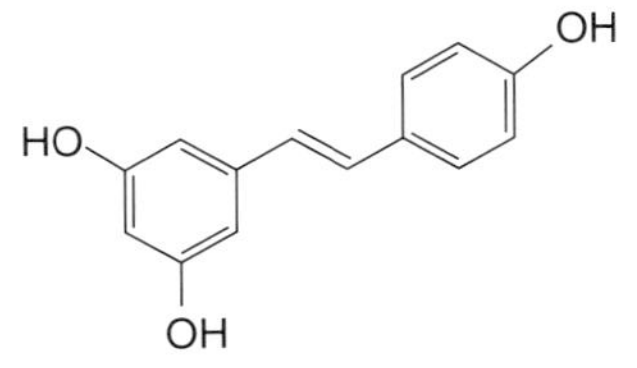

Fig. 14.8. Structure of resveratrol.

exhibited antioxidant, cardioprotective, chemopreventative, anti-inflammatory and oestrogenic properties, as well as interaction with signal transduction pathways. As an antioxidant, resveratrol may delay and/or prevent oxidative stress-induced cellular damage and disease. Excessive damage induced by oxidative stress can induce cells to undergo apoptosis. Resveratrol has been shown to inhibit oxidative-induced apoptosis in a variety of cell lines including Swiss 3T3 mouse fibroblasts, human peripheral blood mononuclear (PBM) and human retinal pigment epithelium (RPE) cells (Jang and Surh, 2001; Kutuk *et al.*, 2004; King *et al.*, 2005). Red wine is one of the few dietary sources of resveratrol and it is believed that this compound is responsible, in part, for the positive cardiovascular effects associated with moderate wine consumption (Constant, 1997). The most accepted mechanism of cardioprotection by resveratrol is the inhibition of platelet aggregation (Bhat *et al.*, 2001).

The antiproliferative activity of resveratrol has been observed in a number of cancer cell lines and may be due, in part, to the induction of apoptosis (Ding and Adrian, 2002). Proliferation inhibition may also be caused by the arrest of the cell cycle (Castello and Tessitore, 2005). Piceatannol, a naturally occurring analogue of resveratrol, has been observed to inhibit the proliferation of cancer cell lines via apoptosis and cell cycle arrest (Wolter *et al.*, 2002; Larrosa *et al.*, 2004). Resveratrol has also been observed to decrease induced COX-2 activity by inhibiting the expression of the enzyme via signal transduction pathways (Kundu *et al.*, 2004). Resveratrol also inhibits the inflammatory actions of cytokines, such as tumour necrosis factor-α (TNF-α) and interleukin-1β (IL-1β) (Culpitt *et al.*, 2003).

The structural similarity of resveratrol to the synthetic oestrogen diethyl-stilbestrol (DES) suggests that it may have oestrogenic activity. The oestrogenic activity of resveratrol has been proposed as a mechanism for cardioprotection and prevention of oestrogen-dependent cancers. The oestrogenic activity of resveratrol may also help prevent bone loss in post-menopausal women (Liu, Z.P. *et al.*, 2005).

Resveratrol has been shown to exert neuroprotective effects (de Ruvo *et al.*, 2000), as well as beneficial effects on the cardiovascular system (Pendurthi *et al.*, 1999). These effects are mostly attributed to its antioxidant properties. It inhibits the proliferation and induces apoptotic cell death in multiple cancer cell types *in vitro* (Kim *et al.*, 2002; Pozo-Guisado *et al.*, 2002); moreover, in animal models of cancer, resveratrol has been shown to inhibit angiogenesis and delay tumour growth (Tseng *et al.*, 2004), impede carcinogenesis (Gusman *et al.*, 2001) and reduce experimental metastasis (Busquets *et al.*, 2007).

14.3.3 Phenolic acids

Phenolic acids (Fig. 14.9) are a group of secondary plant metabolites that encompasses approximately 8000 naturally occurring compounds, all of which possess one common structural feature, a phenol (an aromatic ring bearing at least one hydroxyl group) (Croteau *et al.*, 2000). Current classification divides the broad category of phenolics into polyphenols and simple phenols, based solely on the number of phenol subunits present (Clifford, 1999). Polyphenols possessing at least two phenol subunits include the flavonoids, and those compounds possessing three or more phenol subunits are referred to as the tannins (hydrolysable and non-hydrolysable). The major compounds (Fig. 14.9) are protocatechuic acid, gallic acid, syringic acid, vanillic acid, *p*-cumaric acid, caffeic acid, ferulic acid and chlorogenic acid. Phenolics behave as antioxidants, due to the reactivity of the phenol

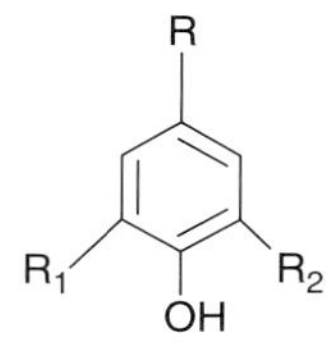

	R	R_1	R_2
Protocatechuic acid	COOH	OH	H
Gallic acid	COOH	OH	OH
Syringic acid	COOH	OCH_3	OCH_3
Vanillic acid	COOH	OCH_3	H
p-Cumaric acid	CH=CH.COOH	H	H
Caffeic acid	CH=CH.COOH	OH	H
Ferulic acid	CH=CH.COOH	OCH_3	H
Chlorogenic acid	CH=CH.COO-quinic acid	OH	H

Fig. 14.9. Structure of phenolic acids: hydroxycinnamic acids.

moiety (hydroxyl substituent on the aromatic ring). Although there are several mechanisms, the predominant mode of antioxidant activity is believed to be radical scavenging via hydrogen atom donation. Other established antioxidant, radical quenching mechanisms are through electron donation and singlet oxygen quenching (Shahidi and Wanasundara, 1992). Substituents on the aromatic ring affect the stabilization and therefore affect the radical-quenching ability of these phenolic acids. Different acids therefore have different antioxidant activity (Chalas *et al.*, 2001).

The naturally occurring phenolic acids contain two distinguishing constitutive carbon frameworks: the hydroxycinnamic and hydroxybenzoic structures. Phenolic acids have received considerable attention due to their various biological activities, including antioxidant, anti-apoptotic and anti-inflammatory capacities (Manach *et al.*, 2004). Phenolic acids have been associated with colour, sensory qualities, and nutritional and antioxidant properties of foods. They play an important role in the organoleptic properties such as flavour, astringency and hardness of foods (Clifford, 1999; Tan, 2000).

Hydroxycinnamic acids

The most common hydroxycinnamic acid (Fig. 14.9) derivatives are *p*-cumaric acid, caffeic acid and ferulic acid, which frequently occur in foods as simple esters with quinic acid or glucose.

p-Cumaric acid is a hydroxycinnamic acid, an organic compound that is a hydroxyderivative of cinnamic acid. There are three isomers, *o*-cumaric acid, *m*-cumaric acid and *p*-cumaric acid that differ by the position of the hydroxy substitution of the phenyl group. *p*-Cumaric acid is the most abundant isomer of the three in nature. *p*-Cumaric acid can be found in a wide variety of edible plants and widely exists in fruits, such as apples and pears, and in vegetables and plant products, such as beans, potatoes, tomatoes and tea (Galvez *et al.*, 1994). It was reported that *p*-cumaric acid *in vitro* can provide antioxidant protection to LDL as a result of its chain-breaking activity (Castelluccio *et al.*, 1996). Diet supplementation with a crude extract of *p*-cumaric acid isolated from pulses resulted in the reduction of ester cholesterol, providing a protective mechanism against the development of atherosclerosis (Sharma, 1979).

The ability of *p*-cumaric acid to prevent excessive lipid peroxidation on the basis of its chain-breaking activity of α-tocopherol oxidation has also been demonstrated (Laranjinha *et al.*, 1996). Castelluccio *et al.* (1996) reported that *p*-cumaric acid was effective in enhancing the resistance of LDL to oxidation. If *p*-cumaric acid is an efficient antioxidant for LDL, it may play a key role in the purported effect of oxidized lipoprotein on platelet activity to inhibit atherogenesis. In addition, the dehydrogenation polymer of *p*-cumaric acid was reported to have anti-human immunodeficiency virus activity (Shimizu *et al.*, 1993).

Caffeic acid (Fig. 14.9) is abundantly found in fruits, leaves and other tissues of numerous dicotyledenous plant species from the families *Caprifoliaceae, Compositae, Cruciferae, Cucurbitaceae, Labiatae, Leguminosae, Polygonaceae, Saxifragaceae, Solanaceae, Theaceae, Umbelliferae* and *Valerianaceae* (Litvinenko *et al.*, 1975). It is abundantly found in plants because it is a key intermediate in the biosynthesis of lignin, one of the principal sources of biomass. It exists in *cis* and *trans* forms; the *trans* form is the predominant one in nature. The uses of caffeic acid for treating asthma and allergies have been extensively investigated (Koshihara *et al.*, 1984; Murota and Koshihara, 1985). Caffeic acid is supposed to catalyse the integration of viral DNA into the host chromatin. Its derivatives (e.g. dicaffeoylquinic and dicaffeolytartaric acids) have been shown to be potent and selective inhibitors of human immunodeficiency virus type 1 (HIV-1) integrase (King *et al.*, 1999). Caffeic acid is a potent antioxidant, metal chelating, free radical scavenger, anti-inflammatory, inhibitor of the lipoxygenase and antidiabetic agent (Psotova *et al.*, 2003; Gulcin, 2006; Chao *et al.*, 2010).

Ferulic acid (Fig. 14.9) is a potent phenolic antioxidant found ubiquitously and at high concentrations in plants (Rice-Evans *et al.*, 1996). Ferulic acid (4-hydroxy-3-methoxy cinnamic acid) is commonly found in fruits such as oranges and in vegetables such as tomato, carrot, sweetcorn and rice bran. Ferulic acid is beneficial for human health due to its antibacterial, anti-inflammatory, hepatoprotective, anticancer, antidiabetic, neuroprotective, anti-atherogenic and antioxidant properties (Srinivasan *et al.*, 2007). Partially, because of its antioxidant and anti-inflammatory activity, ferulic acid is considered as a potential therapeutic agent against various diseases such as cancer, diabetes, cardiovascular dysfunction, inflammatory diseases and neurodegenerative diseases. Furthermore, it has been found that ferulic acid stays in the blood for longer than other antioxidants such as vitamin C and has higher bioavailability than that of other dietary flavonoids and monophenolics studied so far (Beecher, 1998).

Ferulic acid is a potent antioxidant with synergistic interactions with ascorbic acid (Trombino *et al.*, 2004). It readily forms a resonance stabilized phenoxy radical, which accounts for its potent antioxidant potential (Graf, 1992). Ferulic acid protected membranes from lipid peroxidation and neutralized alkoxyl and peroxyl radicals (Trombino *et al.*, 2004). It protected against iron-induced oxidative damage (Hynes and O'Coinceanainn, 2004). Ferulic acid scavenged hydroxyl radical, nitric oxide (Wenk *et al.*, 2004), peroxynitrite (Dinis *et al.*, 2002) and superoxide radical (Kikuzaki *et al.*, 2002). It was antimutagenic (Ferguson *et al.*, 2003), protected against menadione-induced oxidative DNA damage (Burdette *et al.*, 2002) and demonstrated anticarcinogenic effects in animal models of pulmonary and colon carcinoma (Wargovich *et al.*, 2000). Topical application of ferulic acid inhibited UVB-induced erythema (Saija *et al.*, 2000).

Chlorogenic acid constitutes up to 90% of the total phenolic content of potato tubers and its main function is presumably as a defence against phytopathogens (Friedman, 1997). The chlorogenic acid is most likely metabolized by the colonic microflora and it serves as a potent anti-inflammatory agent alternative to conventional chemotherapeutics. Chlorogenic acid is therapeutically useful for mitigating the pathogenic effects of staphylococcal exotoxins (SE). It has been found that chlorogenic acid inhibited SE-induced T-cell proliferation (by 98%) and production of IL-1β tumour necrosis factor, IL-6, interferon-γ, monocyte chemotactic protein I (MCP-l), macrophage inflammatory protein (MIP)-l-α, and MIP-l-β by human peripheral blood mononuclear cells (Kraukauer, 2002).

Hydroxybenzoic acids

Hydroxybenzoic acids (Fig. 14.10), the phenolic derivatives of benzoic acid, are found naturally in cocoa (*Cocos nucifera*) (Dey *et al.*, 2005). They are popular antioxidants and have very low toxicity. Salicylic acid (Fig. 14.10) is a principle hydroxybenzoic acid derivative. Fruits and vegetables are natural sources of salicylic acid (SA), with fruits having large amounts of salicylates, particularly berries. The prime food sources of salicylic acid include radish, tomato, green pepper, olive oil, mushrooms, broccoli, cucumber, spinach and fruits such as blackberries, blueberries, cantaloupe, dates, apricot and guava. A moderate to small quantity is also available in lentil, green peas, beans, celery, cabbage and cauliflower. SA is known for its ability to ease aches and pains and reduce fever. These medicinal properties, particularly fever relief, have been known since ancient times, and it is used as an anti-inflammatory drug (Mackowiak, 2000). In modern medicine, salicylic acid and its derivatives are used as constituents of some rubefacient products. For example, methyl salicylate is used as a liniment to soothe joint and muscle pain, and choline salicylate is used topically to relieve the pain of aphthous ulcers. SA is a key ingredient in many skin-care products for the treatment of seborrhoeic dermatitis, acne, psoriasis, calluses, keratosis and warts (Steele *et al.*, 1988). Salicylic acid is also used as a food preservative, bactericidal and antiseptic.

14.4 Natural Sources of Polyphenols

Polyphenols are considered to be the most effective antioxidants; they can also intensify the activity of other antioxidants. The most popular polyphenols are flavonoids, among which quercetin, kaempferol and apigenin glycosides dominate. Epidemiological and animal studies suggest that the regular consumption of fruits, vegetables and whole grains reduces the risk of chronic diseases associated with oxidative damage (Prakash and Gupta, 2009). A whole variety of phenolic compounds, in addition to flavonoids, are widely distributed in nature (Table 14.1).

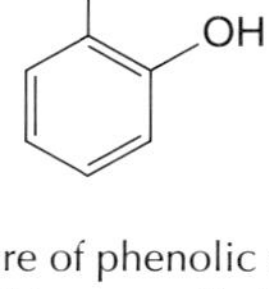

Fig. 14.10. Structure of phenolic acids: hydroxybenzoic acids, e.g. salicylic acid.

14.4.1 Fruits

The majority of fruits are rich sources of vitamin C, carotenoids and polyphenols. Phenolic compounds present in seeds and peel of grapes are mainly anthocyanins and derivatives of hydroxycinnamic acid, flavonols and stilbenoids. Among polyphenols present in grape seeds are gallic acid, catechins and epicatechins, while in peel ellagic acid, myricetin, quercetin, kaempferol and *trans*-resveratrol predominate. In black currant fruits phenols present are mainly anthocyanins (Kammerer *et al.*, 2004). *Aronia melanocarpa* fruits are considered as rich source of polyphenols (40–70 mg g^{-1}) with over 50% share for anthocyanins. The other polyphenols are the derivatives of hydroxycinnamic acid, mainly chlorogenic and neochlorogenic acids and epicatechins. In mature guava leaves, the greatest concentrations of flavonoids found were myricetin (208.44 mg kg^{-1}), quercetin (2883.08 mg kg^{-1}), luteolin (51.22 mg kg^{-1}) and kaempferol (97.25 mg kg^{-1}). The antioxidant phytochemicals of strawberry fruits were phenols (20 mg g^{-1}), with anthocyanins, ellagic acid, their glycoside and ellagitannin derivatives. Among anthocyanins, mainly pelargonidin-3-glucoside and cyanidin-3-glucoside are present. Ellagic acid makes up also over a half of the amount of polyphenols in raspberries. Bilberries, a very popular fruit, contain vitamin C and carotenoids, but the most important antioxidants are phenols (30 mg g^{-1}) with 70% consisting of anthocyans and about 10% derivatives of hydroxycinnamic acid (Aaby *et al.*, 2005; Anttonen and Karjalainen, 2005). The other rich sources of phenolic compounds are cranberries (20 mg g^{-1}). They contain anthocyanins (peonidin and cyanidin), flavanones and procyanidin, and from flavonols, quercetin, myricetin and derivatives of hydroxycinnamic acid are present (Maatta-Riihinen *et al.*, 2004). Crowberry fruits also contain phenols (26–46 mg g^{-1}). Among phenols, flavanols and procyanidins, cinnamic acid, *trans*-resveratrol and *p*-cumaric acid dominate. Polyphenolics (23 mg g^{-1}) of blackberry are mainly responsible for its antioxidant activity (Reyes-Carmona *et al.*, 2005). Besides anthocyanins and flavonols, ellagic acid has the biggest share and the next in turn are procyanidins and epicatechins in seeds (Reyes-Carmona *et al.*, 2005). Citrus fruits, grapes, lemons and oranges are rich sources of antioxidant phenols, among which flavanones (hesperitin, naringenin, eriodictyol) predominate (Cieslik *et al.*, 2006). Apples are a source of phenols (5.0 g kg^{-1}), with about seven-fold higher content in peel than that in pulp. Almost 80% of apple polyphenols contain polymeric procyanidins and monomeric flavanols, with dominating epicatechin and its dimer procyanidin. The other phytochemicals are phenolic acids, dihydrochalcons and flavonols. The main phenolic acid in apples is chlorogenic acid, among dihydrochalcons (phloridzin and phloretin-2-xyloglucoside) and they influence juice quality, especially colour and taste (Lu and Foo, 2000; Cieslik *et al.*, 2006).

14.4.2 Vegetables

Anthocyanin pigments found in vegetables are acyl derivatives of cyanidin (red cabbage, red onion, radish and lettuce), pelargonidin (radish and potatoes) and delphinidin (aubergine). Generally, the flavonoids group dominates among vegetable polyphenols (Bahorun *et al.*, 2004). Among glycosides of flavonol in onion, 4-glucoside of quercetin and 3,4-glucoside of quercetin were identified (Stewart *et al.*, 2000; Marotti and Piccaglia, 2002). Derivatives of quercetin were found also in lettuce. Main polyphenol compounds of broccoli are quercetin-3-sophoroside and kaempferol-3-sophoroside (Marotti and Piccaglia, 2002). Over 20 compounds of quercetin and kaempferol were found in cabbage. In red pepper, two derivatives of quercetin, three derivatives of luteolin and one derivative of apigenin were found (Materska and Perucka, 2005). Chlorogenic acids in potatoes constitute 90% of all phenolic compounds. In the case of carrot, the amount of chlorogenic acid depends on vegetable colour, with the highest amount in carrots with a purple colour and the lower in yellow and white. Chlorogenic acid is also present in aubergine and tomatoes. Neochlorogenic acid is present in a high amount in broccoli. In carrot, besides chlorogenic acid, caffeic acid and its derivatives were identified (Mattila and Hellstrom, 2007). Tomatoes are source of polyphenols such as quercetin and kaempferol (Stewart *et al.*, 2000). *Brassica* vegetables also contain derivatives of hydroxycinnamic acids such as caffeic, chlorogenic, ferulic and sinapic acids and flavonols (Kopsell *et al.*, 2004).

14.4.3 Cereals, legumes and beverages

Among the polyphenols found in cereal grains, phenolic acids play an important role and especially ferulic acid is dominant in grains (wheat and rye). Besides this compound, vanillic and *p*-cumaric acids play an important role, even though they are present in smaller amounts. In the case of oats and buckwheat, avertramidin and rutin were reported, respectively. Phenolic acids are present as ester and glycoside forms. Cereal grains are also a source of catechins; the higher amounts of these compounds were found in seeds of buckwheat, oats, rye and wheat (Peterson *et al.*, 2001; Holasova *et al.*, 2002). Catechins are present in seeds of beans, which contain phenolic acids (ferulic, sinapic), quercetin, tannins, anthocyanins and isoflavones (genistein, daidzein, glycitein), of which soybean is the richest source (Prakash *et al.*, 2007; Sikora *et al.*, 2008).

Extracted oilseed meals consist of phenolic acids, either free or esterified or in condensed forms. Another important antioxidant fraction is flavonoids, which can be detected in nearly all extracted meals, at least in small amounts. Some substances possessing antioxidant activities have been detected in most expeller cakes and extracted meals, such as phenolic acids, like caffeic, dihydrocaffeic acids, ferulic and sinapic acids, or flavonoids. Rapeseed meal is very rich in phenols (77–81 mg kg^{-1}), mainly sinapic acid (Amarowicz *et al.*, 2001). Soybean flour or defatted flour has been used as an antioxidant due to the presence of isoflavones and cinnamic acid derivatives. The antioxidant activity of aqueous extracts is attributed to genistein and glycitein-7-O-monoglucosides. The antioxidant activity of groundnuts is mainly due to phenolic acids. Defatted sunflower meal contained 3.0–3.5 g kg^{-1} phenolics; chlorogenic and caffeic acids constitute about 70% of phenolic antioxidants. Defatted grape seed meal contains a mixture of catechins and procyanidins (Saito *et al.*, 1998). Evening primrose seeds are used for the extraction of oil and extracted meal is rich in phenols such as proanthocyanidins, catechins, polymerized polyphenols and isoflavones (Shahidi *et al.*, 1997). The main precursors of sesame seed antioxidants are lignans, such as sesamolin. The defatted extract of sesame flour contained 41 mg kg^{-1} free phenolic acids, 325 mg kg^{-1} esterified acids and 14 mg kg^{-1} insoluble phenolic acids. Olive fruit is rich in phenolic antioxidants (Ninfali *et al.*, 2001), such as hydroxytyrosol, tyrosol, secoiridoids, such as oleoeuropein and its aglycone, flavonoids and lignans (Shahidi *et al.*, 1997; Ninfali *et al.*, 2001).

Beverages, such as cocoa, coffee, tea, red wine and beer can supply high amount of antioxidants. Phenols are present in high amounts (12–18%) in cocoa seeds and procyanidins consist of about 60%, with quercetin and its glycosides in smaller quantities. These flavonols and procyanidins are also present in chocolate and their contents depend on the kind and colour of chocolate. The contents of phenols in roasted coffee reach 8%, from which chlorogenic acid is dominant. Main phenolic compounds present in tea are catechins and, generally, green tea contains more of these compounds. Red wine is a very good source of antioxidants, and contains resveratrol, a valuable polyphenol. Antioxidant phenols present in beer are mainly derived from barley, malt and hop. The most important phenolic compounds present in beer are phenolic acids such as cinnamic, chlorogenic, vanillic, ferulic, gallic, caffeic, syringic, *o*- and *p*-cumaric acids, derivatives of flavan-3-ol such as catechin, epicatechin, procyanidin, prodelphinidin and flavonoglycosides (Rupasinghe and Clegg, 2007).

14.5 Conclusions

The natural polyphenols are phytochemicals of nutraceutical importance with antioxidant activity that may play a key role to control several diseases induced by oxidative stress and might aid in the design of novel therapies targeting the respective molecular pathways. Antioxidants can also offer suitable answers to the question of the anticancer, antidiabetic, anti-ulcer, anti-inflammatory and antimutagenic effects and other oxidative stress-related health problems. Several clinical evidences, together with epidemiologic observations, suggest that for example tomato consumption may have protective effects on tumour development, tumour dissemination, neurodegenerative and cardiovascular diseases. In a number of clinical trials the importance of daily consumption of whole grains, fruits, vegetables, nuts and vegetable oils that are rich in unsaturated fatty acids in the prevention or treatment of various diseases has been observed. Randomized clinical trials and evidence from epidemiological studies on naturally occurring antioxidants have shown protective effects. Thus the future of polyphenolic antioxidants holds a great promise to ensure a better disease-free lifestyle for mankind by scavenging free radicals and consequently preventing mutagenic changes and associated disorders. More recently, advancement in molecular biology and genomics technology have provided additional understanding of the mechanisms underlying the synthesis of these compounds with special emphasis on the regulation of gene expression. Many interesting findings indicate towards strongly positive correlations between the dietary intake of polyphenol-containing food and the prevention of many chronic diseases. The future of phenolic research will likely include surprising and unexpected advances in the characterization of new structures, new functions and new exploitations in human health.

References

Aaby, K., Skrede, G. and Wrolstad, R.E. (2005) Phenolic composition and antioxidant acivities in flesh and achenes of strawberries (*Fragaria ananassa*). *Journal of Agricultural and Food Chemistry* 53, 4032–4040.

Adom, K.K. and Liu, R.H. (2002) Antioxidant activity of grains. *Journal of Agricultural Food Chemistry* 50, 6182–6187.

Agarwal, B., Swaroop, P., Protiva, P., Raj, S.V., Shirin, H. and Holt, P.R. (2003) Cox-2 is needed but not sufficient for apoptosis induced by Cox-selective inhibitors in colon cancer cells. *Apoptosis* 8, 649–654.

Aggarwal, B.B. and Shishodia, S. (2006) Molecular targets of dietary agents for prevention and therapy of cancer. *Biochemical Pharmacology* 71, 1397–1421.

Akkarachiyasit, S., Charoenlertkul, P., Yibchok-Anun, S. and Adisakwattana, S. (2010) Inhibitory activities of cyanidin and its glycosides and synergistic effect with acarbose against intestinal α-Glucosidase and pancreatic α-Amylase. *International Journal of Molecular Sciences* 11, 3387–3396.

Amarowicz, R., Fornal, J., Karamac, M. and Shahidi, F. (2001) Antioxidant activity of extracts of phenolic compounds from rapeseed oil cakes. *Journal of Food Lipids* 8, 65–74.

Anttonen, M.J. and Karjalainen, R.O. (2005) Environmental and genetic variation of phenolic compounds in red raspberry. *Journal of Food Composition and Analysis* 18, 759–769.

Araujo, C.A.C. and Leon, L.L. (2001) Biological Activities of *Curcuma longa* L. *Memorias do Instituto Oswaldo Cruz Rio de Janeiro* 96, 723–728.

Asai, A. and Miyasawa, T. (2001) Dietary curcuminoids prevent high fat diet induced lipid accumulation in rat liver and epididymal adipose tissue. *Journal of Nutrition* 131, 2932–2935.

Awad, H.M., Boersma, M.G., Vervoort, J. and Rietjens, I.M. (2000) Peroxidase catalyzed formation of quercetin quinone methideglutathione adducts. *Archives of Biochemistry and Biophysics* 378, 224–233.

Azam, S., Hadi, N. and Khan, N.U. (2004) Prooxidant property of green tea polyphenols epicatechin and epigallocatechin-3-gallate: implications for anticancer properties. *Toxicology In Vitro* 18, 555–561.

Azuine, M.A. and Bhide, S.V. (1992) Chemopreventive effect of turmeric against stomach and skin tumors induced by chemical carcinogens in Swiss mice. *Nutrition and Cancer* 17, 77e83.

Babu, P.V. and Liu, D. (2008) Green tea catechins and cardiovascular health: an update. *Current Medicinal Chemistry* 15, 1840–1850.

Bahorun, T., Luximon-Ramma, A., Crozier, A. and Aruoma, O.I. (2004) Total phenol, flavonoid, proanthocyanidin and vitamin C levels and antioxidant activities of Mauritian vegetables. *Journal of the Science of Food and Agriculture* 84, 1553–1561.

Beecher, G.R. (1998) Nutrient content of tomatoes and tomato products. *Proceedings of the Society for Experimental Biology and Medicine* 218, 98–100.

Benavente-Garcia, O., Castillo, J., Marin, F.R., Ortuno, A. and del Rio, J.A. (1997) Uses and properties of citrus flavonoids. *Journal of Agricultural Food Chemistry* 45, 4505–4515.

Bhat, K.P.L., Kosmeder, J.W. II. and Pezzuto, J.M. (2001) Biological effects of resveratrol. *Antioxidants and Redox Signaling* 3, 1041–1064.

Bravo, L. (1998) Polyphenols: chemistry, dietary sources, metabolism, and nutritional significance. *Nutrition Review* 56, 317–333.

Burda, S. and Oleszek, W. (2001) Antioxidant and antiradical activities of flavonoids. *Journal of Agricultural Food Chemistry* 49, 2774–2779.

Burdette, J.E., Chen, S.N., Lu, Z.Z., Xu, H., White, B.E., Fabricant, D.S., Liu, J., Fong, H.H., Farnsworth, N.R., Constantinou, A.I., van Breemen, R.B., Pezzuto, J.M. and Bolton, J.L. (2002) Black cohosh (*Cimicifuga racemosa* L.) protects against menadione-induced DNA damage through scavenging of reactive oxygen species: Bioassay-directed isolation and characterization of active principles. *Journal of Agricultural Food Chemistry* 50, 7022–7028.

Busquets, S., Ametller, E., Fuster, G., Olivan, M., Raab, V., Argiles, J.M. and Lopez-Soriano, F.J. (2007) Resveratrol, a natural diphenol, reduces metastatic growth in an experimental cancer model. *Cancer Letters* 245, 144–148.

Caltagirone, S., Rossi, C., Poggi, A., Ranelletti, F.O., Natali, P.G., Brunetti, M., Aiello, F.B. and Piantelli, M. (2000) Flavonoids apigenin and quercetin inhibit melanoma growth and metastatic potential. *International Journal of Cancer* 87, 595–600.

Castello, L. and Tessitore, L. (2005) Resveratrol inhibits cell cycle progression in U937 cells. *Oncology Reports* 13, 133–137.

Castelluccio, C., Bolwell, G.P., Gerrish, C. and Rice-Evans, C. (1996) Differential distribution of ferulic acid to the major plasma constituents in relation to its potential as an antioxidant. *Biochemical Journal* 316, 691–694.

Chalas, J., Claise, C., Edeas, M., Messaoudi, C., Vergnes, L., Abella, A. and Lindenbaum, A. (2001) Effect of ethyl esterification of phenolic acids on low-density lipoprotein oxidation. *Biomedicine and Pharmacotherapy* 55, 54–60.

Chang, F., Lee, J.T., Navolanic, P.M., Steelman, L.S., Shelton, J.G., Blalock, W.L., Franklin, R.A. and McCubrey, J.A. (2003) Involvement of PI3K/Akt pathway in cell cycle progression, apoptosis, and neoplastic transformation: a target for cancer chemotherapy. *Leukemia* 17, 590–603.

Chao, C.Y., Mong, M.C., Chan, K.C. and Yin, M.C. (2010) Anti-glycative and anti-inflammatory effects of caffeic acid and ellagic acid in kidney of diabetic mice. *Molecular Nutrition and Food Research* 54, 388–395.

Cheng, A.L., Hsu, C.H., Lin, J.K., Hsu, M.M., Ho, Y.F., Shen, T.S., Ko, J.Y., Lin, J.T., Lin, B.R., Ming-Shiang, W., Yu, H.S., Jee, S.H., Chen, G.S., Chen, T.M., Chen, C.A., Lai, M.K., Pu, Y.S., Pan, M.H., Wang, Y.J., Tsai, C.C. and Hsieh, C.Y. (2001) Phase-I clinical trial of curcumin, a chemopreventive agent, in patients with high risk or premalignant leisions. *Anticancer Research* 27, 2895–2900.

Cheplick, S., Kwon, Y.I., Bhowmik, P. and Shetty, K. (2010) Phenolic-linked variation in strawberry cultivars for potential dietary management of hyperglycemia and related complications of hypertension. *Bioresource Technology* 101, 404–413.

Choung, M.G., Baek, I.Y., Kang, S.T., Baek, I.Y., Kang, S.T., Han, W.Y., Shin, D.C., Moon, H.P. and Kang, K.H. (2001) Isolation and determination of anthocyanins in seed coats of black soybean (*Glycine max* (L.) Merr.). *Journal of Agricultural and Food Chemistry* 49, 5848–5851.

Cieslik, E., Greda, A. and Adamus, W. (2006) Contents of polyphenols in fruits and vegetables. *Food Chemistry* 94, 135–142.

Clifford, M.N. (1999) *Appendix 1. A Nomenclature for Phenols with Special Reference to Tea*. CRC Press LLC Boca Raton, Florida, pp. 393–397.

Constant, J. (1997) Alcohol, ischemic heart disease, and the French paradox. *Coronary Artery Disease* 8, 645–649.

Croteau, R., Kutchan, T.M. and Lewis, N.G. (2000) Natural products (secondary metabolites). In: Buchanan, B. Gruissem, W. and Joneas, R. (eds) *Biochemistry and Molecular Biology of Plants*. American Society of Plant Biologists, Rockville, Maryland, pp. 1250–1268.

Cui, L., Miao, J. and Cui, L. (2007) Cytotoxic effect of curcumin on malaria parasite *Plasmodium falciparum*: inhibition of acetylation and generation of reactive oxygen species. *Antimicrobial Agents and Chemotherapy* 51, 488–494.

Culpitt, S.V., Rogers, D.F., Fenwick, P.S., Shah, P., deMatos, C., Russell, R.E.K., Barnes, P.J. and Donnelly, L.E. (2003) Inhibition by red wine extract, resveratrol, of cytokine release by alveolar macrophages in COPD. *Thorax* 58, 942–946.

Davis, J.M. (2010) The dietary flavonoid quercetin increases VO(2max) and endurance capacity. *International Journal of Sport Nutrition and Exercise Metabolism* 20, 56–62.

Davis, J.M., Murphy, E.A. and Carmichael, M.D. (2009) Effects of the Dietary Flavonoid Quercetin upon Performance and Health. *Current Sports Medicine Reports* 8, 206–213.

de Boer, V.C., Dihal, A.A., van der Woude, H., Arts, I.C., Wolffram, S., Alink, G.M., Rietjens, I.M., Keijer, J. and Hollman, P.C. (2005) Tissue distribution of Quercetin in rats and pigs. *Journal of Nutrition* 135, 1718–1725.

de Ruvo, C., Amodio, R., Algeri, S., Martelli, N., Intilangelo, A., D'Ancona, G.M. and Esposito, E. (2000) Nutritional antioxidants as anti-degenerative agents. *International Journal of Developmental Neuroscience* 18, 359–366.

Dey, G., Chakraborty, M. and Mitra, A. (2005) Profiling C6-C3 and C6-C1 phenolic metabolites in *Cocos nucifera*. *Journal of Plant Physiology* 162, 375–381.

Ding, X.Z. and Adrian, T.E. (2002) Resveratrol inhibits proliferation and induces apoptosis in human pancreatic cancer cells. *Pancreas* 25, 71–76.

Dinis, T.C., Santosa, C.L. and Almeida, L.M. (2002) The apoprotein is the preferential target for peroxynitrite-induced LDL damage protection by dietary phenolic acids. *Free Radical Research* 36, 531–543.

Dorai, T., Cao, Y.C., Dorai, B., Buttyan, R. and Katz, A.E. (2001) Therapeutic potential of curcumin in human prostate cancer. III. Curcumin inhibits proliferation, induces apoptosis and inhibits angiogenesis of PC3 prostate cancer cells in vivo. *Prostate* 47, 293–303.

Escarpa, A. and Gonzalez, M.C. (2001) An overview of analytical chemistry of phenolic compounds in foods. *Critical Reviews in Analytical Chemistry* 3, 57–139.

Ferguson, L.R., Lim, I.F., Pearson, A.E., Ralph, J. and Harris, P.J. (2003) Bacterial antimutagenesis by hydroxycinnamic acids from plant cell walls. *Mutation Research* 542, 49–58.

Friedman, M. (1997) Chemistry, biochemistry, and dietary role of potato polyphenols. A review. *Journal of Agricultural and Food Chemistry* 45, 1523–1540.

Galvez, M.C., Barroso, C.G. and Perez-Bustamante, J.A. (1994) Analysis of polyphenolic compounds of different vinegar samples. *Z Lebensm Unters Forsch A* 199, 29–31.

Gerhauser, C., Klimo, K., Heiss, E., Neumann, I., Gamal-Eldeen, A., Knauft, J., Liu, G.Y., Sitthimonchai, S. and Frank, N. (2003) Mechanism-based *in vitro* screening of potential cancer chemopreventive agents. *Mutation Research* 523–524, 163–172.

Ghosh, D. and Konishi, T. (2007) Anthocyanins and anthocyanin-rich extracts: role in diabetes and eye function. *Asia Pacific Journal of Clinical Nutrition* 16, 200–208.

Graf, E. (1992) Antioxidant potential of ferulic acid. *Free Radical Biology and Medicine* 13, 435–448.

Gulcin, I. (2006) Antioxidant activity of caffeic acid (3,4-dihydroxycinnamic acid). *Toxicology* 217, 213–220.

Gusman, J., Malonne, H. and Atassi, G. (2001) A reappraisal of the potential chemo-preventive and chemotherapeutic properties of resveratrol. *Carcinogenesis* 22, 1111–1117.

Hasima, N. and Aggarwal, B.B. (2012) Cancer-linked targets modulated by curcumin. *International Journal of Biochemistry and Molecular Biology* 3, 328–351.

He, M., Zhao, L., Wet, M.J., Yao, W.F., Zhao, H.S. and Chen, F.J. (2009) Neuroprotective effects of (-)- epigallocatechin-3-gallate on aging mice induced by D-galactose. *Biological and Pharmaceutical Bulletin* 32, 55–60.

Heber, D., Berdanier, C.D., Dwyer, J.T. and Feldman, E.B. (eds) (2008) *Plant Foods and Phytochemicals in Human Health*. CRC Press, Boca Raton, Florida, pp. 176–181.

Hidaka, H., Ishiko, T., Furuhashi, T., Kamohara, H., Suzuki, S., Miyazaki, M., Ikeda, O., Mita, S., Setoguchi, T. and Ogawa, M. (2002) Curcumin inhibits interleukin 8 production and enhances interleukin 8 receptor expression on the cell surface impact on human pancreatic carcinoma cell growth by autocrine regulation. *Cancer* 95, 1206–1214.

Holasova, M., Fiedlerova, V., Smrcinova, H., Orsak, M., Lachman, J. and Vavreinova, S. (2002) Buckwheat – the source of antioxidant activity in functional foods. *Food Research International* 35, 207–211.

Hynes, M.J. and O'Coinceanainn, M. (2004) The kinetics and mechanisms of reactions of iron(III) with caffeic acid, chlorogenic acid, sinapic acid, ferulic acid and naringin. *Journal of Inorganic Biochemistry* 98, 1457–1464.

Jang, J.H. and Surh, Y.J. (2001) Protective effects of resveratrol on hydrogen peroxide-induced apoptosis in rat phenochromocytoma. *Mutation Research* 496, 181–190.

Jang, M.S., Cai, E.N. and Udeani, G.O. (1997) Cancer chemopreventive activity of resveratrol, a natural product derived from grapes. *Science* 275, 218–220.

Kaldas, M.I., Walle, U.K. and Walle, T. (2003) Resveratrol transport and metabolism by human intestinal CaCo-2 cells. *Journal of Pharmacy and Pharmacology* 55, 307–312.

Kammerer, D., Claus, A., Carle, R. and Scheiber, A. (2004) Polyphenol screening of pomace from red and white grape varieties (*Vitis vinifera* L.) by HPLC-DAD-MS/MS. *Journal of Agricultural and Food Chemistry* 52, 4360–4367.

Khalatbary, A.R., Tiraihi, T., Beigi, B.M., Ahmadvand, H., Tavafi, M. and Tamjidipoor, A. (2010) Effects of epigallocatechin gallate on tissue protection and functional recovery after contusive spinal cord injury in rats. *Brain Research* 1306, 168–175.

Kikuzaki, H., Hisamoto, M., Hirose, K., Akiyama, K. and Taniguchi, H. (2002) Antioxidant properties of ferulic acid and its related compounds. *Journal of Agricultural and Food Chemistry* 50, 2161–2168.

Kim, K.H., Tsao, R., Yang, R. and Cui, S.W. (2006) Phenolic acid profiles and antioxidant activities of wheat bran extracts and the effect of hydrolysis conditions. *Food Chemistry* 95, 466–473.

Kim, Y.M., Yun, J., Lee, C.K., Lee, H., Min, K.R. and Kim, Y. (2002) Oxyresveratrol and hydroxystilbene compounds. Inhibitory effect on tyrosinase and mechanism of action. *Journal of Biological Chemistry* 277, 16340–16344.

King, P.J., Ma, G., Miao, W., Jia, Q., McDoughall, B.R., Reinecke, M.G., Cornell, C., Kuan, J., Kim, T.R. and Robinson, W.E. Jr (1999) Structure-activity relationships: analogues of the dicaffeoylquinic and dicaffeoyltartaric acids as potent inhibitors of human immunodeficiency virus type 1 integrase and replication. *Journal of Medicinal Chemistry* 42, 497–509.

King, R.E., Kent, K.D. and Bomser, J.A. (2005) Resveratrol reduces oxidation and proliferation of human retinal pigment epithelial cells via extracellular signal-regulated kinase inhibition. *Chemico Biological Interactions* 151, 143–149.

Knekt, P., Kumpulainen, J., Jarvinen, R., Rissanen, H., Heliovaara, M., Reunanen, A., Hakulinen, T. and Aromaa, A. (2002) Flavonoid intake and risk of chronic diseases. *American Journal of Clinical Nutrition* 76, 560–568.

Konatham, S., Kumar, P. and Aukunuru, J. (2010) Synthesis and screening of antidiabetic activity of some novel Curcumin analogues. *International Journal of Pharma and Bio Sciences* 1, 1–12.

Kondratyuk, T.P. and Pezzuto, J.M. (2004) Natural product polyphenols of relevance to human health. *Pharmaceutical Biology* 42, 46–63.

Kopsell, D.A., Kopsell, D.E., Lefsrud, M.G., Curran-Celentano, J. and Dukach, L.E. (2004) Variation in lutein, (beta) β-carotene, and chlorophyll concentrations among *Brassica oleracea* cultigens and seasons. *Horticultural Science* 39, 361–364.

Koshihara, Y., Neichi, T., Murota, S., Lao, A., Fujimoto, Y. and Tatsuno, T. (1984) Caffeic acid is a selective inhibitor for leukotriene biosynthesis. *Biochimica et Biophysica Acta* 792, 92–97.

Kowalski, J., Samojedny, A., Paul, M., Pietsz, G. and Wilczok, T. (2005) Effect of apigenin, kaempferol and resveratrol on the expression of interleukin-1 beta and tumor necrosis factor-alpha genes in J774.2 macrophages. *Pharmacological Reports* 57, 390–394.

Kraukauer, T. (2002) The polyphenol chlorogenic acid inhibits Staphylococcal exotoxin-induced inflammatory cytokines and chemokines. *Immunopharmacology and Immunotoxicology* 24, 113–119.

Kris-etherton, P., Hecker, K., Bonanome, A., Coval, S., Binkoski, A. and Hilpert, K. (2002) Bioactive compounds in foods: their role in the prevention of cardiovascular disease and cancer. *American Journal of Medicine* 113, 71S–78S.

Kundu, J.K., Chun, K.S., Kim, S.O. and Surh, Y.J. (2004) Resveratrol inhibits phorbol ester-induced cyclooxygenase-2 expression in mouse skin: MAPKs and AP-1 as potential molecular targets. *BioFactors* 21, 33–39.

Kutuk, O., Adli, M., Poli, G. and Basaga, H. (2004) Resveratrol protects against 4-HNE induced oxidative stress and apoptosis in Swiss 3T3 fibroblasts. *BioFactors* 20, 1–10.

Labinskyy, N., Csiszar, A., Veress, G., Stef, G., Pacher, P., Oroszi, G., Wu, J. and Ungvari, Z. (2006) Vascular dysfunction in aging: potential effects of resveratrol, an anti-inflammatory phytoestrogen. *Current Medicinal Chemistry* 13, 989–996.

Laranjinha, J., Vierira, O., Almeida, L. and Madeira, V. (1996) Inhibition of metmyoglobin/H_2O_2-dependent low density lipoprotein lipid peroxidation by naturally occurring phenolic acids. *Biochemical Pharmacology* 51, 395–402.

Larrosa, M., Tomas-Barberan, F.A. and Espın, J.C. (2004) The grape and wine polyphenol piceatannol is a potent inducer of apoptosis in human SK-Mel-28 melanoma cells. *European Journal of Nutrition* 43, 275–284.

Lee, J.C., Kinniry, P.A., Arguiri, E., Serota, M., Kanterakis, S., Chatterjee, S., Solomides, C.C., Javvadi, P., Koumenis, C. and Cengel, K.A. (2010) Dietary curcumin increases antioxidant defenses in lung, ameliorates radiation-induced pulmonary fibrosis, and improves survival in mice. *Radiation Research* 173, 590–601.

Leiro, J., Alvarez, E., Arranz, J.A., Laguna, R., Uriarte, E. and Orallo, F. (2004) Effects of *cis*-resveratrol on inflammatory murine macrophages: antioxidant activity and down regulation of inflammatory genes. *Journal of Leukocyte Biology* 75, 1156–1165.

Li, Y. and Sarkar, F.H. (2002) Inhibition of nuclear factor kappa-B activation in PC3 cells by genistein is mediated via Akt signaling pathway. *Clinical Cancer Research* 7, 2369–2377.

Litvinenko, V.I., Popova, T.P., Simonjan, A.V., Zoz, I.G. and Sokolov, V.S. (1975) Tannins and derivatives of hydroxycinnamic acid in Labiatae. *Plant Medicine* 27, 372–380.

Liu, L.Z., Fang, J., Zhou, Q., Hu, X., Shi, X. and Jiang, B.H. (2005) Apigenin inhibits expression of vascular endothelial growth factor and angiogenesis in human lung cancer cells: implication of chemoprevention of lung cancer. *Molecular Pharmacology* 68, 635–643.

Liu, Z.P., Li, W.X., Yu, B., Huang, J., Sun, J., Huo, J.S. and Liu, C.X. (2005) Effects of trans-resveratrol from *Polygonum cuspidatum* on bone loss using the ovariectomized rat model. *Journal of Medicinal Food* 8, 14–19.

Lu, Y. and Foo, L.Y. (2000) Antioxidant and radical scavenging activities of polyphenols from apple pomace. *Food Chemistry* 68, 81–85.

Maatta-Riihinen, K.R., Kamal-Eldin, A., Mattila, P.H., Paramas, G. and Torronen, R. (2004) Distribution and contents of phenolic compounds in eighteen Scandinavian berry species. *Journal of Agricultural and Food Chemistry* 52, 4477–4486.

Mackowiak, P.A. (2000) Brief history of antipyretic therapy. *Clinical Infectious Diseases* 31, 154–156.

Manach, C., Scalbert, A., Morand, C., Remesy, C. and Jimenez, L. (2004) Polyphenols: food sources and bioavailability. *American Journal of Clinical Nutrition* 79, 727–747.

Manjunatha, H. and Srinivasan, K. (2007) Hypolipidemic and antioxidant effects of dietary curcumin and capsaicin in induced hypercholesterolemic rats. *Lipids* 42, 1133–1142.

Marotti, M. and Piccaglia, R. (2002) Characterization of flavonoids in different cultivars of onion (*Allium cepa*). *Journal of Food Science* 67, 1229–1232.

Materska, M. and Perucka, I. (2005) Antioxidant activity of the main phenolic compounds isolated from hot pepper fruit (*Capsicum annuum* L.) *Journal of Agricultural Food and Chemistry* 53, 1750–1756.

Mattila, P. and Hellstrom, J. (2007) Phenolic acids in potatoes, vegetables, and some of their products. *Journal of Food Composition and Analysis* 20, 152–160.

Mimche, P.N. (2010) Modulation of *Plasmodium falciparum* phagocytosis, inflammatory cytokines production and cytoadhesion molecules expression by the natural product curcumin. PhD thesis. EMBL/University of Milan-LSHTM.

Montanari, A., Chen, J. and Widmer, W. (1998) Citrus flavonoids: a review of past biological activity against disease. Discovery of new flavonoids from Dancy tangerine cold pressed peel oil solids and leaves. In: Manthey, J. and Buslig, B. (eds) *Flavonoids in the Living System*. Plenum, New York, pp. 103–116.

Murota, K. and Terao, J. (2003) Antioxidative flavonoid quercetin: implications of its intestinal absorption and metabolism. *Archives of Biochemistry and Biophysics* 417, 12–17.

Murota, S. and Koshihara, Y. (1985) New lipoxygenase inhibitors isolated from Chinese plants. Development of new anti-allergic drugs. *Drugs and Experimental Clinical Research* 11, 641–644.

Nichenametla, S.N., Taruscio, T.G., Barney, D.L. and Exon, J.H. (2006) A review of the effects and mechanism of polyphenolics in cancer. *Critical Reviews in Food Science and Nutrition* 46, 161–183.

Nichols, J.A. and Katiyar, S.K. (2010) Skin photoprotection by natural polyphenols: anti-inflammatory, antioxidant and DNA repair mechanisms. *Archives of Dermatological Research* 302, 71–83.

Ninfali, P., Aluigi, G., Bacchiocca, M. and Magnani, M. (2001) Antioxidant capacity of extra-virgin olive oil. *Journal of the American Oil Chemists' Society* 78, 243–247.

Nishiyama, T., Mae, T., Kishida, H., Tsukagawa, M., Mimaki, Y., Kuroda, M., Sashida, Y., Takahashi, K., Kawada, T., Nakagawa, K. and Kitahara, M. (2005) Curcuminoids and sesquiterpenoids in turmeric (*Curcuma longa* L) suppress and increase in blood glucose level in type 2 diabetic KK-Ay mice. *Journal of Agricultural and Food Chemistry* 53, 959–963.

Pendurthi, U.R., Williams, J.T. and Rao, L.V. (1999) Resveratrol, a polyphenolic compound found in wine, inhibits tissue factor expression in vascular cells: A possible mechanism for the cardiovascular benefits associated with moderate consumption of wine. *Arteriosclerosis, Thrombosis and Vascular Biology* 19, 419–426.

Pérez-Arriaga, L., Mendoza-Magana, M.L., Cortes-Zarate, R., Corona-Rivera, A., Bobadilla-Morales, L., Troyo-Sanroman, R. and Ramirez-Herrera, M.A. (2006) Cytotoxic effects of curcumin on *Giardia lamblia* trophozoites. *Acta Tropica* 98, 152–161.

Peterson, D.M., Emmons, C.L. and Hibbs, A.H. (2001) Phenolic antioxidants and antioxidant activity in pearling fractions of oat groats. *Journal of Cereal Science* 33, 97–103.

Pozo-Guisado, E., Alvarez-Barrientos, A., Mulero-Navarro, S., Santiago-Josefat, B. and Fernandez-Salguero, P.M. (2002) The anti-proliferative activity of resveratrol results in apoptosis in MCF-7 but not in MDAMB- 231 human breast cancer cells: cell-specific alteration of the cell cycle. *Biochemical Pharmacology* 64, 1375–1386.

Prakash, D. and Gupta, K.R. (2009) The antioxidant phytochemicals of nutraceutical importance. *The Open Nutraceuticals Journal* 2, 20–35.

Prakash, D., Upadhyay, G., Singh, B.N. and Singh, H.B. (2007) Antioxidant and free radical scavenging activities of seeds and agri-waste of some varieties of soybean (*Glycine max*). *Food Chemistry* 104, 783–790.

Psotova, J., Lasovsky, J. and Vicar, J. (2003) Metal-chelating properties, electrochemical behavior, scavenging and cytoprotective activities of six natural phenolics. *Biomedical Papers of Medical Faculty of University of Palacky Olomouc Czech Republic* 147, 147–153.

Qin, Y., Xia, M., Ma, J., Hao, Y., Liu, J., Mou, H., Cao, L. and Ling, W. (2009) Anthocyanin supplementation improves serum LDL- and HDL-cholesterol concentrations associated with the inhibition of cholesteryl ester transfer protein in dyslipidemic subjects. *American Journal of Clinical Nutrition* 90, 485–492.

Rahman, I., Biswas, S.K. and Kirkham, P.A. (2006) Regulation of inflammation and redox signaling by dietary polyphenols. *Biochemical Pharmacology* 72, 1439–1452.

Ramachandran, C., Fonseca, H.B., Jhabvala, P., Escalon, E.A. and Melnick, S.J. (2002) Curcumin inhibits telomerase activity through human telomerase reverse transcriptase in MCF-7 breast cancer cell line. *Cancer Letters* 184, 1–6.

Ramadan, A.M., Nadia, H.M. and Dalia, D. (2012) Curcumin reduced potato chips and roasted bread induced chromosomal aberrations and micronuclei formation in albino rats. *Life Science Journal* 9, 330–336.

Reyes-Carmona, J., Youseg, G.G., Martinez-Peniche, R.A. and Lila, M.A. (2005) Antioxidant capacity of fruit extracts of blackberry (*Rubus* sp.) produced in different climatic regions. *Journal of Food Science* 70, 497–503.

Rice-Evans, C.A., Miller, N.J. and Paganga, G. (1996) Structure-antioxidant activity relationships of flavonoids and phenolic acids. *Free Radical Biology and Medicine* 20, 933–956.

Rios, L.Y., Gonthier, M.P., Remesy, C., Mila, I., Lapierre, C., Lazarus, S.A., Williamson, G. and Scalbert, A. (2003) Chocolate intake increases urinary excretion of polyphenol derived phenolic acids in healthy human subjects. *American Journal of Clinical Nutrition* 77, 912–918.

Riva, D.A., Fernandez-Larrosa, P.N., Dolcini, G.L., Martinez-Peralta, L.A., Coulombie, F.C. and Mersich, S.E. (2008) Two immunomodulators, curcumin and sulfasalazine, enhance IDV antiretroviral activity in HIV-1 persistently infected cells. *Archives of Virology* 153, 561–565.

Rolfs, C.H. and Kindl, H. (1984) Stilbene synthase and chalcone synthase-two different constitutive enzymes in cultured cells of *Picea exelsa*. *Plant Physiology* 75, 489–492.

Rupasinghe, V.H.P. and Clegg, S. (2007) Total antioxidant capacity, total phenolic content, mineral elements, and histamine concentrations in wine of different fruit sources. *Journal of Food Composition and Analysis* 20, 133–137.

Saija, A., Tomaino, A., Trombetta, D., de Pasquale, A., Uccella, N., Barbuzzi, T., Paolino, D. and Bonina, F. (2000) *In vitro* and *in vivo* evaluation of caffeic and ferulic acids as topical photo-protective agents. *International Journal of Pharmaceutics* 199, 39–47.

Saito, M., Hosoyoma, H., Ariga, T., Kataoka, S. and Yamaji, N. (1998) Antiulcer activity of grape seed extract and procyanidins. *Journal of Agricultural Food and Chemistry* 46, 1450–1464.

Scalbert, A. and Williamson, G. (2000) Dietary intake and bioavailability of polyphenols. *Journal of Nutrition* 130, 2073S–2085S.

Scalbert, A., Manach, C., Morand, C. and Remesy, C. (2005) Dietary polyphenols and the prevention of diseases. *Critical Reviews in Food Science and Nutrition* 45, 287–306.

Shahidi, F. and Wanasundara, J.P.D. (1992) Phenolic antioxidants. *Critical Reviews in Food Science and Nutrition* 32, 67–103.

Shahidi, F., Amarowicz, R., He, Y. and Wettasinghe, M. (1997) Antioxidant activity of phenolic extracts of evening primrose (*Oenothera bienis*). A preliminary study. *Journal of Food Lipids* 4, 75–86.

Sharma, R.D. (1979) Isoflavones and hypercholesterolemia in rats. *Lipids* 14, 535–539.

Shimizu, N., Naoe, T., Kawazoe, Y., Sakagami, H., Nakashima, H., Murakami, T. and Yamamoto, N. (1993) Lignified materials as medicinal resources. VI. Anti-HIV activity of dehydrogenation polymer of p-coumaric acid, a synthetic lignin, in a quasi-in-vivo assay system as an intermediary step to clinical trials. *Biological and Pharmaceutical Bulletin* 16, 434–436.

Shukla, Y., Arora, A. and Taneja, P. (2003) Antigenotoxic potential of certain dietary constituents. *Teratogenesis, Carcinogenesis and Mutagenesis* 23, 323–335.

Siegelin, M.D., Reuss, D.E., Habel, A., Rami, A. and von Deimling, A. (2009) Quercetin promotes degradation of survivin and thereby enhances death-receptor-mediated apoptosis in glioma cells. *Neuro-Oncology* 11, 122–131.

Sikora, E., Ewa, C. and Kinga, T. (2008) The sources of natural antioxidants. *Acta Scientiarum Polonorum Technologia Alimentaria* 7, 5–17.

Smith, M.R., Gangireddy, S.R., Narala, V.R., Hogaboam, C.M., Standiford, T.J., Christensen, P.J., Kondapi, A.K. and Reddy, R.C. (2010) Curcumin inhibits fibrosis-related effects in IPF fibroblasts and in mice following bleomycin-induced lung injury. *American Journal of Physiology – Lung Cellular and Molecular Physiology* 298, 616–625.

Sobolev, V.S., Horn, B.W., Potter, T.L., Deyrup, S.T. and Gloer, J.B. (2006) Production of stilbenoids and phenolic acids by the peanut plant at early stages of growth. *Journal of Agricultural and Food Chemistry* 54, 3505–3511.

Srinivasan, M., Sudheer, A.R. and Menon, V.P. (2007) Ferulic Acid: therapeutic potential through its antioxidant property. *Journal of Clinical Biochemistry and Nutrition* 40, 92–100.

Steele, K., Shirodaria, P., O'Hare, M., Merrett, J.D., Irwin, W.G., Simpson, D.I.H. and Pfister, H. (1988) Monochloroacetic acid and 60% salicylic acid as a treatment for simple plantar warts: effectiveness and mode of action. *British Journal of Dermatology* 118, 537–543.

Stewart, A.J., Bozonnet, S., Mullen, W., Jenkins, G.I., Lean, M.E. and Crozier, A. (2000) Occurence of flavonols in tomatoes and tomato-based products. *Journal of Agricultural Food and Chemistry* 48, 2663–2669.

Tan, S.C. (2000) Determinants of eating quality in fruits and vegetables. *Proceedings of the Nutrition Society of Australia* 24, 183–190.

Tang, F.Y., Nguyen, N. and Meydani, M. (2003) Green tea catechins inhibit VEGF-induced angiogenesis *in vitro* through suppression of VE-cadherin phosphorylation and inactivation of Akt molecule. *International Journal of Cancer* 106, 871–878.

Tiwari, A.K. (2001) Imbalance in antioxidant defense and human diseases: multiple approach of natural antioxidants therapy. *Current Science* 81, 1179–1187.

Trock, B.J., Hilakivi-Clarke, L. and Clarke, R. (2006) Meta-analysis of soy intake and breast cancer risk. *Journal of National Cancer Institute* 98, 459–471.

Trombino, S., Serini, S., di Nicuolo, F., Celleno, L., Ando, S., Picci, N., Calviello, G. and Palozza, P. (2004) Antioxidant effect of ferulic acid in isolated membranes and intact cells: synergistic interactions with alpha-tocopherol, beta-carotene, and ascorbic acid. *Journal of Agricultural and Food Chemistry* 52, 2411–2420.

Tseng, S.H., Lin, S.M., Chen, J.C., Su, Y.H., Huang, H.Y., Chen, C.K., Lin, P.Y. and Chen, Y. (2004) Resveratrol suppresses the angiogenesis and tumor growth of gliomas in rats. *Clinical Cancer Research* 10, 2190–2202.

Wargovich, M.J., Jimenez, A., McKee, K., Steele, V.E., Velasco, M., Woods, J., Price, R., Gray, K. and Kelloff, G.J. (2000) Efficacy of potential chemopreventive agents on rat colon aberrant crypt formation and progression. *Carcinogenesis* 21, 1149–1155.

Watjen, W., Michels, G., Steffan, B., Niering, P., Chovolou, Y., Kampkotter, A., Tran-Thi, Q.H. and Kahl, P.P. (2005) Low concentrations of flavonoids are protective in rat H4IIE cells whereas high concentrations cause DNA damage and apoptosis. *Journal of Nutrition* 135, 525–531.

Weisberg, S.P., Leibel, R. and Tortoriello, D.V. (2008) Dietary curcumin significantly improves obesity-associated inflammation and diabetes in mouse models of diabesity. *Endocrinology* 149, 3549–3558.

Wenk, G.L., McGann, G.K., Hauss-Wegrzyniak, B., Ronchetti, D., Maucci, R., Rosi, S., Gasparini, L. and Ongini, E. (2004) Attenuation of chronic neuroinflammation by a nitric oxide-releasing derivative of the antioxidant ferulic acid. *Journal of Neurochemistry* 89, 484–493.

Wilken, R., Veena, M.S., Wang, M.B. and Srivatsan, E.S. (2011) Curcumin: A review of anti-cancer properties and therapeutic activity in head and neck squamous cell carcinoma. *Molecular Cancer* 10, 12.

Wolter, F., Clausnitzer, A., Akoglu, B. and Stein, J. (2002) Piceatannol, a natural analog of resveratrol, inhibits progression through the Sphase of the cell cycle in colorectal cancer cell lines. *Journal of Nutrition* 132, 298–302.

Wu, S., Lin, Y., Chu, C., Tsai, Y. and Chao, J.C. (2008) Curcumin or Saikosaponin a Improves Hepatic Antioxidant Capacity and Protects Against CCl4-Induced Liver Injury in Rats. *Journal of Medicinal Food* 11, 224–229.

Xia, X., Ling, W., Ma, J., Xia, M., Hou, M., Wang, Q., Zhu, H. and Tang, Z. (2006) An anthocyanin-rich extract from black rice enhances atherosclerotic plaque stabilization in apolipoprotein E-deficient mice. *Journal of Nutrition* 136, 2220–2225.

Xiao, J., Chen, T., Cao, H., Chen, L. and Yang, F. (2010) Molecular Property-Affinity Relationship of Flavanoids and Flavonoids for Human Serum Albumin *In Vitro*. *Molecular Nutrition and Food Research* 55, 310–317.

Yoshida, H., Takamura, N., Shuto, T., Ogata, K., Tokunaga, J., Kawai, K. and Kai, H. (2010) The citrus flavonoids herceptin and naringenin block the lipolytic actions of TNF-α in mouse adipocytes. *Biochemical and Biophysical Research Communications* 394, 728–732.

Zafra-Stone, S., Yasmin, T., Bagchi, M., Chatterjee, A., Vinson, J.A. and Bagchi, D. (2007) Berry anthocyanins as novel antioxidants in human health and disease prevention. *Molecular Nutrition and Food Research* 51, 675–683.

Zhang, D., Huang, C., Yang, C., Liu, R.J., Wang, J., Niu, J. and Bromme, D. (2011) Antifibrotic effects of curcumin are associated with overexpression of cathepsins K and L in bleomycin treated mice and human fibroblasts. *Respiratory Research* 12, 154.

Zingg, J.M., Hasan, S.T., Cowan, D., Ricciarelli, R., Azzi, A. and Meydani, M. (2012) Regulatory effects of curcumin on lipid accumulation in monocytes/macrophages. *Journal of Cellular Biochemistry* 113, 833–840.

15 Antioxidant Phytochemicals in Cancer Chemoprevention

Narendra Singh,[1] Dhanir Tailor,[1] Raosaheb K. Kale[1,2] and Rana P. Singh[1,2*]

[1]School of Life Sciences, Central University of Gujarat, Gandhinagar, India; [2]School of Life Sciences, Jawaharlal Nehru University, New Delhi, India

15.1 Introduction

Cancer is one of the leading causes of death around the world. In 2008, 13 million people lost their lives due to cancer (Borek, 2004). According to the WHO, 'cancer is an uncontrolled growth and spread of cells which can affect almost any part of the body. A cancer cell often invades surrounding tissues and can metastasize to distant sites'. Many factors such as smoking, chemicals, hereditary inheritance, alcohol consumption, radiation and life style contribute to cancer development. Available treatments for cancer include removal by surgery, radiotherapy, chemotherapy, immunotherapy, gene therapy and bone marrow transplantation. These approaches of treatment are widely used but they have shown limitations and ineffectiveness at several points of time. To overcome some limitations like cytotoxicity to normal cells and other side effects, the use of dietary phytochemicals/antioxidants have been utilized to prevent cancer development. They have been a part of household and traditional practices. Recently, these phytochemicals have caught the attention of many investigators because of their relatively fewer side effects, natural production, affordability and easy availability.

People have used plant products for preventive or therapeutic purposes to prevent or treat various abnormalities and diseases from ancient times. The normal diet includes many components that have various properties such as chemopreventive and anticancer activities. These plant products, also known as 'phytochemicals', have been used successfully for the management of many diseases, including cancer. Antioxidant phytochemicals prevent cell damage resulting from free radicals, which are the major events in the initiation of carcinogenesis. Large numbers of animal studies have reported the role of antioxidants as anticancer agents. Thus, the higher intake of antioxidants in diet may be a good approach to protect us from many deadly diseases. Many of these phytochemicals interact with the signal transduction pathways, alter hormonal/growth factor activities by activation/inhibition of regulatory genes, restore the immune system, induce terminal differentiation and apoptosis, inhibit angiogenesis and inflammation (Manson, 2003; Pan and Ho, 2008). The plant polyphenolics, which form a major group of dietary components, function as regulators of detoxifying enzymes including haem-oxygenase, GST, NADH, quinine oxidoreductase and manganese superoxide dismutase (MnSOD).

* E-mail: rana_singh@mail.jnu.ac.in; ranaps@hotmail.com

Recently, it has been reported that the antioxidant response elements (ARE) and electrophile response elements (EpRe) regulate the expression of genes such as cyclooxygenase 2 (COX2) and apo-lipoprotein A-1. Many pathways including MAPK, PI3K, JNK and PKC can lead to the activation of ARE/EpRe. Hence, it might be possible that these phytochemicals would be able to regulate these response elements by modulating those pathways.

Phytochemicals can act at multiple levels of carcinogenesis by inhibition of one or more than one step of initiation, promotion and progression of the cancer. This may be brought about by induction of cell cycle arrest, and inhibition of proliferation, inflammation, angiogenesis and metastasis, and induction of apoptosis, epigenetic changes and regulation of biochemical pathways. They arrest cell cycle progression via controlling the expression of cyclins and cyclin-dependent kinases (CDKs) or through CDK inhibitors (CDKI). For example, in prostate cancer treatment, silymarin causes cell cycle arrest at G1 phase by inducing the expression of CDKI Cip1/p21 and Kip1/p27. These CDKIs inhibit the CDK2 and CDK4 and related kinase activities (Deep and Agarwal, 2007). Genistein arrests the cell cycle at G2/M phase by targeting the activity of CDC2 and cyclin B1 activity in both prostate and lung cancer (Banerjee *et al.*, 2008). The treatment of prostate cancer cell lines by IP6 causes the hypophosphorylation of retinoblastoma (Rb) protein at serine 780, 807 and 811 sites or Rb-related proteins and enhances the interaction of E2F1, E2F4, and E2F5 with Rb or Rb-related protein, and these transcription factors became unavailable for transcription of genes (Agarwal *et al.*, 2004). Epigallocatechin-gallate (EGCG) and curcumin promote apoptosis by activation of caspase 3, caspase 8, caspase 9 and by decreasing the expression of anti-apoptotic compounds like Bcl2, both of which increase the cleavage of PARP leading to cell death (Khan *et al.*, 2009; Shehzad *et al.*, 2010). Inhibition of angiogenesis and metastasis are promoted by resveratrol through decrease in the expression of VEGF, MMP-2, MMP-9, ERK1/2 and HIF-1α (Bishayee, 2009). Carnosol shows anti-inflammatory action through suppression of iNOS, which leads to the inhibition of NF-κB, p38, ERK1/2; apart from this carnosol hinders the protein kinase C signalling pathway and prevents binding of AP1 to COX-2 promoter (Johnson, 2011). Phytochemicals can also inhibit the growth of a tumour by hampering the different biochemical pathways, including NF-κB, Akt, PI3K, MAPK, Ras and GFR (Ramos, 2008). EGCG shows its effect on the epigenetic regulation through suppression of the expression of hTERT (Berletch *et al.*, 2008), and increases the methylation of p15 (INK), p16 (INK) and oestrogen receptor-I (Berner *et al.*, 2010).

However, most of the studies on phytochemicals have been focused on their antioxidant and anticancer activities, but the major concerns include lower efficacy, potency, specificity, bioavailability and solubility in water (Siddiqui *et al.*, 2010). Many phytochemicals have been shown to induce cytotoxic effects on normal cells at higher concentrations; after metabolism in the liver they become more toxic, interact with other dietary components, become less biocompatible, have a destructive effect on intestinal microflora and also induce cell survival pathways, which may promote tumour growth rather than inhibiting it. Apart from this, most of the dietary phytochemicals are present in conjugated forms or are effective only when present with other plant components, which may have synergistic effects. The properties of phytochemicals have been tested mostly on cancer cell lines or other *in vitro* systems and *in vivo* in animals, which are given a high dose of these phytochemicals/antioxidants, which normally does not happen in normal human diet. The administration timing also differs between experimental and normal conditions. Exposure of carcinogens and treatment with phytochemicals is usually performed simultaneously in experimental conditions, but it does not happen in normal human situation, where carcinogens may enter the system before the phytochemicals are administered (Hodek *et al.*, 2009).

15.2 Antioxidants as Anticancer Agents

Higher intake of fruits and vegetables is believed to have a beneficial effect on our health; the fruits and vegetables have dietary polyphenols, which are known to act as antioxidant and anticancer agents and, along with these activities, they also show activity against cardiovascular and neurodegenerative diseases (Surh, 2003; Ouédraogo *et al.*, 2011). Antioxidants are molecules formed through oxidation that may prevent cells from free radical damage in our body. Oxidation is a reaction in which oxygen is transferred to a higher unstable ionization state from molecules such as proteins, DNA, lipids and carbohydrates. These free radicals initiate many chain reactions, resulting in cell damage or cell death (Sies, 1997). Antioxidants inhibit these chain reactions thus preventing cells from damage; they prevent oxidation reactions by oxidizing themselves, so they are also known as reducing agents (Sies, 1997; Hurrell, 2003). Oxidation reactions are critical to human life; plants and animals maintain very complicated systems with antioxidants/phytochemicals, including vitamin E, vitamin C and glutathione and with many other enzymes such as superoxide dismutase (SOD), catalase and peroxidases. Antioxidants have been studied extensively for pharmaceutics, particularly in heart stroke and neurodegenerative diseases (Bjelakovic *et al.*, 2007). Reactive oxygen species (ROS) are a by-product of normal oxygen metabolism, which is important in cell signal transduction and homeostasis. During environmental stress such as UV and heat exposure, the ROS level increases; it leads to severe damage to the cells, which is cumulatively known as oxidative stress. ROS can also be generated by exogenous sources such as ionization radiations (Devasagayam *et al.*, 2004). Usually cells try to defend against oxidizing agents/ROS with the help of several enzymes and other small molecules, which act as antioxidants, such as ascorbic acid, glutathione, uric acid, carotenoids, polyphenols and tocopherols. ROS are not only involved in apoptosis but also have some positive effects, such as induction of the defence system and mobilization of the ion transport systems. Thus, they are also known as redox signalling or oxidative signalling. Platelets, which have been found to be involved in wound healing and blood homeostasis, release ROS to recruit more platelets. The immune system is also believed to be linked with ROS in order to recruit leucocytes (Conner *et al.*, 2002; Rada and Leto, 2008). ROS has been associated with many inflammatory responses, cardiovascular diseases and hearing impairment, and has also been implicated in apoptosis, specifically in stroke and heart attack. General oxidation results in DNA damage, oxidation of proteins and oxidation of the polyunsaturated fatty acid of lipids and oxidation of enzymes/cofactors (Bergström *et al.*, 2012).

15.3 Sources of Antioxidant Phytochemicals

The distribution of antioxidants in nature is very diverse and ranges from dietary to non-dietary compounds. Antioxidants are found in large amount in fruits and vegetables, including agricultural by-products, herbal tea, coffee, some beverages, cold-pressed vegetable oil, table olives, sesame seed, nuts and grains, which are particularly rich in polyphenolic compounds (Surh, 2003). Major component of polyphenols are flavonoids, and total polyphenolic intake is estimated as 1 g day^{-1} from fruit juice, green tea, coffee, chocolate and beer and to a lesser extent from dry legumes, cereals and vegetables (Scalbert and Williamson, 2000; Ramos, 2007). Epidemiological studies have shown that people who are more dependent on plant products for diet have a higher intake of polyphenolic compounds. Several *in vitro* and *in vivo* studies have shown the potential of polyphenols against many diseases including cancer. Notable anticancer polyphenolic examples are resveratrol, green tea catechins, curcumin, silibinin and genistein (Scalbert and Williamson, 2000; Ramos, 2007; Vauzour *et al.*, 2010).

15.4 Mode of Action

15.4.1 ROS-independent action

There are several small molecules or antioxidants that show their anticancer activity independent of ROS mechanism, including EGCG, silibinin, curcumin, gallic acid and resveratrol. Chemotherapy in combination with these antioxidants increases the beneficial effect of the therapies (Borek, 2004). Antioxidants possess a scavenger effect, which plays important roles in cancer prevention. They can alter cancer signal transduction pathways: modulate antioxidant enzyme activities, such as up-regulation of GST, NQO1, catalase, GPx and/or phase II enzymes. In addition, they show cell cycle arrest and induction of apoptosis (Pan and Ho, 2008; Vauzour *et al.*, 2010). Polyphenolic antioxidants also modulate Nrf2 and NF-κB pathways. MAPK and PI3K signalling pathways are important targets in anticancer approaches, because many cancers become dependent for growth and survival on these pathways (Surh, 2008). Studies have shown that polyphenols can modulate MAPK and PI3K signalling pathways. However, the dietary polyphenols at different concentrations may show an opposite effect, e.g. curcumin, EGCG, quercetin and green tea at low concentration may activate the MAPK, c-Jun N-terminal kinase (JNK), and can lead to the activation of survival genes such as *c-fos* and *c-jun* (Yu *et al.*, 1997); on the other hand at higher concentrations quercetin and EGCG can activate caspase-dependent apoptosis (Spencer *et al.*, 2003).

Apoptosis (programmed cell death) also plays an important role in cancer prevention. It includes intrinsic and extrinsic pathways. The intrinsic pathway involves mitochondria, whereas extrinsic pathways involve activation of death receptors. Both pathways promote the activation of caspases, which can be classified as the initiator caspases (caspases 2, 8, 9 and 10) and effector caspases (caspases 3, 6 and 7). Intrinsic and extrinsic pathways in combination induce the activation of caspase 3 leading to apoptosis (Thornberry, 1998; Thornberry and Lazebnik, 1998). Various *in vitro* and *in vivo* studies have reported that phytochemicals including resveratrol, genistein, luteolin, quercetin and apigenin induce apoptosis in cancer cells and in animal models (Manson, 2003; Vauzour *et al.*, 2010). For example, EGCG induces apoptosis by inducing the expression of Fas, caspase-3, caspase-8 and caspase-9 and down-regulating the expression of anti-apoptotic proteins, including Bcl-2, Bcl-xL and BH3 (Nishikawa *et al.*, 2006). Ellagic acid was found to induce apoptosis in a caspase-independent manner and decreased the expression of Bcl-xL and release of cytochrome-c (Mertens-Talcott and Percival, 2005; Larrosa *et al.*, 2006).

15.4.2 ROS-dependent action

Antioxidants are reducing agents that can also act as pro-oxidants. Pro-oxidants induce oxidative stress, either creating reactive oxygen species or inhibiting antioxidant systems. Many polyphenols show pro-oxidant activities, especially at higher doses. The pro-oxidant activity of polyphenols may be due to chemical instability of polyphenols, mobilization of cellular copper ions and deletion of cellular glutathione (GSH) (Dai and Mumper, 2010). Pro-oxidant activities are more prominent *in vitro*, such as at high pH, at higher concentration of transition metals and oxygen molecules. Small molecules of polyphenols like quercetin and gallic acid become oxidized easily, and are good pro-oxidants; on the other hand, high molecular weight polyphenols like tannins do not get oxidized easily, and are less pro-oxidant (Halliwell, 2008). Some polyphenolic phytochemicals are unstable in structure and spontaneously become oxidized enzymatically in the presence of metal ions, particularly in cell culture condition, and form ROS species (Halliwell, 2003). For example, EGCG and green tea produce H_2O_2 in cell culture conditions and are toxic to the cells (Akagawa *et al.*, 2003; Chai *et al.*, 2003). Cell culture media contain many transition metals; for example, DMEM medium has iron, usually in the form of $Fe(NO_3)_3$, that results in higher pro-oxidant activity. In addition, the cells grow in culture in higher oxygen content, 95% air, 5% CO_2 and about 150 mm Hg pressure and higher concentrations of vitamin E, vitamin C and selenium (Halliwell, 2008).

Many polyphenols are believed to have a role in depletion of cellular glutathione, which results in the induction of apoptosis in cancer cells. Chrysin and 2′, 5′ DHC (a kinase inhibitor) have shown cytotoxicity due to their pro-oxidant activity, leading to dysfunction of mitochondrial membrane potential, depletion of mitochondrial glutathione and an increase in the mitochondrial cytochrome-c release. Luteolin and quercetin act as pro-oxidant by depleting GSH. Green tea catechins and phenolic acid cause mitochondrial dysfunction and formation of ROS species in rat hepatocytes (Kachadourian and Day, 2006). Polyphenols including resveratrol and caffeic acid show cytotoxicity to cancer cells, via DNA damage by the immobilization of endogenous copper ions, probably chromatin-bound copper ions, which results in ROS production (Ullah *et al.*, 2011). The copper accumulation mechanism is not well known so far, however, it has been observed that copper transporter 1 is highly efficient in humans and it becomes overexpressed in malignant cells, causing accumulation of copper in the cancer cells. In addition, it has been proposed that copper accumulation may be required for ceruloplasmin, a copper binding protein, highly expressed in cancer cells (Hrgovcic *et al.*, 1973). There are studies that show that curcumin induces lipid detoxification, probably through copper ion interaction in rats (Nair *et al.*, 2005).

15.4.3 Phytochemicals and their efficacy

Compounds having antioxidant activity show different potency to different cancer types (Table 15.1). It has been shown that 50 μM of EGCG is required for inhibition of lung, cervical, and head and neck cancer, whereas 100 μM is required for pancreatic and prostate cancer and 200 μM for breast cancer cells *in vitro* (Table 15.1). On the other hand, *in vivo* activity of EGCG requires relatively high concentrations of the drug in experimental animal models (Li *et al.*, 2010). Efficacy may decrease further in natural conditions. Clinical studies have shown that oral intake of 400 mg day^{-1} of EGCG had maximum plasma level, i.e. 111.8 ± 98.6 (ng ml^{-1}), with half-life ($T_{1/2}$) of 162.3 ± 84.3 min. It was found in its derivative form in urine but not in intact form (Nakagawa *et al.*, 1997; Chow *et al.*, 2001; Wang *et al.*, 2008; Williamson and Renouf, 2011).

Curcumin is a very well-known chemopreventive as well as chemotherapeutic agent. Curcumin concentration ranges from 10 μM for ovary cancer to 50 μM for other cancers types for its *in vitro* activity. These variations in the requirement of concentration of curcumin also vary *in vivo* models (Table 15.1). Curcumin has not been approved as a therapeutic because of less bioavailability. The bioavailability of curcumin in an *in vivo* study is reported to be approximately 40% from 400 mg kg^{-1} oral administration to rats, and the remaining was eliminated via faeces in intact unchanged form, even though traces were found in kidneys and portal blood (Ravindranath and Chandrasekhara, 1980). Efficient action of the drug requires the proper administration and bioavailability of the drug into the body. In another study, it was reported that 10 mg kg^{-1} delivery of curcumin intravenously into the rats showed 0.36 ± 0.05 mg ml^{-1}, whereas 50 times higher dose given orally showed only 0.06 ± 0.01 mg ml^{-1} maximum serum level (Maiti *et al.*, 2007; Yang *et al.*, 2007). The clinical studies of curcumin in humans reported that after taking a 3.6 g capsule orally, the C_{max} value was about 8.9 ± 0.7 nM l^{-1} and about 1.3 μM intact curcumin was observed in urine (Sharma *et al.*, 2004; Dhillon *et al.*, 2008). Requirement of the concentration of silibinin for its action ranges from 10 μM to 200 μM *in vitro* and varies from 80 mg kg^{-1} to 742 mg kg^{-1} *in vivo* (Ramasamy *et al.*, 2011).

The clinical studies of resveratrol found 538.8 ± 72.5 (ng ml^{-1}) plasma level after taking 5 g day^{-1} oral tablets, with $T_{1/2}$ 511.2 ± 95.8 min; approximately 0.4% intact resveratrol along with its derivatives were found in urine (Boocock *et al.*, 2007). Oral administration of genistein in clinical studies with 11.3 mg/capsule, showed 261.84 ± 110.68 (ng ml^{-1}) plasma level of the drug, with $T_{1/2}$ 477.6 ± 136.8 min, and around 1.2 μM of intact genistein found in urine along with its other derivatives (Watanabe *et al.*, 1998; Anupongsanugool *et al.*, 2005). These studies show that there is plenty of scope to improve the bioavailability and stability of these compounds to increase their efficacy.

Table 15.1. Phytochemicals and their mechanisms of action in cancers *in vitro* and *in vivo*.

Phytochemicals	Cancer type	*In vitro*			*In vivo*				References
		Cell line	Dose	Mechanisms	Animal model	Dose	Route of administration	Mechanisms	
Epigallocatechin-gallate (EGCG) 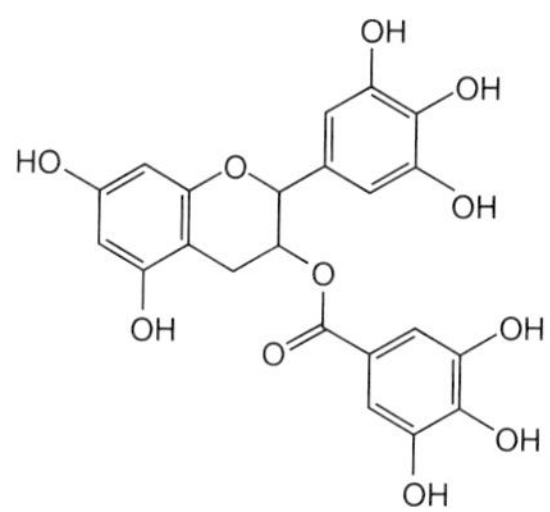 Molar mass: 458.37 g mol^{-1} Soluble in water	Lung	H1299	50 μM	↑p53; ↑P21/ Cip1; ↑P27/ Kip1; ↑ROS	Male NCr nu/nu mice with H1299 cells xenograft	30 mg kg^{-1}	Intraperitoneal injection	↑8-OHdG; ↑8-H2AX; ↑Caspase-3	Qin *et al.*, 2008; Li *et al.*, 2010
	Breast	MCF-7	200 μM	↓HSP70 & 90; ↑P21/ Cip1; ↑P27/ Kip1; ↓Skp2; ↓Cyclin D1; ↓Bcl-2	BALB/c mice with MCF-7 cells xenograft	10 mg kg^{-1}	Intraperitoneal injection	↓HSP70, 90; ↑P21/ Cip1; ↑P27/ Kip1	Huang *et al.*, 2008; Hsieh and Wu, 2008; Tran *et al.*, 2010
	Colorectal	HCT-116 HT 29	35 mg l^{-1}	↓Telomerase; ↓TROP-2; ↓NF-κB; ↓COX-2; ↑P21/ Cip1; ↑P27/ Kip1	Female BALB/c nude mice with HCT-L2 and HCT-S2R cells xenograft	0.04% w/v ECGC as sole source of drinking water	Through drinking water	↓Telomerase; ↑P21/ Cip1; ↑P27/ Kip1	Naasani *et al.*, 2003; Park *et al.*, 2009; Sukhthankar *et al.*, 2010
	Pancreas	AsPC-1 BxPC-3	100 μM	↓Focal Adhesion Kinase, ↓IGFR –I, ↓MAPK, ↓Akt, ↓N-cadherin	–	–	–	–	Vu *et al.*, 2010
	Liver	BEL7404/ADM, BEL7402/5-FU	150 mg l^{-1}	↓172 genes; ↑38 genes	–	–	–	–	Tang *et al.*, 2008
	Prostate	TRAMP-C1	100 μM	↓Cell Proliferation; ↓MMP-2,9; ↑P21/ Cip1; ↑P27/ Kip1;↓IGF-1; ↓COX-2; ↓iNOS	C57/Bl Mice with TRAMP- C1 xenograft	0.06% EGCG in drinking water	Oral	↓Cell Proliferation; ↓MMP-2,9; ↑P21/ Cip1; ↑P27/ Kip1;	Sartor *et al.*, 2004; Harper *et al.*, 2007
	Skin	A431	60 μM	↓Ezh2; ↓Suz12; ↓Bim-1; ↑P21/ Cip1; ↑P27/ Kip1; ↓CDK4; ↓CDK2; ↓Cyclin D1	–	–	–	–	Bhatia *et al.*, 2001; Balasubramanian *et al.*, 2010
	Ovary	SKOV-3	80 μg ml^{-1}	↑P21/ Cip1; ↑P27/ Kip1; ↓CDK2; ↓Cyclin D1; ↑p53; ↓PCNA; ↓Bcl-x_L; ↓E2F-1/4; ↓Rb	–	–	–	–	Kim *et al.*, 2004; Rao and Pagidas, 2010

Curcumin 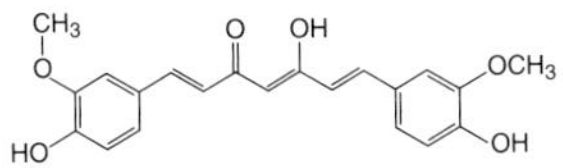 Molar mass: 368.38 g mol^{-1} Insoluble in water	Cervical	HeLa	50 μM	↓TFIIIB subunit Brf1,2; ↑p53; ↓NF-κB; ↓Akt; ↑Cytochrome-c ; ↑Caspase-3/9	–	–	–	–	Jacob *et al.*, 2007; Singh *et al.*, 2011
	Head and Neck	Tu686/M4e	30 μM	↓MMP-2,9; ↑RECK; ↑P21/ Cip1; ↑P27/ Kip1; ↑p53; ↑Bim; ↓NF-κB; ↓p65	–	–	–	–	Kato *et al.*, 2008; Amin *et al.*, 2009
	Lung	A549 H1299	20 μM	↑P21/ Cip1; ↑P27/ Kip1; ↓MMP-2/9/14; ↑JNK; ↑AP-1	SCID mice with CL1-5, A549, H1975 cell xenograft	1g kg^{-1}	Diet supplement	↓MMP-2/9/14; ↓Integrinα6/β4; ↓MEKK3; ↓ERK; ↓NCAM; ↑HLJ1; ↑E-cadherin; ↑JNK; ↑AP-1	Lin *et al.*, 2009; Lee *et al.*, 2011; Ko *et al.*, 2011
	Breast	MDA-MB-231	50 μM	↓VEGF; ↓MMP-2/9; ↓b-FGF; ↑TIMP-I; ↑P21/ Cip1; ↑P27/ Kip1	Female CD1 athymic nude mice with MDA-MB-231 cell xenograft	200 mg kg^{-1}	Oral	↓MMP-2/9; ↓b-FGF; ↑TIMP-I; ↑P21/ Cip1; ↑ROS; ↑JNK	Shao *et al.*, 2002; Somers-Edgar *et al.*, 2008
	Colorectal	HT-29	80 μM	↑Bax; ↓Bcl-2; ↓Bcl-xL; ↑Bad; ↑Caspase-3; ↓PARP; ↓Survivin; ↓COX-2	Male athymic nude mice with HCT-116 cell xenograft	30 mg kg^{-1}	Intraperitoneal injection	↓COX-2; ↓Akt	Subramaniam *et al.*, 2008; Wang *et al.*, 2009; Lee *et al.*, 2009
	Pancreas	Panc28 L3.6pL	50 μM	↓NF-κB; ↓Cyclin D1; ↓c-myc; ↓bcl-2; ↓Cytochrome-c; ↓Survivin; ↓VEGF; ↓ROS	Female athymic nude mice with PaCa-2 cell xenograft	1 g kg^{-1}	Oral	↓NF-κB; ↓Survivin; ↓VEGF; ↓Bcl-2; ↓Bcl-xL; ↓Procaspase -3/9; ↓COX-2	Kunnumakkara *et al.*, 2007; Jutooru *et al.*, 2010
	Liver	SK-Hep-1	40 μM	↑P21/ Cip1; ↑P27/ Kip1; ↓Survivin; ↓Notch1; ↓Bcl-xL; ↓MMP-2,9	Nu/nu mice with SK-Hep-1 cell xenograft	100 mg kg^{-1}	Intraperitoneal injection	↑P21/ Cip1; ↑P27/ Kip1; ↓Notch1	Lin *et al.*, 1998; Ning *et al.*, 2009;Wang *et al.*, 2010
	Prostate	LNCaP PC-3	50 μM	↓VEGF; ↓Cyclin D1; ↓uPA, ↓MMP-2/9; ↓Bcl-2; ↓Bcl-xL; ↑P21/ Cip1; ↑P27/ Kip1; ↓PCNA; ↑TRAIL-R1/ DR4; ↑TRAIL-R2/DR5; ↑Bax; ↑Bak; ↓Akt	Male TRAMP mice	2% w/w	Diet supplement	↓VEGF; ↓Akt; ↓MMP-2/9; ↓Bcl-2; ↓Bcl-xL; ↑P21/ Cip1; ↑P27/ Kip1; ↓PCNA	Dorai *et al.*, 2001; Deng *et al.*, 2008; Shankar *et al.*, 2008

Continued

Table 15.1. Continued.

Phytochemicals	Cancer type	*In vitro*			*In vivo*				References
		Cell line	Dose	Mechanisms	Animal model	Dose	Route of administration	Mechanisms	
	Skin	Keratinocytes from human foreskin	50 μM	↓Bcl-x_L; ↑P21/ Cip1; ↑p16; ↓Procaspase-9/8/3; ↓Cyclin D1	–	–	–	–	Balasubramanian and Eckert, 2007
	Ovary	HeyA8	10 μM	↑P21/ Cip1; ↑P27/ Kip1;↓VEGF; ↓IL-8; ↓MMP-9; ↓NF-κB; ↓COX-2; ↓Bcl-2; ↓Bcl-x_L	Female athymic nude mice with HeyA8 cells	500 mg kg^{-1}	Gavage	↑P21/ Cip1; ↑P27/ Kip1, ↓VEGF; ↓IL-8; ↓MMP-9; ↓NF-κB; ↓COX-2; ↓Bcl-2; ↓Bcl-x_L	Lin *et al.*, 2007
	Cervical	HeLa	50 μM	↓PCNA; ↓Cyclin D1; ↓Telomerase; ↑p53; ↑Survivin; ↓NF-κB	–	–	–	–	Singh and Singh, 2011; Bava *et al.*, 2011
	Head and Neck	CCL 23 CAL 27 UM-SCC1	50 μM	↓IKKβ; ↓NF-κB; ↑p53; ↓Cyclin D1; ↓c-Myc; ↓Cyclooxygenase-2; ↓MMP-9; ↓Bcl-2; ↓Bcl-x_L; ↓Mcl-1L/1S	–	–	–	–	Wang *et al.*, 2008; Duarte *et al.*, 2010; Meyer *et al.*, 2011
Silibinin 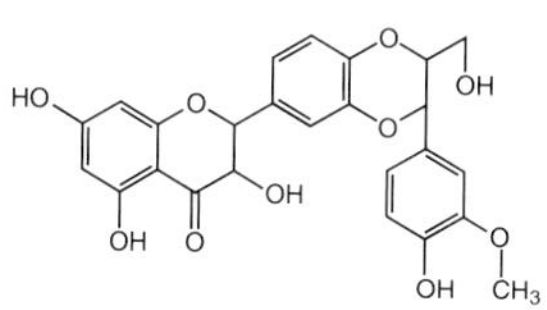	Lung	H460 H1299	75 μM	↑P18/INK4C; ↑P21/ Cip1; ↑P27/ Kip1; ↓CDK2; ↓CDK4; ↓pRb	Male B6/129-Nos2^{tm1Lau} (iNOS$^{-/-}$) &B6/129PF2 WT mice	742 mg kg^{-1} body weight	Dietary supplement	↓VEGFR2; ↓STAT3; ↓NF-κB	Mateen *et al.*, 2010; Ramasamy *et al.*, 2011
	Breast	MCF-7	200 μM	↑p53; ↑RNS; ↑ROS; ↓MMP-9; ↓COX-2; ↓VEGF; ↓Raf/MEK/ ERK; ↓IGF-1; ↓SIRT1	–	–	–	–	Kim *et al.*, 2009; Noh *et al.*, 2011

Molar mass: 482.44 g mol^{-1} Poorly water soluble	Colorectal	SW480, HCT116	200 μM	↓β-catenin; ↓Cyclin D1; ↓c-Myc; ↓CDK8; ↑Caspase-3;8;10; ↓VEGF; ↓iNOS; ↓Survivin	Athymic (nu/nu) male nude mice with SW480 cell xenograft	200 mg kg^{-1}	Dietary supplement	↓β-catenin; ↓Cyclin D1; ↓c-Myc; ↓CDK8; ↑Apoptosis	Kaur *et al.*, 2010; Velmurugan *et al.*, 2010
	Pancreas	No study	–	–	–	–	–	–	–
	Liver	BEL-7402/ 5-FU	200 μM	↓CDK2; ↓CDK4; ↓CDC2; ↑P21/ Cip1; ↓NF-κB; ↑P27/ Kip1; ↓pRb	Nude mice with HCC cell xenograft	80 mg kg^{-1}	Gavage	↑PTEN; ↓p-Akt; ↓p-ERK; ↓PI3K/Akt; ↑P27/ Kip1; ↓NF-κB	Varghese *et al.*, 2005; Cui *et al.*, 2009
	Prostate	PC-3 DU145	20–100 μM	↓CDK2;4;6; ↓CDC2; ↓Cyclin D1; D2; A; ↑P21/ Cip1;↑P27/ Kip1; ↓JNK-1/2; ↓Akt; ↓STAT1; ↓STAT3; ↓STAT5; ↑Caspase-3; ↑ERK-1/2; ↓IGF-1; ↓NF-κB	Athymic mice with PC-3 and DU145 cell xenograft	100 mg kg^{-1}	Gavage	↓CDK2;4;6; ↓CDC2; ↓Cyclin D1; D2; A; ↓JNK-1/2; ↓Akt; ↓STAT1; ↓STAT3; ↓STAT5; ↑Caspase-3; ↑ERK-1/2	Singh and Agarwal, 2006; Singh *et al.*, 2009
	Skin	HaCaT keratino-cytes	50 μM	↑Caspase-3; ↑P21/ Cip1; ↑P27/ Kip1; ↑p53; ↓Akt; ↓COX-2; ↓HIF-1α; ↓iNOS; ↓NF-κB; ↓E2F1,2,3; ↓p65; ↓VEGF; ↓ROS	SKH-1 hairless mice	9 mg/200 μM acetone/ mouse	Topical application	↓VEGF; ↓HIF-1α; ↓iNOS; ↓COX-2; ↓p65; ↓NF-κB; ↓E2F1,2,3; ↓Akt; ↑P21/ Cip1; ↑P27/ Kip1; ↑p53	Gu *et al.*, 2006, 2007; Svobodová *et al.*, 2007
	Ovary	No study	–	–	–	–	–	–	–
	Cervical	HeLa	250–500 μM	↓HIF-1α; ↓mTOR; ↓PI3K/Akt; ↓VEGF	–	–	–	–	García-Maceira and Mateo, 2009
	Head and Neck	SCC-4	100 μM	↓MMP-2; ↓ERK-1/2; ↓TIMP-2; ↓Survivin	–	–	–	–	Chen *et al.*, 2006

Arrows indicate an increase (↑) or decrease (↓) in the levels of protein expression/activation, genetic and epigenetic expression.

15.5 Applications of Antioxidant Phytochemicals in Therapeutics

Usually, the higher intake of antioxidants in diet increases the lifespan of humans by neutralizing/stabilizing free radicals in the body. The continuous intake or exposure of oxidation-inducing foods may lead to the generation of free radicals, such as intake of over-cooked and re-used oil shows a higher amount of free radicals (Halliwell, 2008). These cause oxidation to important cellular macromolecules such as proteins, DNA, lipid, carbohydrates and associated tissues, which can lead to health hazards. For example, smoking and chronic alcoholism increase the free radical formation and decrease the level of antioxidants in serum, which ultimately can cause severe damage to the body (Diplock *et al.*, 1998). Thus dietary antioxidants play a major role in chemoprevention. They are also used in body lotion, as they protect skin from UV radiation in the sunlight, decrease skin roughness and also protect from UV-induced skin cancer and swelling of the skin (Borek, 2004).

15.6. Limitations of Uses of Phytochemicals/Antioxidants

15.6.1 Bioavailability and physico-chemical property

Bioavailability is the degree to which a drug or other substance becomes available to the target tissue after administration. There are several factors that affect the bioavailability of the drugs, such as physico-chemical properties including hydrophobicity, pKa value and solubility, etc. (Moorthi *et al.*, 2011). The other factors include different metabolic rates, chemical degradation and enzyme induction/inhibition by other drugs/food and inter-individualistic differences in metabolism (Hodek *et al.*, 2009), formulation of drugs and effect of the intestinal microflora. Some phytochemicals have big polycyclic structures, due to which the homogeneous distributions of drugs become difficult. For instance, *in vitro* studies with resveratrol have shown inhibition of proliferation, induction of apoptosis and hindrance in cell cycle of human cancer cell lines, including those of colon, breast, lung, prostate, liver and pancreatic cancers. But most of the drug (resveratrol) is excreted in the urine in metabolic state, which causes less amount of drug to be available to the targeted tissues (Walle *et al.*, 2004). Physico-chemical and molecular intricacy of drugs and their low availability to target tissues provide a huge challenge to investigators. Many phytochemicals are hydrophobic in nature, thus show low bioavailability. Thus to overcome these obstacles, formulation of new delivery systems is required, which should also be less toxic to normal cells (Moorthi *et al.*, 2011).

Several approaches have been exploited to overcome the problem of bioavailability, including solubilization, inclusion and complexation of compounds. Complexation includes the encapsulation of the drug with hydrophilic compounds such as poly-acetic acid, lactic acid and glycolic acid, and relatively hydrophobic compounds such as albumin-based nanoparticles, liposome formulation, polymeric micelles cyclodextrin and chitosan-based nanoparticles (Wischke and Schwendeman, 2008). The ionization state of a drug can alter its behaviour, for example an increase in the solubility of a drug in water will decrease its lipophilicity and vice-versa. PKa value plays an important role in the partition coefficient and directly affects the hydrophobicity and drug distribution (Avdeef, 2001, 2003; Jia, 2005).

On the other hand, the problem of excess accumulation of drugs in the body can be overcome by using nano-carriers, which are biodegradable and biocompatible including polylactic acid (PLA), poly-DL-lactide-co-glycolide acid (PLGA), starch, chitosan and protein-based carriers such as apotransferrin and lactoferrin nanoparticles (Peer *et al.*, 2007; Golla *et al.*, 2012). Some of the nano-carriers have been proved safe, biocompatible and provide the flexibility to control the sustainable release and dose formulation of the drugs, i.e. protein-based nanoparticles with doxorubicin has been found safe *in vivo* and *in vitro* (Siddiqui *et al.*, 2010; Golla *et al.*, 2012). It has been proved that the nano-EGCG, at IC_{50} value 2.5 μM, inhibits proliferation, induces apoptosis in 22Rv1 prostate carcinoma cell lines, in comparison with IC_{50} value 35 μM of EGCG alone (Siddiqui *et al.*, 2009; Siddiqui and Mukhtar, 2010).

For proper exposure of targeted tissues the drug should have a slow metabolic rate in the body and higher half-life, so that it will have sufficient time to reach its target site. Thus, faster metabolism of the drug leads to less bioavailability at the target site and decreases the efficacy of the drug. Liver and kidney metabolism play a major role in the phytochemical degradation, which includes phase I reactions such as oxidation and reduction of the drug, hydrolysis and phase II reaction such as conjugation (Gibson and Skett, 2001a, b). Curcumin, which is an anticancer and anti-inflammatory agent, has less amount of availability at the target site *in vivo* studies and most of the drug becomes metabolized through conjugation including glucuronidation, sulfation and reduction (Pfeiffer *et al.*, 2007; Vareed *et al.*, 2008). This limitation of curcumin has been solved with the help of nanoparticles such as curcumin tagged with human serum albumin nanoparticles (CCM-HSA-NPs).When the efficacy of this conjugate was tested, it was found that the drug was 14 times more effective against tumours than that of curcumin alone. CCM-HSA-NPs showed 5.5 times more anti-angiogenic activity than curcumin alone, and it was observed that the nanoparticle conjugate shows 32% higher anticancer activity in comparison to curcuminin HCT116 tumour xenograft model (Kim *et al.*, 2011). These studies indicate that such strategies may help curcumin to escape from metabolic degradation and become more available to the target sites.

15.6.2 Interaction of phytochemicals and drugs

Pure phytochemicals frequently show different potency to cancer to that of phytochemicals in combination, such as extract of *Curcuma longa* (which includes curcuminoids, demethoxycurcumin, bisdemethoxycurcumin and α-turmerone) shows higher anticancer activity than curcumin alone (Yue *et al.*, 2010). A combination study of indol-3-carbinol (300 μM) and genistein (40 μM) on human colon cancer HT-29 cells showed reduced cell viability up to 87% after 24 h of treatment. Both of them were relatively less effective at these concentrations when cells were treated with either of the phytochemicals alone (Nakamura *et al.*, 2009). In combination with doxorubicin and silibinin, a flavonoid, it inhibited the growth of lung cancer and reduced the systemic toxicity of doxorubicin with an enhanced therapeutic efficacy most likely via an inhibition of doxorubicin-induced chemo-resistance involving NF-κβ signalling *in vivo* (Singh *et al.*, 2004).

15.6.3 Interaction of phytochemicals with microflora

The mode of administration of most of the phytochemicals and antioxidants is oral as many of these are dietary components. The intestinal microflora plays a major role in the metabolism of these compounds by their secretory enzymes and affects their bioavailability. Reduction, hydrolysis and ring fission are three common mechanisms by which flavonoids are metabolized in the GI by microflora (Gao and Hu, 2010). Flavonoids, which occur as glycosides, undergo bacterial degradation by glucosidases and glucuronidases, which convert them to toxic aglycones. Flavonoids that have antimicrobial activity can be detrimental to normal microbiota, and may also provide advantage to the harmful bacteria. Moreover, the biochemical transformation by microbiota will increase the intestinal absorption of phytochemicals (Woting *et al.*, 2010). For example, about 50% of the resveratrol was removed after metabolic degradation by the microflora (Walle, 2011). Furthermore, some functional food compounds influence the growth and metabolic activity of the gut microbiota, and thereby influence the behaviour and functions of the drug.

15.6.4 Effect of phytochemicals on cytochrome P450 metabolizing enzymes

Cytochrome P450 (CYPs) contributes up to 80% of all phase I xenobiotic metabolizing enzymes. Expressions of CYPs play a major role in metabolism of phytochemicals. Many compounds induce or suppress the expression of CYPs, which affects the overall chemopreventive efficacy of phytochemicals. The expression of CYPs may be beneficial or detrimental depending on time and tissue-specific expression (Nebert and Dalton, 2006). Benzopyrene (B[a]P), a known carcinogen, when metabolized by CYP1A1 gets converted into 7,8-epoxy-7,8-dihydro-B[a]P, which acts as carcinogen. Natural polyphenolic compounds inhibit the activity of CYP1A1, which reduces B[a]P carcinogenicity (Schwarz and Roots, 2003). Some CYPs are activated, while others are inhibited by the same compound. Dietary curcumin suppresses oesophageal and gastric cancer by activating CYP2B1/2 and CYP2E1, but enhances large intestinal carcinogenesis. Induction of CYPs expression is also dependent on the dose and time of phytochemical administration. Certain flavonoids act as phytoestrogens and inhibit CYP19 involved in oestrogen biosynthesis, shifting the overall hormonal balance (Hodek *et al.*, 2009). Thus cytochrome P450 and phase II enzyme systems play an important role in determining the biological activity of phytochemicals.

15.7 Solubility of Phytochemicals

The solubility of phytochemicals is a major concern in their chemopreventive efficacy. Since phytochemicals are orally administered their absorption and effective bioavailability depends upon their solubility in water. For examples, ellagic acid and resveratrol are poorly absorbed because of their low solubility in aqueous solutions (Lei *et al.*, 2003; López-Nicolás and García-Carmona, 2008). Many bioactive phytochemicals possess polyaromatic rings with one or more hydroxyl groups. The bulkiness and lipophilicity of these aromatic rings make these compounds less soluble. However, researchers are trying to develop analogues of these compounds that will improve their solubility in aqueous environments with the same biological activity, and likely the solubility of resveratrol can be increased by the addition of the salt moiety to the resveratrol structure (Anderson *et al.*, 2001). It has been reported that various delivery systems can increase the solubility of phytochemicals, e.g. phytosome complexes have been used to increase the solubilization of phytochemicals in aqueous medium (Semalty *et al.*, 2010).

15.8 Stability of Phytochemicals

The stability of phytochemicals is greatly influenced by their micro-environment such as metabolic enzyme concentrations, pH and temperature. It has been shown that anthocyanins are unstable in alkaline conditions. Catechins in green tea and flavonoids were also shown to have lower half-life in alkaline conditions than in acidic conditions. Temperature-dependent degradation has been also observed for some compounds. For example, lutein and γ-tocopherol get degraded at room temperature faster than at 4°C and −80°C. Thus, the micro-environment of phytochemicals or antioxidants can significantly affect the overall biological activity of phytochemicals and antioxidants.

15.9 Conclusions

Cancer and other chronological diseases have become prevalent, causing unmanageable suffering and loss of human life worldwide. Many strategies have been employed to treat cancer, but the current available approaches have many limitations. This drives investigators to explore other efficient methods that are more effective and less expensive. Many phytochemicals have both chemopreventive as well as chemotherapeutic properties, which can target several stages of carcinogenesis, including cell proliferation, angiogenesis, metastasis and apoptosis for cell death. Further, the use of antioxidant phytochemicals should be done carefully in different pathological conditions, for example antioxidant agents should not be combined with radiotherapy. But there are several limitations with phytochemicals, which need to be studied, such as poor bioavailability, problem of pharmacokinetics, biotransformation by microflora and their structural and physical properties. The application of structural, biochemical and material sciences including nanotechnology should be collectively employed to tackle these limitations.

References

Agarwal, C., Dhanalakshmi, S., Singh, R.P. and Agarwal, R. (2004) Inositol hexaphosphate inhibits growth and induces G1 arrest and apoptotic death of androgen-dependent human prostate carcinoma LNCaP cells. *Neoplasia* 6, 646–659.

Akagawa, M., Shigemitsu, T. and Suyama, K. (2003) Production of hydrogen peroxide by polyphenols and polyphenol-rich beverages under quasi-physiological conditions. *Bioscience, Biotechnology and Biochemistry* 67, 2632–2640.

Amin, A.R., Khuri, F.R., Chen, Z.G. and Shin, D.M. (2009) Synergistic growth inhibition of squamous cell carcinoma of the head and neck by erlotinib and epigallocatechin-3-gallate: the role of p53-dependent inhibition of nuclear factor-kappaB. *Cancer Prevention Research (Philadelphia, Pa.)* 2, 538–545.

Anderson, A., Belelli, D., Bennett, D., Buchanan, K.I., Casula, A. and Al, A.C. (2001) Alpha-amino acid phenolic ester derivatives: novel water-soluble general anesthetic agents which allosterically modulate GABA (A) receptors. *Journal of Medicinal Chemistry* 44, 3582–3591.

Anupongsanugool, E., Teekachunhatean, S., Rojanasthien, N., Pongsatha, S. and Sangdee, C. (2005) Pharmacokinetics of isoflavones, daidzein and genistein, after ingestion of soy beverage compared with soy extract capsules in postmenopausal Thai women. *BMC Clinical Pharmacology* 5, 2.

Avdeef, A. (2001) Physicochemical profiling (solubility, permeability and charge state). *Current Topics in Medicinal Chemistry* 1, 277–351.

Avdeef, A. (2003) *Absorption and Drug Development: solubility, permeability, and charge state*. Wiley, New York.

Balasubramanian, S. and Eckert, R.L. (2007) Curcumin suppresses AP1 transcription factor-dependent differentiation and activates apoptosis in human epidermal keratinocytes. *The Journal Of Biological Chemistry* 282, 6707–6715.

Balasubramanian, S., Adhikary, G. and Eckert, R.L. (2010) The Bmi-1 polycomb protein antagonizes the (-)-epigallocatechin-3-gallate-dependent suppression of skin cancer cell survival. *Carcinogenesis* 31, 496–503.

Banerjee, S., Li, Y., Wang, Z. and Sarkar, F.H. (2008) Multi-targeted therapy of cancer by genistein. *Cancer Letters* 269, 226–242.

Bava, S.V., Sreekanth, C.N., Thulasidasan, A.K., Anto, N.P., Cheriyan, V.T., Puliyappadamba, V.T., Menon, S.G., Ravichandran, S.D. and Anto, R.J. (2011) Akt is upstream and MAPKs are downstream of NF-κB in paclitaxel-induced survival signaling events, which are down-regulated by curcumin contributing to their synergism. *The International Journal of Biochemistry and Cell Biology* 43, 331–341.

Bergström, T., Ersson, C., Bergman, J. and Möller, L. (2012) Vitamins at physiological levels cause oxidation to the DNA nucleoside deoxyguanosine and to DNA-alone or in synergism with metals. *Mutagenesis* 27, 511–517.

Berletch, J.B., Liu, C., Love, W.K., Andrews, L.G., Katiyar, S.K. and Tollefsbol, T.O. (2008) Epigenetic and genetic mechanisms contribute to telomerase inhibition by EGCG. *Journal of Cellular Biochemistry* 103, 509–519.

Berner, C., Aumüller, E., Gnauck, A., Nestelberger, M., Just, A. and Haslberger, A.G. (2010) Epigenetic control of estrogen receptor expression and tumor suppressor genes is modulated by bioactive food compounds. *Annals of Nutrtion and Metabolism* 57, 183–189.

Bhatia, N., Agarwal, C. and Agarwal, R. (2001) Differential responses of skin cancer-chemopreventive agents silibinin, quercetin, and epigallocatechin 3-gallate on mitogenic signaling and cell cycle regulators in human epidermoid carcinoma A431 cells. *Nutrtion and Cancer* 39, 292–299.

Bishayee, A. (2009) Cancer prevention and treatment with resveratrol: from rodent studies to clinical trials. *Cancer Prevention Research (Philadelphia, Pa.)* 2, 409–418.

Bjelakovic, G., Nikolova, D., Gluud, L.L., Simonetti, R.G. and Gluud, C. (2007) Mortality in randomized trials of antioxidant supplements for primary and secondary prevention: systematic review and meta-analysis. *The Journal of The American Medical Association* 297, 842–857.

Boocock, D.J., Faust, G.E., Patel, K.R., Schinas, M., Brown, V.A., Ducharme, M.P., Booth, T.D., Crowell, J.A., Perloff, M., Gescher, A.J., Steward, W.P. and Brenner, D.E. (2007) Phase I dose escalation pharmacokinetic study in healthy volunteers of resveratrol, a potential cancer chemopreventive agent. *Cancer Epidemiology, Biomarkers and Prevention* 16, 1246–1252.

Borek, C. (2004) Dietary antioxidants and human cancer. *Integrative Cancer Therapies* 4, 333–341.

Chai, P.C., Long, L.H. and Halliwell, B. (2003) Contribution of hydrogen peroxide to the cytotoxicity of green tea and red wines. *Biochemical and Biophysical Research Communications* 304, 650–654.

Chen, P.N., Hsieh, Y.S., Chiang, C.L., Chiou, H.L., Yang, S.F. and Chu, S.C. (2006) Silibinin inhibits invasion of oral cancer cells by suppressing the MAPK pathway. *Journal of Dental Research* 83, 220–225.

Chow, H.H., Cai, Y., Alberts, D.S., Hakim, I., Dorr, R., Shahi, F., Crowell, J.A., Yang, C.S. and Hara, Y. (2001) Phase I pharmacokinetic study of tea polyphenols following single-dose administration of epigallocatechin gallate and polyphenon E. *Cancer Epidemiology, Biomarker and Prevention* 10, 53–58.

Conner, G.E., Salathe, M. and Forteza, R. (2002) Lactoperoxidase and hydrogen peroxide metabolism in the airway. *American Journal of Respiratory and Critical Care Medicine* 166(12 Pt 2), S57–61.

Cui, W., Gu, F. and Hu, K.Q. (2009) Effects and mechanisms of silibinin on human hepatocellular carcinoma xenografts in nude mice. *World Journal of Gastroenterology* 15, 1943–1950.

Dai, J. and Mumper, R.J. (2010) Plant phenolics: extraction, analysis and their antioxidant and anticancer properties. *Molecules* 15, 7313–7352.

Deep, G. and Agarwal, R. (2007) Chemopreventive efficacy of silymarin in skin and prostate cancer. *Intergrative Cancer Therapies* 6, 130–145.

Deng, G., Yu, J.H., Ye, Z.Q. and Hu, Z.Q. (2008) Curcumin inhibits the expression of vascular endothelial growth factor and androgen-independent prostate cancer cell line PC-3 *in vitro*. *Zhonghua Nan Ke Xue* 14, 116–121.

Devasagayam, T.P., Tilak, J.C., Boloor, K.K., Sane, K.S., Ghaskadbi, S.S. and Lele, R.D. (2004) Free radicals and antioxidants in human health: current status and future prospects. *The Journal of The Association of Physicians of India* 52, 794–804.

Dhillon, N., Aggarwal, B.B., Newman, R.A., Wolff, R.A., Kunnumakkara, A.B., Abbruzzese, J.L., Ng, C.S., Badmaev, V. and Kurzrock, R. (2008) Phase II trial of curcumin in patients with advanced pancreatic cancer. *Clinical Cancer Research* 14, 4491–4499.

Diplock, A.T., Charleux, L., Crozier-Willi, G., Kok, F.J., Rice-Evans, C., Roberfroid, M., Stahl, W. and Vina-Ribes, J. (1998) Functional food science and defence against reactive oxidative species. *The British Journal of Nutrition* 80, S77–112.

Dorai, T., Cao, Y.C., Dorai, B., Buttyan, R. and Katz, A.E. (2001) Therapeutic potential of curcumin in human prostate cancer. III. Curcumin inhibits proliferation, induces apoptosis, and inhibits angiogenesis of LNCaP prostate cancer cells in vivo. *The Prostate* 47, 293–303.

Duarte, V.M., Han, E., Veena, M.S., Salvado, A., Suh, J.D., Liang, L.J., Faull, K.F., Srivatsan, E.S. and Wang M.B. (2010) Curcumin enhances the effect of cisplatin in suppression of head and neck squamous cell carcinoma via inhibition of IKKβ protein of the NFκB pathway. *Molecular Cancer Therapeutics* 9, 2665–2675.

Gao, S. and Hu, M. (2010) Bioavailability challenges associated with development of anti-cancer phenolics. *Mini Reviews in Medicinal Chemistry* 10, 550–567.

García-Maceira, P. and Mateo, J. (2009) Silibinin inhibits hypoxia-inducible factor-1alpha and mTOR/p70S6K/4E-BP1 signalling pathway in human cervical and hepatoma cancer cells: implications for anti-cancer therapy. *Oncogene* 28, 313–324.

Gibson, G.G. and Skett, P. (2001a) Enzymology and molecular mechnism of drug metaboloism reaction. In: Gibson, G.G. and Skett, P. (eds) *Introduction to Drug Metabolism*, 3rd edn. Nelson Thornes, London, pp. 37–84.

Gibson, G.G. and Skett, P. (2001b) Pathways of drug metabolism. In: Gibson, G.G. and Skett, P. (eds) *Introduction to Drug Metabolism*, 3rd edn. Nelson Thornes, London, pp. 1–34.

Golla, K., Cherukuvada, B., Ahmed, F. and Kondapi, A.K. (2012) Efficacy, safety and anticancer activity of protein nanoparticle-based delivery of doxorubicin through intravenous administration in rats. *PLoS One* 7(12), e51960.

Gu, M., Singh, R.P., Dhanalakshmi, S., Agarwal, C. and Agarwal, R. (2007) Silibinin inhibits inflammatory and angiogenic attributes in photocarcinogenesis in SKH-1 hairless mice. *Cancer Research* 67, 3483–3491.

Gu, M., Singh, R.P., Dhanalakshmi, S., Mohan, S. and Agarwal, R. (2006) Differential effect of silibinin on E2F transcription factors and associated biological events in chronically UVB-exposed skin versus tumors in SKH-1 hairless mice. *Molecular Cancer Therapeutics* 5, 2121–2129.

Halliwell, B. (2003) Oxidative stress in cell culture: an under-appreciated problem? *FEBS Letters* 540, 3–6.

Halliwell, B. (2008) Are polyphenols antioxidants or pro-oxidants? What do we learn from cell culture and *in vivo* studies? *Archives of Biochemistry and Biophysics* 476, 107–112.

Harper, C.E., Patel, B.B., Wang, J., Eltoum, I.A. and Lamartiniere, C.A. (2007) Epigallocatechin-3-Gallate suppresses early stage, but not late stage prostate cancer in TRAMP mice: mechanisms of action. *Prostate* 67, 1576–1589.

Hodek, P., Krízková, J., Burdová, K., Sulc, M., Kizek, R., Hudecek, J., Wahed, A.S., Belle, S.H., Afdhal, N.H., Navarro, V.J., Berman, J., Liu, Q.Y., Doo, E., Fried, M.W. and SyNCH Trial Group (2009) Chemopreventive compounds – view from the other side. *Chemico-Biological Interactions* 180, 1–9.

Hrgovcic, M., Tessmer, C.F., Thomas, F.B., Ong, P.S., Gamble, J.F. and Shullenberger, C.C. (1973) Serum copper observations in patients with malignant lymphoma. *Cancer* 32, 1512–1524.

Hsieh, T.C. and Wu, J.M. (2008) Suppression of cell proliferation and gene expression by combinatorial synergy of EGCG, resveratrol and gamma-tocotrienol in estrogen receptor-positive MCF-7 breast cancer cells. *International Journal of Oncology* 33, 851–859.

Huang, H.C., Way, T.D., Lin, C.L. and Lin, J.K. (2008) EGCG stabilizes p27kip1 in E2-stimulated MCF-7 cells through down-regulation of the Skp2 protein. *Endocrinology* 49, 5972–5983.

Hurrell, R.F. (2003) Influence of vegetable protein sources on trace element and mineral bioavailability. *The Journal of Nutrition* 133, 2973S–2977S.

Jacob, J., Cabarcas, S., Veras, I., Zaveri, N. and Schramm, L. (2007) The green tea component EGCG inhibits RNA polymerase III transcription. *Biochemical and Biophysical Research Communications* 360, 778–783.

Jia, L. (2005) Nanoparticle Formulation Increases Oral Bioavailability of Poorly Soluble Drugs: Approaches Experimental Evidences and Theory. *Current Nanosciences* 1, 237–243.

Johnson, J.J. (2011) Carnosol: a promising anti-cancer and anti-inflammatory agent. *Cancer Letter* 305, 1–7.

Jutooru, I., Chadalapaka, G., Lei, P. and Safe, S. (2010) Inhibition of NFkappaB and pancreatic cancer cell and tumor growth by curcumin is dependent on specificity protein down-regulation. *The Journal of Biological Chemistry* 285, 25332–25344.

Kachadourian, R. and Day, B.J. (2006) Flavonoid-induced glutathione depletion: potential implications for cancer treatment. *Free Radical Biology and Medicine* 41, 65–76.

Kato, K., Long, N.K., Makita, H., Toida, M., Yamashita, T., Hatakeyama, D., Hara, A., Mori, H. and Shibata, T. (2008) Effects of green tea polyphenol on methylation status of RECK gene and cancer cell invasion in oral squamous cell carcinoma cells. *British Journal of Cancer* 99, 647–654.

Kaur, M., Velmurugan, B., Tyagi, A., Agarwal, C., Singh, R.P. and Agarwal, R. (2010) Silibinin suppresses growth of human colorectal carcinoma SW480 cells in culture and xenograft through down-regulation of beta-catenin-dependent signaling. *Neoplasia* 12, 415–424.

Khan, N., Adhami, V. and Mukhtar, H. (2009) Green tea polyphenols in chemoprevention of prostate cancer: preclinical and clinical studies. *Nutrition and Cancer* 61, 836–841.

Kim, S., Choi, J.H., Lim, H.I., Lee, S.K., Kim, W.W., Kim, J.S., Kim, J.H., Choe, J.H., Yang, J.H., Nam, S.J. and Lee, J.E. (2009) Silibinin prevents TPA-induced MMP-9 expression and VEGF secretion by inactivation of the Raf/MEK/ERK pathway in MCF-7 human breast cancer cells. *Phytomedicine* 16, 573–580.

Kim, T.H., Jiang, H.H., Youn, Y.S., Park, C.W., Tak, K.K., Lee, S., Kim, H., Jon, S., Chen, X. and Lee, K.C. (2011) Preparation and characterization of water-soluble albumin-bound curcumin nanoparticles with improved antitumor activity. *International Journal of Pharmaceutics* 403, 285–291.

Kim, Y.W., Bae, S.M., Lee, J.M., Namkoong, S.E., Han, S.J., Lee, B.R., Lee, I.P., Kim, S.H., Lee, Y.J., Kim, C.K., Kim, Y.W. and Ahn, W.S. (2004) Activity of green tea polyphenol epigallocatechin-3-gallate against ovarian carcinoma cell lines. *Cancer Research and Treatment* 36, 315–323.

Ko, J.C., Tsai, M.S., Weng, S.H., Kuo, Y.H., Chiu, Y.F. and Lin, Y.W. (2011) Curcumin enhances the mitomycin C-induced cytotoxicity via downregulation of MKK1/2-ERK1/2-mediated Rad51 expression in non-small cell lung cancer cells. *Toxicology and Applied Pharmacology* 255, 327–338.

Kunnumakkara, A.B., Guha, S., Krishnan, S., Diagaradjane, P., Gelovani, J. and Aggarwal, B.B. (2007) Curcumin potentiates antitumor activity of gemcitabine in an orthotopic model of pancreatic cancer through suppression of proliferation, angiogenesis, and inhibition of nuclear factor-kappaB-regulated gene products. *Cancer Research* 67, 3853–3861.

Larrosa, M., Tomás-Barberán, F.A. and Espín, J.C. (2006) The dietary hydrolysable tannin punicalagin releases ellagic acid that induces apoptosis in human colon adenocarcinoma Caco-2 cells by using the mitochondrial pathway. *The Journal of Nutritional Biochemistry* 17, 611–625.

Lee, J.Y., Lee, Y.M., Chang, G.C., Yu, S.L., Hsieh, W.Y., Chen, J.J., Chen, J.J., Chen, H.W. and Yang, P.C. (2011) Curcumin Induces EGFR Degradation in Lung Adenocarcinoma and Modulates p38 Activation in Intestine: The Versatile Adjuvant for Gefitinib Therapy. *PLoS One* 6(8), e23756.

Lee, Y.K., Park, S.Y., Kim, Y.M. and Park, O.J. (2009) Regulatory effect of the AMPK-COX-2 signaling pathway in curcumin-induced apoptosis in HT-29 colon cancer cells. *Annals of the New York Academy of Sciences* 1171, 489–494.

Lei, F., Xing, D.M., Xiang, L., Zhao, Y.N., Wang, W., Zhang, L.J. and Du, L.J. (2003) Pharmacokinetic study of ellagic acid in rat after oral administration of pomegranate leaf extract. *Journal of Chromatography, B. Analytical Technologies Biomedical and Life Sciences* 796, 189–194.

Li, G.X., Chen, Y.K., Hou, Z., Xiao, H., Jin, H., Lu, G., Lee, M.J., Liu, B., Guan, F., Yang, Z., Yu, A. and Yang, C.S. (2010) Pro-oxidative activities and dose-response relationship of (-)-epigallocatechin-3-gallate in the inhibition of lung cancer cell growth: a comparative study *in vivo* and *in vitro*. *Carcinogenesis* 31, 902–910.

Lin, L.I., Ke, Y.F., Ko, Y.C. and Lin, J.K. (1998) Curcumin inhibits SK-Hep-1 hepatocellular carcinoma cell invasion *in vitro* and suppresses matrix metalloproteinase-9 secretion. *Oncology* 55, 349–353.

Lin, S.S., Lai, K.C., Hsu, S.C., Yang, J.S., Kuo, C.L., Lin, J.P., Ma, Y.S., Wu, C.C and Chung, J.G. (2009) Curcumin inhibits the migration and invasion of human A549 lung cancer cells through the inhibition of matrix metalloproteinase-2 and -9 and Vascular Endothelial Growth Factor (VEGF). *Cancer Letters* 285, 127–133.

Lin, Y.G., Kunnumakkara, A.B., Nair, A., Merritt, W.M., Han, L.Y., Armaiz-Pena, G.N., Kamat, A.A., Spannuth, W.A., Gershenson, D.M., Lutgendorf, S.K., Aggarwal, B.B. and Sood, A.K. (2007) Curcumin inhibits tumor growth and angiogenesis in ovarian carcinoma by targeting the nuclear factor-kappaB pathway. *Clinical Cancer Research* 13, 3423–3430.

López-Nicolás, J.M. and García-Carmona, F. (2008) Aggregation state and pKa values of (E)-resveratrol as determined by fluorescence spectroscopy and UV-visible absorption. *Journal of Agricultural and Food Chemistry* 56, 7600–7605.

Maiti, K., Mukherjee, K., Gantait, A., Saha, B.P. and Mukherjee, P.K. (2007) Curcumin-phospholipid complex: preparation, therapeutic evaluation and pharmacokinetic study in rats. *International Journal of Pharmaceutics* 330, 155–163.

Manson, M.M. (2003) Cancer prevention – the potential for diet to modulate molecular signalling. *Trends In Molecular Medicine* 9, 11–18.

Mateen, S., Tyagi, A., Agarwal, C., Singh, R.P. and Agarwal, R. (2010) Silibinin inhibits human nonsmall cell lung cancer cell growth through cell-cycle arrest by modulating expression and function of key cell-cycle regulators. *Molecular Carcinogenesis* 49(3), 247–258.

Mertens-Talcott, S.U. and Percival, S.S. (2005) Ellagic acid and quercetin interact synergistically with resveratrol in the induction of apoptosis and cause transient cell cycle arrest in human leukemia cells. *Cancer Letters* 218, 141–151.

Meyer, C., Pries, R. and Wollenberg, B. (2011) Established and novel NF-κB inhibitors lead to downregulation of TLR3 and the proliferation and cytokine secretion in HNSCC. *Oral Oncology* 47, 818–826.

Moorthi, C., Manavalan, R. and Kathiresan, K. (2011) Nanotherapeutics to Overcome Conventional Cancer Chemotherapy. *Journal of Pharmacy and Pharmaceutical Sciences* 14, 67–77.

Naasani, I., Oh-Hashi, F., Oh-Hara, T., Feng, W., Johnston, J., Chan, K. and Tsuruo, T. (2003) Blocking telomerase by dietary polyphenols is a major mechanism for limiting the growth of human cancer cells *in vitro* and *in vivo*. *Cancer Research* 63, 824–830.

Nair, J., Strand, S., Frank, N., Knauft, J., Wesch, H., Galle, P.R. and Bartsch, H. (2005) Apoptosis and age-dependant induction of nuclear and mitochondrial etheno-DNA adducts in Long-Evans Cinnamon (LEC) rats: enhanced DNA damage by dietary curcumin upon copper accumulation. *Carcinogenesis* 26, 1307–1315.

Nakagawa, K., Okuda, S. and Miyazawa, T. (1997) Dose-dependent incorporation of tea catechins, (-)-epigallocatechin-3-gallate and (-)-epigallocatechin, into human plasma. *Bioscience, Biotechnology and Biochemistry* 61, 1981–1985.

Nakamura, Y., Yogosawa, S., Izutani, Y., Watanabe, H., Otsuji, E. and Sakai, T. (2009) A combination of indol-carbinol and genistein synergistically induces apoptosis in human colon cancer HT29 cells by inhibiting Akt phosphorylation and progression of autophagy. *Molecular Cancer* 12, 100.

Nebert, D.W. and Dalton, T.P. (2006) The role of cytochrome P450 enzymes in endogenous signalling pathways and environmental carcinogenesis. *Nature Reviews Cancer* 6(12), 947–960.

Ning, L., Wentworth, L., Chen, H. and Weber, S.M. (2009) Down-regulation of Notch1 signaling inhibits tumor growth in human hepatocellular carcinoma. *American Journal of Translational Research* 1, 358–366.

Nishikawa, T., Nakajima, T., Moriguchi, M., Jo, M., Sekoguchi, S., Ishii, M., Takashima, H., Katagishi, T., Kimura, H., Minami, M., Itoh, Y., Kagawa, K. and Okanoue, T. (2006) A green tea polyphenol, epigalocatechin-3-gallate, induces apoptosis of human hepatocellular carcinoma, possibly through inhibition of Bcl-2 family proteins. *Journal of Hepatology* 44, 1074–1082.

Noh, E.M., Yi, M.S., Youn, H.J., Lee, B.K., Lee, Y.R., Han, J.H., Yu, H.N., Kim, J.S. and Jung, S.H. (2011) Silibinin enhances ultraviolet B-induced apoptosis in mcf-7 human breast cancer cells. *Journal of Breast Cancer* 14, 8–13.

Ouédraogo, M., Charles, C., Ouédraogo, M., Guissou, I.P., Stévigny, C. and Duez, P. (2011) An overview of cancer chemopreventive potential and safety of proanthocyanidins. *Nutrition and Cancer* 63, 1163–1173.

Pan, M.H. and Ho, C.T. (2008) Chemopreventive effects of natural dietary compounds on cancer development. *Chemical Society Reviews* 37, 2558–2574.

Park, I.J., Lee, Y.K., Hwang, J.T., Kwon, D.Y., Ha, J. and Park, O.J. (2009) Green tea catechin controls apoptosis in colon cancer cells by attenuation of H2O2-stimulated COX-2 expression via the AMPK signaling pathway at low-dose H2O2. *Annals of the New York Academy of Sciences* 1171, 538–544.

Peer, D., Karp, J.M., Hong, S., Farokhzad, O.C., Margalit, R. and Langer, R. (2007) Nanocarriers as an emerging platform for cancer therapy. *Nature Nanotechnology* 2, 751–760.

Pfeiffer, E., Hoehle, S.I., Walch, S.G., Riess, A., Sólyom, A.M., and Metzler, M. (2007) Curcuminoids form reactive glucuronides *in vitro*. *Journal of Agricultural and Food Chemistry* 55, 538–544.

Qin, J., Chen, H.G., Yan, Q., Deng, M., Liu, J., Doerge, S., Ma, W., Dong, Z. and Li, D.W. (2008) Protein phosphatase-2A is a target of epigallocatechin-3-gallate and modulates p53-Bak apoptotic pathway. *Cancer Research* 68, 4150–4162.

Rada, B. and Leto, T.L. (2008) Oxidative innate immune defenses by Nox/Duox family NADPH oxidases. *Contributions to Microbiology* 15, 164–187.

Ramasamy, K., Dwyer-Nield, L.D., Serkova, N.J., Hasebroock, K.M., Tyagi, A., Raina, K., Singh, R.P., Malkinson, A.M. and Agarwal, R. (2011) Silibinin prevents lung tumorigenesis in wild-type but not in iNOS-/- mice: potential of real-time micro-CT in lung cancer chemoprevention studies. *Clincal Cancer Research* 17, 753–761.

Ramos, S. (2007) Effects of dietary flavonoids on apoptotic pathways related to cancer chemoprevention. *The Journal of Nutritional Biochemistry* 18, 427–442.

Ramos, S. (2008) Cancer chemoprevention and chemotherapy: dietary polyphenols and signalling pathways. *Molecular Nutrition and Food Research* 52, 507–526.

Rao, S.D. and Pagidas, K. (2010) Epigallocatechin-3-gallate, a natural polyphenol, inhibits cell proliferation and induces apoptosis in human ovarian cancer cells. *Anticancer Research* 30, 2519–2523.

Ravindranath, V. and Chandrasekhara, N. (1980) Absorption and tissue distribution of curcumin in rats. *Toxicology* 16, 259–265.

Sartor, L., Pezzato, E., Donà, M., Dell'Aica, I., Calabrese, F., Morini, M., Albini, A. and Garbisa, S. (2004) Prostate carcinoma and green tea: (-)epigallocatechin-3-gallate inhibits inflammation-triggered MMP-2 activation and invasion in murine TRAMP model. *International Journal of Cancer* 112, 823–829.

Scalbert, A. and Williamson, G. (2000) Dietary intake and bioavailability of polyphenols. *The Journal of Nutrition* 130, 2073S–2085S.

Schwarz, D. and Roots, I. (2003) *In vitro* assessment of inhibition by natural polyphenols of metabolic activation of procarcinogens by human CYP1A1. *Biochemical and Biophysical Research Communications* 303, 902–907.

Semalty, A., Semalty, M., Rawat, M.S. and Franceschi, F. (2010) Supramolecular phospholipids-polyphenolics interactions: the PHYTOSOME strategy to improve the bioavailability of phytochemicals. *Fitoterapia* 81, 306–314.

Shankar, S., Ganapathy, S., Chen, Q. and Srivastava, R.K. (2008) Curcumin sensitizes TRAIL-resistant xenografts: molecular mechanisms of apoptosis, metastasis and angiogenesis. *Molecular Cancer* 7, 16.

Shao, Z.M., Shen, Z.Z., Liu, C.H., Sartippour, M.R., Go, V.L., Heber, D. *et al.* (2002) Curcumin exerts multiple suppressive effects on human breast carcinoma cells. *International Journal of Cancer* 98(2), 234–240.

Sharma, R.A., Euden, S.A., Platton, S.L., Cooke, D.N., Shafayat, A., Hewitt, H.R., Marczylo, T.H., Morgan, B., Hemingway, D., Plummer, S.M., Pirmohamed, M., Gescher, A.J. and Steward, W.P. (2004) Phase I Clinical Trial of Oral Curcumin: Biomarkers of Systemic Activity and Compliance. *Clinical Cancer Research* 10, 6847–6854.

Shehzad, A., Wahid, F. and Lee, Y.S. (2010) Curcumin in cancer chemoprevention: molecular targets, pharmacokinetics, bioavailability, and clinical trials. *Archiv Der Pharmazei (Weinheim)* 343, 489–499.

Siddiqui, I.A. and Mukhtar, H. (2010) Nanochemoprevention by bioactive food components: a perspective. *Pharmaceutical Research* 26, 1054–1060.

Siddiqui, I., Adhami, V., Bharali, D., Hafeez, B.B., Asim, M., Khwaja, S.I., Ahmad, N., Cui, H., Mousa, S.A. and Mukhtar, H. (2009) Introducing nanochemoprevention as a novel approach for cancer control: proof of principle with green tea polyphenol epigallocatechin-3-gallate. *Cancer Research* 69(5), 1712–1716.

Siddiqui, I.A., Adhami, V.M., Ahmad, N. and Mukhtar, H. (2010) Nanochemoprevention: sustained release of bioactive food components for cancer prevention. *Nutrition and Cancer* 62, 883–890.

Sies, H. (1997) Oxidative stress: oxidants and antioxidants. *Experimental Physiology* 82, 291–295.

Singh, M. and Singh, N. (2011) Curcumin counteracts the proliferative effect of estradiol and induces apoptosis in cervical cancer cells. *Molecular and Cellular Biochemistry* 347, 1–11.

Singh, M., Singh, R., Bhui, K., Tyagi, S., Mahmood, Z. and Shukla, Y. (2011) Tea polyphenols induce apoptosis through mitochondrial pathway and by inhibiting nuclear factor-kappaB and Akt activation in human cervical cancer cells. *Oncology Research* 19, 245–257.

Singh, R.P. and Agarwal, R. (2006) Prostate cancer chemoprevention by silibinin: bench to bedside. *Molecular Carcinogenesis* 45, 436–442.

Singh, R.P., Mallikarjuna, G.U., Sharma, G., Dhanalakshmi, S., Tyagi, A.K., Chan, D.C., Agarwal, C. and Agarwal, R. (2004) Oral silibinin inhibits lung tumor growth in athymic nude mice and forms a novel chemocombination with doxorubicin targeting nuclear factor kappa B-mediated inducible chemoresistance. *Clinical Cancer Research* 10, 8641–8647.

Singh, R.P., Raina, K., Deep, G., Chan, D. and Agarwal, R. (2009) Silibinin suppresses growth of human prostate carcinoma PC-3 orthotopic xenograft via activation of ERK1/2 and inhibition of STAT signaling. *Clinical Cancer Research* 15, 613–621.

Somers-Edgar, T.J., Scandlyn, M.J., Stuart, E.C., Nedelec, M.L., Valentine, S.P. and Rosengren, R.J. (2008) The combination of epigallocatechin gallate and curcumin suppresses ER alpha-breast cancer cell growth *in vitro* and *in vivo*. *International Journal of Cancer* 122, 1966–1971.

Spencer, J.P.E., Rice-Evans, C. and Williams, R.J. (2003) Modulation of pro-survival Akt/protein kinase B and ERK1/2 signaling cascades by quercetin and its *in vivo* metabolites underlie their action on neuronal viability. *The Journal of Biological Chemistry* 278, 34783–34793.

Subramaniam, D., May, R., Sureban, S.M., Lee, K.B., George, R., Kuppusamy, P., Ramanujam, R.P., Hideg, K., Dieckgraefe, B.K., Houchen, C.W. and Anant, S. (2008) Diphenyl difluoroketone: a curcumin derivative with potent *in vivo* anticancer activity. *Cancer Research* 68, 1962–1969.

Sukhthankar, M., Alberti, S. and Baek, S.J. (2010) (-)-Epigallocatechin-3-gallate (EGCG) post-transcriptionally and post-translationally suppresses the cell proliferative protein TROP2 in human colorectal cancer cells. *Anticancer Research* 30, 2497–2503.

Surh, Y.J. (2003) Cancer chemoprevention with dietary phytochemicals. *Nature Reviews Cancer* 3, 768–780.

Surh, Y.J. (2008) NF-kappa B and Nrf2 as potential chemopreventive targets of some anti-inflammatory and antioxidative phytonutrients with anti-inflammatory and antioxidative activities. *Asia Pacific Journal of Clinical Nutrition* 17, 269–272.

Svobodová, A., Zdarilová, A., Walterová, D. and Vostálová, J. (2007) Flavonolignans from *Silybum marianum* moderate UVA-induced oxidative damage to HaCaT keratinocytes. *Journal of Dermatological Science* 48, 213–224.

Tang, H.H., Zhou, M. and Liang, G. (2008) Impact of epigallocatechin gallate on gene expression profiles of human hepatocellular carcinoma cell lines BEL7404/ADM and BEL7402/5-FU. *Ai Zheng* 27, 325–332.

Thornberry, N.A. (1998) Caspases: key mediators of apoptosis. *Chemistry and Biology* 5, R97–103.

Thornberry, N.A. and Lazebnik, Y. (1998) Caspases: enemies within. *Science* 281, 1312–1316.

Tran, P.L., Kim, S.A., Choi, H.S., Yoon, J.H. and Ahn, S.G. (2010) Epigallocatechin-3-gallate suppresses the expression of HSP70 and HSP90 and exhibits anti-tumor activity *in vitro* and *in vivo*. *BMC Cancer* 10, 276.

Ullah, M.F., Ahmad, A., Zubair, H., Khan, H.Y., Wang, Z., Sarkar, F.H. and Hadi, S.M. (2011) Soy isoflavone genistein induces cell death in breast cancer cells through mobilization of endogenous copper ions and generation of reactive oxygen species. *Molecular Nutrition and Food Research* 55, 553–559.

Vareed, S.K., Kakarala, M., Ruffin, M.T., Crowell, J.A., Normolle, D.P., Djuric, Z. and Brenner, D.E. (2008) Pharmacokinetics of curcumin conjugate metabolites in healthy human subjects. *Cancer Epidemiology, Biomarkers and Prevention* 17, 1411–1417.

Varghese, L., Agarwal, C., Tyagi, A., Singh, R.P. and Agarwal, R. (2005) Silibinin efficacy against human hepatocellular carcinoma. *Clinical Cancer Research* 11, 8441–8448.

Vauzour, D., Rodriguez-Mateos, A., Corona, G., Oruna-Concha, M.J. and Spencer, J.P.E. (2010) Polyphenols and human health: prevention of disease and mechanisms of action. *Nutrients* 2, 1106–1131.

Velmurugan, B., Gangar, S.C., Kaur, M., Tyagi, A., Deep, G. and Agarwal, R. (2010) Silibinin exerts sustained growth suppressive effect against human colon carcinoma SW480 xenograft by targeting multiple signaling molecules. *Pharmaceutical Research* 27, 2085–2097.

Vu, H.A., Beppu, Y., Chi, H.T., Sasaki, K., Yamamoto, H., Xinh, P.T. *et al.* (2010) Green tea epigallocatechin gallate exhibits anticancer effect in human pancreatic carcinoma cells via the inhibition of both focal adhesion kinase and insulin-like growth factor-I receptor. *Journal of Biomedicine and Biotechnology* 2010, 290516.

Walle, T. (2011) Bioavailability of resveratrol. *Annals of the New York Academy of Sciences* 1215, 9–15.

Walle, T., Hsieh, F., DeLegge, M.H., Otis, J.E. Jr and Walle, U.K. (2004) High absorption but very low bioavailability of oral resveratrol in humans. *Drug Metabolism and Disposition: The Biological Fate of Chemicals* 32, 1377–1382.

Wang, J.B., Qi, L.L., Zheng, S.D. and Wu, T.X. (2009) Curcumin induces apoptosis through the mitochondria-mediated apoptotic pathway in HT-29 cells. *Journal of Zhejiang University Science B* 10, 93–102.

Wang, J.S., Luo, H., Wang, P., Tang, L., Yu, J., Huang, T., Cox, S. and Gao, W. (2008) Validation of green tea polyphenol biomarkers in a phase II human intervention trial. *Food and Chemical Toxicology* 46, 232–240.

Wang, W., Zhang, B., Chen, H. and Zhang, L. (2010) Anticancer activities of curcumin on human hepato-carcinoma cell line Sk-hep-1. *Zhongguo Zhong Yao Za Zhi* 35, 485–488.

Watanabe, S., Yamaguchi, M., Sobue, T., Takahashi, T., Miura, T., Arai, Y., Mazur, W., Wahala, K. and Adlercreutz, H. (1998) Pharmacokinetics of soybean isoflavones in plasma, urine and feces of men after ingestion of 60 g baked soybean powder (kinako). *The Journal of Nutrition* 128, 1710–1715.

Williamson G.F.D. and Renouf, M. (2011) Flavanols from green tea and phenolic acids from coffee: critical quantitative evaluation of the pharmacokinetic data in humans after consumption of single doses of beverages. *Molecular Nutrition and Food Research* 55, 864–873.

Wischke, C. and Schwendeman, S.P. (2008) Principles of encapsulating hydrophobic drugs in PLA/PLGA microparticles. *International Journal of Pharmaceutics* 364, 298–327.

Woting, A., Clavel, T., Loh, G. and Blaut, M. (2010) Bacterial transformation of dietary lignans in gnotobiotic rats. *FEMS Microbiology Ecology* 72, 507–514.

Yang, K.Y., Lin, L.C., Tseng, T.Y., Wang, S.C. and Tsai, T.H. (2007) Oral bioavailability of curcumin in rat and the herbal analysis from *Curcuma longa* by LC-MS/MS. *Journal of Chromatography. B, Analytical Technologies in the Biomedical and Life Sciences* 853, 183–189.

Yu, R., Jiao, J.J., Duh, J.L., Gudehithlu, K., Tan, T.H. and Kong, A.N. (1997) Activation of mitogen-activated protein kinases by green tea polyphenols: potential signaling pathways in the regulation of antioxidant-responsive element-mediated phase II enzyme gene expression. *Carcinogenesis* 18, 451–456.

Yue, G.G., Chan, B., Hon, P.M., Lee, M.Y., Fung, K.P., Leung, P. and Lau, C.B. (2010) Evaluation of *in vitro* anti-proliferative and immunomodulatory activities of compounds isolated from *Curcuma longa. Food and Chemical Toxicology* 48, 2011–2020.

16 Antioxidants: Their Health Benefits and Plant Sources

R.L. Singh,[1]* Sapna Sharma[2] and Pankaj Singh[1]
[1]Department of Biochemistry, Dr RML Avadh University, Faizabad, India; [2]Division of Nephrology, Department of Medicine, University of Chicago Medical Center, Chicago, USA

16.1 Introduction

An important field of research today is the control of 'free radicals' generation or redox status with the properties of food and food components. Reactive oxygen species (ROS) may interact with cellular macromolecules and modify several cellular proteins, lipids and DNA, which results in altered target cell functions. Oxidative stress occurs in a cell or tissue when the ROS generation level exceeds the antioxidant capability of that cell (Kumar *et al.*, 2011). ROS can be produced both endogenously and exogenously. Endogenous oxidative stress can be the result of normal cellular metabolism and oxidative phosphorylation. Exogenous sources of ROS can also impact on the overall oxidative status of a cell. Drugs, hormones and other xenobiotic chemicals can produce ROS by either direct or indirect mechanisms (Kakkar and Singh, 2007). Several human chronic disease states, including cancer, have been associated with oxidative stress produced through either an increased free radical generation and/or a decreased antioxidant level in the target cells and tissues (Rice-Evans and Burdon, 1993). Natural antioxidants present in the diet increase the resistance toward oxidative damages and they may have a substantial impact on human health. It has been reported that a diet rich in antioxidant phytochemicals, such as polyphenolics, carotenoids, terpenoids and flavonoids, protects against cellular damage due to ability to quench oxygen-derived free radicals (Dhakarey *et al.*, 2005; Singh, P., 2008; Singh, B.N., 2009a). If antioxidant defence systems are not sufficiently present in critical situations like oxidative stress, contamination, UV exposure etc., the production of free radicals increases significantly (Singh, U. *et al.*, 2008). Non-enzymatic (vitamin E, vitamin C, glutathione (GSH), etc.) and enzymatic (superoxide dismutase, GSH peroxides, glutathione-S-transferase and catalase) antioxidant levels in the cell can be decreased through modification in gene expression, decreased antioxidant uptake in the diet, or can be overloaded in ROS production, which creates a net increase in the amount of oxygen free radicals present in the cell. It has been reported that with the administration of antioxidants, cells are protected against carcinogen-induced damage (Kumar *et al.*, 2011). Mechanisms of protection could be effective against a wide

* E-mail: drrlsingh@rediffmail.com

range of dietary carcinogens possibly influencing several cancer sites. Antioxidant enzymes are detoxification/biotransformation enzymes that are involved in the detoxification of toxic substances such as xenobiotics, carcinogens, free radicals and peroxides by conjugating these substances with GSH (Tripathi *et al.*, 2010).

Traditional medicine all over the world is nowadays being revalued by an extensive amount of research on different plant species and their therapeutic principles. Experimental evidence suggests that free radicals (FR) and ROS can be involved in a high number of diseases (Richards and Sharma, 1991). As plants produce a lot of antioxidants to control the oxidative stress caused by sunlight and oxygen, they can represent a source of new compounds with antioxidant activity. One of the clinical specialities of Ayurveda is Rasayana. Rasayana is not only a drug therapy but is a specialized procedure practised in the form of rejuvenating recipes and dietary regimen promoting good habit. The purpose of Rasayana is two-fold: prevention of disease and counteraction of ageing processes which result from optimization of homeostasis. The meaning of the word Rasayana (rasa: essence, water; ayana: going) essentially refers to nutrition and its acquisition, movement, circulation and perfusion in the body tissues (Singh, 1992). With regard to Rasayana drug therapy, Sharma *et al.* (1992) reported the strong antioxidant activity of any Rasayana: these compounds were found to be 1000 times more potent than ascorbic acid, α-tocopherol and probucol.

16.2 Antioxidants

In living cells, two antioxidant defence system are present against free radical damage. The first line of defence includes antioxidant enzymes (such as superoxide dismutase, catalase, GSH peroxidase), whereas the second defence system includes low molecular non-enzymatic antioxidants (thioredoxin, GSH, vitamins A, C, E, lycopene, lutein, quercetin etc.). These antioxidants inhibit the formation of FRs by breaking the chain reaction or can reduce the concentration of FR by donating hydrogen and an electron. They also act as peroxide decomposers (vitamin E), enzyme inhibitors, singlet oxygen quenchers (vitamin E), synergists and metal-chelating agents (tranferritin). To provide maximum intracellular protection, antioxidants are strategically compartmentalized throughout the cell. As FR are produced intracellularly and extracellularly during metabolism, both enzymatic and non-enzymatic antioxidants are able to detoxify FRs.

Certain antioxidant enzymes (superoxide dismutase, catalase and glutathione peroxidase) are produced within the body. Other antioxidant agents are found in foods, such as green leafy vegetables, and it is believed that diets rich in antioxidant (such as β-carotene and vitamins A, C and E) are beneficial to human health (Halliwell and Gutteridge, 1989). Therefore, antioxidant naturally present in the body or supplied in the form of diet (phytonutrients) plays an important role to control various diseases resulting from oxidative stress. Fresh fruits and vegetables are of more importance than cooked, because of the high concentration and maximum absorption of antioxidants. In recent years, researchers have been researching the relationship between antioxidants and prevention of some diseases, such as cardiovascular disease and cancer (Kubola and Siriamornpun, 2008).

As soon as these FRs are generated in the body, they are trapped by antioxidants present in extracellular and intracellular defence system. If the generation of free radicals is much more than the concentration of antioxidants then oxidative stress arises. As a result of oxidative stress, arthritis in joints, emphysema and bronchitis in lungs, atherosclerosis or heart disease in the blood vessels, peptic ulcer in the stomach, ageing and wrinkling in the skin are caused. In the nucleus, it also alters the sequence of nucleotide base pairs, strand breaks etc. in the DNA resulting in transformed and mutated DNA. Mutated DNA will produce diseases like cancer, leukaemia and lymphoma (Prakash *et al.*, 2012).

16.2.1 Antioxidant enzymes

Three groups of enzymes play significant roles in protecting cells from oxidative stress.

Superoxide dismutase

Superoxide dismutase (SOD) has been recognized to play an important role in the body defence mechanism against the deleterious effect of superoxide FR in the biological system. It acts on two superoxide molecules and converts them into hydrogen peroxide and oxygen. The beneficial aspect of this reaction is that it produces less toxic hydrogen peroxide. The organisms that resist oxygen toxicity must have the SOD enzyme. On the basis of metal cofactor, the organism has three distinct types of SOD. In eukaryotes, cytosol has the copper- and zinc-containing form of SOD while mitochondria and bacterial cells have the manganese-containing form of SOD (Table 16.1). Iron-containing SOD is found in bacteria, cyanobacteria and some plants. Newly discovered forms of SOD, also found in bacteria, contain nickel as a cofactor. Interestingly, SODs are inducible enzymes, i.e. with the increase in the concentration of oxygen in the environment of the cell, the concentration of SOD enzyme also increases. The main source of naturally occurring SOD enzyme is green vegetables such as in broccoli, Brussels sprouts and cabbage, as well as barley, wheat and most green plants (Gassen and Youdim, 1999).

Catalase

The catalase activity of mammalian tissue varies greatly. It is highest in liver and kidney and low in connective tissue. In the cell, it is mainly particle bound (in mitochondria and peroxisomes) whereas in erythrocytes it exists in a soluble state. Catalase activity has received much attention for its role in oxidative metabolism as well as protective function by acting as a H_2O_2 scavenger. Catalase located in the organelles acts as a regulator of H_2O_2 levels and, on the other hand, in erythrocytes, catalase and GSH peroxidase jointly exert a protective function for haemoglobin and other SH-protein. It degrades hydrogen peroxide to water and oxygen, and hence finishes the detoxification reaction started by SOD (Gassen and Youdim, 1999).

Glutathione peroxidase

GSH peroxidase is a member of family of GPx enzymes, whose function is to detoxify

Table16.1. Important enzymatic and non-enzymatic physiological antioxidants.

Antioxidants	Location	Properties
Enzymatic		
Superoxide dismutase	Mitochondria, cytosol	Dismutase superoxide radicals
Glutathione peroxidase	Mitochondria, cytosol	Removes hydrogen peroxide and organic hydroperoxides
Catalase	Mitochondria, cytosol	Removes hydrogen peroxide
Non-enzymatic		
Vitamin E	Cell membrane	Chain-breaking antioxidant in cell membrane
Vitamin C	Aqueous phase of cell Sap	Acts as free radical scavenger and recycles vitamin E
α-Lipoic acid	Endogenous thiol	Effective in recycling vitamin C, may also be an effective glutathione substitute
Carotenoids	Membrane tissue	Scavengers of reactive oxygen species, singlet oxygen quencher
Bilirubin	Blood	Extracellular antioxidant
Ubiquinones	Mitochondria	Reduced forms are efficient antioxidants
Metals ions sequestration: transferrin, ferritin, lactoferrin		Chelating metals ions, responsible for Fenton reactions
Nitric oxide		Free radical scavenger, inhibitor of LP

peroxide in the cell. Peroxides decompose to form highly reactive free radicals, which can damage the macromolecules like protein, DNA and lipid. GPx enzyme plays an important role in the protection of cells from this damage, particularly lipid peroxidation. GSH peroxidase contains selenium as a cofactor. The synthesis of GSH peroxidase in humans appears to be very important in scavenging H_2O_2 (Cheng *et al.*, 2003).

16.2.2 Antioxidant phytochemicals

There are more than a thousand phytochemicals that have been identified with antioxidant properties. Plants produce these chemicals to protect themselves from microorganisms and oxidative stress, but now several evidences suggest that these phytochemicals also protect humans against various diseases caused by FRs. Some of the well-known phytochemicals are lycopene (tomatoes), isoflavones (in soy), flavanoids (in fruits, vegetables), allyl sulfides (onions, leeks, garlic), carotenoids (fruits, carrots) and polyphenols (tea, grapes). Medicinal plant parts are commonly rich in phenolic compounds, such as flavonoids, phenolic acids, stilbenes, tannins, cumarins, lignans and lignins. These compounds have multiple biological effects including antioxidant activity (Shukla *et al.*, 2009). The antioxidant activity of phytochemicals is mainly due to their redox properties, which can play an important role in adsorbing and neutralizing free radicals, quenching oxygen, or decomposing peroxides.

Flavonoids

Flavonoids are the most common secondary metabolites in higher plants, and can directly scavenge the superoxide ion, hydroxyl radical and H_2O_2. These include more than 4000 phenolic compounds that occur naturally in plants.

Flavonols

The main flavonol is quercetin, followed by myricetin, kaempferol, laricitrin, isorhamnetin and syringetin. The main sources of flavonols are onion, kale, broccoli, lettuce, tomato, apple, grape, berries, tea and red wine. High contents of flavonols are present in greener leaves (Manach *et al.*, 2004). Flavonols have multiple biological health benefits. They reduce risk of cardiovascular diseases, cancer, improve endothelial function and reduce platelet activity. This property is mainly attributed to their antioxidant properties (Patel, 2008). Furthermore, flavonols also help to prevent oxidative damage to cells, lipids and DNA. The antioxidant properties of flavonols are drawn from the presence of aromatic rings of the flavonoid molecule, which allows the donation and acceptance of electrons from FR species.

Anthocyanins

Anthocyanins are violet, blue and purple pigments, which are mainly present in fruits, berries and flowers. The major dietary anthocyanins include cyanidin, delphinidin, malvidin, pelargonidin, peonidin and petunidin (Manach *et al.*, 2004). Anthocyanins and their derivatives have the capacity to scavenge FRs through a number of mechanisms, thereby reducing the oxidative stress. Anthocyanins present in red cabbage reduce the oxidative stress caused by the toxin paraquat (Igarashi *et al.*, 2000). Tsuda (2000) reported that cyanidin, which is found in most fruit sources, has potential antioxidant activity under *in vivo* conditions. In another animal study, Tsuda (1998) reported that cyanidins protect cell membrane lipids from oxidation by a variety of harmful substances.

Tannins

Tannins are commonly present in fruits (grapes, persimmon, blueberry, etc.), tea, chocolate, legume forages and legume trees (*Acacia* sp., *Sesbania* spp. etc.) and grasses (sorghum, maize, etc.). Tannins include proanthocyanidins, gallotannins and ellagitannins. At high temperatures in alcohol solutions or in a strong mineral acid, proanthocyanidins release anthocyanidins, which have antioxidant properties. Gallotannins and ellagitannins are both hydrolysable tannins. Gallotannins constitute galloyl esters of glucose or quinic acid whereas ellagitannins are derivatives of hexahydroxydiphenic acid (HHDP). Another form of tannin is phloroglucinols, which are subunits of phlorotannins and present in marine brown algae only. Tannins give an astringent or bitter taste to foods and beverages (e.g. some red wines, teas and unripe fruits). The basic function of tannin is not as a primary antioxidant (i.e. they donate hydrogen atom or electrons) but they act as secondary antioxidants (i.e. interfere with the chain reaction or by chelating the metal ions such as Fe(II) thereby retarding oxidation or Fenton reaction). Zhang *et al.* (2004) showed that the inhibition of lipid peroxidation by tannin constituents can act via the inhibition of cyclooxygenase.

Phenolic acids

Phenolic acids are a major class of phenolic compounds, widely occurring in the plant kingdom. Predominant phenolic acids include hydroxybenzoic acids (e.g. gallic acid, *p*-hydroxybenzoic acid, protocatechuic acid, vanillic acid and syringic acid) and hydroxycinnamic acids (e.g. ferulic acid, caffeic acid, *p*-cumaric acid, chlorogenic acid and sinapic acid) (Wrigstedt *et al.*, 2010). Ferulic, caffeic and *p*-cumaric acid are present in many medicinal herbs and dietary spices, fruits, vegetables and grains. Wheat bran is a good source of ferulic acids. Hydroxycinnamic acids (non-flavonoid phenolics) are characterized by the C6–C3 structure. Plants use these compounds in both structural and chemical defence strategies against microbial flora as well as oxidative stress (Cartea *et al.*, 2011). Naturally occurring hydroxycinnamic acids possess greater antioxidant activity in comparison to hydroxybenzoic acid due to increased possibilities for delocalization of the phenoxy radical (Beer *et al.*, 2002). Phenolic compounds have the potential to function as antioxidants by scavenging the superoxide anion, hydroxyl radical, peroxy radical or quenching singlet oxygen and inhibiting lipid peroxidation in biological systems (Izunya *et al.*, 2010). At low temperatures during the maturity of leaves, the leaves have been shown to increase the phenols and flavonoids content (Singh, P. *et al.*, 2008; Singh, B.N., 2009c).

16.2.3 Antioxidant nutrients

Vitamin E

Vitamin E is the main lipid-soluble antioxidant and plays a vital role in protecting membranes from lipid peroxidation. Primary function of vitamin E is to trap peroxy radical formation during lipid peroxidation in cellular membranes. It is mainly present in nuts, seeds, vegetables, fish oils, whole grains (especially wheat germ), fortified cereals and apricots (Glenville, 2006). Current recommended daily allowance (RDA) is 15 IU day^{-1} for men and 12 IU day^{-1} for women.

Vitamin C or ascorbic acid

Vitamin C or ascorbic acid is a water-soluble antioxidant that can reduce a variety of free radicals. It acts as a synergist for tocopherol by converting the oxidized tocopherols back to their reduced status. Ascorbic acid can also act as a pro-oxidant under certain circumstances and helps regeneration of membrane-bound oxidized vitamin E. Vitamin C reacts with the α-tocopheroxyl radical and is oxidized to dehydroascorbic acid. Humans lack L-gulono-γ-lactone oxidase, which is a key enzyme in ascorbic acid synthesis, hence it cannot be synthesized in the body and must be acquired from dietary sources. Ascorbic acid is mainly present in citrus fruits and juices, kiwi, cabbage, green peppers, spinach, broccoli, kale, cantaloupe and strawberries. The RDA for vitamin C is 60 mg day^{-1}. If taken in high dosages it may be excreted out due to its water-soluble nature but may cause adverse side effects in some individuals. The efficiency of ascorbic acid as scavenger of superoxide in mammalian tissue is not less than the SOD enzyme. The ascorbic acid level in extracellular fluids is higher than those of glutathione. So, ascorbate probably plays a predominant role in extracellular antioxidant protection. Vitamin C reacts with the superoxide radical to form dehydroascorbic acid and it returns to its original state (vitamin C) with the help of gluthathione (Prakash *et al.*, 2012).

Glutathione

Glutathione, a tripeptide (glutamyl-cysteinyl-glycine) antioxidant, is the most important intracellular defence against damage by ROS. It is widely distributed among living cells and apparently involved in many biological functions. Glutathione present in the oxidized (GSSH) form is converted to the reduced GSH by enzyme glutathione reductase. It has been reported that reduced GSH is mainly present in tissue. The free sulfhydryl (SH) is a very reactive group in cysteine, providing a target for radical attack. Reduced glutathione is oxidized when it reacts with free radicals and it gets back to the reduced state by redox cycle involving GSH reductase and the electron acceptor NADPH (Gassen and Youdim, 1999).

Selenium

Selenium, an essential element for antioxidation reactions, is required only in small amounts in humans and animals (Thomson, 2004). Selenoproteins (proteins containing selenium) are important antioxidant enzymes. There are nearly 30 known selenoproteins, mainly containing selenocysteine. The active site of GSH peroxidase (the most abundant selenoprotein in mammals) and thioredoxin reductase enzyme has selenocysteine. Thioredoxin reductase not only maintains cell proteins in a reduced state but also provides deoxyribonucleases required for DNA synthesis (Holmgren, 1989). At low concentrations it acts as an antioxidant, inhibiting lipid peroxidation, whereas at higher concentrations it behaves as pro-oxidant, enhancing the accumulation of lipid peroxidation products. The antioxidant properties of selenoproteins help to regulate thyroid function, play an important role in the immune system and prevent cellular damage from free radicals (Corvilain *et al.*, 1993). Selenium deficiency may cause a form of heart disease, hypothyroidism and a weakened immune system (Zimmerman and Kohrle, 2002).

β-Carotene

β-carotene (precursor to vitamin A, retinol) is present in liver, egg yolk, butter, milk, spinach, squash, carrots, broccoli, tomato, yams, cantaloupe, peaches and grains. β-carotene is converted to vitamin A by the body. The carotenoids (fat-soluble antioxidant) are one of the most common pigments found in nature (Daun, 1988). β-carotene (one of the best known carotenoids) is necessary for the synthesis of vitamin A. Some other related pigments include α-carotene, lutein, lycopene and astaxanthin. There is evidence that a diet containing fruit and vegetables is associated with lower incidences of cancer (Giovannucci, 1999). β-carotene has the capacity to quench reactive oxygen (stop oxidative mechanisms), making them chemoprotective against cancer. There is strong evidence that β-carotene increases the detoxification of carcinogens present in the liver, thereby reducing the development of cancer (Solomons, 2001).

Metal-binding protein

Transition metals are tightly bound to various proteins that prevent them from reacting with peroxides to form free radicals. These include the following.

Ceruloplasmin

Ceruloplasmin is an effective antioxidant with potent peroxidase property. It decomposes hydrogen peroxide in the presence of reduced glutathione. Ceruloplasmin is expressed mainly in the liver but has been found to be expressed in the lungs (Fleming *et al.*, 1991) and mammary glands. The role of ceruloplasmin as antioxidant is against organic and inorganic oxygen radicals from iron and ascorbate. It contains 90–95% of the circulating copper in normal mammals. The concentration of ceruloplasmin increases by a factor of 2 to 3 during pregnancy and hormonal conditions. It also inhibits lipid peroxidation induced by ferrous ion by way of decomposing lipid peroxides (Verma *et al.*, 2005).

Lactoferrin

Lactoferrin belongs to the iron transporter or transferrin family of glycoproteins and is mainly present in whey and exocrine secretions from mammals and is released from neutrophil granules during inflammation. Human breast milk may contain as much as 15% lactoferrin while cow's milk may have only 0.5% to 1.0%. It has two important roles: (i) it shows antibacterial, antiviral, antifungal, anti-inflammatory, antioxidant and immunomodulatory activities; and (ii) lactoferrin plays an important role in the uptake and absorption of iron through the intestinal mucosa. Its ability to bind iron probably contributes to both its antioxidant properties and its antibacterial action (Gupta *et al.*, 2012).

Metallothionein

Metallothionein (MT) consists of four low-molecular-weight (6000–7000), metal-binding proteins with high cysteine content. Metallothioneins (MTs) are sulfhydryl-rich proteins, which specifically neutralize hydroxyl radicals (Viarengo *et al.*, 2000). Antioxidant properties of MTs are mainly due to sulfhydryl nucleophilicity. *In vitro* studies have revealed that it reacts directly with ROS including superoxide and hydroxyl radicals and hydrogen peroxide. Binding of transition metals (Fe, Cu) to the protein reduce the Fenton reactivity, resulting in reduced oxidative stress.

Transferrin

Transferrin (iron-binding blood plasma glycoprotein) has a molecular weight of approximately 80 kDa and binds iron very tightly but reversibly and hence controls the level of free iron in biological fluids (Crichton and Charloteaux-Wauters, 1987). It has two specific high-affinity Fe(III) binding sites. Iron present in the body is always found in protein-bound form and never in a free state. If iron is being transported or stored it must be chelated in very specific ways by transferrin or ferritin. Transferrin is mainly present in serum, but it is also found in other body fluids at lower concentrations (Chauhan *et al.*, 2004). The antioxidant activity of transferrin is due to its reducing properties. It reduces the concentration of free ferrous ion that catalyses the conversion of hydrogen peroxide to highly toxic hydroxyl radicals by Fenton reaction. Transferrin is a universal iron carrier and is able to deliver iron to cells without formation of free radicals.

Ferritin

Ferritin (a globular protein complex consisting of 24 protein subunits) is a ubiquitous intracellular protein that stores iron and releases it in a controlled fashion. Ferritin is synthesized by almost all living organisms, including algae, bacteria, higher plants and animals. Intracellular iron is stored in the ferritin in both prokaryotes and eukaryotes and released into cells when needed; hence it acts as buffer against iron deficiency. Ferritin that is not combined with iron is called apoferritin. Ferritin converts ferrous (Fe^{2+}) to ferric (Fe^{3+}) form by ferroxidase activity, thereby reducing the chance of the deleterious reaction that occurs between ferrous iron and hydrogen peroxide known as the Fenton reaction, which produces the highly damaging hydroxyl radical (Sarma *et al.*, 2010).

16.3 Some Commonly Measured Analytes with Antioxidant and Pro-oxidant Activities

16.3.1 Gamma-glutamyltransferase

Gamma-glutamyl transpeptidase (also known as γ-glutamyltransferase, GGT, GGTP, gamma-GT) (EC 2.3.2.2) is an enzyme that transfers γ-glutamyl functional groups. It is the first enzyme of the γ-glutamyl cycle that regulates the antioxidant glutathione; hence it is a critical enzyme in glutathione homeostasis. GGT is present in the cell membrane of many tissues, including the kidney, bile duct, pancreas, gallbladder, spleen, heart, brain and seminal vesicle (Sarma *et al.*, 2010).

16.3.2 Uric acid

Uric acid, the end product of purine metabolism, works as an antioxidant. It is the most abundant aqueous antioxidant in humans and contributes as much as two-thirds of all free-radical scavenging capacity in plasma. It is particularly effective in quenching hydroxyl, superoxide and singlet oxygen and peroxynitrite radicals and may play a protective physiological role by preventing lipid peroxidation. The major antioxidant role of uric acid is its ability to bind and inactivate peroxynitrite. At physiological concentrations, urate protects erythrocyte ghosts against lipid peroxidation leading to lysis of erythrocytes. Urate is found to be about as effective an antioxidant as ascorbate in these experiments. Urate is much more easily oxidized than deoxynucleosides by singlet oxygen and is destroyed by hydroxyl radicals at a comparable rate (Nieto *et al.*, 2000).

16.3.3 Bilirubin

Bilirubin, the end product of haem metabolism, has the ability to function as an antioxidant in the brain, scavenging free radicals and reducing oxidative damage. It is reported that bilirubin protects oxidation of lipids such as linoleic acid and vitamin A. Stocker *et al.* (1987) demonstrated that bilirubin has more of an antioxidant effect than vitamin E towards lipid peroxidation. It has also been experimentally proved that a higher concentration of serum bilirubin increases its antioxidant capacity.

16.3.4 High-density lipoprotein

High-density lipoprotein (HDL) has long been known as the 'good cholesterol', protecting against heart disease and atherosclerosis. It has been experimentally found that HDL has powerful antioxidant properties, similar to vitamin C and vitamin E. An enzyme related to synthesis of HDL cholesterol, lecithin-cholesterol acyltransferase, is a powerful antioxidant enzyme that blocks the oxidization of low-density lipoprotein (LDL) cholesterol. Cholesterol is beneficial if it is not oxidized. Barter *et al.* (2007) suggested that a low level of HDL increases the risk of diseases even in people with very low LDL levels. Jafri *et al.* (2010) suggested that there is an inverse relationship between high HDL and cancer occurrence.

16.3.5 Nitric oxide

Nitric oxide is an uncharged lipophilic molecule that behaves like an amphoteric molecule, i.e. NO could function as an electron donor (oxidant) or an electron acceptor (antioxidant) (Drew and Leeuwenburgh, 2002). It contains a single unpaired electron (NO•), which reacts with other molecules, such as oxygen, GSH and superoxide radicals. They prevent free radicals from stealing electrons from other molecules.

16.4 Sources of Natural Antioxidants

Dietary antioxidants include ascorbate, tocopherols, carotenoids and bioactive plant phenols. The health benefits of fruits and vegetables are largely due to the antioxidant vitamins supported by the large number of phytochemicals, some with greater antioxidant properties. Sources of tocopherols, carotenoids and ascorbic acid are well known and there are plenty of publications related to their roles in health. Exogenous dietary antioxidants capable of scavenging free radicals are of great interest in combating oxidative stress-induced cell damage. Plants containing a high content of polyphenols and flavonoids are considered as potential antioxidants and can be used as adjuvant therapy. These plant polyphenols and flavonoids are multifunctional and can act as reducing agents, hydrogen donors, singlet oxygen quenchers and metal ion chelators (Gassen and Youdim, 1999).

Several natural antioxidants such as silymarin, grape seed extract, resveratrol, curcumin etc., are known to reduce oxidative stress and protect from hepatic damage. Ulusoy *et al.* (2012) reported antioxidant and anti-apoptotic effects of proanthocyanidine from grape seed extract. Silymarin, a flavonoid complex from *Silybum marianum*, has been used in the treatment of hepatitis, liver cirrhosis, viral hepatitis and fatty liver. It has been shown to have antioxidant, antilipid peroxidative, anti-inflammatory and liver regenerative effects. Lupeol, a pentacyclic triterpenoid, found in many plants such as crataeva, mango, olive etc., received much attention due to its wide spectrum of medicinal properties that include antiprotozoal, anti-inflammatory, anticarcinogenic, cardioprotective and antimicrobial activities. Hepatoprotective action of lupeol against aflatoxin B1-induced toxicity has been reported by Preetha *et al.* (2006).

Cymbopogon citratus D. Stapf., commonly known as lemongrass, contains volatile oil consisting of citral, a monoterpene (a mixture of two isomeric aldehydes, neral and geranial in the ratio of 2:3), as a major component, which is used in various perfume and cosmetic industries (Rauber *et al.*, 2005). The plant is reported to possess antifungal, mosquito repellent, insecticidal, antidiabetic, antiseptic, antimutagenic and anticarcinogenic activity (Masuda *et al.*, 2008).

Fumaria parviflora Lam. (Fp) is used for dermatological diseases, stimulation of liver function and gall bladder, as antiscabies, antiscorbite, antibronchite, diuretic, expectorant, antipyretic, diaphoretic, appetizer and antineoplastic agent. Its antinoceceptive effect has also been worked out (Heidari *et al.*, 2004). Phytochemical analysis of Fp indicated presence of organic acids and isoquinoline alkaloids, namely: fumaric acid, protropine, cryptopine, sinactine, stylopine, dihydrofumariline, per-fumidine and dihydrosanguirine (Suau *et al.*, 2002). Acetylcholinesterase and butyrylcholinesterase inhibitory activity of Fp has also been reported (Orhan *et al.*, 2004). Significant oral antipyretic activity has been shown by hexane-chloroform and water-soluble extracts of Fp in rabbits (Akhtar *et al.*, 1984). A 50% ethanolic extract of Fp was also tested to discover the role of mitochondria and ROS/oxidative stress in cytoprotective and anti-apoptotic effects against nimesulide-induced hepatotoxicity (Tripathi *et al.*, 2010).

Glycyrrhiza glabra (liquorice) possesses triterpene, saponins, glycyrrhizin/glycyrrhizic acid and glycyrrhetic acid. Glycyrrhizic acid (GA), a biologically active constituent of liquorice root with a structure of 20b-carboxy-11-oxo-30-norolean-12en-3-b-yl-2-o-b-D-glucopyranosiduronic acid, is believed to be partly responsible for antiulcer, anti-inflammatory, antidiuretic, anti-epileptic, anti-allergic, antidote, antitumour, antiviral, antihypotensive and several other properties of the plant (Baltina, 2003). Hypocholesterolaemic and hypoglycaemic activities have also been reported (Sitohy *et al.*, 1991).

Bacopa monnieri Linn. (syn. *Herpestis monnieri* Linn. H.B. and K) is used as a nerve tonic, brain tonic, memory enhancer, laxative, astringent, antipyretic, anti-inflammatory and leprosy healer. It is also useful in renal disorders, blood diseases, cough, anaemia and poisoning. The plant also finds various applications in central nervous system depressant activity. Its major constituents including two saponins (bacoside A and bacoside B) have been isolated and characterized (Chowdhuri *et al.*, 2002).

Geraniol, an acyclic monoterpenoid, is an important constituent of essential oils of ginger, lemon, lime, lavender, nutmeg, orange, rose and palmarosa. It is reported to prevent cancer. Camphene, another component, is a bicyclic monoterpene with a pungent smell. It constitutes a minor part of many essential oils including turpentine oil, cypress oil, citronella oil, ginger oil etc., and is known to possess antilithic and expectorant properties. Camphene is also present in apricot, carrots, cinnamon, ginger, cumin seed, nutmeg, cardamom and turmeric. It is used as a food additive for flavouring as well as in the preparation of fragrances, plasticizers for resins and lacquers (Verschueren, 2001).

Free radicals generated in diabetes may lead to several kinds of diabetic complications including nephropathy, neuropathy, cardiopathy and many more. Many herbal medicines as single agents or in different oral formulations have been recommended for diabetes mellitus due to the fact that they are less toxic than oral hypoglycaemic agents such as sulfonylureas, metformin, etc. (Ponnachan *et al.*, 1993).

Anthocyanins have been shown to be natural anti-inflammatory agents and pain relievers. Chronic inflammation has also been associated with an increased risk of cancer, but anti-inflammatory drugs are not effective for reducing this type of inflammation (Singh, B.N. *et al.*, 2009b). Some important sources of antioxidants are presented in Table 16.2.

16.5 Roles of Antioxidants in the Prevention of Diseases

Plants have numerous natural antioxidants to control the oxidative stress induced by these free radicals (Pacher *et al.*, 1997; Sarma *et al.*, 2010). Free radicals have been implicated in the pathogenesis of over 100 human diseases such as cancer, heart disease, stroke, Alzheimer's disease, diabetes, premature ageing, high blood pressure and sepsis, to name a few.

16.5.1 Cancer

Antioxidants protect DNA thereby reducing the oxidative DNA damage caused by the free radical and ultimately control the increased abnormal cell division, the main characteristic of carcinogenesis. Experimental evidence using cell culture and animal models indicate that antioxidants either slow or prevent the development of cancer through their action as free-radical scavengers (Rock *et al.*, 1996). Using *in vitro* and an animal model system, it was experimentally found that plant-derived phytochemicals, such as allyl sulfides, isothiocyanates and sulforaphene, inhibit the various steps of tumour development (Milner, 1994). Blot *et al.* (1993) and Sardas (2003) reported that a combination of β-carotene, vitamin E and selenium significantly reduces the chance of cancer development especially in the case of gastric cancer. Experimental evidence also suggests that β-carotene with α-tocopherol/retinol significantly reduced the chance of lung cancer (Omenn *et al.*, 1994).

16.5.2 Alzheimer's disease

Alzheimer's disease (AD) is characterized by progressive loss of memory as the major clinical manifestation. Studies on free radicals suggest that oxidative stress causes neurodegenerative disorders, including AD. Metal ions also play an important role in the development of AD. Nutraceutical antioxidants such as β-carotene, curcumin, lutein, lycopene, turmerin etc., showed positive effects by reducing oxidative stress, mitochondrial dysfunction and various forms of neural degeneration (Glenville, 2006). It has been observed that a lower activity of antioxidant enzyme such as superoxide dismutase is related to occurrence of Alzheimer's disease in humans (Thome *et al.*, 1997). Kontush *et al.* (2001) reported that supplementation with vitamins E and C to the patient significantly increases the concentration of vitamins in plasma and decreases the oxidation of lipoprotein, while vitamin E alone does not have any significant effects. High intake of nutraceutical postpones the development of dementias such as AD (Haider and Bhutta, 2006).

16.5.3 Atherosclerosis

Atherosclerosis is a common cardiovascular disease, which occurs due to deposition of oxidized fatty acid to the arteries in the form of plaque. Approximately two-thirds of the serum cholesterol pool in a normal subject is low-density lipoprotein-cholesterol (LDL-C), which is believed to play an important role in the development of atherosclerosis (Shukla *et al.*, 2011).

Flavonoids and other plant-derived polyphenols, present in fresh fruits and vegetables, have been shown to be powerful antioxidants capable of preventing LDL oxidation induced by free radicals. Recommended daily allowance for the flavonoids is 1 g in an ordinary diet, which is sufficient for the antioxidant defence system. Interestingly, it has been found that the antioxidant activity of some of flavonoids synergistically increases when they are supplemented with ascorbic acid to prevent LDL oxidation. The beneficial properties

Table 16.2. Some important sources of antioxidants.

Plant	Antioxidants	References
Medicinal plants		
Terminalia chebula (Bahera)	Casuarinin, chebulanin and chebulinic acid	Cheng *et al.*, 2003
Cassia fistula (Amaltas)	Lupeol, β-sitosterol, hexacosanol, kaempferol, proanthocyanidin, bianthraquinone glycoside, anthraquinones, flavonoids, flavan-3-ol derivatives, sennoside A, sennoside B	Akiremi *et al.*, 2000
Withania somnifera (Ashwagandha)	Withanolides, cuscohygrine, anahygrine, tropane, pseudotropine, anaferine, dl-iso-pllatierine, withanine, withasominine, withaninine, somniferin, pseudowithanine, tropanol, pseudotropanol, cuscokygrene, 3-tigioyloxytropana, isopelletierine	Sangwan, 2004; Mohammad and Elisabeth, 2009; Kushwaha and Karanjekar, 2011
Fruits		
Berries (Sarashphal)	Flavanols, hydroxycinammic acids, hydroxybenzoic acids, anthocyanins	Wang and Lin, 2000; Yanishlieva-Maslarova and Heinonen, 2001
Citrus fruits	Flavanones, flavonols, phenolic acids	Yanishlieva-Maslarova and Heinonen, 2001; Manach *et al.*, 2004
Black grapes	Anthocyanins, flavonols	Belitz and Grosch, 1999; Yanishlieva-Maslarova and Heinonen, 2001
Cherries	Hydroxycinnamic acids, anthocyanins	Belitz and Grosch, 1999; Yanishlieva-Maslarova and Heinonen, 2001
Plums (Jamun), apples, pears	Hydroxycinnamic acids, catechin	Belitz and Grosch, 1999; Yanishlieva-Maslarova and Heinonen, 2001
Vegetables		
Allium sativum (Garlic)	Aliin, allicin, ajoene, allylpropyl disulfide, diallyl trisulfide, sallylcysteine, vinyldithiines, S-allylmercaptocystein, S-allylcysteine, S-allyl mercaptocysteine, saponins	Kemper, 2000; Amagase, 2006
Allium cepa (Onion)	Phenolic acids, flavonoids, cepaenes, thiosulfinates, anthocyanins, sulfur compounds, saponins, quercetrin	Singh, B.N. *et al.*, 2009a; Panduranga Murthy *et al.*, 2011
Trigonella foenum-graecum (Fenugreek)	Cumarin, fenugreekine, nicotinic acid, sapogenins, phytic acid, scopoletin, trigonelline, L-tryptophan-rich proteins and saponins	Yoshikawa *et al.*, 1997
Daucus carota (Carrot)	Carotol, daucene, germacrene D, bergamotene, selinene, carotol, daucol, copaenol	Ozcan and Chalchat, 2007
Sweet potato leaves	Flavonols, flavones,	Chu *et al.*, 2000
Yellow onion	Flavonols	Manach *et al.*, 2004
Beans	Flavanols	Manach *et al.*, 2004
Spinach	Flavonoids, *p*-cumaric acid	Bergman *et al.*, 2001

Flours		
Oats, wheat, rice	Caffeic, ferulic acids	Yanishlieva-Maslarova and Heinonen, 2001
Drinks		
Orange juice	Flavanols	Manach *et al*., 2004
Coffee	Hydroxycinnamic acids	Manach *et al*., 2004
Chocolate	Flavanols	Manach *et al*., 2004
Red wine	Flavan-3-ols, flavonols, anthocyanins	Manach *et al*., 2004
Herbs and spices		
Sage, carnosol	Carnosic acid, lateolin, rosmanul, rosmarinic acid	Yanishlieva-Maslarova and Heinonen, 2001
Foeniculum vulgare (Fennel)	Essential oil (trans-anethole, α-phellandrene, α-pinene), dipentene, methyl chavicol, feniculun, anisaldehyde and anisic acid	Piccaglia and Marotti, 2001; Mimica-Dukic *et al*., 2003; Araque *et al*., 2007
Rosemary	Carnosic acid, carnosol, rosmarinic acid, rosmanol	Yanishlieva-Maslarova and Heinonen, 2001; Ibanez *et al*., 2003
Thyme	Thymol, carvacrol, flavonoids, lubeolin	Exarchou *et al*., 2002
Ginger	Gingerol and related compounds	Moure *et al*., 2001; Yanishlieva-Maslarova and Heinonen, 2001

of certain plants may be explained by the presence of some especially effective flavonoids like resveratrol, which has also been found in red wines. Probucol, a hypocholesterolaemic drug, has significant antioxidant activity and an *in vivo* study on rabbit showed that it has protective effects against atherosclerosis. In animal studies, aspirin has also been shown to prevent atherosclerosis (Jaichander *et al.*, 2008).

16.5.4 Heart diseases

There are several factors such as high cholesterol level, hypertension, diabetes, cigarette smoking etc. that provide a platform for the development of heart disease. Oxidation of low density lipoprotein (LDL-cholesterol) causes deposition of fatty acid in arteries leading to development of atherosclerosis, which ultimately causes heart disease (Anderson *et al.*, 1995). Heart disease is acquired with age because oxidized fatty acid gets more 'sticky' and easier to adhere to the artery walls. It is believed that high intake of ascorbic acid reconstitutes the endothelial dysfunctions (Ting *et al.*, 1997) and protects the circulating lipoprotein from free radicals.

16.5.5 Diabetes

Diabetes mellitus (DM) is characterized by hyperglycaemia (Grill and Bjorklund, 2000). Oxidative stress due to lack of antioxidant defences may also cause diabetes (Cross *et al.*, 1987; Maxwell *et al.*, 1997; Keaney and Loscalzo, 1999; Bonnefont-Rousselot *et al.*, 2000; West, 2000). It is hypothesized that if ROS are involved in the genesis of diabetes, then antioxidants may be an effective approach in prevention of diabetes (Giugliano *et al.*, 1996). Reaven (1995) revealed that supplementation of vitamin E reduces the sensitivity of LDL to *in vitro* oxidation and availability of oxidized LDL in type-2 diabetics as well as in healthy subjects (Liao *et al.*, 1995). It is hypothesized that imbalance between generation and scavenging of free radicals is the main cause associated with diabetes. Insulin increases the uptake of vitamin C in to the cell but in hyperglycaemic conditions this process is inhibited resulting in a condition known as 'tissue scurvy'. Supplementation of vitamin C alone controls the blood glucose level, improves endothelium-dependent vasodilation and increases the resistance of lipoprotein towards oxidation in the patient with either type-1 or type-2 diabetes mellitus (Ting *et al.*, 1996; Timimi *et al.*, 1998; Kawano *et al.*, 1999).

16.5.6 Parkinson's disease

Parkinson's disease (PD) results from damage in neuronal cells in certain regions of the brain, and is characterized by muscle rigidity, shaking and difficulty in walking (Losso, 2003). Latif *et al.* (2007) reported that vitamin E in food may be protective against PD. Glutathione has also shown some promising results in preliminary studies to treat PD but appropriate long-term dosing, side-effects and the most effective method of administration are not yet clear.

16.6 Conclusions

Antioxidants may be a promising source for the prevention and or treatment of free radical-generated diseases such as atherosclerosis, hypertension, diabetes, cancer, Parkinson's and Alzheimer's diseases etc. Evidence also indicates that antioxidants protect/cure the diseases by involving a number of biological processes, including signal transduction pathways, activation of antioxidant defences, cell proliferation, cell survival-associated gene expression, differentiation and preservation of mitochondrial integrity. To protect the cells and organ systems of the body against reactive oxygen species, humans have evolved a highly sophisticated and complex antioxidant protection system. It involves a variety of antioxidant components, both endogenous and exogenous in origin, that function interactively and synergistically to neutralize free radicals. Increasing dietary intake of antioxidants may help to maintain an adequate antioxidant status and, therefore, the normal physiological function of human beings.

References

Akhtar, M.S., Khan, Q.M. and Khaliq, T. (1984) Effects of *Euphorbia prostrata* and *Fumaria parviflora* in normoglycemic and alloxan-treated hyperglycaemic rabbits. *Planta Medica* 50, 138–142.

Akiremi, A.A., Omobuwajo, O.R. and Elujoba, A.A. (2000) Pharmcopieal standards for the fruits of *Cassia fistula* and *Cassia podocarpa*. *Nigerian Journal of Natural Products and Medicine* 4, 23–26.

Amagase, H. (2006) Clarifying the real bioactive constituents of garlic. *Journal of Nutrition* 136, 716S–725S.

Anderson, T.J., Meredith, I.T., Yeung, A.C., Frei, B., Selwyn, A.P. and Ganz, P. (1995) The effect of cholesterol-lowering and antioxidant therapy on endothelium-dependent coronary vasomotion. *The New England Journal of Medicine* 332, 488–493.

Araque, M., Rojas, L.B. and Usubillaga, A.D. (2007) Antibacterial activity of essential oil of *Foeniculum vulgare* Miller against multi resistant gram-negative bacilli from nosocomial infections. *Ciencia* 15, 366–370.

Baltina, L.A. (2003) Chemical modification of glycyrrhizic acid as a route to new bioactive compounds for medicine. *Current Medicinal Chemistry* 10, 155–171.

Barter, P., Gotto, A.M., LaRosa, J.C., Maroni, J., Szarek, M., Grundy, S.M., Kastelein, J.J.P., Bittner, V., Fruchart, J.C. and Treating to New Targets Investigators (2007) HDL cholesterol, very low levels of LDL cholesterol, and cardiovascular events. *New England Journal of Medicine* 357, 1301–1310.

Beer, D.D., Joubert, E., Gelderblom, W.C.A. and Mandey, M. (2002) Phenolic compounds: a review of their possible role as *in vivo* antioxidant of wine. *South African Journal of Enology* 23, 48–61.

Belitz, H.D. and Grosch, W. (1999) Phenolic compounds. In: *Food Chemistry*, 2nd edn. Springer, Berlin, pp. 764–775.

Bergman, M., Vershavsky, L., Gottlieb, H.E. and Grossman, S. (2001) The antioxidant activity of aqueous spinach extract: chemical identification of active fractions. *Phytochemistry* 58, 143–152.

Blot, W.J., Li, J.Y., Taylor, P.R., Guo, W., Dawsey, S., Qingang, G., Yang, C.S., Zheng, S.F., Gail, M., Li, G.Y., Yu, Y., Liu, B., Tangrea, J., Sun, Y.H., Liu, F., Fraumeni, J.F., Zhang, Jr Y.H. and Li, B. (1993) Nutrition intervention trials in Linxian, China: supplementation with specific vitamin/mineral combinations, cancer incidence, and disease-specific mortality in the general population. *Journal of the National Cancer Institute* 85, 1483–1491.

Bonnefont-Rousselot, D., Bastard, J.P., Jaudon, M.C. and Delattre, J. (2000) Consequences of the diabetic status on the oxidant/antioxidant balance. *Diabetes & Metabolism* 26, 163–176.

Cartea, M.E., Francisco, M., Soengas, P. and Velasco, P. (2011) Phenolic Compounds in *Brassica* Vegetables. *Molecules* 16, 251–280.

Chauhan, A., Chauhan, V., Brown, W.T. and Cohen, I. (2004) Oxidative stress in autism: Increased lipid peroxidation and reduced serum levels of ceruloplasmin and transferrin – the antioxidant proteins. *Life Sciences* 75, 2539–2549.

Cheng, H.Y., Lin, T.C., Yu, K.H., Yang, C.M. and Lin, C.C. (2003) Antioxidant and Free Radical Scavenging activities of *Terminalia chebula*. *Biological and Pharmaceutical Bulletin* 26, 1331–1335.

Chowdhuri, D.K., Parmar, D., Kakkar, P., Shukla, R., Seth, P.K. and Srimal, R.C. (2002) Antistress effects of bacosides of *Bacopa monnieri*: modulation of Hsp70 expression, superoxide dismutase and cytochrome P450 activity in rat brain. *Phytotherapy Research* 16, 639–645.

Chu, Y.H., Chang, C.L. and Hsu, H.F. (2000) Flavonoid content of several vegetables and their antioxidant activity. *Journal of the Science of Food and Agriculture* 80, 561–566.

Corvilain, B., Contempre, B., Longombe, A.O., Goyens, P., Gervy-Decoster, C., Lamy, F., Vanderpas, J.B. and Dumont, J.E. (1993) Selenium and the thyroid: How the relationship was established. *American Journal of Clinical Nutrition* 57, 244–248.

Crichton, R.R. and Charloteaux-Wauters, M. (1987) Iron transport and storage. *European Journal of Biochemistry* 164, 485–506.

Cross, C.E., Halliwell, B. and Borish, E.T. (1987) Oxygen radicals and human disease. *Annals of Internal Medicine* 107, 526–545.

Daun, H. (1988) The chemistry of carotenoids and their importance in food. *Clinical Nutrition* 7, 97.

Dhakarey, R., Uppadhyay, G., Singh, B.N., Singh, H.B., Prakash, D., Kumar, S., Singh, K.K. and Singh, R.L. (2005) Phenolic content and antioxidant potential of Rhododendron species. *Indian Society of Agricultural Biochemists* 18, 40–43.

Drew, B. and Leeuwenburgh, C. (2002) Aging and the role of reactive nitrogen species. *Annals of the New York Academy of Sciences* 959, 66–81.

Exarchou, V., Nenadis, N., Tsimidou, M., Gerothanassis, I.P., Troganis, A. and Boskou, D. (2002) Antioxidant activities and phenolic composition of extracts from Greek oregano, Greek sage and summer savory. *Journal of Agricultural and Food Chemistry* 50, 5294–5299.

Fleming, R.E., Whitman, I.P. and Gitlin, J.D. (1991) Induction of ceruloplasmin in rat lung during hyperoxia. *American Journal of Physiology* 260, 68–74.

Gassen, M. and Youdim, M.B. (1999) Free radical scavengers: chemical concept and chemical relevance. *Journal of Neural Transmission. Supplementa* 56, 193–210.

Giovannucci, E. (1999) Tomatoes, tomato-based products, lycopene, and cancer: review of the epidemiologic literature. *Journal of the National Cancer Institute* 91, 317.

Giugliano, D., Ceriello, A. and Paolisso, G. (1996) Oxidative stress and diabetic vascular complications. *Diabetes Care* 19, 257–267.

Glenville, M. (2006) Nutritional supplements in pregnancy: commercial push or evidence based. *Current Opinion in Obstetrics and Gynecology* 18, 642–647.

Grill, V. and Bjorklund, A. (2000) Dysfunctional insulin secretion in type 2 diabetes: role of metabolic abnormalities. *Cellular and Molecular Life Sciences* 57, 429–440.

Gupta, C., Prakash, D., Garg, A.P. and Gupta, S. (2012) Whey Proteins: A novel source of Bioceuticals. *Middle-East Journal of Scientific Research* 12, 365–375.

Haider, B.A. and Bhutta, Z.A. (2006) Multiple-micronutrient supplementation for women during pregnancy. *Cochrane Database of Systematic Reviews* 18, CD004905.

Halliwell, B. and Gutteridge, J.M.C. (1989) *Free Radicals in Biology and Medicine*, 2nd edn. Clarendon Press, Oxford, UK.

Heidari, R.M., Mandgary, A. and Enayati, M. (2004) Antinociceptive effects and toxicity of *Fumaria parviflora* Lam. in mice and rats. *DARU* 12, 136–140.

Holmgren, A. (1989) Thioredoxin and glutaredoxin systems. *Journal of Biological Chemistry* 264, 13963–13966.

Ibanez, E., Kubatova, A., Senorans, F.J., Cavero, S., Regiero, G. and Hawthorn, S.B. (2003) Subcritical water extraction of antioxidant compounds from rosemary plants. *Journal of Agricultural and Food Chemistry* 571, 375–382.

Igarashi, K., Kimura, Y. and Takenaka, A. (2000) Preventive effects of dietary cabbage acylated anthocyanins on paraquat-induced oxidative stress in rats. *Bioscience Biotechnology and Biochemistry* 64, 1600–1607.

Izunya, A.M., Nwaopara, A.O., Aigbiremolen, A., Odike, M.A.C., Oaikhena, G.A., Bankole, J.K. and Ogarah, P.A. (2010) Morphological and biochemical effects of crude aqueous extract of *mangifera indica* l. (mango) stem bark on the liver in wistar rats. *Research Journal of Applied Sciences, Engineering and Technology* 2, 460–465.

Jafri, H., Alsheikh-Ali, A.A. and Karas, R.H. (2010) Baseline and on-treatment high-density lipoprotein cholesterol and the risk of cancer in randomized controlled trials of lipid-altering therapy. *Journal of the American College of Cardiology* 55, 2846–2854.

Jaichander, P., Selvarajan, K., Garelnabi, M. and Parthasarathy, S. (2008) Induction of paraoxonase 1 and apolipoprotein A1 gene expression by aspirin. *Journal of Lipid Research* 49, 2142–2148.

Kakkar, P. and Singh, B.K. (2007) Mitochondria: a hub of redox activities and cellular distress control. *Molecular and Cellular Biochemistry* 305, 235–253.

Kawano, H., Motoyama, T., Hirashima, O., Hirai, N., Miyao, Y., Sakamoto, T., Kugiyama, K., Ogawa, H. and Yasue, H. (1999) Hyperglycemia rapidly suppresses flow-mediated endothelium-dependent vasodilation of brachial artery. *Journal of the American College of Cardiology* 34, 146–154.

Keaney, J.F. and Loscalzo, J. (1999) Diabetes, Oxidative Stress, and Platelet Activation. *Circulation* 99, 89–191.

Kemper, K.J. (2000) Garlic (*Allium sativum*). Available at: http://www.mcp.edu/herbal/default.htm (accessed 25 April 2013).

Kontush, A., Mann, U., Arlt, S., Ujeyl, A., Lührs, C., Müller-Thomsen, T. and Beisiegel, U. (2001) Influence of vitamin E and C supplementation on lipoprotein oxidation in patients with Alzheimer's disease. *Free Radical Biology and Medicine* 31, 345–354.

Kubola, J. and Siriamornpun, S. (2008) Phenolic contents and antioxidant activities of bitter gourd (*Momordica charantia* L.) leaf, stem and fruit fraction extracts in vitro. *Food Chemistry* 110, 881–890.

Kumar, M., Kumar, S. and Kaur, S. (2011) Investigations on DNA protective and antioxidant potential of chloroform and ethyl acetate fractions of *Koelreuteria paniculata* Laxm. *African Journal of Pharmacy and Pharmacology* 5, 421–427.

Kushwaha, R. and Karanjekar, S. (2011) Standardization of Ashwagandharishta formulation by TLC Method. *International Journal of Chem Tech Research* 3, 1033–1036.

Latif, S., Anwar, F., Ashraf, M. and Gilani, A.H. (2007) *Moringa oleifera*: a food plant with multiple medicinal uses. *Phytotherapy Research* 21, 17–25.

Liao, J.K., Shin, W.S., Lee, W.Y. and Clark, S. (1995) Oxidized LDL decreases the expression of eNOS. *Journal of Biological Chemistry* 270, 319–324.

Losso, J.N. (2003) Targeting excessive angiogenesis with functional foods and nutraceuticals. *Trends in Food Science & Technology* 14, 455–468.

Manach, C., Scalbert, A., Morand, C., Remesy, C. and Jimenez, L. (2004) Polyphenols: food sources and bioavailability. *American Journal of Clinical Nutrition* 79, 727–747.

Masuda, T., Odaka, Y., Ogawa, N., Nakamoto, K. and Kuninaga, H. (2008) Identification of geranic acid, a tyrosinase inhibitor in lemongrass *Cymbopogon citratus*. *Journal of Agricultural and Food Chemistry* 56, 597–601.

Maxwell, S.R., Thomason, H., Sandler, D., Leguen, C., Baxter, M.A., Thorpe, G.H., Jones, A.F. and Barnett, A.H. (1997) Antioxidant status in patients with uncomplicated insulin-dependent and non-insulin-dependent diabetes mellitus. *European Society for Clinical Investigation* 27, 484–490.

Milner, J.A. (1994) Reducing the Risk of Cancer. In: Goldberg, I. (ed.) *Functional Foods*. Chapman and Hall, New York, pp. 39–70.

Mimica-Dukic, N., Kujundzic, S., Sokovic, M. and Couladis, M. (2003) Essential oil composition and antifungal activity of *Foeniculum vulgare* Mill obtained by different distillation conditions. *Phytotherapy Research* 17, 368–371.

Mohammad, H.M. and Elisabeth, M. (2009) Steroidal Lactones from *Withania somnifera*, an Ancient Plant for Novel Medicine. *Molecules* 14, 2373–2393.

Moure, A., Cruz, J.M., Franco, D., Dominguez, J.M., Sineiro, J., Dominguez, H., Numez, M.J. and Parajo, J.C. (2001) Natural antioxidants from residual sources. *Food Chemistry* 72, 145–171.

Nieto, F.J., Iribarren, C., Gross, M.D., Comstock, G.W. and Cutler, R.G. (2000) Uric acid and serum antioxidant capacity: a reaction to atherosclerosis? *Atherosclerosis* 148, 131–139.

Omenn, G.S., Goodman, G., Thornquist, M., Grizzle, J., Rosenstock, L., Barnhart, S., Balmes, J., Cherniack, M.G., Cullen, M.R. and Glass, A. (1994) The beta-carotene and retinol efficacy trial (CARET) for chemoprevention of lung cancer in high risk populations: smokers and asbestos-exposed workers. *Cancer Research* 54, 2038s–2043s.

Orhan, I., Sener, B., Choudhary, M.I. and Khalid, A. (2004) Acetylcholinesterase and butyrylcholinesterase inhibitory activity of some Turkish medicinal plants. *Journal of Ethnopharmacology* 91, 57–60.

Ozcan, M.M. and Chalchat, J.C. (2007) Chemical composition of carrot seeds (*Daucus carota* L.) cultivated in Turkey: characterization of the seed oil and essential oil. *Grasasy Aceites* 58, 359–365.

Pacher, P., Beckman, J.S. and Liaudet, L. (1997) Nitric oxide and peroxynitrite in health and disease. *Physiological Reviews* 87, 315–424.

Panduranga Murthy, G., Mamtharani, D.R., Tejas, T.S. and Suarlikerimath, N.M. (2011) Phytochemical analysis, *in vitro* anti-bacterial and antioxidant activities of wild onion sps. *International Journal of Pharma and Bio Sciences* 2, 230–237.

Patel, J.M. (2008) A Review of Potential Health Benefits of Flavonoids. Lethbridge Undergraduate Research Journal. Vol. 3, Number 2. Available at: http://www.lurj.org/article.php/vol3n2/flavonoids.xml (accessed 25 April 2013).

Piccaglia, R. and Marotti, M. (2001) Characterization of some Italian types of wild fennel (*Foeniculum vulgare* Mill.). *Journal of Agricultural and Food Chemistry* 49, 239–244.

Ponnachan, P.T.C., Paulose, C.S. and Panikar, K.R. (1993) Effect of leaf extract of *Aegle marmelose* in diabetic rats. *Indian Journal of Experimental Biology* 31, 345–347.

Prakash, D., Upadhyay, G., Gupta, C., Pushpangadan, P. and Singh, K.K. (2012) Antioxidant and Free Radical scavenging activities of some promising wild edible fruits. *International Food Research Journal* 19, 1109–1116.

Preetha, S.P., Kanniappan, M., Selvakumar, E., Nagaraj, M. and Varalakshmi, P. (2006) Lupeol ameliorates aflatoxin B1-induced peroxidative hepatic damage in rats. *Comparative Biochemistry and Physiology Part C: Toxicology & Pharmacology* 143, 333–339.

Rauber, C.S., Guterrs, S.S. and Schapoval, E.E.S. (2005) LC determination of citral in *Cymbopogon citratus* volatile oil. *Journal of Pharmaceutical and Biomedical Analysis* 37, 597–601.

Reaven, P. (1995) Dietary and pharmacologic regimens to reduce lipid peroxidation in noninsulin-dependent diabetes mellitus. *American Journal of Clinical Nutrition* 62, 1483S–1489S.

Rice-Evans, C. and Burdon, R. (1993) Free radical-lipid interactions and their pathological consequences. *Progress in Lipid Research* 32, 71–110.

Richards, R.T. and Sharma, H.M. (1991) Free radicals in health and disease. *Indian Journal of Clinical Practice* 2, 15–26.

Rock, C.L., Jacob, R.A. and Bowen, P.E. (1996) Update on the biological characteristics of the antioxidant micronutrients: vitamin C, vitamin E, and the carotenoids. *Journal of the American Dietetic Association* 96, 693–702.

Sangwan, R.S. (2004) Photochemical variability in commercial herbal products and preparations of *Withania somnifera. Current Science* 10, 461.

Sardas, S. (2003) The role of antioxidants in cancer prevention and treatment. *Indoor Built Environment* 12, 401–404.

Sarma, A.D., Mallick, A.R. and Ghosh, A.K. (2010) Free Radicals and Their Role in Different Clinical Conditions: An Overview. *International Journal of Pharma Sciences and Research* 1, 185–192.

Sharma, H.M., Hanna, A.N., Kauffman, E.M. and Newman, H.A.I. (1992) Inhibition of human low-density lipoprotein oxidation in vitro by Maharishi Ayurveda herbal mixtures. *Pharmacology Biochemistry and Behavior* 43, 1175–1182.

Shukla, M., Singh, U., Singh, P. and Singh, R.L. (2009) Nutraceutical properties of agrowaste part of some citrus plants. *Journal of Ecophysiology and Occupational Health* 9, 97–103.

Shukla, M., Singh, S.V., Singh, P., Singh, U., Vishwakerma, S.P., Khanna, A.A., Sexana, J.K. and Singh, R.L. (2011) Anti-dyslipidimic and antioxidant activity of hydro-ethanolic fruit extract of *Ficus glomerota. Asian Journal of Pharmaceutical and Clinical Research* 4, 145–148.

Singh, B.N., Singh, B.R., Singh, R.L., Prakash, D., Dhakarey, R., Uppadhyay, G. and Singh, H.B. (2009a) Oxidative DNA damage protective activity, antioxidant and antiquorum sensing potential of *Moringa olifera. Food Chemistry and Toxicology* 47, 1109–1116.

Singh, B.N., Singh, B.R., Singh, R.L., Prakash, D., Sarma, B.K. and Singh, H.B. (2009b) Antioxidant antiquorum sensing activities of green pod of *Acacia nilotica* L. *Food Chemistry and Toxicology* 47, 778–786.

Singh, B.N., Singh, B.R., Singh, R.L., Prakash, D., Singh, D.P., Sarma, B.K., Uppadhyay, G. and Singh, H.B. (2009c) Polyphenolics from various extracts/fractions of red onion (*Allium cepa*) peel with potent antioxidant and antimutagenic activities. *Food Chemistry and Toxicology* 47, 1161–1167.

Singh, P., Singh, U., Shukla, M. and Singh, R.L. (2008) Antioxidant activity imparting biomolecules in *Cassia fistula. Advances in Life Sciences* 2, 23–28.

Singh, R.H. (1992) Rasayana and Vajikarana. In: Sharma, P.V. (ed.) *History of Medicine in India*. Indian National Science Academy, New Delhi.

Singh, U., Singh, P., Shukla, M. and Singh, R.L. (2008) Antioxidant activity of vegetables belonging to Papilionaceae family. *Advances in Life Sciences* 2, 31–36.

Sitohy, M.Z., el-Massry, R.A., el-Saadany, S.S. and Labib, S.M. (1991) Metabolic effect of licorice roots (*Glycyrrhiza glabra*) on lipid distribution pattern, liver and renal functions of albino rats. *Nahrung* 35, 799–806.

Solomons, N.W. (2001) Vitamin A and carotenoids. In: *Present Knowledge in Nutrition*, 8th edn. ISLI Press, Washington, DC.

Stocker, R., Glazer, A.N. and Ames, B.N. (1987) Antioxidant activity of albumin-bound bilirubin. *Proceedings of the National Academy of Sciences of the United States of America* 84, 5918–5922.

Suau, R., Cabezudo, B., Rico, R., Nájera, F. and López-Romero, J.M. (2002) Direct determination of alkaloid contents in *Fumaria* species by GC–MS. *Phytochemical Analysis* 13, 363–367.

Thome, J., Gsell, W., Rösler, M., Kornhuber, J., Frölich, L., Hashimoto, E., Zielke, B., Wiesbeck, G.A. and Riederer, P. (1997) Oxidative-stress associated parameters (lactoferrin, superoxide dismutases) in serum of patients with Alzheimer's disease. *Life Science* 60, 13–19.

Thomson, C.D. (2004) Assessment of requirements for selenium and adequacy of selenium status: a review. *European Journal of Clinical Nutrition* 58, 391–402.

Timimi, F.K., Ting, H.H., Haley, E.A., Roddy, M.A., Ganz, P. and Creager, M.A. (1998) Vitamin C improves endothelium-dependent vasodilation in patients with insulin-dependent diabetes mellitus. *Journal of the American College of Cardiology* 31, 552–557.

Ting, H.H.,Timimi, F.K., Boles, K.S., Creager, S.H.J., Gans, P. and Creager, M.A. (1996) Vitamin C improves endothelium-dependent vasodilation in patients with non-insulin-dependent diabetes mellitus. *Journal of Clinical Investigation* 97, 22–28.

Ting, H.H., Timimi, F.K., Haley, E.A., Roddy, M.A., Ganz, P. and Creager, M.A. (1997) Vitamin C improves endothelium-dependent vasodilation in forearm resistance vessels of humans with hypercholesterolemia. *Circulation* 95, 2617–2622.

Tripathi, M., Singh, B.K., Mishra, C., Raisuddin, S. and Kakkar, P. (2010) Involvement of mitochondria mediated pathways in hepatoprotection conferred by *Fumaria parviflora* Lam. extract against nimesulide induced apoptosis *in vitro*. *Toxicology In Vitro* 24, 495–508.

Tsuda, T. (1998) Dietary cyanidin 3-0-beta-D-glucoside increases *ex vivo* oxidative resistance of serum in rats. *Lipids* 33, 583–588.

Tsuda, T. (2000) The role of anthocyanins as an antioxidant under oxidative stress in rats. *Biofactors* 13, 133–139.

Ulusoy, S., Ozkan, G., Yucesan, F.B., Ersoz, S., Orem, A., Alkanat, M., Yulug, E., Kaynar, K. and Al, S. (2012) Anti-apoptotic and antioxidant effects of grape seed proanthocyanidin extract (GSPE) in preventing cyclosporine A-induced nephropathy. *Nephrology* http://dx.doi.org/10.1111/j.1440-1797.2012.01565.x.

Verma, V.K., Ramesh, V., Tewari, S., Gupta, R.K., Sinha, N. and Pandey, C.M. (2005) Role of bilirubin, vitamin c and ceruloplasmin as antioxidants in coronary artery disease [CAD]. *Indian Journal of Clinical Biochemistry* 20, 68–74.

Verschueren, K. (2001) *Handbook of Environmental Data on Organic Chemicals*, 4th edn, vols 1–2. John Wiley and Sons, New York, 419 pp.

Viarengo, A., Burlando, B., Ceratto, N. and Panfoli, I. (2000) Antioxidant role of metallothioneins: a comparative overview. *Cellular and Molecular Biology (Noisy-le-grand)* 46, 407–417.

Wang, S.Y. and Lin, H.S. (2000) Antioxidant activity in fruits and leaves of blackberry, raspberry, and strawberry varies with cultivar and development stage. *Journal of Agricultural and Food Chemistry* 48, 140–146.

West, I.C. (2000) Radicals and oxidative stress in diabetes. *Diabetic Medicine* 17, 171–180.

Wrigstedt, P., Kylli, P., Pitkanen, L., Nousiainen, P., Tenkanen, M. and Sipila, J. (2010) Synthesis and antioxidant activity of hydroxycinnamic acid xylan esters. *Journal of Agricultural and Food Chemistry* 58, 6937–6943.

Yanishlieva-Maslarova, N.N. and Heinonen, M. (2001) Sources of natural antioxidants. In: Pokorny, J., Yanishlieva, N. and Gordon, M. (eds) *Antioxidants in Food*. CRC Press, Boca Raton, Florida, pp. 210–249.

Yoshikawa, M., Murakami, T., Komatsu, H., Murakami, N., Yamahara, J. and Matsuda, H. (1997) Medicinal Foodstuffs: IV. Fenugreek seeds (1): structures of trigoneosides Ia, Ib, IIb, IIIa and IIIb new furostanol saponins from the seeds of Indian *Trigonella foenum- graecum* L. *Chemistry and Pharmacology Bulletin* 45, 81–87.

Zhang, Y.J., DeWitt, D.L., Murugesan, S. and Nair, M.G. (2004) Novel lipid-peroxidation and cyclooxygenase inhibitory tannins from *Picrorhiza kurrora* seeds. *Chemistry & Biodiversity* 1, 426–441.

Zimmerman, M.B. and Kohrle, J. (2002) The impact of iron and selenium deficiencies on iodine and thyroid metabolism: biochemistry and relevance to public health. *Thyroid* 12, 867–878.

17 Phytochemicals of Nutraceutical Importance from *Curcuma longa* L. and their Role in Human Health

Dhan Prakash* and Charu Gupta
Amity Institute for Herbal Research and Studies, Amity University, Noida, India

17.1 Introduction

Curcuma longa L. syn. *C. domestica* Val. ('Haldi') family *Zingiberaceae* is extensively cultivated for its rhizomes, which are dried, powdered and used as turmeric. It is a perennial herb distributed throughout tropical and subtropical regions of the world like India, Pakistan, Sri Lanka, China, Bangladesh and Indonesia (Reema *et al.*, 2006). The pharmacologic safety of curcumin (Fig. 17.1) is well demonstrated by the fact that people in certain countries have been consuming curcumin as a dietary spice for centuries (Ammon and Wahl, 1991). Its rhizomes are harvested, washed and boiled in mild alkaline water to soften and then dried in the sun or in electric driers. It is used as colouring matter in pharmacy, food industry and confectionery, and for dyeing wool, silk, cotton and in combination with other natural dyes to obtain different shades (Lawhavinit *et al.*, 2011). Rhizomes are used as cosmeceutical, expectorant, antiseptic, blood purifier, in leprosy, spleen disorders, rheumatism, bronchitis, cough and cold, insecticide, spasmolytic, hypotensive, cholera and syphilis (Yu *et al.*, 2002; Gayatri and Rajani, 2011). It is a minor ingredient of the 'Ayurvedic' drug for malarial fever; indigenous antifatigue drug Geriforte (Geri Care/Stress Care), Unani drug Majnoon-E-Falsfa, Vitafix, Opthacare, Purime (Hemo Care), V-Gel, Fem Care Gel, Acne-n-Pimple Cream, Anti-Wrinkle Cream, Blood purifier Capsules and Syrup, Foot care Cream, Dibecon (Gluco Care), Curcumin-97 and Curcumin 900 MG; in modern pharmaceutical products it is an ingredient of 'Geriforte', which is effective in senile pruritus, insecticide, spasmolytic, hypotensive, antifungal, anti-inflammatory, antibacterial and to fight decaying metabolism to prevent cancer (Johnson and Mukhtar, 2007; Goel *et al.*, 2008; Jurenka, 2009; Ronita *et al.*, 2009; Lee *et al.*, 2010).

17.2 Chemical Constituents of *Curcuma longa* L.

Turmeric contains protein (6.3%), fat (5.1%), minerals (3.5%), carbohydrates (69.4%) and moisture (13.1%). It contains several polyphenolic compounds within the range of 3–6% beneficial for health, collectively known as curcuminoids, which is a mixture of curcumin (Fig. 17.1), demethoxycurcumin (Fig. 17.2) and bisdemethoxycurcumin (Fig. 17.3) (Satyavati

* E-mail: dprakash_in@yahoo.com

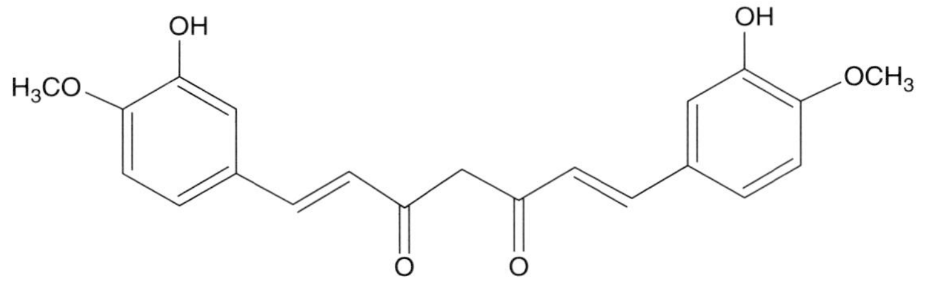

Fig. 17.1. Curcumin.

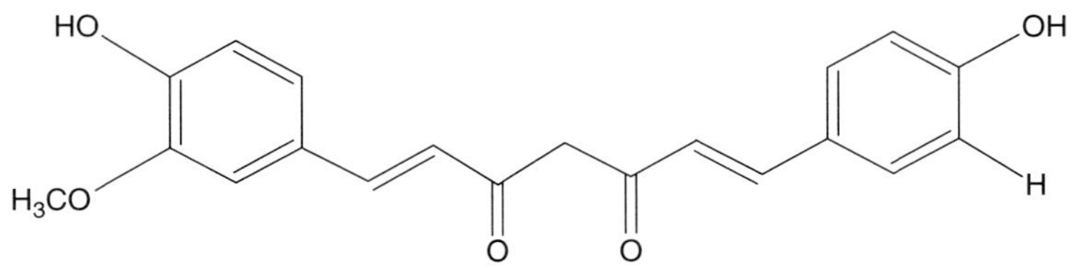

Fig. 17.2. Demethoxycurcumin.

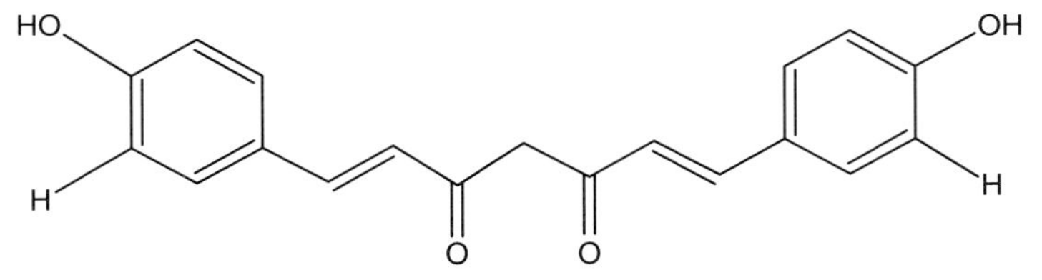

Fig. 17.3. Bisdemethoxycurcumin.

et al., 1976; Ravindranath and Satyanarayana, 1980). The main colouring principle of turmeric rhizomes is curcumin, which has a brilliant yellow hue at pH 2.5–7 and takes on a red hue at pH >7 (Ishita *et al.*, 2004). Other phenolic compounds present in rhizomes are 1-hydroxy,1,7-bis(4-hydroxy-3-methoxyphenyl)-(6E)-6-heptene-3,5-dione (Fig. 17.4), 1-(4-hydroxy-3,5-dimethoxyphenyl)-7-(4-hydroxy-3-methoxyphenyl)-(1E,6E)-1,6-heptadiene-3,4-dione (Fig. 17.5), 1,5-bis (4-hydroxy-3-methoxyphenyl)-penta-(1E,4E)-1,4-dien-3-one (Fig. 17.6), 1-(4-hydroxy-3-methoxyphenyl-5-(4-hydroxyphenyl)-penta-(1E,4E)-1,4-dien-3-one (Fig. 17.7), 1-(4-hydroxy-3-methoxyphenyl)-7-(3,4-dihydroxyphenyl)-1,6-heptadiene-3,5-dione (Fig. 17.8) and 1,7-bis(4-hydroxyphenyl)-1,4,6-heptatrien-3-one (Fig. 17.9) (Nakayama *et al.*, 1993). The pale yellow to orange-yellow volatile oil (4–6%) obtained from *C. longa* consists of a number of mono- and sesquiterpenes. The sesquiterpenes were named as curcumenone (Fig. 17.10), dehydrocurdione (Fig. 17.11), (4S,5S)-germacrone 4,5-epoxide (Fig. 17.12), bisabola 3,10-diene 2-one (Fig. 17.13), ar-turmerone (Fig. 17.14) (Roth *et al.*, 1998), bisacumol (Fig. 17.15), bisacurone (Fig. 17.16), curcumenol (Fig. 17.17), isoprocurcumenol (Fig. 17.18), zedoaronediol (Fig. 17.19), procurcumenol (Fig. 17.20), epiprocurcumenol (Fig. 17.21), germacrone-13-al (Fig. 17.22), 4-hydroxybisabola-2,10-diene-9-one (Fig. 17.23), 4,5-dihydroxybisabola-2,10-diene (Fig. 17.24), 4-methoxy-5-hydroxybisabola-2,10-diene-9-one (Fig. 17.25), 2,5-dihydroxybisabola-3,10-diene (Fig. 17.26) and procurcumadiol (Fig. 17.27) (Ohshiro *et al.*, 1990). Some other

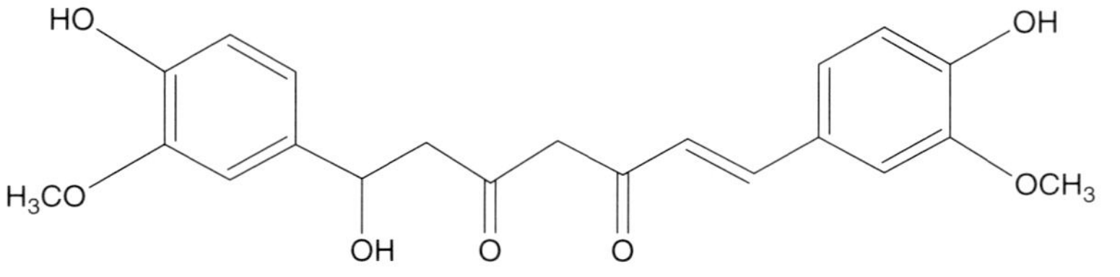

Fig. 17.4. 1-hydroxy-1,7-bis(4-hydroxy-3-methoxyphenyl)-(6E)-6-heptene-3,5-dione.

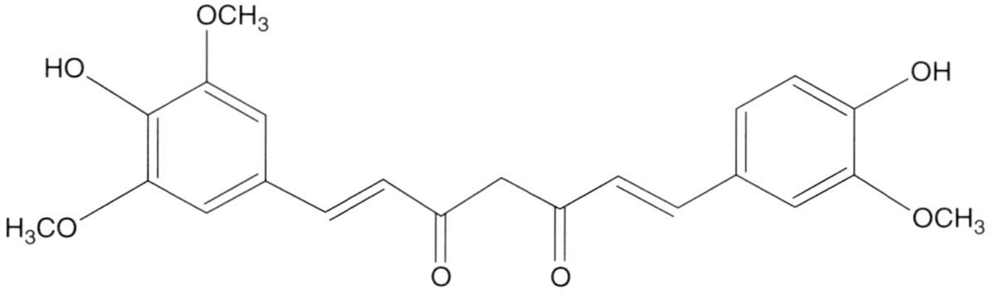

Fig. 17.5. 1-(4-hydroxy-3,5-dimethoxyphenyl)-7-(4-hydroxy-3-methoxyphenyl)-(1E,6E)-1,6-heptadiene-3,4-dione.

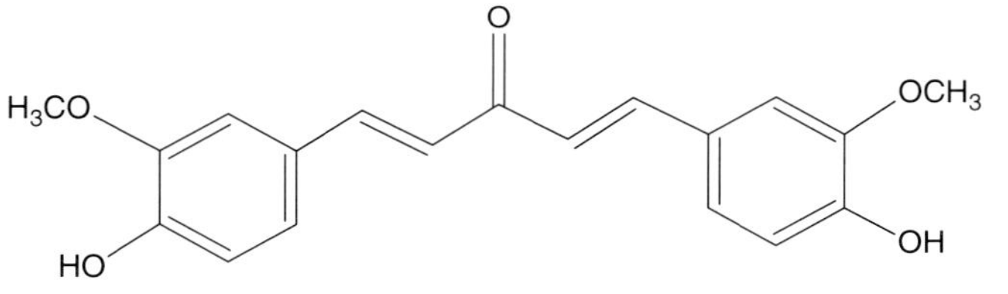

Fig. 17.6. 1,5-bis(4-hydroxy-3-methoxyphenyl)-penta-(1E,4E)-1,4-dien-3-one.

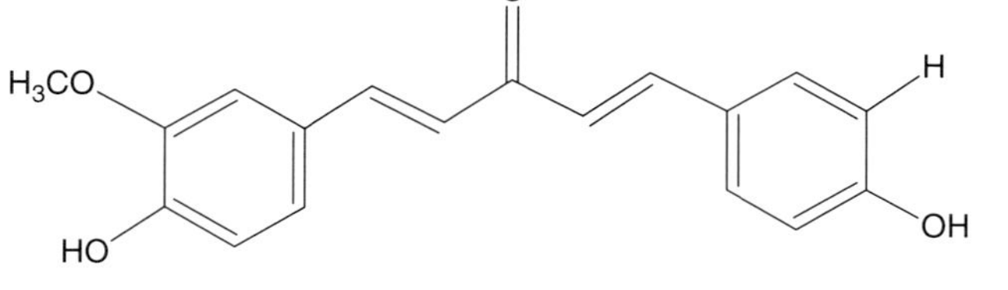

Fig. 17.7. 1-(4-hydroxy-3-methoxyphenyl-5-(4-hydroxyphenyl)-penta-(1E,4E)-1,4-dien-3-one.

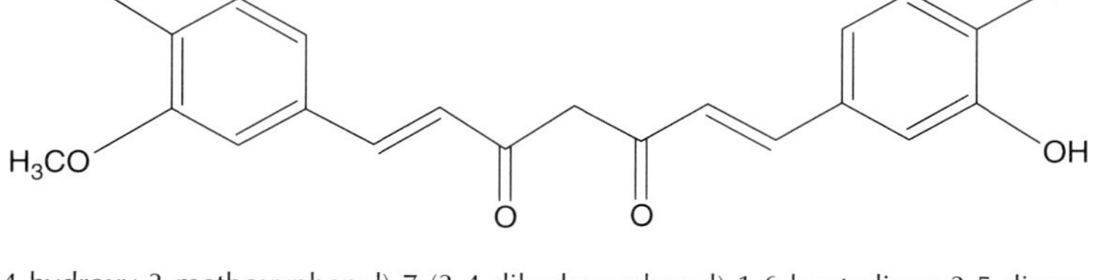

Fig. 17.8. 1-(4-hydroxy-3-methoxyphenyl)-7-(3,4-dihydroxyphenyl)-1,6-heptadiene-3,5-dione.

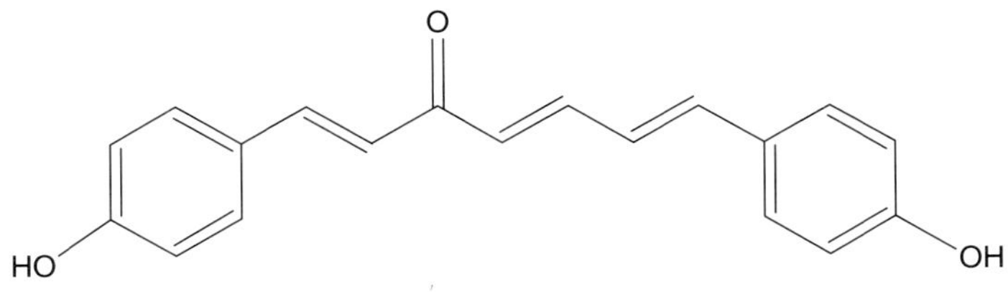

Fig. 17.9. 1,7-bis(4-hydroxyphenyl)-1,4,6-heptatrien-3-one.

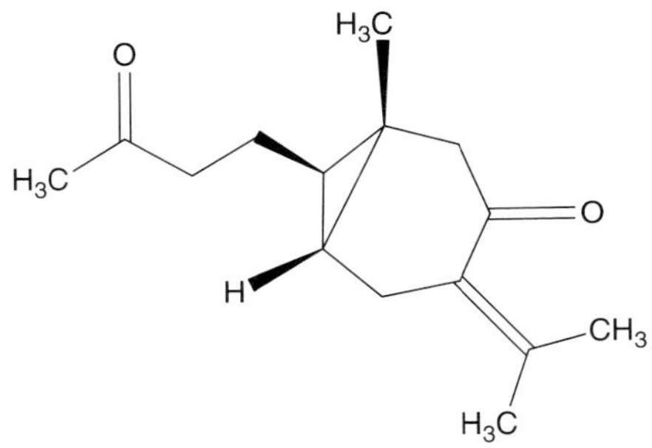

Fig. 17.10. Curcumenone.

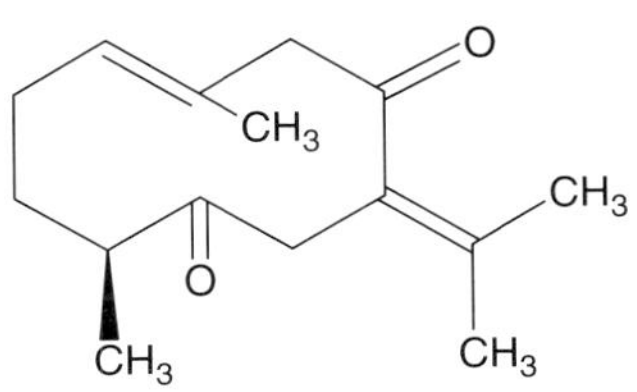

Fig. 17.11. Dehydrocurdione.

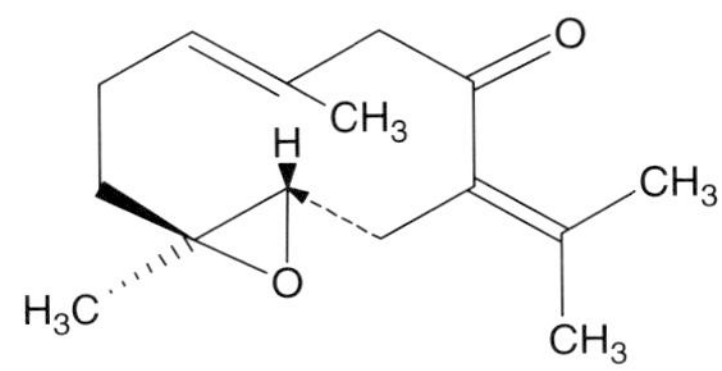

Fig. 17.12. (4 S, 5 S)-Germacrone 4,5-epoxide.

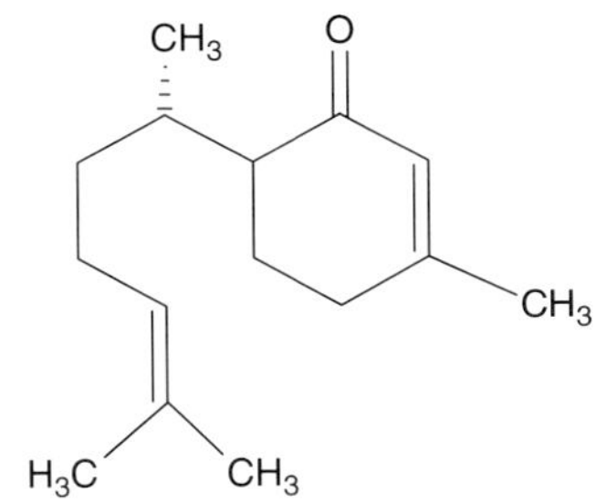

Fig. 17.13. Bisabola 3,10-diene 2-one.

compounds named as curlone (Fig. 17.28) (Kiso *et al.*, 1983), α-turmerone (Fig. 17.29), β-turmerone (Fig. 17.30), terpinolene (Fig. 17.31), α-phellandrene (Fig. 17.32), curcumadiol (Fig. 17.33), lambda-8(17)-diene-15,16-dial (Fig. 17.34) and three acidic polysaccharides named as Ukon A, B and C. They were composed of L-arabinose, D-xylose, D-galactose, D-glucose, L-rhamnose and D-galacturonic acid in the molar ratio 12:4:12:1:4:10, (Ukon A), 12:4:12:1:2:4 (Ukon B) and 8:3:6:14:2:3 (Ukon C). The polysaccharide named as Ukon D was composed of L-arabinose, D-galactose, D-glucose and D-mannose in the molar ratio of 1:1:12:2. The water soluble peptide was named as turmerin with an amino acid composition as aspartic acid/asparagine, glutamic acid/glutamine, serine, glycine, arginine, proline, alanine, tyrosine, valine, methionine, leucine, isoleucine and phenylalanine in the ratio: 1:2:3:8:1:1:1:3:2:6:3:4:5:3 (Rastogi and Mehrotra, 1990, 1991, 1993, 1995, 1998; Srinivas *et al.*, 1992; Rohman, 2012).

17.3 Biological Activity of *Curcuma longa* L.

A great variety of pharmacological activities of *C. longa* has been reported. Curcumin is one of its major components being responsible for its various biological actions. *In vitro*, it exhibits antiparasitic, antispasmodic, anti-inflammatory, anticarcinogenic and gastrointestinal, antifungal, antiviral, antiprotozoal and nematocidal

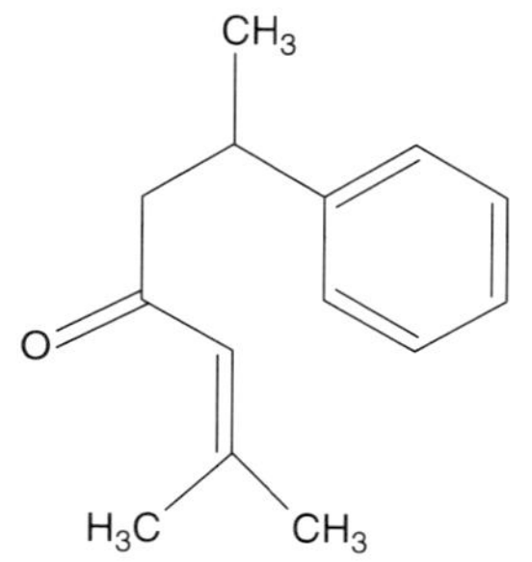

Fig. 17.14. ar-Turmerone.

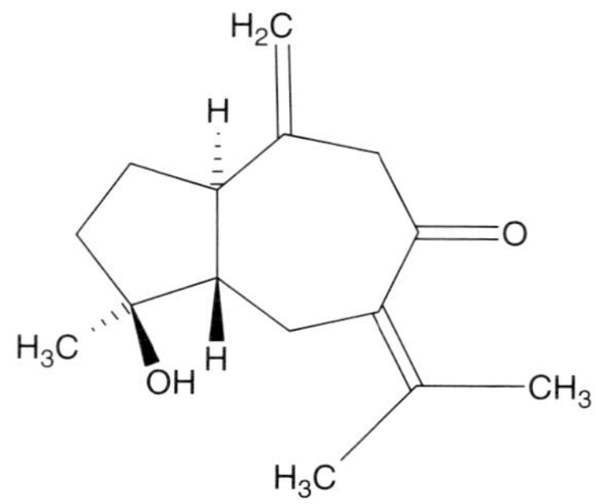

Fig. 17.18. Isoprocurcumenol.

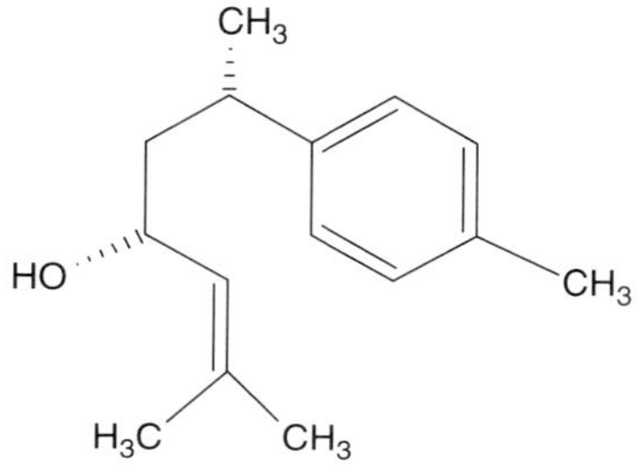

Fig. 17.15. Bisacumol.

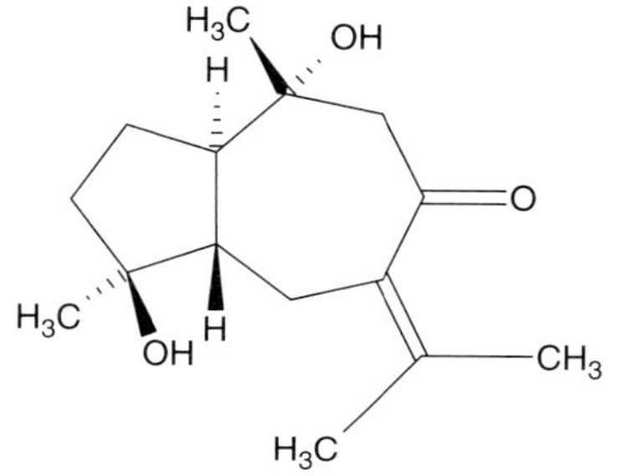

Fig. 17.19. Zedoaronediol.

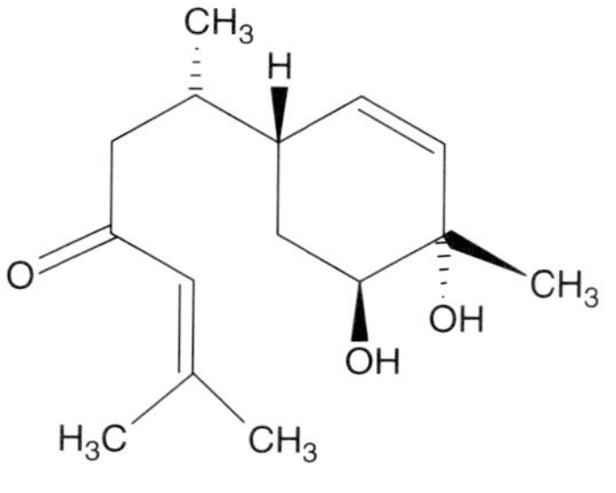

Fig. 17.16. Bisacurone.

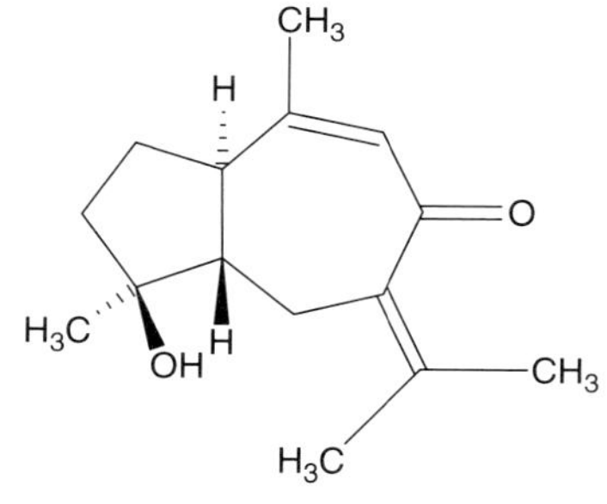

Fig. 17.20. Procurcumenol.

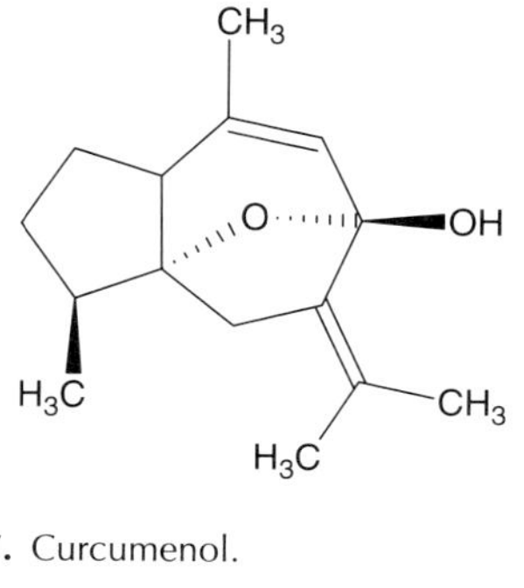

Fig. 17.17. Curcumenol.

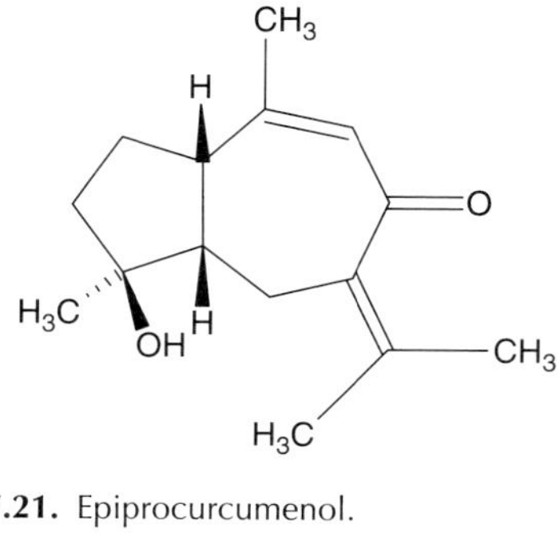

Fig. 17.21. Epiprocurcumenol.

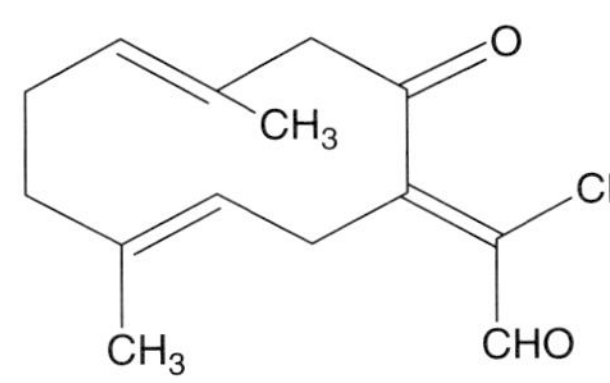

Fig. 17.22. Germacrone-13-al.

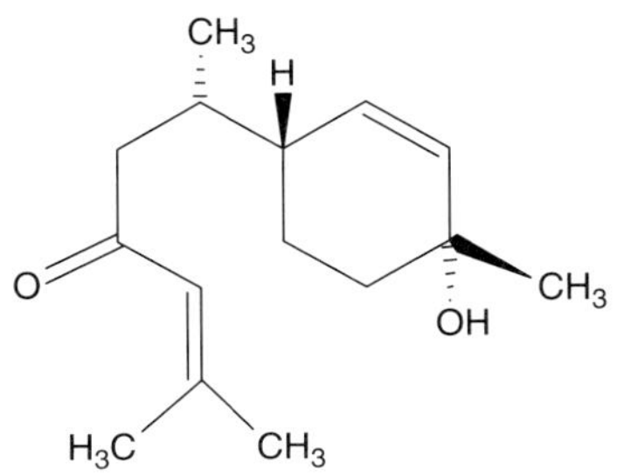

Fig. 17.23. 4-Hydroxybisabola-2,10-diene-9-one.

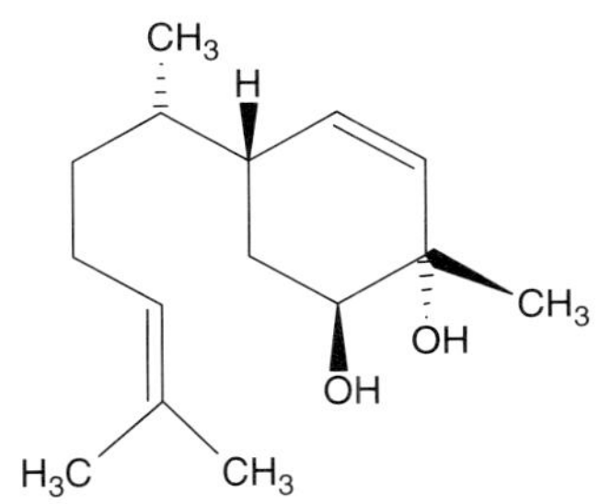

Fig. 17.24. 4, 5-Dihydroxybisabola-2, 10-diene.

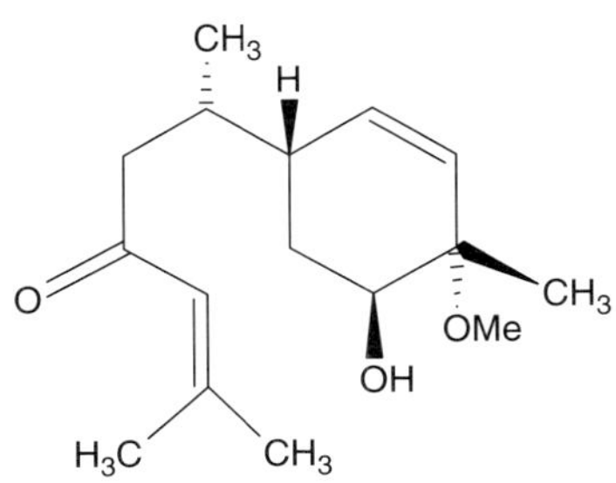

Fig. 17.25. 4-Methoxy-5-hydroxybisabola-2,10-diene-9-one.

properties (Cui *et al.*, 2007). *In vivo*, it has shown antiparasitic and anti-inflammatory activity through oral application in animal models (Araujo and Leon, 2001; Pérez-Arriaga *et al.*, 2006). As the principle active compound of *C. longa*, curcumin has been shown to interact with a wide variety of proteins, modifying

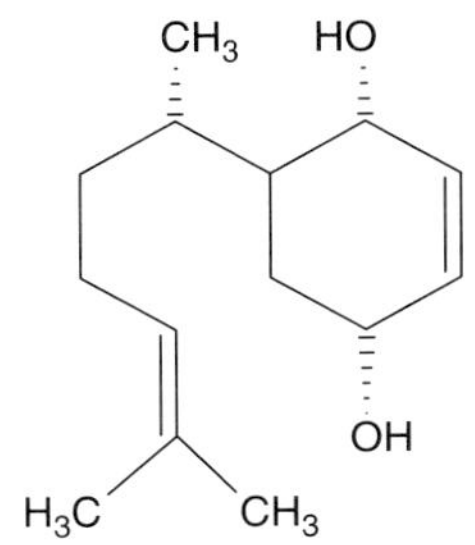

Fig. 17.26. 2,5-Dihydroxybisabola-3,10-diene.

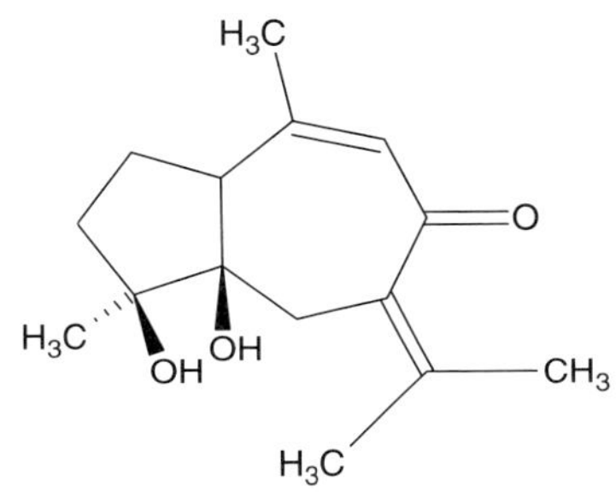

Fig. 17.27. Procurcumadiol.

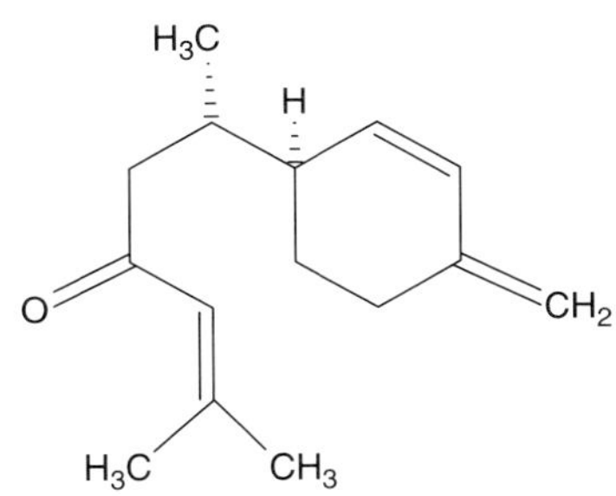

Fig. 17.28. Curlone.

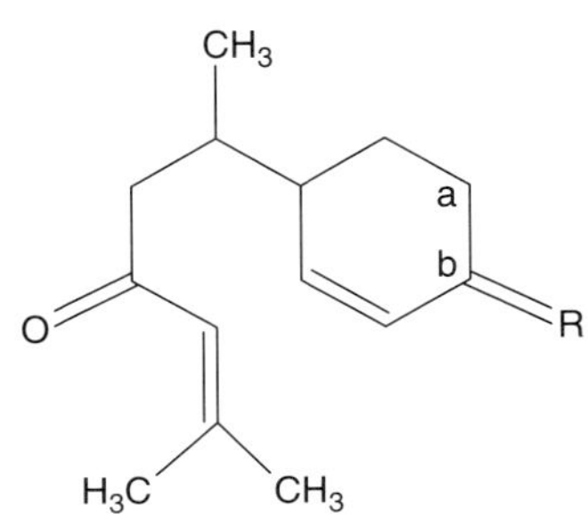

Fig. 17.29. α-Turmerone a b = Δ, R=Me.

their expression and regulating their functions (Chen *et al.*, 2011). Curcumin alone has poor oral bioavailability due to glucuronidation in the small intestine. Piperine from black pepper

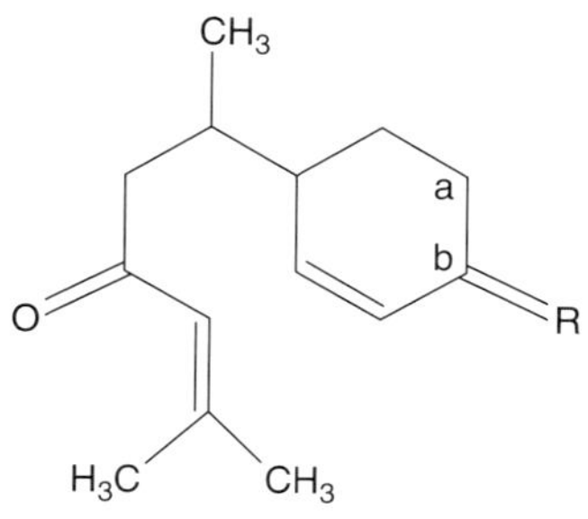

Fig. 17.30. β-Turmerone R=CH_2.

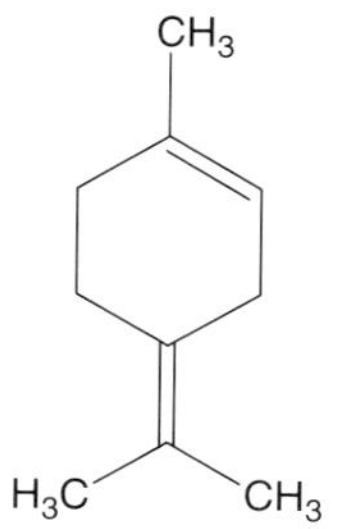

Fig. 17.31. Terpinolene.

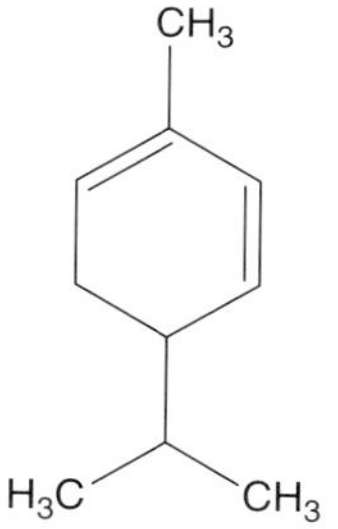

Fig. 17.32. α-Phellandrene.

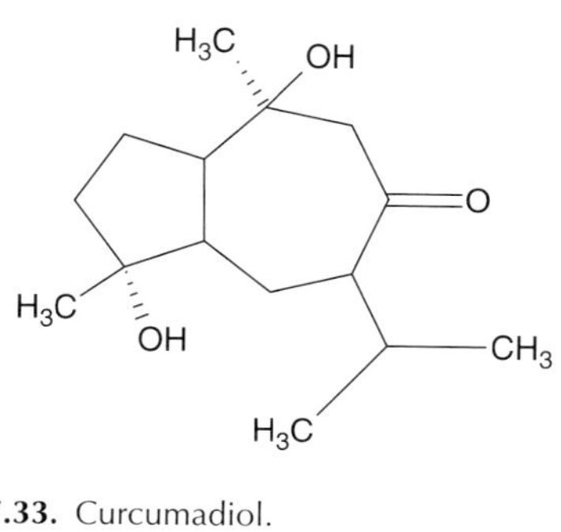

Fig. 17.33. Curcumadiol.

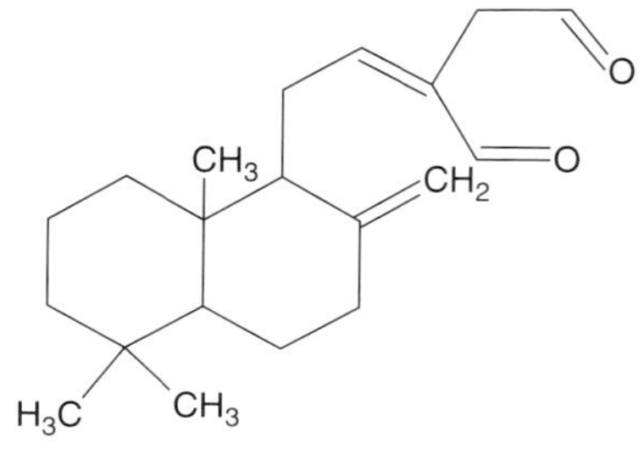

Fig. 17.34. Lambda-8(17)-diene-15,16-dial.

(*Piper nigrum* seeds) enhances the bioavailability of curcumin by 2000% in humans, due to an inhibition of this glucuronidation and slowing the gastrointestinal transit (Shoba *et al.*, 1998; Rohman, 2012).

17.3.1 Anti-inflammatory activity

Curcumin is a highly pleiotropic molecule capable of interacting with numerous molecular targets involved in inflammation. Chronic inflammation leads to destruction of normal tissue injury. Production of inflammation mediators through up-regulation of several inducible gene products, such as inducible nitric oxide synthase (iNOS) and cyclooxygenase-2 (COX-2), contributes to inflammatory responses and tissue damage. Lee *et al.* (2005) have prepared four diaryl-heptanoid and a series of new diaryl-heptalamine analogue derivatives from curcumin and evaluated anti-inflammatory activity by using LPS-stimulated macrophages for induction of iNOS and COX-2 as a model system. Curcumin modulates the inflammatory response by down-regulating the activity of COX-2, lipoxygenase and iNOS enzymes; inhibits the production of the inflammatory cytokines tumour necrosis factor-alpha (TNF-α), interleukin (IL)-1, -2, -6, -8 and -12, monocyte chemo-attractant protein (MCP) and migration inhibitory protein; and down-regulates mitogen-activated and Janus kinases (Goel *et al.*, 2008). *In vitro* studies indicate that the curcumin regulates activation of certain transcription factors such as activating protein-1 (AP-1) and NF-κB in stimulated monocytes and alveolar macrophages, thereby blocking expression of cytokine gene expression. Down-regulation of intercellular signalling proteins, such as protein kinase C, may be another way in which curcumin inhibits cytokine production (Jurenka, 2009). Venkatesan *et al.* (2007) studied the anti-inflammatory activity of curcumin, diacetyl curcumin, triethyl curcumin, tetrahydrocurcumin and phenylbutazone by using carrageenan-induced rat paw oedema. The curcumin analogues decreased the carrageenan-induced paw oedema. Yamamoto *et al.* (1997) hypothesized that the anti-inflammatory action of curcumin is partly due to the inhibition of the enzyme G protein mediated phospholipase D (PLD). Curcumin inhibits 12-O-tetradeconoylphorbel-13 in a dose dependent manner that suggests that PLD inhibition may contribute to the mechanism of chemopreventive action of it. Xu *et al.* (1997) evaluated the effects of curcumin on chemotactic cytokines or chemokine expression in bone marrow cells. Regulatory effects on chemokine expression by IL-1α was measured and was found that it lowered mRNA levels by inhibition of the transcription of chemokine genes, affecting immunomodulation. Awasthi (1996) found that local application of turmeric in eye inflammatory condition and bacterial conjunctivitis showed significant improvement. Curcumin has also been extensively studied as a potential drug for the treatment of lung fibrosis (Zhang *et al.*, 2011). Recent research indicates that the mechanism of blocking fibrosis by curcumin is related to decreasing collagen accumulation in the lungs (Smith *et al.*, 2010), which is attributed to its antioxidant (Li *et al.*, 2008; Guzel *et al.*, 2009; Lee *et al.*, 2010) and anti-inflammatory activities (Venkatesan *et al.*, 2007).

17.3.2 Antioxidant activity

Unnikrishnan and Rao (1995) reported protection of haemoglobin from oxidation, inhibition of lipid peroxidation, erythrocyte membranes and brain homogenates. Curcumin is capable of scavenging oxygen free radicals such as superoxide anions and hydroxyl radicals, which are important to inhibit the lipid peroxidation (Selvem *et al.*, 1995; Sreejayan and Rao, 1996, 1997; Stano *et al.*, 2000). Curcumin has been shown to scavenge various reactive oxygen species (ROS) produced by macrophages (including superoxide anions, hydrogen peroxide and nitrite radicals) both *in vitro* as well as *in vivo* using rat peritoneal macrophages as a model. Inducible nitric oxide synthase (iNOS) is an enzyme found in macrophages that generates large amounts of NO to provide the 'oxidative burst' necessary for defence against pathogens. iNOS is induced in response to an oxidative environment, and the NO generated can react with superoxide radicals to form peroxynitrite, which is highly toxic to cells. Curcumin down-regulates the iNOS activity in macrophages, thus reducing the amount of ROS generated in response to oxidative stress (Chan *et al.*, 1998). Jayaprakasha *et al.* (2006) and Ramsewak *et al.* (2000) reported antioxidant activity of individual curcuminoids by *in vitro* model systems. Its capacities as ascorbic acid equivalent (μmol g^{-1}) were in the order of curcumin > demethoxycurcumin > bisdemethoxycurcumin. The phenolic and the methoxy group on the phenyl ring and the 1,3-diketone systems seem to play a major role and potent ligand for metals. The antioxidant activity increases when phenolic and methoxy groups are on the ortho- position. Masuda *et al.* (1999) and Venkatesan and Rao (2000) showed that the phenolic group is the most important for the free radical reaction of curcumin that enhances antioxidant properties to a significant extent. There was direct evidence for involvement of curcumin in reducing arsenic- and lead-induced oxidative stress in Swiss albino mice by virtue of its antioxidant potential and trapping of free radicals (Jaydip *et al.*, 2010; Murphy *et al.*, 2010). Curcumin reduces pathogenesis of opisthorchiasis and prevents oxidative and nitrative stress (Kaewsamut *et al.*, 2007).

17.3.3 Anticancer activity

Ample evidence exists to support curcumin's use in cancer prevention through its antiproliferative and anticarcinogenic properties or as an adjunct in overall cancer treatment. Curcumin is generally regarded as safe in a phase I clinical trial of cancer patients and marketed as a dietary supplement (Cheng *et al.*, 2001). Curcumin has been shown to possess apoptotic activity against human colon cancer cells (Agarwal *et al.*, 2003), stomach and skin tumours (Azuine and Bhide, 1992), breast cancer cells (Ramachandran *et al.*, 2002) and prostate cancer cells (Dorai *et al.*, 2001). A recent study reported that curcumin inhibited proliferation of colon cancer cell lines (HT-29 and HCT-15) by accumulating cells in G2-M phase (Hanif *et al.*, 1997). Curcumin also induced apoptosis in NIH3T3 and leukaemic cell line HL-60 (Jiang *et al.*, 1996). Huang *et al.* (1988, 1991, 1992) reported that turmeric inhibits the epidermal ornithine decarboxylase (ODC) and epidermal DNA synthesis on tumour promotion in mouse skin. According to Lin *et al.* (2000), curcumin was found to enhance cytotoxicity of chemotherapeutic agents in prostate cancer cells and has the ability to block colon tumour initiation. Synergistic effects were observed when curcumin was combined with standard chemotherapeutic agents. It effectively inhibits UV radiation-induced damage and thereby reduces the incidence of skin cancer by virtue of its free radical quenching action. It was found to inhibit the proliferation of human breast cancer cells *in vitro* (Soudamini, 1988; Azuin and Bhide, 1994). The development of cancerous and precancerous lesions in the glandular stomach was decreased by exposure to pure curcumin as compared to controls and was found to exert chemopreventive effects in rats (Lal *et al.*, 2000). Curcumin is a potent inhibitor of the transcriptional factors activated protein 1 and NF-κB. These factors are known to play important functional roles in the survival of osteoclast. Curcumin was shown to stimulate osteoclast apoptosis in a dose-dependent and time-dependent manner. It also inhibited osteoclastic bone resorption, supporting the results that it stimulates osteoclasts apoptosis (Ozaki *et al.*, 2000; Hasima and Aggarwal, 2012).

The mechanisms by which curcumin exerts its anticancer effects are comprehensive and diverse, targeting many levels of regulation in the processes of cellular growth and apoptosis. Curcumin's potent antioxidant and free radical quenching properties play an important role in the inhibitory effects of the compound on the initial stages of carcinogenesis. It has been shown that curcumin has the ability to suppress UV irradiation-induced DNA mutagenesis and induction of cellular SOS functions (Wilken *et al.*, 2011). Curcumin's inhibitory effect on carcinogenesis has been demonstrated in several animal models of various tumour types including oral cancer, mammary carcinoma and intestinal tumours (Collett *et al.*, 2001; Maheshwari *et al.*, 2006; Johnson and Mukhtar, 2007; Sharma *et al.*, 2007; Goel *et al.*, 2008). Bachmeier *et al.* (2007) found that curcumin induces apoptosis and inhibits the formation of breast cancer metastasis while Li *et al.* (2008) showed that it prevents lung cancer from Quartz particles in rat lung epithelial cell lines.

17.3.4 Anti-HIV activity

Vlietinck (1998) reported the inhibition of the virus cell fusion stage in the replication cycle of HIV. Mazumber *et al.* (1995, 1997) and De Clercq (2002) demonstrated the antiviral activity, being a HIV-1 integrase inhibitor (IC_{50}=40 μM) and suggested that its analogues such as dicaffeoylmethane and rosmarinic acid may be developed as anti-AIDS drugs. Rafael *et al.* (1993), Jorden and Drew (1996) and Eigner and Scholz (1999) showed that curcumin was claimed for anti-HIV-1 and HIV-2 activities. In a study, Riva *et al.* (2008) characterized the action of curcumin on HIV-1 persistently infected CD4+ T-cells as a model for HIV cell reservoirs. The results presented in this study suggest that curcumin interferes with viral production. Since the appearance of resistance to antiretroviral treatment is unavoidable, the use of an immunomodulator with a different mechanism of action could help to reduce the persisting replication observed in the presence of antiviral therapy and the selection of resistant HIV-1 variants.

Curcumin has been reported to inhibit HIV-1 integrase (Mazumber *et al.*, 1995). Dicaffeoylquinic acids inhibit HIV-1 integrase at submicromolecular concentrations (Robinson *et al.*, 1996a). The dicaffeoylquinic and L-chicoric acids were reported to inhibit HIV-1 integrase and HIV-1 replication in cell cultures (Robinson *et al.*, 1996b). Curcumin analogues such as dicaffeoylmethane and rosmarinic acid both inhibit activities of integrase with IC_{50} values below 10 mM (Mazumber *et al.*, 1997). In follow-up studies, dicaffeoylquinic acids (DCQAs) and dicaffeoyltartaric acids (DCTAs) were found to inhibit HIV-1 integrase and HIV replication; no inhibition of gp120 binding to CD4 was noted. Likewise, no inhibition of reverse transcription or RNase H was noted, and it was concluded that the DCQAs and DCTAs act as specific integrase inhibitors, and that their activity against integrase is consistent with their observed anti-HIV activity in cell cultures (McDougall *et al.*, 1998). That integrase would be an excellent target for combination chemotherapy of HIV infection was further ascertained by combination experiments where L-chicoric acid, the putative integrase inhibitor, was combined with a protease inhibitor (AG1350) and zidovudine. Arguing for an integrase-targeted action was the finding that introduction of the mutant integrase containing a single Gly-Ser substitution at position 140 into the native, L-chicoric acid-sensitive virus was found to be sufficient to confer resistance to L-chicoric acid (King and Robinson, 1998).

17.3.5 Antimutagenic potential

Li *et al.* (1998) and Shukla (2002) evaluated the antimutagenic effects *in vitro* using chromosomal aberration assay in Wister rats, induced by cyclophosphamide, a known carcinogen. When curcumin was given at a dose of 100 and 200 mg kg^{-1} body weight through gastric intubation for 7 consecutive days before cyclophosphamide treatment, the incidence of aberrant cells was found to be reduced with both doses of curcumin when compared to a control group treated with phosphamide alone. It was concluded that curcumin has antigenotoxic potential against cyclophosphamide-induced chromosomal mutations. Curcumin is able to inhibit the genotoxic and histochemical changes induced in the experimental animals by various chemical agents, as it reduced the percentages of micronucleated polychromatic erythrocytes in bone marrow cells of mice and inhibited chromosomal aberrations, micronuclei formation and sister chromatid exchanges (SCEs) incidences in mouse bone marrow cells induced by benzo(a)pyrene (Shukla *et al.*, 2003) and lead acetate (El-Ashmawy *et al.*, 2006; Ramadan *et al.*, 2012).

The antimutagenic potential of curcumin has been widely reported. It has been suggested that the hydroxy groups on the benzene rings, double bonds in the alkene portion of the molecule and/or the central diketone moeity could be responsible for the high biological activity of curcumin (Huang *et al.*, 1991). Curcumin inhibited the mutagenic activity of 2-acetamidofluorene and prevented crotean oil-induced skin tumour and papilloma formation in mice (Anto *et al.*, 1996). They significantly reduced tumour size in Swiss albino mice implanted with solid tumours (Ruby *et al.*, 1995).

17.3.6 Antibacterial, antifungal and antiprotozoal activity

Curcuma oil showed inhibition of *Staphylococcus aureus* growth in the concentration of 1 to 5000 ppm. Activity of turmeric against some intestinal bacteria *in vitro* and a total inhibition of *Lactobacilli* growth were found. Antifungal effect against *Candida albicans*, *C. kruseii*, *C. parapsilosis*, isolates of dermatophytes, yeast and pathogenic moulds was reported (Apisariakul *et al.*, 1995). Nose (1998) found the weak growth and toxigenesis of selected *Aspergillus flavus* strains on curcumin, indicating its antifungal effects and inhibition of aflatoxin production. It was found to have marked antiparasitic activity, showing cytotoxicity against African trypanosomes *in vitro*. Saleheen (2002) found the IC_{50} of 5.3 μM against prostagotes of various leishmanial strains, which was much lower as compared to pentamidine. Antiprotozoal activity was reported in the ethanolic extract of *C. longa* against *Plasmodium falciparum* and *Leishmania major*, which was able to inhibit the *in vitro* growth of these parasites (Rasmussen *et al.*, 2000). There are reports on synthesis of mono-carbonyl analogues of curcumin (Liang *et al.*, 2008) or preparation of bioactive conjugates of curcumin (Dubey *et al.*, 2008) in order to increase antimicrobial activity. Ronita *et al.* (2009) conducted a study to evaluate the antimicrobial activity of curcumin against *Helicobacter pylori* isolates from India. Their study highlighted the potential antibacterial activity of curcumin against *H. pylori in vitro*, as curcumin was highly effective in inhibiting *H. pylori* growth irrespective of the genetic makeup of the strains. Cytotoxic and parasiticidal effects of curcumin on protozoan parasites have been demonstrated in cultures against *Leishmania, Trypanosoma, Giardia* and *Plasmodium falciparum* (Koide *et al.*, 2002; Pérez-Arriaga *et al.*, 2006). *In vivo*, curcumin has displayed potent activity against *Plasmodium berghei* (Nandakumar *et al.*, 2006).

17.3.7 Antimalarial activity

In vitro and *in vivo* studies indicated that curcumin possesses a moderate antimalarial activity (Cui *et al.*, 2007) while its synthetic derivatives have demonstrated an increased activity (Mishra *et al.*, 2008). It was found that this compound inhibits histone acetyltransferase (HAT) and increases the production of ROS in the malaria parasite (Cui *et al.*, 2007). With extensive research efforts on-going to explore the clinical applications of curcumin in chronic inflammatory disorders, diabetes and cancer, the development of oral and parenteral curcumin formulations or curcumin analogues with improved bioavailability while retaining their immunomodulatory properties and more potent antimalarial activity have been investigated.

Curcumin is reported to have direct antimalarial activity and turmeric is reported as a component of traditional remedies for malaria and fever (Reddy *et al.*, 2005). In combination with artemisinin, curcumin prevents recrudescence of malaria parasites and death in animal models (Nandakumar *et al.*, 2006). In combination with *Andrographis paniculata* and *Hedyotis corymbosa* extracts, curcumin displayed a clear synergistic effect *in vitro* and *in vivo* in rodent malaria models (Mishra *et al.*, 2009).

17.3.8 Antidiabetic activity

Administration of turmeric or curcumin to diabetic rats reduced the blood sugar and glycosylated haemoglobin level significantly (Arun and Nalini, 2002; Hussain, 2002). Oxidative stress was also reduced as determined by the standard TBARS test. The authors postulated that this could be due to decrease of glucose into the polyol pathway, leading to an increased NADPH/NADP ratio and elevated activity of the antioxidant enzyme glutathione peroxidase. Konatham *et al.* (2010) reported curcumin to reduce hyperlipidaemia, delay the development of cataracts, ameliorate renal lesions, and reduce the cross-linking of collagen in a streptozotocin-treated diabetic animal model. Curcumin has also been shown to lower blood glucose levels in type-2 diabetic KK-Ay mice (Nishiyama *et al.*, 2005). Jang *et al.* (2008) reported curcumin to have antidiabetic and antihyperlipidaemic activities. The mechanism of antidiabetic activity of turmeric was explained due to the beneficial effects of curcumin on the liver of diabetic animals. An important enzyme that converts glucose into glycogen was found in higher concentration in diabetic mice treated with curcumin compared to control mice. This enzyme was thought to inhibit the post-meal rise of glucose level.

The anti-inflammatory and antioxidant properties of turmeric also have been proposed to lessen insulin resistance and prevent type-2 diabetes in a mice model by dampening the inflammatory response caused by obesity. It was also found that dietary curcumin could increase the expression of adiponectin, which in turn improves insulin sensitivity in insulin-resistant animal models (Weisberg *et al.*, 2008). *Curcuma longa* rhizomes have also been reported to possess blood glucose-lowering activity in alloxan-induced diabetic rats (Shankar *et al.*, 1980).

17.3.9 Anti-obesity activity

On a global scale, obesity has reached epidemic proportions and is a major contributor to several chronic diseases. At present, because of dissatisfaction with high costs and potentially hazardous side-effects, the potential of natural products for treating obesity may be an excellent alternative strategy for developing future effective, safe anti-obesity drugs (Mayer *et al.*, 2009). Curcumin obtained from *C. longa* has been reported with potential anti-obesity activities. It is known to decrease ATP biosynthesis resulting in an increase of AMP:ATP ratio and then activation of 5′-AMP kinase (AMPK). Activated AMPK would inhibit the synthesis of fatty acid and cholesterol, which explains the anti-obesity effect of curcumin (Lim *et al.*, 2009). It was also found to increase LDL receptor (LXR), which plays a role in elimination of LDL from blood (Peschel *et al.*, 2007). Curcumin has a significant effect on adiposity and lipid metabolism through several mechanisms, including modulation of energy metabolism, inflammation, and suppression of angiogenesis. It has been well established that angiogenesis plays pivotal roles in the growth and expansion of adipose tissue (Hausman and Richardson, 2004; Lijnen, 2008). Therefore, curcumin may contribute to the prevention of adipogenesis through suppression of angiogenesis into the adipose tissue (Rupnick *et al.*, 2002). Several other studies in animal models of obesity have reported the beneficial effects of curcumin on body weight and fat, adiposity, and energy metabolism (Ruderman *et al.*, 2003).

17.3.10 Antifibrinogen activity

Kang (2002) reported the effects of curcumin on the production of collagen and smooth muscle alpha proteins and of alpha collagen mRNA *in vivo* and *in vitro* and found reduced DNA synthesis *in vitro* and down-regulated smooth muscle alpha action, type I collagen expression and alpha collagen mRNA expression. It was concluded that curcumin might therefore prove to be a valuable antifibrinogenic agent. Ramirez *et al.* (2000) found a significant drop in elevated fibrinogen with *Curcuma longa* extract without any adverse side-effect. This new pharmacological activity of *C. longa* indicates that it may be an ideal drug for the treatment of atherosclerosis and cardiovascular diseases, without altering coagulation parameters.

17.3.11 Wound healing activity

Tissue repair and wound healing are complex processes that involve inflammation, granulation and remoulding of the tissue. Sidhu *et al.* (1998) evaluated that the localization of transforming growth factor beta and fibronectin, which are important criteria in wound healing, shows increase in curcumin-treated wounds. Phan (2001) investigated the effects of curcumin on hydrogen peroxide and hypoxanthine-xanthine oxidase-induced damage to cultured human keratinocytes and fibroblasts, in an effort to elucidate the mechanism of wound healing action of curcumin. It was observed that exposure of human keratinocytes to curcumin (10 μg ml^{-1}) offered significant protection against hydrogen peroxide. However, no protective effects were observed against hypoxanthine-xanthine oxidase injury. The authors concluded that curcumin is a powerful inhibitor of damage to human keratinocytes and fibroblasts by hydrogen peroxide. Significant wound healing activity of curcumin has been reported (Panchatcharam *et al.*, 2006; Sundarananthavalli *et al.*, 2011). The curcumin-treated wounds showed a faster rate of wound contraction compared with controls, which was further supported by histopathological studies. The gel formulations produced better healing compared with the emulsifying ointment formulations (Bhat *et al.*, 2007).

17.3.12 Lipid lowering activity

Turmeric's protective effects on the cardiovascular system include lowering cholesterol and triglyceride levels, decreasing susceptibility of LDL to lipid peroxidation and inhibiting platelet aggregation. These effects have been noted even with low doses of turmeric. In atherosclerotic rabbits, turmeric extract demonstrated decreased susceptibility of LDL to lipid peroxidation, in addition to lower plasma cholesterol and triglyceride levels. The effect on cholesterol levels may be due to decreased cholesterol uptake in the intestines and increased conversion of cholesterol to bile acids in the liver. Inhibition of platelet aggregation by *C. longa* constituents is thought to be via potentiation of prostacyclin synthesis and inhibition of thromboxane synthesis (Akram *et al.*, 2010).

The effects of diet-supplemented curcuminoids (commercial grade curcumin: a mixture of curcumin (73.4%), demethoxycurcumin (16.1%) and bisdemethoxycurcumin (10.5%)) showed encouraging effects on triacylglycerol and cholesterol concentrations with lipid-lowering potency (Shalini and Srinivas, 1987; Asai and Miyasawa, 2001). Sreejayan and Rao (1996) reported the ability of curcumin to scavenge electrophilic reactive intermediates found by metabolic activation of drug and lipid peroxidation. Venkatesan (1998) studied the protective effect against acute adriamycin myocardial toxicity. Curcumin treatment before and after adriamycin treatment significantly inhibits lipid peroxidation and increased the levels of endogenous antioxidants. Asai and Miyasawa (2001) reported that the dietary curcuminoids prevent high-fat diet-induced lipid accumulation in rat liver and epididymal adipose tissue. Wu *et al.* (2008) reported that curcumin significantly reduced the plasma and hepatic cholesterol and triglyceride levels in rats. Manjunatha and Srinivasan (2007) have reported lowering of serum and liver cholesterol levels in induced hypercholesterolaemic rats. In a recent study conducted by Zingg *et al.* (2012), curcumin showed a trend for reduction of lipid levels in peritoneal macrophages in LDL receptor knockout mice fed a high fat diet for 4 months.

17.3.13 Hepatoprotective effects

Turmeric has been found to have a hepatoprotective characteristic similar to silymarin. Animal studies have demonstrated turmeric's hepatoprotective effects from a variety of hepatotoxic insults, including carbon tetrachloride (CCl_4), galactosamine (Donatus *et al.*, 1990), acetaminophen (paracetamol) and *Aspergillus* aflatoxin (Soni *et al.*, 1992). Turmeric's hepatoprotective effect is mainly a result of its antioxidant properties, as well as its ability to decrease the formation of pro-inflammatory cytokines. In rats with CCl_4-induced acute and subacute liver injury, curcumin administration significantly decreased liver injury in test animals compared to controls. Turmeric extract inhibited fungal aflatoxin production by 90% when given to ducklings infected with *Aspergillus parasiticus*. Turmeric and curcumin also reversed biliary hyperplasia, fatty changes and necrosis induced by aflatoxin production. Sodium curcuminate also exerts choleretic effects by increasing biliary excretion of bile salts, cholesterol and bilirubin, as well as increasing bile solubility, therefore possibly preventing and treating cholelithiasis (Akram *et al.*, 2010).

17.3.14 Radioprotective activity

Curcumin was found to be effective in inhibiting radiation-induced protein kinase C (PKC) activity and was potentially useful as a chemopreventive agent. It is beneficial in reducing the risk of developing cancer, and provides protection from radiation-induced toxicity and harmful effects of organochlorine pesticides (Thresiamma, 1996; Cheng *et al.*, 2001). Varadkar (2001) reported curcumin to be potentially useful in preventing the development of radio-resistance following radiotherapy. Curcumin was found to be effective in inhibiting radiation-induced PKC activity. Activation of PKC is reported to be one of the means of conferring radio-resistance on a tumour cell. Therefore, suppression of PKC by curcumin may be a means of preventing the development of radio-resistance following radiotherapy. In a study conducted by Shabon (2008), female albino rats were treated with oral administration of curcumin for 14 days before γ-irradiation. Serum total protein, albumin, globulins, cholesterol, triglycerides, serum glutamic oxalo-acetic transaminase (SGOT) and serum glutamic pyruvic transaminase (SGPT) were determined. Curcuma was found to elevate the protein profile (total protein, albumin and globulin) and ameliorate the hyperlipidaemic effects of γ-radiation. Curcuma also improved the liver functions affected by γ-irradiation, concluding that curcuma can be used as a radio-protector for occupationally exposed individuals.

17.3.15 Immunomodulating activity

Curcumin has been found to modulate the growth and cellular response of various cell types of the immune system. Numerous lines of evidence suggest that curcumin can modulate both the proliferation and the activation of T cells. It inhibited the proliferation induced by concanavalin A (Con A), phyto-haemagglutinin (PHA) and phorbol-12-myristate-13-acetate (PMA) of lymphocytes derived from fresh human spleen (Ranjan *et al.*, 2004). Curcumin not only plays an important role in the immunomodulation of normal but also transformed T cells, where it adversely affects the cell proliferation of these cells by suppression of IL-2 gene expression and by inhibiting the activation of NF-κ-light-chain-enhancer of activated B cells (NF-κB) (Gertsch *et al.*, 2003). In addition to affecting T cells, curcumin can also influence the proliferation of B cells and B lymphocyte-mediated immune function. This effect of curcumin appears to be mediated through down-regulation of oxidative stress induced by cyclosporine and hydrogen peroxide. Thus, post-transplant lympho-proliferative disorder (PLTD) associated with the use of cyclosporine during organ transplantation, can be reversed by curcumin (Ranjan *et al.*, 1998). Many studies have shown curcumin's ability to modulate the activation of macrophages. For example, curcumin seems to regulate the immune function of mice in a dose-dependent fashion, as curcumin treatment enhanced the phagocytosis of peritoneal macrophages and differentially regulates the proliferation of splenocytes (Li and Liu, 2005). Apart from cell proliferation, a daily diet of curcumin (30 mg kg^{-1} body weight day^{-1}) for 2 weeks in rats, reportedly attenuated the ability of macrophages to generate free radicals and secrete lysosomal enzymes collagenase, elastase and hyaluronidase (Joe and Lokesh, 2000). Curcumin can also apparently modulate the activation of natural killer (NK) cells. Studies by South and his colleagues in rats showed that curcumin at a dose of 1 and 20 mg kg^{-1} body weight could not enhance the IgG levels in the NK cells, whereas a higher dose (40 mg kg^{-1}) did elevate IgG levels significantly. More importantly, none of the three doses of curcumin significantly enhanced either delayed-type hypersensitivity or NK cell activity (South, 1997).

Kim *et al.* (2005) found that curcumin efficiently blocked the lipopolysaccharide-induced expression of IL-12 and inflammatory cytokines including IL-1β, IL-6 and TNF-α. Curcumin treatment enhanced the antigen capturing ability of dendritic cells via mannose receptor-mediated endocytosis. However, their Th1 and normal cell-mediated immune response was very poor. Further studies showed that treatment of dendritic cells with curcumin before lipopolysaccharide stimulation completely suppressed the lipopolysaccharide-induced phosphorylation of mitogen-activated protein kinase (MAPK) and NF-κB nuclear translocation. The direct suppression of these activities by curcumin in dendritic cells may lead to the attenuated T cell-mediated immune responses by interfering with handling and presentation of antigens by dendritic cells.

17.4 Structure–Activity Relationships

Curcumin belongs to a class of curcuminoids and is very similar to diaryl-heptanoids. The anti-inflammatory activity of curcumin and its derivatives is due to the presence of hydroxyl and phenol groups in the molecule that are essential for the inhibition of prostaglandins (PG synthetase) and leucotrienes (LT) (Araujo and Leon, 2001). On the other hand, some authors suggested that the anti-inflammatory action associated to the existence of the α-dicarbonylic system having the conjugated double bonds (dienes) is responsible for this activity. This system seems to be responsible for its anti-inflammatory and antiparasitic activity (Barclay *et al.*, 2000). The presence of diene and keto system provides lipophilicity to the compounds and thus probably has better skin penetration. The presence of the diketone moiety in the curcumin molecule seems to be essential for the inhibitory activity (Kiuchi *et al.*, 1993; Simon *et al.*, 1998). The highest antioxidant activity was obtained when the phenolic group was stearically hindered by the introduction of two methyl groups at the ortho- position. The phenolic group is essential for free radical scavenging activity, and the presence of the methoxy group further increases the activity (Sreejayan and Rao, 1996). Studies have shown that both antioxidant and pro-oxidant effects are determined by the same structural moieties of the curcuminoids (Ahsan *et al.*, 1999).

17.5 Pharmacodynamic and Pharmacokinetics

Curcumin having low oral bioavailability in humans may undergo intestinal metabolism (Sharma *et al.*, 2001). Curcumin when added to isolated hepatocytes is quickly metabolized and the major biliary metabolites are glucuronides of tetrahydrocurcumin and hexahydrocurcumin. Pharmacokinetic studies in animals demonstrated that 40–85% of an oral dose of curcumin passes through the gastrointestinal tract unchanged, with most of the absorbed flavonoid being metabolized in the intestinal mucosa and liver. After metabolism in the liver it is mainly excreted through bile. In another study (Sharma *et al.*, 2007), preclinical data from animal models and phase I clinical studies performed with human volunteers and patients with cancer have demonstrated low systemic bioavailability following oral dosing. Efficient first-pass metabolism and some degree of intestinal metabolism, particularly glucuronidation and sulfation of curcumin, might explain its poor systemic availability when administered via the oral route. A dose escalation pilot study revealed that *Curcuma* extract could be administered safely to patients at doses of up to 2.2 g daily, equivalent to approximately 180 mg of curcumin. Most of the biological activities of curcumin can be attributed to its potent antioxidant capacity at neutral and acidic pH, its inhibition of cell signalling pathways at multiple levels, its diverse effects on cellular enzymes and its effects on cell adhesion and angiogenesis. In particular, curcumin's ability to alter gene transcription and induce apoptosis in preclinical models advocates its potential utility in cancer chemoprevention and chemotherapy.

17.6 Safety and Dosage

Small doses of turmeric (curcumin) are taken daily as a spice by the population in many Asian countries. In one epidemiologic survey, in terms of its dietary use in Nepal, turmeric consumption was found to be up to 1500 mg per person per day, equivalent to approximately 50 mg day^{-1} of curcumin (Eigner and Scholz, 1999). In India, where the average intake of turmeric can be as high as 2000–2500 mg day^{-1} (approx. 100 mg of curcumin), no toxicities or adverse effects have been reported (Chainani-Wu, 2003). However, the doses administered in clinical trials are expected to be rather higher than those normally consumed in the diet. Based on repeated studies, turmeric is Generally Recognized as Safe (GRAS) by the US FDA, and curcumin has been granted an acceptable daily intake level of 0.1–3 mg kg^{-1} body weight by the Joint FAO/WHO Expert Committee on Food Additives. No significant toxicity has been reported following either acute or chronic administration of turmeric extracts at standard doses or very high doses (100 mg kg^{-1} body weight). Curcumin may be ulcerogenic in animals, as evidenced by one rat study (Ammon and Wahl, 1991). In phase I clinical trials, curcumin with doses up to 3600–8000 mg day^{-1} for 4 months did not result in discernible toxicities except mild nausea and diarrhoea (Hsu and Cheng, 2007).

17.7 Conclusions

Curcuma longa has been used in the Ayurveda, Unani and Siddha systems of medicines from ancient times, and is known for various biological applications. A wide spectrum of pharmacological activities of turmeric either in the form of powder, extracts or in its isolated compounds with minimum side effects has been reported. The methoxy group on the phenyl ring, phenolic and 1,3-diketone systems in curcumin play an important role for various pharmacological activities. Several products fortified with curcumin or turmeric have been launched in national and international markets for various diseases. Changing of route and medium of curcumin administration, blocking of metabolic pathways by concomitant administration with other agents, developing novel delivery systems and structural modifications are the main strategies now being explored in attempts to improve the bioavailability of curcumin and avoiding acid degradation of the drug due to stomach acids. Enhanced bioavailability of curcumin and avoidance of stomach acid degradation in the near future is likely to bring this promising natural product to the forefront of therapeutic agents for treatment of human disease. Curcumin is considered to be a safe, non-toxic, strong natural antioxidant in comparison to other phytochemicals.

References

Agarwal, B., Swaroop, P., Protiva, P., Raj, S.V., Shirin, H. and Holt, P.R. (2003) Cox-2 is needed but not sufficient for apoptosis induced by Cox-selective inhibitors in colon cancer cells. *Apoptosis* 8, 649–654.

Ahsan, H., Parveen, N. and Khan, N.U. (1999) Pro-oxidant, anti-oxidant and cleavage activities on DNA of curcumin and its derivatives demethoxycurcumin and bisdemethoxycurcumin. *Chemico-Biological Interactions* 121, 161–175.

Akram, M., Uddin, S., Ahmed, A., Usmanghani, K., Hannan, A., Mohiuddin, E. and Asif, F. (2010) *Curcuma longa* and Curcumin: A review article. *Romanian Journal of Biology - Plant Biology* 55, 65–70.

Ammon, H.P. and Wahl, M.A. (1991) Pharmacology of *Curcuma longa*. *Planta Medica* 57, 1–7.

Anto, R.J., George, J., Babu, K.V., Rajasekharan, K.N. and Kuttan, R. (1996) Antimutagenic and anti-carcinogenic activity of natural and synthetic curcuminoids. *Mutation Research* 370, 127–131.

Apisariyakul, A., Vanittanakom, N. and Buddhasukh, D. (1995) Antifungal activity of turmeric oil extracted from *Curcuma longa* (Zingiberaceae). *Journal of Ethnopharmacology* 49, 163–169.

Araujo, C.A.C. and Leon, L.L. (2001) Biological activities of *Curcuma longa* L. *Memorias do Instituto Oswaldo Cruz Rio de Janeiro* 96, 723–728.

Arun, N. and Nalini, N. (2002) Efficacy of turmeric on blood sugar and polyol pathway in diabetic albino rats. *Plant Foods for Human Nutrition* 57, 41–52.

Asai, A. and Miyasawa, T. (2001) Dietary curcuminoids prevent high fat diet induced lipid accumulation in rat liver and epididymal adipose tissue. *Journal of Nutrition* 131, 2932–2935.

Awasthi, S. (1996) Curcumin protects against 4-hydroxy-2-transnonenal-induced cataract formation in rat lenses. *American Journal of Clinical Nutrition* 64, 761–766.

Azuine, M.A. and Bhide, S.V. (1992) Chemopreventive effect of turmeric against stomach and skin tumors induced by chemical carcinogens in Swiss mice. *Nutrition and Cancer* 17, 77–83.

Azuine, M.A. and Bhide, S.V. (1994) Adjuvant chemoprevention of experimental cancer: catechin and dietary turmeric in forestomach and oral cancer models. *Journal of Ethanopharmacology* 44, 211–217.

Bachmeier, B., Nerlich, A.G. and Iancu, C.M. (2007) The chemopreventive polyphenol Curcumin prevents hematogenous breast cancer metastases in immunodeficient mice. *Cellular Physiology and Biochemistry* 19, 137–152.

Barclay, L.R., Vinqvist, M.R., Mukai, K., Goto, H., Hashimoto, Y., Yokunaga, A. and Uno, H. (2000) On the antioxidant mechanism of curcumin: classical methods are needed to determine antioxidant mechanism and activity. *Organic Letters* 2, 2841–2843.

Bhat, R.S., Shankrappa, J. and Shivakumar, H.G. (2007) Formulation and evaluation of polyherbal wound treatments. *Asian Journal of Pharmaceutical Sciences* 2, 11–17.

Chainani-Wu, N. (2003) Safety and anti-inflammatory activity of curcumin: a component of turmeric (*Curcuma longa*). *Journal of Alternative and Complementary Medicine* 9, 161–168.

Chan, M.M., Huang, H.I., Fenton, M.R. and Fong, D. (1998) *In vivo* inhibition of nitric oxide synthase gene expression by curcumin, a cancer preventive natural product with anti-inflammatory properties. *Biochemical Pharmacology* 55, 1955–1962.

Chen, Xi., Liang, S., Zhu, L., Liu, J., Zhang, G. and Chen, Z. (2011) High-sensitivity determination of curcumin in human urine using gemini zwitterionic surfactant as a probe by resonance light scattering technique. *Phytochemical Analysis* DOI: 10.1002/pca.1380.

Cheng, A.L., Hsu, C.H. and Lin, J.K. (2001) Phase-I Clinical trail of curcumin, a chemopreventive agent, in patients with high risk or premalignant leisions. *Anticancer Research* 27, 2895–2900.

Collett, G.P., Robson, C.N., Mathers, J.C. and Campbell, F.C. (2001) Curcumin modifies Apc (min) apoptosis resistance and inhibits 2-amino 1-methyl-6- phenylimidazo[4,5-b]pyridine (PhIP) induced tumour formation in Apc (min) mice. *Carcinogenesis* 22, 821–825.

Cui, L., Miao, J. and Cui, L. (2007) Cytotoxic effect of curcumin on malaria parasite *Plasmodium falciparum*: inhibition of acetylation and generation of reactive oxygen species. *Antimicrobial Agents Chemotherapy* 51, 488–494.

De Clercq, E. (2002) New Anti-HIV Agents and Targets. *Medicinal Research Reviews* 22(6), 531–565.

Donatus, I.A., Sardjoko and Vermeulen, N.P. (1990) Cytotoxic and cytoprotective activities of curcumin. Effects on paracetamol-induced cytotoxicity, lipid peroxidation and glutathione depletion in rat hepatocytes. *Biochemical Pharmacology* 39, 1869–1875.

Dorai, T., Cao, Y.C., Dorai, B., Buttyan, R. and Katz, A.E. (2001) Therapeutic potential of curcumin in human prostate cancer. III. Curcumin inhibits proliferation, induces apoptosis, and inhibits angiogenesis of LNCaP prostate cancer cells *in vivo*. *Prostate* 47, 293–303.

Dubey, S.K., Sharma, A.K., Narian, U., Misra, K. and Pati, U. (2008) Design, synthesis and characterization of some bioactive conjugates of curcumin with glycine, glutamic acid, valine and demethylenated piperic acid and study of their antimicrobial and antiproliferative properties. *European Journal of Medicinal Chemistry* 43, 1837–1846.

Eigner, D. and Scholz, D. (1999) *Ferula asa-foetida* and *Curcuma longa* in traditional medical treatment and diet in Nepal. *Journal of Ethnopharmacology* 67, 1–6.

El-Ashmawy, I.M., Ashry, K.M., El-Nahasa, F. and Salama, O.M. (2006) Protection by turmeric and myrrh against liver oxidative damage and genotoxicity induced by lead acetate in mice. *Basic and Clinical Pharmacology and Toxicology* 98, 32–37.

Gayatri, N. and Rajani, K.S. (2011) Evaluation of antioxidant activity in ethanolic extracts of five *Curcuma* species. *International Research Journal of Pharmacy* 2, 243–248.

Gertsch, J., Guttinger, M., Heilmann, J. and Sticher, O. (2003) Curcumin differentially modulates mRNA profiles in Jurkat T and human peripheral blood mononuclear cells. *Bioorganic and Medicinal Chemistry* 11, 1057–1063.

Goel, A., Kunnumakkara, A.B. and Aggarwal, B.B. (2008) Curcumin as 'Curecumin': from kitchen to clinic. *Biochemical Pharmacology* 75, 787–809.

Guzel, A., Kanter, M., Aksu, B., Basaran, U.N., Yalcin, O., Uzun, H., Konukoglu, D. and Karasalihoglu, S. (2009) Preventive effects of curcumin on different aspiration material-induced lung injury in rats. *Pediatric Surgery International* 25, 83–92.

Hanif, R., Qiao, L., Shiff, S.J. and Rigas, B. (1997) Curcumin a natural plant phenolic food additive, inhibits cell proliferation and induces cell cycle changes in colon adenocarcinoma cell lines by a prostaglandin-independent pathway. *Journal of Laboratory and Clinical Medicine* 130, 576–584.

Hasima, N. and Aggarwal, B.B. (2012) Cancer-linked targets modulated by curcumin. *International Journal of Biochemistry and Molecular Biology* 3, 328–351.

Hausman, G.J. and Richardson, R.L. (2004) Adipose tissue angiogenesis. *Journal of Animal Science* 82, 925–934.

Hsu, C. and Cheng, A. (2007) Clinical studies with curcumin. *Advances in Experimental Medicine and Biology* 595, 471–480.

Huang, H.C., Jan, T.R. and Yeh, S.F. (1992) Inhibitory effect of curcumin, an anti-inflammatory agent, on vascular smooth muscle cell proliferation. *European Journal of Pharmacology* 221, 381–384.

Huang, M.T., Smart, R.C., Wong, C.Q. and Conney, A.H. (1988) Inhibitory effect of curcumin, chlorogenic acid, caffeic acid and ferulic acid on tumor promotion in mouse skin by 12-O-tetradecanoylphorbol-13-acetate. *Cancer Research* 48, 5941–5946.

Huang, M.T., Lysz, T., Ferrara, T., Abidi, T.F., Laskin, J.D. and Conney, A.H. (1991) Inhibitory effects of curcumin on *in vitro* lipoxygenase and cyclooxygenase activities in mouse epidermis. *Cancer Research* 51, 813–819.

Hussain, H.E.M. (2002) Hypoglycemic, hypolipidemic and antioxidant properties of combination of curcumin from *Curcuma longa* Linn and partially purified product from *Abroma auguesta* Linn in streptozotocin induced diabetes. *Indian Journal of Clinical Biochemistry* 17, 33–43.

Ishita, C., Kaushik, B., Uday, B. and Ranajit, K.B. (2004) Turmeric and curcumin: biological actions and medicinal applications. *Current Science* 87, 44–53.

Jang, E.M., Choi, M.S., Jung, U.J., Kim, M.J. and Kim, H.J. (2008) Beneficial effects of curcumin on hyperlipidemia and insulin-resistance in high-fat-fed hamster. *Metabolism-Clinical and Experimental* 57, 1576–1583.

Jayaprakasha, G.K., Rao, L.J. and Sakariah, K.K. (2006) Antioxidant activities of curcumin, demethoxycurcumin and bisdemethoxycurcumin. *Food Chemistry* 98, 720–724.

Jaydip, B., Soumi, R., Sutapa, M., Dona, S. and Madhumita, R. (2010) Indian spice curcumin may be an effective strategy to combat the genotoxicity of arsenic in swiss albino mice. *Asian Pacific Journal of Cancer Prevention* 11, 239–248.

Jiang, M.C., Young, Y.H.F., Yen, J.I. and Lin, J.K. (1996) Curcumin induces apoptosis in immortalized NIH3T3 and malignant cancer cell lines. *Nutrition and Cancer* 26, 111–120.

Joe, B. and Lokesh, B.R. (2000) Dietary *n*-3 fatty acids, curcumin and capsaicin lower the release of lysosomal enzymes and eicosanoids in rat peritoneal macrophages. *Molecular and Cellular Biochemistry* 203, 153–161.

Johnson, J. and Mukhtar, H. (2007) Curcumin for chemoprevention of colon cancer. *Cancer Letters* 255, 170–181.

Jorden, W.C. and Drew, C.R. (1996) Curcumin – a natural herb with anti-HIV. *Journal of the National Medical Association* 88, 333.

Jurenka, J.S. (2009) Anti-inflammatory properties of curcumin, a major constituents of *Curcuma longa*: a review of preclinical and clinical research. *Alternative Medicine Review* 14, 141–153.

Kaewsamut, B., Pinlaor, S., Boonmars, T., Srisawangwong, T. and Yongvanit, P. (2007) Effect of curcumin on the inducible nitric oxide synthase (iNOS) and antioxidant enzyme expression in hamsters infected with *Opisthorchis viverrini*. *Southeast Asian Journal of Tropical Medicine and Public Health* 38, 66–73.

Kang, H.C. (2002) Curcumin inhibits collagen synthesis and hepatic stellate cell activation *in vivo* and *in vitro*. *Journal of Pharmacy and Pharmacology* 54, 119–126.

Kim, G.Y., Kim, K.H., Lee, S.H., Yoon, M.S., Lee, H.J., Moon, D.O., Lee, C.M., Ahn, S.C., Park, Y.C. and Park, Y.M. (2005) Curcumin inhibits immunostimulatory function of dendritic cells:MAPKs and translocation of NF-kappa B as potential targets. *Journal of Immunology* 174, 8116–8124.

King, P.J. and Robinson, W.E. Jr (1998) Resistance to the anti-human immunodeficiency virus type 1 compound l-chicoric acid results from a single mutation at amino acid 140 of integrase. *Journal of Virology* 72, 8420–8424.

Kiso, Y., Suzuki, Y., Oshima, Y. and Hikino, H. (1983) Stereostructure of curlone a sesquiterpenoid of *Curcuma longa* rhizomes. *Phytochemistry* 22, 596–597.

Kiuchi, F., Goto, Y., Sugimoto, N., Akao, N., Kondo, K. and Tsuda, Y. (1993) Nematocidal activity of Turmeric: synergistic action of curcuminoids. *Chemical and Pharmaceutical Bulletin* 41, 1640–1643.

Koide, T., Nose, M., Ogihara, Y., Yabu, Y. and Ohta, N. (2002) Leishmanicidal effect of curcumin *in vitro*. *Biological and Pharmaceutical Bulletin* 25, 131–133.

Konatham, S., Kumar, P. and Aukunuru, J. (2010) Synthesis and screening of antidiabetic activity of some novel curcumin analogues. *International Journal of Pharma and Bio Sciences* 1(2), 1–12.

Lal, B., Kapoor, A.K., Agarwal, P.K., Asthana, O.P. and Srimal, R.C. (2000) Role of curcumin in idiopathic inflammatory orbital pseudotumours. *Phytotherapy Research* 14, 443–447.

Lawhavinit, O., Sincharoenpokai, P. and Sunthornandh, P. (2011) Effects of ethanol tumeric (*Curcuma longa* Linn.) extract against shrimp pathogenic *Vibrio* spp. and on growth performance and immune status of white shrimp (*Litopenaeus vannamei*). *Kasetsart Journal (Natural Science)* 45, 70–77.

Lee, J.C., Kinniry, P.A., Arguiri, E., Serota, M., Kanterakis, S., Chatterjee, S., Solomides, C.C., Javvadi, P., Koumenis, C. and Cengel, K.A. (2010) Dietary curcumin increases antioxidant defenses in lung, ameliorates radiation-induced pulmonary fibrosis, and improves survival in mice. *Radiation Research* 173, 590–601.

Lee, S.L., Huang, W.J., Lin, W.W., Lee, S.S. and Chen, C.H. (2005) Preparation and anti-inflammatory activities of diarylheptanoid and diarylheptylamine analogs. *Bioorganic and Medicinal Chemistry* 13, 6175–6181.

Li, H., van Berlo, D. and Shi, T. (2008) Curcumin protects against cytotoxic and inflammatory effects of quartz particles but causes oxidative DNA damage in a rat lung epithelial cell line. *Toxicology and Applied Pharmacology* 227, 115–124.

Li, X. and Liu, X. (2005) Effect of curcumin on immune function of mice. *Journal of Huazhong University Science and Technology (Medical Science)* 25, 137–140.

Li, X., Song, Q. and Chen, B. (1998) Study on the antimutagenicity of curcumin. *Wei Sheng Yan Jiu* 27, 163–165.

Liang, G., Yang, S., Jiang, L., Zhao, Y., Shao, L., Xiao, J., Ye, F., Li, Y. and Li, X. (2008) Synthesis and antibacterial properties of mono-carbonyl analogues of curcumin. *Chemical and Pharmaceutical Bulletin* 56, 162–167.

Lijnen, H.R. (2008) Angiogenesis and obesity. *Cardiovascular Research* 78, 286–293.

Lim, H.W., Lim, H.Y. and Wong, K.P. (2009) Uncoupling of oxidative phosphorylation by curcumin: implication of its cellular mechanism of action. *Biochemical and Biophysical Research Communications* 389, 187–192.

Lin, J.K., Pan, M.H. and Lin-Shiau, S.Y. (2000) Recent studies on the biofunctions and biotransformations of curcumin. *Biofactors* 13, 153–158.

Maheshwari, R., Singh, A.K., Gaddipati, J. and Srimal, R.C. (2006) Multiple biological activities of curcumin: a short review. *Life Sciences* 78, 2081–2087.

Manjunatha, H. and Srinivasan, K. (2007) Hypolipidemic and antioxidant effects of dietary curcumin and capsaicin in induced hypercholesterolemic rats. *Lipids* 42, 1133–1142.

Masuda, T., Hidaka, S.A., Maikawa, T., Takeda, Y. and Yamaguchi, H. (1999) Chemical studies on antioxidant mechanism of curcuminoid: analysis of radical reaction products from curcumin. *Journal of Agricultural and Food Chemistry* 47, 71–77.

Mayer, M.A., Hocht, C., Puyo, A. and Taira, C.A. (2009) Recent advances in obesity pharmacotherapy. *Current Clinical Pharmacology* 4, 53–61.

Mazumber, A., Raghavan, K., Weinstein, J., Kohn, K.W. and Pommer, Y. (1995) Inhibition of human immunodeficiency virus type-1 integrase by curcumin. *Biochemical Pharmacology* 49, 1165–1170.

Mazumber, A., Neamati, N., Sunder, S., Schulz, J., Pertz, H., Eich, E. and Pommier, Y. (1997) Curcumin analogs with altered potencies against HIV-I integrase as probes for biochemical mechanisms of drug action. *Journal of Medicinal Chemistry* 40, 3057–3063.

McDougall, B., King, P.J., Wu, B.W., Hostomsky, Z., Reinecke, M.G. and Robinson, W.E. Jr (1998) Dicaffeoylquinic and dicaffeoyltartaric acids are selective inhibitors of human immunodeficiency virus type 1 integrase. *Antimicrobial Agents Chemotherapy* 42, 140–146.

Mishra, K., Dash, A.P., Swain, B.K. and Dey, N. (2009) Anti-malarial activities of *Andrographis paniculata* and *Hedyotis corymbosa* extracts and their combination with curcumin. *Malaria Journal* 8, 26.

Mishra, S., Karmodiya, K., Surolia, N. and Surolia, A. (2008) Synthesis and exploration of novel curcumin analogues as anti-malarial agents. *Bioorganic and Medicinal Chemistry* 16, 2894–2902.

Murphy, E.A., Davis, J.M., McClellan, J.L., Gordon, B.T. and Carmichael, M.D. (2010) Curcumin's effect on intestinal inflammation and tumorigenesis in the Apc mouse. *Journal of Interferon & Cytokine Research* 15, 512–519.

Nakayama, R., Tamura, Y., Yamanaka, H., Kikuzaki, H. and Nakatani, N. (1993) Two curcuminoid pigments from *Curcuma domestica*. *Phytochemistry* 33, 501–502.

Nandakumar, D.N., Nagaraj, V.A., Vathsala, P.G., Rangarajan, P. and Padmanaban, G. (2006) Curcumin-artemisinin combination therapy for malaria. *Antimicrobial Agents and Chemotherapy* 50, 1859–1860.

Nishiyama, T., Mae, T., Kishida, H., Tsukagawa, M., Mimaki, Y., Kuroda, M., Sashida, Y., Takahashi, K., Kawada, T., Nakagawa, K. and Kitahara, M. (2005) Curcuminoids and sesquiterpenoids in turmeric (*Curcuma longa* L) suppress and increase in blood glucose level in type 2 diabetic KK-Ay mice. *Journal of Agricultural and Food Chemistry* 53, 959–963.

Nose, M. (1998) Trypanocidal effects of curcumin *in vitro*. *Biological and Pharmaceutical Bulletin* 21, 643–645.

Ohshiro, M., Kuroyanagi, M. and Ueno, A. (1990) Structures of sesquiterpenes from *Curcuma longa*. *Phytochemistry* 29, 2201–2205.

Ozaki, K., Kawata, Y., Amano, S. and Hanazawa, S. (2000) Stimulatory effect of curcumin on osteoclast apoptosis. *Biochemical Pharmacology* 59, 1577–1581.

Panchatcharam, M., Miriyala, S., Gayathri, V.S. and Suguna, L. (2006) Curcumin improves wound healing by modulating collagen and decreasing reactive oxygen species. *Molecular and Cellular Biochemistry* 290, 87–96.

Pérez-Arriaga, L., Mendoza-Magana, M.L., Cortes-Zarate, R., Corona-Rivera, A., Bobadilla-Morales, L., Troyo-Sanroman, R. and Ramirez-Herrera, M.A. (2006) Cytotoxic effects of curcumin on *Giardia lamblia* trophozoites. *Acta Tropica* 98, 152–161.

Peschel, D., Koerting, R. and Nass, N. (2007) Curcumin induces changes in expression of genes involved in cholesterol homeostasis. *Journal of Nutritional Biochemistry* 18, 113–119.

Phan, T.T. (2001) Protective effects of curcumin against oxidative damage on skin cells *in vitro*: its implication for wound healing. *Journal of Trauma* 51, 327–331.

Rafael, Z., Salto, R., Li, J., Craik, C. and de Montellano, P.R.O. (1993) Inhibition of HIV-1 and HIV-2 proteases by curcumin and curcumin boron complexes. *Bioorganic and Medicinal Chemistry* 1, 415–422.

Ramachandran, C., Fonseca, H.B., Jhabvala, P., Escalon, E.A. and Melnick, S.J. (2002) Curcumin inhibits telomerase activity through human telomerase reverse transcriptase in MCF-7 breast cancer cell line. *Cancer Letters* 184, 1–6.

Ramadan, A.M., Nadia, H.M. and Dalia, D. (2012) Curcumin Reduced Potato Chips and Roasted Bread Induced Chromosomal Aberrations and Micronuclei Formation in Albino Rats. *Life Science Journal* 9, 330–336.

Ramirez, B.A., Soler, A., Carrión-Gutiérrez, M.A., Pamies, M.D., Pardo, Z.J., Diaz-Alperi, J., Bernd, A., Quintanilla, A.E. and Miquel, J. (2000) An hydroalcoholic extract of *Curcuma longa* lowers the abnormally high values of human-plasma fibrinogen. *Mechanisms of Ageing and Development* 114, 207–210.

Ramsewak, R.S., De Witt, D.L. and Nair, M.G. (2000) Cytotoxicity, antioxidant and anti inflammatory activities of curcumins I-III from *Curcuma longa. Phytomedicine* 7, 303–308.

Ranjan, D., Siquijor, A., Johnston, T.D., Wu, G. and Nagabhuskahn, M. (1998) The effect of curcumin on human B-cell immortalization by Epstein-Barr virus. *American Journal of Surgery* 64, 47–51.

Ranjan, D., Chen, C., Johnston, T.D., Jeon, H. and Nagabhushan, M. (2004) Curcumin inhibits mitogen stimulated lymphocyte proliferation, NF-κB activation, and IL-2 signaling. *Journal of Surgical Research* 121, 171–177.

Rasmussen, H.B., Christensen, S.B., Kvist, L.P. and Karazmi, A. (2000) A simple and efficient separation of the curcumins, the antiprotozoal constituents of *Curcuma longa. Planta Medica* 66, 396–398.

Rastogi, R.P. and Mehrotra, B.N. (ed.) Central Drug Research Institute, Lucknow and National Institute of Science Communication, New Delhi. Vol 1: (1990), Vol 2: (1991), Vol 3: (1993), Vol 4: (1995), Vol 5: (1998).

Ravindranath, V. and Satyanarayana, M.N. (eds) (1980) Compendium of Indian Medicinal Plants. An Unsymmetrical diarylheptanoid from *Curcuma longa. Phytochemistry* 19, 2031–2032.

Reddy, R.C., Vatsala, P.G., Keshamouni, V.G., Padmanaban, G. and Rangarajan, P.N. (2005) Curcumin for malaria therapy. *Biochemical and Biophysical Research Communications* 326, 472–474.

Reema, F.T., Dennis, D.H., Wael, K.A. and Cheryl, L.R. (2006) Curcumin Content of Turmeric and Curry Powders. *Nutrition and Cancer* 55, 126–131.

Riva, D.A., Fernandez-Larrosa, P.N., Dolcini, G.L., Martinez-Peralta, L.A., Coulombie, F.C. and Mersich, S.E. (2008) Two immunomodulators, curcumin and sulfasalazine, enhance IDV antiretroviral activity in HIV-1 persistently infected cells. *Archives of Virology* 153, 561–565.

Robinson, W.E. Jr, Cordeiro, M., Abdel-Malek, S., Jia, Q., Chow, S.A., Reinecke, M.G. and Mitchell, W.M. (1996a) Dicaffeoylquinic acid inhibitors of human immunodeficiency virus integrase: inhibition of the core catalytic domain of human immunodeficiency virus integrase. *Molecular Pharmacology* 50, 846–855.

Robinson, W.E. Jr, Reinecke, M.G., Abdel-Malek, S., Jia, Q. and Chow, S.A. (1996b) Inhibitors of HIV-1 replication that inhibit HIV integrase. *Proceedings of National Academy of Sciences USA* 93, 6326–6331.

Rohman, A. (2012) Analysis of curcuminoids in food and pharmaceutical products. *International Food Research Journal* 19, 19–27.

Ronita, D., Kundu, P., Swarnakar, S., Ramamurthy, T., Chowdhury, A., Balakrish, N.G. and Mukhopadhyay, A.K. (2009) Antimicrobial activity of curcumin against *Helicobacter pylori* isolates from India and during infections in mice. *Antimicrobial Agents and Chemotherapy* 53, 1592–1597.

Roth, G.N., Chandra, A. and Nair, M.G. (1998) Novel bioactivities of *Curcuma longa* constituents. *Journal of Natural Products* 61, 542–545.

Ruby, A.J., Kuttan, G., Babu, K.D., Rajasekharan, K.N. and Kuttan, R. (1995) Anti-tumour and antioxidant activity of natural curcuminoids. *Cancer Letters* 94, 79–83.

Ruderman, N.B., Park, H., Kaushik, V.K., Dean, D., Constant, S., Prentki, M. and Saha, A.K. (2003) AMPK as a metabolic switch in rat muscle, liver and adipose tissue after exercise. *Acta Physiologica Scandinavica* 178, 435–442.

Rupnick, M.A., Panigrahy, D., Zhang, C.Y., Dallabrida, S.M., Lowell, B.B., Langer, R. and Folkman, M.J. (2002) Adipose tissue mass can be regulated through the vasculature. *Proceedings of National Academy of Sciences USA* 99, 10730–10735.

Saleheen, D. (2002) Latent activity of curcumin against leshmaniasis *in vitro. Biological and Pharmaceutical Bulletin* 25, 386–389.

Satyavati, G.V., Raina, M.K. and Sharma, M. (1976) *Medicinal Plants of India.* (ed.) Indian Council of Medical Research, New Delhi, India.

Selvem, R., Subramanium, L., Gayathri, R. and Angayarkanni, N. (1995) The antioxidant activity of turmeric (*Curcuma longa*). *Journal of Ethnopharmacology* 47, 59–67.

Shabon, M.H. (2008) Use of curcuma as a radioprotector. *Proceedings of the 3rd Environmental Physics Conference*. Aswan, Egypt, pp. 19–23.

Shalini, V.K. and Srinivas, L. (1987) Lipid peroxide induced DNA damage: protection by turmeric (*Curcuma longa*). *Molecular and Cellular Biochemistry* 77, 3–10.

Shankar, T.N.B., Shanta, N.V., Ramesh. H.P., Murthy, I.A.S. and Murthy, V.S. (1980) Toxicity studies on turmeric (*Curcuma longa*): Acute toxicity studies in rats, Guinea pigs and Monkeys. *Indian Journal of Experimental Biology* 18, 73–75.

Sharma, R.A., McLelland, H.R., Hill, K.A., Ireson, C.R., Euden, S.A., Manson, M.M., Pirmohamed, M., Marnett, L.J., Gescher, A.J. and Steward, W.P. (2001) Pharmacodynamic and pharmacokinetic study of oral curcuma extract in patients with colorectal cancer. *Clinical Cancer Research* 7, 1894–1900.

Sharma, R.A., Steward, W.P. and Gescher, A.J. (2007) Pharmacokinetics and pharmacodynamics of curcumin. *Advances in Experimental Medicine and Biology* 595, 453–470.

Shoba, G., Joy, D., Joseph, T., Majeed, M., Rajendran, R. and Srinivas, P.S. (1998) Influence of piperine on the pharmacokinetics of curcumin in animals and human volunteers. *Planta Medica* 64, 353–356.

Shukla, Y. (2002) Antimutagenic potential of curcumin on chromosomal aberrations in wistar rats. *Mutation Research* 575, 197–202.

Shukla, Y., Arora, A. and Taneja, P. (2003) Antigenotoxic potential of certain dietary constituents. *Teratogenesis, Carcinogenesis and Mutagenesis* 1, 323–335.

Sidhu, G.S., Singh, A.K., Thaloor, D., Banaudha, K.K., Patniak, G.K., Srimal, R.C. and Maheshwari, R.K. (1998) Enhancement of wound healing by curcumin in animals. *Wound Repair and Regeneration* 6, 167–177.

Simon, A., Allais, D.P. and Duroux, J.L. (1998) Inhibitory effect of curcuminoids on MCF-7 cell proliferation and structure-activity relationships. *Cancer Letters* 129, 111–116.

Smith, M.R., Gangireddy, S.R., Narala, V.R., Hogaboam, C.M., Standiford, T.J., Christensen, P.J., Kondapi, A.K. and Reddy, R.C. (2010) Curcumin inhibits fibrosis-related effects in IPF fibroblasts and in mice following bleomycin-induced lung injury. *American Journal of Physiology – Lung Cellular and Molecular Physiology* 298, 616–625.

Soni, K.B., Rajan, A. and Kuttan, R. (1992) Reversal of aflatoxin induced liver damage by turmeric and curcumin. *Cancer Letters* 66, 115–121.

Soudamini, K.R. (1988) Cytotoxic and tumor reducing properties of curcumin. *Indian Journal of Pharmacology* 20, 95–101.

South, E.H. (1997) Dietary curcumin enhances antibody response in rats. *Immunopharmacology and Immunotoxicology* 19, 105–119.

Sreejayan, N. and Rao, M.N. (1996) Free radical scavenging activity of curcuminoids. *Drug Research* 46, 169–171.

Sreejayan, N. and Rao, M.N. (1997) Nitric acid oxide scavenging by curcuminoids. *Journal of Pharmacology* 49, 105–107.

Srinivas, L., Shalini, V.K. and Shylaja, M. (1992) Turmerin: a water soluble antioxidant peptide from turmeric. *Archives of Biochemistry and Biophysics* 292, 617–623.

Stano, J., Grancai, D., Neubert, K. and Kresanek, J. (2000) Curcumin as a potential antioxidant. *Ceska Slov Farm* 49, 168–170.

Sundarananthavalli, S., Kulandaisamy, A. and Christopher, C.C. (2011) Synthesis, characterization, analgesic, anti-inflammatory, anti-ulcer, wound healing and antimicrobial effects of curcuminoids. *International Journal of Chemistry and Technology Research* 3, 2040–2046.

Thresiamma, K.C. (1996) Protective effect of curcumin, ellagic acid and bixin on radiation induced toxicity. *Indian Journal of Experimental Biology* 34, 845–847.

Unnikrishnan, M.K. and Rao, M.N. (1995) Inhibition of nitrite induced oxidation of heamoglobin by curcuminoids. *Pharmazie* 50, 490–492.

Varadkar, P. (2001) Modulation of radiation induced protein kinase and activity by phenolics. *Journal of Radiological Protection* 21, 261–370.

Venkatesan, N. (1998) Curcumin attenuation of acute adriyamycin myocardial toxicity in rats. *British Journal of Pharmacology* 124, 425–427.

Venkatesan, N., Punithavathi, D. and Babu, M. (2007) Protection from acute and chronic lung diseases by curcumin. *Advances in Experimental Medicine and Biology* 595, 379–405.

Venkatesan, P. and Rao, M.N. (2000) Structure activity relationships for the inhibition of lipid peroxidation and the scavenging of free radicals by synthetic symmetric curcumin analogues. *Journal of Pharmacy and Pharmacology* 52, 1123–1128.

Vlietinck, A.J. (1998) Plant-derived leading compounds for chemotherapy of HIV infection. *Planta Medica* 64, 97–109.

Weisberg, S.P., Leibel, R. and Tortoriello, D.V. (2008). Dietary curcumin significantly improves obesity-associated inflammation and diabetes in mouse models of diabesity. *Endocrinology* 149, 3549–3558.

Wilken, R., Veena, M.S., Wang, M.B. and Srivatsan, E.S. (2011) Curcumin: a review of anti-cancer properties and therapeutic activity in head and neck squamous cell carcinoma. *Molecular Cancer* 10, 12, doi, 10.1186/1476-4598-10-12.

Wu, S., Lin, Y., Chu, C., Tsai, Y. and Chao, J.C. (2008) Curcumin or Saikosaponin improves Hepatic Antioxidant Capacity and Protects Against CCl_4-Induced Liver Injury in Rats. *Journal of Medicinal Food* 11, 224–229.

Xu, Y.X., Pindolia, K.R., Janakiraman, N., Noth, C.J., Chapman, R.A. and Gautam, S.C. (1997) Curcumin, a compound with anti-inflammatory and antioxidant properties downregulates chemokine expression in bone marrow stromal cells. *Experimental Hematology* 25, 413–422.

Yamamoto, H., Mizutami, T. and Nomura, H. (1997) Inhibitory effects of curcurmin on mammalian phospholipase D activity. *FEBS Letters* 417, 196–198.

Yu, Z.F., Kong, L.D. and Chen, Y. (2002) Anti-depressant activity of aqueous extracts of *Curcuma longa* in mice. *Journal of Ethnopharmacology* 83, 161–165.

Zhang, D., Huang, C., Yang, C., Liu, R.J., Wang, J., Niu, J. and Bromme, D. (2011) Antifibrotic effects of curcumin are associated with over-expression of cathepsins K and L in bleomycin treated mice and human fibroblasts. *Respiratory Research* 12, 154, doi:10.1186/1465-9921-12-154.

Zingg, J.M., Hasan, S.T., Cowan, D., Ricciarelli, R., Azzi, A. and Meydani, M. (2012) Regulatory effects of curcumin on lipid accumulation in monocytes/macrophages. *Journal of Cellular Biochemistry* 113, 833–840.

18 Phytochemistry of Plants Used in Traditional Medicine

Armando Enrique González-Stuart,[1] Dhan Prakash[2] and Charu Gupta[2]*
[1]*College of Health Sciences, University of Texas at El Paso, Texas, USA;*
[2]*Amity Institute for Herbal Research and Studies, Amity University, Noida, India*

18.1 Introduction

There is an increasing interest in natural plant products as a source of new pharmaceuticals and other biologically active phytochemicals. The major screens for biological activities of plant extracts have been carried out in the search for new anticancer, antiviral and antifertility drugs. The development of the rapid screening tests now in use in industry has meant that many more plants can be evaluated for a wide range of biological activities. There still remains an urgent need to develop new clinical drugs and this can be exemplified by the numerous diseases like cancer, hypertension, obesity, diabetes and other age-related disorders. Natural products already have a proven track record for various activities and it is possible that there are further such drugs still to be found from nature. Unfortunately the results of such tests do not necessarily reach the public domain and are kept in locked industrial files. The present chapter deals with the latest advances and trends in a field that is becoming a commercially significant area of investigation for the pharmaceutical industry.

18.2 *Aloe barbadensis* (*Aloe vera*, Ghritkumari, Family *Liliaceae*)

The name *Aloe vera* derives from the Arabic word 'Alloeh' meaning 'shining bitter substance', while 'vera' in Latin means 'true'. It belongs to *Asphodelaceae* (*Liliaceae*) family, and is a shrubby or arborescent, perennial, xerophytic, succulent, pea-green colour plant. There has been partial historic documentation on the evolution of *A. vera* and its use in past times. It has been rumoured that Egyptian queens, such as Cleopatra and Nefertiti, used *A. vera* in their routine beauty regimens. Originating in Africa, *A. barbadensis* then spread to the Americas after the expeditions of Columbus and Vespucci. It grows mainly in the dry regions of Africa, Asia, Europe and America. There are over 400 species of *Aloe* plants in the lily family. *Aloe vera* has succulent, fleshy leaves of a mottled light green colour with delicate edges sometimes punctuated with a pink hue during the winter months. These leaf-like structures have spiky edges that protect the plant from being consumed easily.

There are two parts of *A. vera* that are commonly used. The bitter exudate is used as

* E-mail: charumicro@gmail.com

a natural drug for its cathartic effect and is widely employed as a bittering agent in alcoholic beverages and as a laxative. The inner gel, or 'pure gel,' is the more readily known part of the *A. vera* plant. This is the section of the plant that is most commonly known to be used for treatments of sunburns. Many of the health benefits associated with *A. vera* have been attributed to the polysaccharides contained in the gel of the leaves. These biological activities include promotion of wound healing, antifungal activity, hypoglycaemic or antidiabetic effects, anti-inflammatory, anticancer, immunomodulatory and gastroprotective properties (Hamman, 2008).

The predominant substance in this gel is the muco-polysaccharide acemannen, a complex carbohydrate involved mainly in the processes of immune-modulation, wound healing and anti-inflammatory reactions. The aloin contained in the plant, an anthraquinone, has numerous actions, i.e. laxative, blood purifying and diuretic. *Aloe vera* contains 75 potentially active constituents: vitamins, enzymes, minerals, sugars, lignin, saponins, salicylic acids and amino acids (Shelton, 1991; Atherton, 1998).

Aloe vera is used in the food industry as a source of functional foods and as an ingredient in other food products, for the production of gel-containing health drinks and beverages. In the cosmetic and toiletry industry, it has been used as base material for the production of creams, lotions, soaps, shampoos, facial cleansers and other products (Hamman, 2008). In the pharmaceutical industry, it has been used for the manufacture of topical products such as ointments and gel preparations, as well as in the production of tablets and capsules (Eshun and He, 2004; He *et al.*, 2005). *Aloe vera* gel and whole leaf extract has the ability to improve the bioavailability of co-administered vitamins in human subjects (Vinson *et al.*, 2005).

18.3 *Terminalia arjuna* (Arjun, Family *Combretaceae*)

Terminalia arjuna is a deciduous tree, and is found throughout India growing to a height of 60–90 feet. The thick, white to pinkish-grey bark has been used in India's native Ayurvedic medicine for over three centuries, primarily as a cardiac tonic.

Clinical evaluation of this botanical medicine indicates that it can be used in the treatment of coronary artery disease, heart failure, and possibly hypercholesterolaemia. It is believed to have the ability to cure hepatic, urogenital, venereal and viral diseases (Kumar and Prabhakar, 1987). It also possesses antilipidaemic, antioxidant (Chander *et al.*, 2004), anti-inflammatory, antinociceptive and immunomodulatory activities (Halder *et al.*, 2009).

Terminalia's active constituents include tannins, triterpenoid saponins (arjunic acid, arjunolic acid, arjungenin, arjunglycosides), flavonoids (arjunone, arjunolone, luteolin), gallic acid, ellagic acid, oligomeric proanthocyanidins (OPCs), phytosterols, calcium, magnesium, zinc and copper (Kapoor, 1990; Bone, 1996). Improvement of cardiac muscle function and subsequent improved pumping activity of the heart seems to be the primary benefit of *Terminalia*. It is thought that saponin glycosides might be responsible for inotropic effects of *Terminalia*, while the flavonoids and OPCs provide free radical antioxidant activity and vascular strengthening. A dose-dependent decrease in heart rate and blood pressure was noted in dogs given *Terminalia* intravenously.

Terminalia arjuna is effective in many cardiac disorders such as angina, myocardial infarction, hypertension, hypercholesteraemia, cardiac arrest etc. Experimental studies have revealed that its bark possesses significant inotropic and hypotensive effects, increasing coronary artery flow (Rose and Treadway, 2000; Khan and Balick, 2001). The ethanol extract of the bark has been found to be effective in lowering low-density lipoprotein (LDL) cholesterol levels significantly (Alpana *et al.*, 1997). *Terminalia arjuna* is reported to possess antimutagenic activities. The bark of the plant is rich in polyphenols (60–70%) including flavones, flavanols and tannins. The high contents of tannins and polyphenols are responsible for anticancer activities (Kaur *et al.*, 2002). *Terminalia arjuna* bark extract has also been shown to have antioxidant effects on N-nitrosodiethylamine (DEN)-induced hepatocellular carcinoma in rats induced by a single intraperitonial injection of DEN (200 mg kg^{-1}). Hence, the protective effect is against DEN-induced liver cancer (Ramnath *et al.*, 2007). *Terminalia arjuna* bark extract is known to have antibacterial activity against *Escherichia coli, Plasmodium vulgaris* and *Plasmodium aerogenes* (Aggarwal and Dutt, 1936).

18.4 *Withania somnifera* (Ashwagandha, Family *Solanaceae*)

Withania somnifera, commonly known in Sanskrit as Ashwagandha, is a perennial shrubby plant cultivated in India, parts of East Asia and Africa that offers tremendous potential as an energizing medicinal herb. Ayurvedic practitioners have used the roots of this plant for centuries with success as a tonic to increase vitality and longevity, as well as to treat health conditions as diverse as tumours and arthritis. Its leaves are used in Ayurvedic and Unani systems for treatment of tumours and tubercular glands (Chopra *et al.*, 1992). The herb is termed a rasayana in Ayurvedic practice, which means it acts as a tonic for vitality and longevity.

Recent laboratory studies have begun to confirm what Ayurvedic practitioners have known for years – that *W. somnifera* deserves attention as a herbal therapy to ease or even eliminate many of today's common health problems. Sometimes referred to as Indian ginseng because of its stimulating effects, Ashwagandha is used to calm the mind, relieve weakness and nervous exhaustion, build sexual energy and promote healthy sleep. The pharmacological effects of the roots of *W. somnifera* are attributed to the presence of withanolides, a group of steroidal lactones. A number of withanolide steroidal lactones have been isolated from the leaves of *W. somnifera* and exhibit antibacterial, antifungal and antitumour properties (Devi *et al.*, 1993). There are a number of reports elucidating the chemical and pharmacological properties of *W. somnifera* (Kandil *et al.*, 1994). The roots of *W. somnifera* consist primarily of compounds known as withanolides, which are believed to account for its extraordinary medicinal properties. Withanolides are steroidal and bear a resemblance, both in their action and appearance, to the active constituents of Asian ginseng (*Panax ginseng*) known as ginsenosides. Ashwagandha's withanolides have been researched in a variety of animal studies examining their effect on numerous conditions, including immune function and even cancer (Grandhi *et al.*, 1994).

Chemical analysis of Ashwagandha shows its main constituents to be alkaloids and steroidal lactones. Among the various alkaloids, withanine is the main constituent. The other alkaloids are somniferine, somnine, somniferinine, withananine, pseudowithanine, tropine, pseudotropine, 3-α-gloyloxytropane, choline, cuscohygrine, isopelletierine, anaferine and anahydrine. Research results showed that both *W. somnifera* and *P. ginseng* decreased the frequency and severity of stress-induced ulcers, reversed stress-induced inhibition of male sexual behaviour and inhibited the effects of chronic stress on retention of learned tasks. Both botanicals also reversed stress-induced immunosuppression, but only the *Withania* extract increased peritoneal macrophage activity. Ashwagandha is reported to have anticarcinogenic effects. Research on animal cell cultures has shown that the herb decreases the levels of the nuclear factor kappa B (NF-κB), suppresses the intercellular tumour necrosis factor and potentiates apoptotic signalling in cancerous cell lines (Prakash *et al.*, 2002). One of the most exciting of the possible uses of Ashwagandha is its capacity to fight cancers by reducing tumour size (Jayaprakasam *et al.*, 2003).

18.5 *Acacia arabica* (Babul, Family *Fabaceae*)

Acacia species are commonly known as 'Babul' in India and ethno-medicinally have long been used for the treatment of skin, sexual, stomach and tooth problems. Commonly known as babul, kikar or Indian gum Arabic tree, it has been recognized worldwide as a multipurpose tree. It is widely distributed throughout arid and semi-arid zones of the world.

Acacia arabica has been proved as effective medicine in treatment of malaria, sore throat (aerial part) and toothache (bark) (Chowdhury *et al.*, 1983; Kubmarawa *et al.*, 2007). Researchers have tested the antifertility activity of *A. arabica* pods and nuts. The methanolic extracts of *A. arabica* pods have been claimed against HIV-PR (Hussein *et al.*, 1999; Bessong and Obi, 2006). One study has reported the antiplasmodial activity of *Acacia* ethyl acetate extract against different chloroquine-resistant and -sensitive strains of *Plasmodium falciparum* (El-Tahir *et al.*, 1999). The fresh plant parts of this species have been reported to be most active against hepatitis C virus (Hussein *et al.*, 2000). It is an important multipurpose tree that has been used extensively for the treatment of various diseases, e.g. colds, bronchitis, diarrhoea, dysentery, biliousness, bleeding piles and leucoderma.

18.6 *Phyllanthus niruri* (Bhumi Amla, Family *Euphorbiaceae*)

The annual herb *Phyllanthus niruri* is commonly known as Bhumi Amla. It grows 50 to 70 cm tall and bears ascending herbaceous branches. The bark is smooth and light green. *Phyllanthus niruri* is an important plant of the Indian Ayurvedic system of medicine and is used for problems of the stomach, genito-urinary system, liver, kidney and spleen (Patel *et al.*, 2011).

More than 50 compounds were identified in *P. niruri* including alkaloids, flavanoids, lignans and triterpenes (Bagalkotkar *et al.*, 2006). Among these substances, the triterpenes have been found to inhibit the cytotoxicity induced by calcium oxalate (Malini *et al.*, 2000) as well as to reduce excretion of stone-forming constituents (Vidya *et al.*, 2002) and the markers of crystal deposition in the kidneys (Murugaiyah and Chan, 2006). According to Calixto *et al.* (1998), alkaloids extracted from plants of the genus *Phyllanthus* present an antispasmodic activity leading to smooth muscle relaxation, mostly evidenced in the urinary tract, which would facilitate the elimination of urinary calculi. These data strongly suggest that *P. niruri* may be a potential source of many substances with antilithiasic properties.

A clinical study with *P. niruri* (Micali *et al.*, 2006) indicated that it may reduce the levels of urinary calcium (Nishiura *et al.*, 2004). A subsequent study of 150 patients over a 6 month period indicated that an extract of this herb reduces the incidence of stone formation, and concluded that:

> Regular self-administration of *P. niruri* after extracorporeal shock wave lithotripsy for renal stones results in an increased stone-free rate that appears statistically significant for lower caliceal location. Its efficacy and the absolute lack of side effects make this therapy suitable to improve overall outcomes after extracorporeal shock wave lithotripsy for lower pole stones.

Experimental and clinical studies performed by several groups have produced interesting and hopeful data concerning the potential therapeutic use of *P. niruri* to treat and/or to prevent stone formation. A more recent rat study found that *P. niruri* has been shown to interfere with many stages of stone formation, reducing crystals aggregation, modifying their structure and composition as well as altering the interaction of the crystals with tubular cells leading to reduced subsequent endocytosis (Boim *et al.*, 2010).

18.7 *Vaccinium* spp. (Blueberry, Family *Ericaceae*)

Blueberries are perennial flowering plants with indigo-coloured berries. Blueberries are a native to North America. They are usually erect, but sometimes prostrate shrubs varying in size from 10 cm to 4 m. In commercial blueberry production, smaller species are known as 'low bush blueberries' (synonymous with 'wild') and the larger species are known as 'high bush blueberries'. The fruit is a berry 5–16 mm (0.20–0.63 in) in diameter with a flared crown at the end; they are pale greenish at first, then reddish purple and finally dark blue when ripe. They have a sweet taste when mature, with variable acidity. Blueberry bushes typically bear fruit in the middle of the growing season: fruiting times are affected by local conditions such as altitude and latitude, so the height of the crop can vary from May to August depending upon these conditions.

Blueberries are among the fruits that are best recognized for their anthocyanin and flavonoid content, and for their potential health benefits. Natural plant-produced anthocyanin pigments, the substances that are largely responsible for the intense colours of blueberries, are considered to be responsible for a range of unique and broad-spectrum health benefits (Serafini *et al.*, 1998). Although the reasons behind these disease-preventive properties are not fully understood, ample evidence suggests that these polyphenolic compounds are readily absorbed *in vivo*, are involved in antioxidant defences, and may play similar roles in combating both heart disease and various forms of cancer (de Groot and Rauen, 1998; Parthasarathy, 1998).

Blueberries are one of the richest sources of anthocyanins and flavonoid compounds, making them an exceptional whole food antioxidant. Given the possibility that blueberries may reverse short-term memory loss and forestall other effects of ageing, their potential may be even greater. The anthocyanins that give the fruit its blue hue are the major contributors to its antioxidant activity (Seeram *et al.*, 2008). The abundance of vitamin C is also a big factor for this as well. The studies have found that animals fed on a blueberry extract diet showed fewer age-related motor changes and outperformed their non-blueberry-consuming peers on memory tests. The antioxidant activity was measured by a standard test that measures fruit or vegetables' ability to quench free radicals *in vivo*, called ORAC (oxygen radical absorbency capacity), which was used to measure both lipophilic and hydrophilic antioxidant capacity of blueberries. The higher the ORAC value of an item, the higher 'anti-ageing' capacity it has (Wang *et al.*, 2008). The thought is that the blueberries containing an abundant quantity of antioxidants may act to protect the body against damage from 'oxidative stress', one of several biological processes associated with ageing and neurological diseases. In addition to anti-ageing, wild blueberries also have been shown to have anti-inflammatory, pro-heart and pro-vision properties (Seeram *et al.*, 2008).

18.8 *Caulophyllum thalictroides* (Blue Cohosh Root, Family *Berberidaceae*)

Caulophyllum thalictroides is a small woodland perennial plant, native to the American northeast. The medicinal effects of blue cohosh are derived from its root and rhizomes. Blue cohosh is also referred to as 'papoose root' or 'squaw root', which reflects the use of this herbal medicine by Native American women who brewed blue cohosh as a tea to relieve menstrual cramps and to ease the pains associated with childbirth. Between 1882 and 1905, blue cohosh was listed in the United States Pharmacopoeia as a labour inducer (McFarlin *et al.*, 1999). Blue cohosh is often part of a combination of herbal medicines that have been traditionally used in the third trimester to prepare a woman for delivery; this preparation is called 'mother's cordial' or 'partus preparatus'. In a 1999 survey of Certified Nurse Midwives in the USA, 64% claimed to use blue cohosh during labour (McFarlin *et al.*, 1999).

Blue cohosh enhances oestradiol binding to oestrogen receptors and increases oestradiol-induced transcription activity in oestrogen-responsive cells (Jellin *et al.*, 2002). Blue cohosh decreases luteinizing hormone (LH) levels and increases serum ceruloplasmin oxidase activity, which are measures of oestrogenic activity in the liver (Jellin *et al.*, 2002). Pharmacological studies reported glycosides in blue cohosh have significant oxytocic activity (hastening childbirth) by acting as smooth muscle stimulant. Furthermore, the aglycone obtained from acid hydrolysis of the glycosides matched with those obtained from caulosaponin. However, the glycosides also exert a toxic effect on cardiac muscle by constricting the coronary blood vessels. In addition, it is well known that blue cohosh has a number of toxic alkaloids including quinolizidine alkaloids implicated as teratogens (Woldemariam *et al.*, 1997). Among them, *N*-methylcytisine showed teratogenic activity in the REC (rat embryo culture) (Kennelly *et al.*, 1999).

18.9 *Bacopa monnieri* (Brahmi, Family *Plantaginaceae*)

Bacopa monnieri Linn. is a herb that occurs naturally in India and has a long history of use in the Ayurvedic medicine tradition in the treatment of a number of disorders, particularly those involving anxiety, intellect and poor memory (Singh and Dhawan, 1997).

Bacopa monnieri is used as a nerve tonic in the traditional medicinal system in India.

It is currently being marketed in Western countries as a memory-enhancing agent. Studies have shown that the herb contains many active constituents, including a number of alkaloids and saponins, however, the major constituents are the steroidal saponins, bacosides A and B. There are no published scientific studies of the effects of Brahmi on memory in humans; however, there are some behavioural studies with rats. These studies have shown that it improves the rate of learning in a brightness discrimination task and a conditioned avoidance task, that it improves retention, as demonstrated by savings in relearning, and that it attenuates amnesia induced by immobilization, electroconvulsive shock and scopolamine (Singh and Dhawan, 1997). This later finding involved administration of the extracted bacosides, A and B, and suggests that they influence cholinergic systems. Recently, however, it has been reported that Brahmi has an antioxidant effect in the rat frontal cortex, striatum and hippocampus (Bhattacharya *et al.*, 2000).

Its ethanolic extract contains a mixture of triterpenoid and steroidal saponins (Roodenrys *et al.*, 2002; Jyoti and Sharma, 2006). Bacoside A comprises a mixture of three saponins, bacogenin A1, A2 and A3, with A3 being a major constituent (Russo and Borrelli, 2005). Several other types of saponins have been isolated and characterized in the last few years (Russo and Borrelli, 2005). *Bacopa monnieri* (BM) extract has earlier been reported to augment both the cognitive functions and mental retention capacity in different behaviour studies (Singh and Dhawan, 1997).

Ethanolic extract of *B. monnieri* has been found to increase the activity of antioxidative enzymes in different brain regions of the rat. This exhibits its antioxidative potential (Bhattacharya *et al.*, 2000). It has been reported that ethanolic extract was also found to inhibit the amnesic effects of scopolamine, electroshock and immobilization stress (Singh and Dhawan, 1997) and can significantly improve the speed of visual processing, learning rate and memory consolidation (Russo and Borrelli, 2005). This extract also reduces the different stress effect in rat brain by Hsp70 expression, superoxide dismutase activity and P450 enzyme activity (Russo and Borrelli, 2005).

18.10 *Rhamnus cathartica* (Buckthorn Bark, Family *Rhamnaceae*)

Rhamnus cathartica is a shrub known from the 14th century; *R. cathartica* is especially administrated as a laxative or purgative. Only the buckthorn bark is used for medicinal purposes. The external surface is covered with irregular longitudinal cracks. The internal part is finely striated longitudinally. On a section it can be seen that the cut is smooth towards the exterior and fibrous towards the interior. The maximum thickness of the bark is 2 mm and the colour differs from the exterior where it is brown, to the interior, where it is light-orange or yellow-brown. If the external surface is scratched a red layer appears immediately inside. The taste is first mucilaginous, then bitter and astringent, giving off a weak smell.

It is also a good cholagogue and choleretic. Its laxative action manifests within 10 h from the moment of administration by stimulating mobility of the large intestine. The bark contains a mixture of anthraquinone derivatives (anthranoids) of which the majority is present as glycosides. The total content of anthranoids is from 2 to 6%. It also contains glycosides of emodin such as glucofrangulin A and B and frangulin A and B. The free aglycones emodin, chrysophanol and physcion are also present in varying concentrations (Lichtensteiger *et al.*, 1997). Buckthorn has also choleretic effects over the bile. It can relax the intestinal muscles and has vermifuge properties. The fruits contain vitamins, mineral substances and fatty acids and have an antioxidant, protective and regenerator action. Chronic and acute constipations can be treated by administrating buckthorn powder (1–3 g over 24 h) or warm tea before bedtime. The dose can be repeated in the morning, on the empty stomach, if the problems persist. The quantity used depends on the constipation level, but it should always be low at the beginning of the cure. Buckthorn is also recommended for liver insufficiency in association with other herbs such as dandelion (Lichtensteiger *et al.*, 1997).

It is used in the treatment of liver disorders and especially for the treatment of constipations caused by bile insufficiency.

In cases of obesity, the administration of buckthorn powder each morning can produce very good results. The buckthorn bark can increase bile secretion. It is recommended to combine it with chicory or dandelion in order to increase its effect. Other afflictions that can be treated with buckthorn are: *Giardia*, rheumatism, headaches followed by constipation, allergies and hepatitis. It can also be used for intestinal worms. Used as compresses, buckthorn helps in the treatment of skin diseases associated with constipation (acne, allergic eczema, psoriasis, infections) (Lichtensteiger *et al.*, 1997).

18.11 *Cinnamomum cassia* (Cinnamon, Family *Lauraceae*)

Cinnamomum cassia has been a favorite spice around the world not only because of its health benefits but also because it flavours and preserves food. Cinnamon is native to southern Asia and South America. It is also now cultivated in many tropical countries such as India, China, Madagascar, Brazil, Mexico and the Caribbean. Cinnamon is also known as sweet wood (Chaudhary and Tariq, 2006).

It contains medicinally important essential oil in leaves, fruits, inner and outer bark. Much of cinnamon's bioactivity resides in its oil content, which is about 90% cinnamaldehyde. It is used mainly in medicine, foods and cosmetics (Bown, 1995), and is employed in aromatherapy as a rub to promote blood circulation. It also contains both antifungal and antibacterial principles that can be used to prevent food spoilage due to bacterial contamination (Fabio *et al.*, 2003). Research interest has focused on cinnamon that possesses chemopreventive, antispasmodic, anti-ulcer, choleretic, sedative, hypothermic, antifungal, antibacterial, antiviral, antipyretic, lipolytic, antiseptic, anaesthetic, anodyne, cytotoxic, hypolipidaemic and antiplatelet properties and also stimulates the immune system, which may be useful adjuncts in helping to reduce the risk of cardiovascular disease and cancer (Cralg, 1999).

18.12 *Coleus forskohlii* (Coleus, Family *Lamiaceae*)

Coleus forskohlii is part of the mint family and has long been cultivated in India, Thailand and parts of South-east Asia as a spice and as a condiment for heart ailments and stomach cramps. This species is a perennial herb with fleshy, fibrous roots that grows wild in the warm subtropical temperate areas in India, Burma and Thailand. In Ayurvedic medicine *Coleus* species have been used to treat heart disease, convulsions, spasmodic pain and painful urination. In traditional Ayurvedic systems of medicine, *C. forskohlii* has been used for treating heart diseases, abdominal colic, respiratory disorders, insomnia, convulsions, asthma, bronchitis, intestinal disorders, burning sensation, constipation, epilepsy and angina (Ammon and Muller, 1985). The roots are also used in treatment of worms and to alleviate burning in festering boils. When mixed with mustard oil, the root extract is applied to treat eczema and skin infections. The roots of the plant are a natural source of a diterpene alkaloid called forskolin, the only plant-derived compound known to directly stimulate the enzyme adenylate cyclase and subsequently cyclic AMP. Forskolin is 7β-acetoxy-8,13-epoxy-1α,6 β,9α-trihydroxylabd-14-en11-one (a diterpenoid compound), which directly activates adenylate cyclase (Ammon and Muller, 1985; De Souza and Shah, 1988).

The biological activities are antiglaucoma, antiplatelet, broncho-spasmolytic, cardiotonic, hypotensive, anti-ageing, anti-allergic, smooth muscle and arterial relaxant and anti-asthmatic.

Coleus also aids in weight loss due to its ability to break down stored fat as well as inhibit the synthesis of adipose tissue. Additionally, it increases thyroid hormone production thereby increasing metabolism. Forskolin is used for the treatment of eczema, asthma, psoriasis, cardiovascular disorders and hypertension, where decreased intracellular cAMP level is believed to be a major factor in the development of the disease process (Rupp *et al.*, 1986). It is being developed as a drug for hypertension, glaucoma, asthma, congestive heart failures and certain types of cancers. Forskolin is in great demand in Japan and European countries for its medicinal use and related research purposes. The plant is also used for veterinary purposes (De Souza and Shah, 1988). Forskolin is also used in the preparation of medicines preventing hair greying and restoring grey hair to its normal colour. Though grouped as a medicinal plant, it also contains essential oil in tubers, which has a very attractive and delicate odour with spicy notes. The essential oil has potential uses in the food-flavouring industry and can be used as an antimicrobial agent.

Forskolin is reported to be antiglaucoma, antiplatelet, broncho-spasmolytic, cardiotonic, hypotensive, anti-ageing, and anti-allergic, smooth muscle and arterial relaxant, and anti-asthmatic (Ammon and Muller, 1985). The constituents of *C. forskohlii*-like alkaloids (forskolin and its derivatives), phenols and tannins have been reported to exhibit some biological activities such as stimulating adenyl cyclase, inhibition of platelet aggregation, mast cell degranulation, relaxation of the arteries, increasing the insulin secretion and thyroid function, decreasing adipose accumulation, reduction of body weight, treating skin diseases, cardiovascular disease and asthma, stimulating the secretion of digestive enzymes and absorption of nutrients in the small intestine etc. (Badmaev *et al.*, 2002; Murugesan *et al.*, 2012). The diterpenoids in *Coleus* have attracted interest on account of their antibacterial activity (Murugesan *et al.*, 2012).

18.13 *Eucalyptus globulus* (Eucalyptus, Family *Myrtaceae*)

Eucalyptus globulus is a member of one of the world's important and most widely planted genera (Akin *et al.*, 2010). It is a tall, evergreen tree, native to Australia and Tasmania, successfully introduced worldwide, now extensively planted in several countries (Mubita *et al.*, 2008). *Eucalyptus* species are well known as medicinal plants because of their biological and pharmacological properties.

In the International Pharmacopeia, the most important and represented species is *E. globulus*, which is the main furnisher of essential oils (Bajaj, 1995). These essential oils are in great demand because of their applications as anaesthetic, anodyne, antiseptic, astringent, deodorant, expectorant, fumigant, insect repellent and vermifuge, as a folk remedy for arthritis, boils, cancer, diabetes, diarrhoea, dysentery, encephalitis, inflammation, leprosy, malaria and wounds (Elliot and Jones, 1986). Sometimes their demand is also high in the soap and cosmetic industries (Bajaj, 1995).

Gende *et al.* (2010) reported the biological activity of *E. globulus* essential oils derived from plant material obtained from different geographic areas. They carried out *in vitro* experiments on *Paenibacillus* larvae, *Varroa destructor* and *Apis mellifera*. The *E. globulus* essential oils tested in this study featured high efficiency against *V. destructor*, yet their antimicrobial activity against *Paenibacillus* larvae proved to be lower, and was innocuous to bees. The physico-chemical properties, composition, antimicrobial and bioactivity of essential oils were studied. They observed that the essential oils differed in their composition, albeit their physico-chemical properties were similar. The bioautography method determined that limonene accounted for the greatest antimicrobial activity with respect to other compounds (Gende *et al.*, 2010).

18.14 *Zingiber officinale* (Ginger, Family *Zingiberaceae*)

Zingiber officinale is one of the most widely used species and is a common condiment for various foods and beverages. Ginger has a long history of medicinal use dating back 2500 years in China and India for conditions such as headaches, nausea, rheumatism and colds (Grant and Lutz, 2000). Ginger contains a number of pungent constituents and active ingredients.

Steam distillation of powdered ginger produces ginger oil, which contains a high proportion of sesquiterpene hydrocarbons, predominantly zingiberene (Govindarajan, 1982). The major pungent compounds in ginger, from studies of the lipophilic rhizome extracts, have yielded potentially active gingerols, which can be converted to shogaols, zingerone and paradol (Govindarajan, 1982).

The compound 6-gingerol appears to be responsible for its characteristic taste. Zingerone and shogaols are found in small amounts in fresh ginger and in larger amounts in dried or extracted products.

The mechanism underlying ginger's anti-emetic activity is not clearly understood, but the aromatic, spasmolytic, carminative and absorbent properties of ginger suggest it has direct effects on the gastrointestinal tract (Tyler, 1986). The compounds 6-gingerol and 6-shogaol have been shown to have a number of pharmacological activities, including antipyretic, analgesic, antitussive and hypotensive effects (Suekawa *et al.*, 1984). Ginger has long been used as a remedy to decrease nausea and vomiting associated with several conditions. A randomized, double-blind, placebo-controlled study was performed to assess the effects of ginger extracts on motion sickness and gastric slow-wave dysrhythmias induced by circular vection (Lien *et al.*, 2003). No drug interactions are known; however, due to ginger's apparent effect on platelets, it should be used cautiously in individuals using anticoagulants (Anonymous, 2003).

18.15 *Gymnema sylvestre* (Gudmar, Family *Asclepiadaceae*)

Gymnema sylvestre is distributed throughout India in dry forests up to 600 m and also in Asia, tropical Africa, Malaysia and Sri Lanka. On account of its property of abolishing the taste of sugar it has been given the name of Gudmar, meaning sugar destroying, and it neutralizes the excess of sugar present in the body in diabetes mellitus. In the Indian traditional medicine system it has been used mainly for its antidiabetic properties (Warren *et al.*, 1969).

The plant is also reported to be bitter, astringent, acrid, thermogenic, anti-inflammatory, anodyne, digestive, liver tonic, emetic, diuretic, stomachic, stimulant, antihelmenthic, laxative, cardiotonic, expectorant, antipyretic and uterine tonic. It is useful in dyspepsia, constipation, jaundice, haemorrhoids, renal and vesical calculi, cardiopathy, asthma, bronchitis, amenorrhoea, conjunctivitis and leucoderma (Prakash *et al.*, 1986). Its leaves contain triterpene saponins belonging to oleanane and dammarene classes. Oleanane saponins are gymnemic acids and gymnema saponins, while dammarene saponins are gymnemasides. Besides this, other plant constituents are flavones, anthraquinones, quercitol, lupeol, β-amyrin, stigmasterol, hentri-acontane, pentatriacontane, phytin, resins, quercitol, lupeol, related glycosides and stigmasterol. The plant extract also tests positive for alkaloids. Leaves of this species yield acidic glycosides and anthraquinones and their derivatives (Dateo *et al.*, 1973). Gymnemic acid (GA) is reported to have antidiabetic, antisweetener and anti-inflammatory activities. GA I, II, III and IV are antisweet substances from the leaves of *G. sylvestre*. They all contain a glucuronic acid moiety, and the gymnemagenin aglycone esterified at position C-21 and C-28. A second series of gymnemic acid V–VII has also been reported. GA VII is the 3-O-glucuronide of gymnemagenin and GA V is the O-3-glycuronyl-22,21-*bis*-O-tigloyl substitution pattern. GA VIII–IX are also esters of saponin, have an oxoglycoside moiety attached to the glucuronic acid residue. Gurmarin, another bioactive constituent of the leaves, and gymnemic acid have been shown to block sweet taste in humans. Some researchers have suggested gymnemic acid(s) and gurmarin as possible candidates responsible for antidiabetic activity. These are considered as main bioactive constituents.

Gymnema sylvestre yields gymnemic acid, a glycoside isolated from its leaves, which is a destroyer of madhumeha (glycosuria) and other urinary disorders. It is believed that it neutralizes the excess of sugar present in the body in diabetes mellitus (Warren *et al.*, 1969). It is useful in dyspepsia, constipation, jaundice, haemorrhoids, renal and vesical calculi, cardiopathy, asthma, bronchitis, amenorrhoea, conjunctivitis and leucoderma (Prakash *et al.*, 1986). The leaves are also noted for lowering serum cholesterol and triglycerides. The primary chemical constituents of *Gymnema* include gymnemic acid, gurmarin, stigmasterol, betaine and choline. The water-soluble acidic fractions reportedly provide the hypoglycaemic action. Gurmarin and gymnemic acid have been reported to block sweet taste in humans (Flier, 2001; Steppan *et al.*, 2001; Ramachandran *et al.*, 2003). Mosquito larvicidal effects of *Gymnema* were reported by Khanna and Kannabiran (2007). Sathya *et al.* (2008) provided experimental evidence for the herbal plant *G. sylvestre* in the prevention and curing of alloxan-induced diabetic rats without any side effects.

18.16 *Panax ginseng* (Ginseng, Family *Araliaceae*)

In Asia, ginseng has a long history of traditional medicinal use as a general health-promoting tonic (Xiang *et al.*, 2008; Jia and Zhao, 2009; Lu *et al.*, 2009). Ginseng is found only in the northern hemisphere, in North America and in eastern Asia. In the Korean tradition, several different ways of preparing and manufacturing *P. ginseng* exist. Fresh ginseng is less than 4 years old; white ginseng is 4–6 years old and is dried after peeling; red ginseng is harvested when it is 6 years old, subsequently it is not skinned but steamed and then dried.

The subject is further complicated by the fact that, according to several laboratory investigations, commercially available ginseng products are sometimes of less than optimal quality (Sievenpiper *et al.*, 2004). There are extensive reports that have determined that ginseng has many pharmacological effects on the immune, cardiovascular, endocrine and central nervous systems (Nah *et al.*, 1995; Attele *et al.*, 1999). A blood glucose-lowering effect of ginseng root has also been found (Sotaniemi *et al.*, 1995; Chung *et al.*, 2001). However, despite the various reported functions of ginseng, no studies have yet reported the effects of ginseng on skin ageing. Chemically, the constituents of ginseng can be divided into saponin and non-saponin fractions.

Panax ginseng was shown to possess a potent antisepticaemic activity through nitric oxide via cytokine production in stimulated macrophage (Zhao *et al.*, 1995; Cohen, 2000; Muller-Kobold *et al.*, 2000). These results suggested that the polysaccharide from *P. ginseng* augments the production of the cytokines (TNF-α, IL-1, IL-6 and IFN-α). Since cytokines such as TNF-α, IL-1, IL-6 and IFN-α are known to be potent macrophage activators as well as immunomodulating agents, it was, therefore, possible that the *P. ginseng* polysaccharide activated macrophages by up-regulating the synthesis and production of these cytokines (Corradin *et al.*, 1991). When activated by cytokines, macrophages show enhanced ability to kill both invading extracellular as well as intracellular pathogens residing within these cells (Corradin *et al.*, 1991), one of the primary and important pathways by which intracellular killing may be achieved.

18.17 *Commiphora weightii* (Guggul, Family *Burseraceae*)

Commiphora weightii is one of the very ancient Ayurvedic drugs, having been first described in Atharva Veda (2000 BC). *Commiphora weightii* is found extensively in the dry regions of the Indian subcontinent, mainly India, Pakistan and Bangladesh (Satyavati, 1988). According to Sushrut Samhita, *C. weightii* is, when taken orally, curative of obesity, liver dysfunction, internal tumours, malignant sores and ulcers, urinary complaints, fistula-in-ano, intestinal worms, leucoderma, sinus, oedema and sudden paralytic seizures. Guggul is the dry gum resin obtained from the bark of the *Commiphora* tree. It is a mixture of diterpenes, sterols, steroids, esters and higher alcohols.

The active components of the plant are the guggulsterones, specifically the stereoisomers, guggulsterone E and guggulsterone Z. These are plant sterols with a high degree of human bioactivity, which have been shown to affect many biological processes. It is also considered a cardiac tonic (Satyavati, 1991). The oleo-gum-resin of the Guggul tree is a very complex mixture of gum, minerals, essential oils, terpenes, sterols, ferrulates, flavanones and sterones; several other unknown compounds have also been isolated. The resin yields two fractions upon ethyl acetate extraction. The ethyl acetate-soluble fraction contains 45% of the gum resin. The insoluble fraction consists of the carbohydrate gum, which is about 55% of the gum resin. The bioactive components have been found in the ethyl acetate-soluble fraction, whereas the insoluble carbohydrate fraction is devoid of any hypolipidaemic effects (Nityanand and Kapoor, 1975). The ethyl acetate-soluble fraction consists of diterpenoids, triterpenoids, steroids, lignans and fatty tetrol esters (Dev, 1989). Pharmacological studies revealed that the pure guggulsterone isomers had pronounced hypolipidaemic activity (Satyavati, 1988).

Guggulsterone has been shown to induce apoptosis and suppress proliferation, invasion, angiogenesis and metastasis of tumour cells. Various mechanisms have been suggested to explain the anticarcinogenic effects of guggulsterone, including inhibition of ROI, suppression of inflammation and inhibition of nuclear receptors, transcription factors, inflammatory cytokines, anti-apoptotic proteins, cell survival pathways, COX-2, MMP-9, iNOS and cell cycle-related proteins.

Oxidative stress plays an important part in many human diseases. Although it is unknown whether oxidative stress is the cause or a consequence of disease, antioxidants are widely used for maintaining health and preventing diseases. Guggulipid suppresses formation of lipid peroxides (Singh *et al.*, 1994) and prevents oxidation of LDL *in vitro* (Singh *et al.*, 1997; Wang *et al.*, 2004). In more recent studies, guggulsterone at concentrations of 5–20 μM effectively inhibited LDL peroxidation and generation of free oxygen radicals (Chander *et al.*, 2002, 2003). This finding indicates that guggulsterone may be of therapeutic benefit in diseases associated with oxidative stress, such as myocardial ischaemia and neurodegenerative diseases. Several studies have reported the cardioprotective activity of guggulsterone (Kaul and Kapoor, 1989).

18.18 *Myristica fragrans* (Mace, Nutmeg, Family *Myristicaceae*)

Myristica fragrans is commonly known as 'mace', and produces two spices: mace and nutmeg. Nutmeg is the seed kernel inside the fruit and mace is the red lacy covering (aril) on the kernel. *Myristica* species are natives of the Moluccas, indigenous to India, Indonesia and Sri Lanka and now cultivated in many tropical countries of both hemispheres as well as in South Africa (Pal *et al.*, 2011). *Myristica fragrans* is a spreading aromatic evergreen tree usually growing to about 5 to 13 m high. When the fruits are ripe, the succulent yellow fruit coat splits into two halves revealing a purplish brown, shiny seed (nutmeg) surrounded by a red aril (mace). When fresh, the aril (mace) is bright scarlet becoming more horny, brittle and with a yellowish brown colour when dried.

Mace is popular as a spice and also possesses various therapeutic properties. Mace has a characteristic pleasant fragrance and a slightly warm taste. It is used to flavour many kinds of baked foods, confections, puddings, meats, sausages, sauces, vegetables and beverages. It is also used as a component of curry powder, teas and soft drinks or mixed in milk and alcohol (Olaleye *et al.*, 2006). For a long time, *M. fragrans* has been used as a folklore medicine for treating diarrhoea, mouth sores and insomnia (Somani and Singhai, 2008). Since the Middle Ages, mace has been used as a stomachic, stimulant and carminative as well as for intestinal catarrh and colic, to stimulate appetite, to control flatulence and has a reputation as an emmenagogue and abortifacient (Min *et al.*, 2011). The essential oil of nutmeg is used externally for rheumatism and possesses analgesic and anti-inflammatory properties (Olajide *et al.*, 1999).

Compounds isolated from the seeds of this plant have been reported to possess strong platelet anti-aggregatory activity (Somani and Singhai, 2008). Mace also prevents hypercholesterolaemia and atherosclerosis (Sharma *et al.*, 1996). It has also been found to be useful as tonic for the heart and brain and also in sexual and general debility (Olaleye *et al.*, 2006). The presence of two compounds, myristicin and elemicin, is often related to intoxication and hallucinogenic action of nutmeg, while safrole has been suspected to be carcinogenic. However, the mechanism by which these compounds act is still a subject of extensive research (Jukic *et al.*, 2006).

18.19 *Trigonella foenum-graecum* (Fenugreek, Family *Fabaceae*)

Trigonella foenum-graecum is an annual plant, cultivated worldwide as a semi-arid crop and is a common ingredient in dishes from the Indian subcontinent, where it is known as Methi or fenugreek. The largest producer of fenugreek in the world is India, where the major fenugreek-producing states are Rajasthan, Gujarat, Uttar Pradesh, Madhya Pradesh, Maharashtra, Haryana, Uttarakhand and Punjab. Rajasthan produces the largest share of India's production, accounting for over 80% of the nation's total fenugreek output (Parthasarathy *et al.*, 2008). *Trigonella foenum-graecum* is an important medicinal plant and its leaves and seeds have been used in various ailments and as a health tonic.

Trigonella foenum-graecum is a well-known spicy agent, which prevents ageing, labour pain, imparts immunity, improves mental function and adds vitality to the body and it is also used in nervous disorders, dyspepsia, inflammation, tumours, cholesterolaemic, hyperglycaemic and ulcer conditions (Subhashini *et al.*, 2011). Its seeds are also used as herbal medicine in many parts of the world for their carminative, tonic and aphrodisiac effects. Various reports have demonstrated that *T. foenum-graecum* seeds can lower blood glucose and cholesterol in type 1 and type 2 diabetics and experimental diabetic animals (Kumar *et al.*, 2005). Reports indicate that the pharmacological activities of *T. foenum-graecum* include antidiabetic, antifertility, antifungal, analgesic, anti-inflammatory, antipyretic and immunomodulatory activities (Ahmadiani *et al.*, 2001; Bin-Hafeez *et al.*, 2003).

18.20 *Punica granatum* (Pomegranate, Family *Lythraceae*)

Punica granatum originated from Persia (Iran) and has been cultivated in central Asia, Georgia, Armenia and the Mediterranean region for several millennia, eventually making its way to other parts of the world. Pomegranate (*P. granatum*) is used in the traditional medicine of different Asian cultures for the treatment of a variety of ailments. In Ayurvedic medicine the plant, described under its Sanskrit name 'dalima' (fruit), is considered as a 'blood purifier' and used to cure parasitic infections, aphthae (mouth ulcers), diarrhoea and ulcers (Jurenka, 2008).

Pomegranates are high in polyphenolic compounds, making its juice higher in antioxidant activity than red wine and green tea (Malik *et al.*, 2005). The most abundant of these compounds is ellagic acid, which has been shown in research to be the antioxidant responsible for the free-radical scavenging ability of pomegranate juice. In animal research, pomegranate extract has also been shown to protect the antioxidant enzymes catalase, peroxidase and superoxide dismutase from the effects of toxic chemicals (Chidambara Murthy *et al.*, 2002).

According to some researchers (Lansky and Newman, 2007), the actions of pomegranate's components suggest a wide range of clinical applications for the treatment and prevention of cancer, as well as other diseases where chronic inflammation is believed to play an essential developmental role. Pomegranate extract has been shown to inhibit the growth of human prostate cancer cells cultured in laboratory dishes, as well as slow prostate cancer growth in mice (Malik *et al.*, 2005). An advantage that pomegranate has is that it seems to be capable of intervening at more than one critical pathway in the process of carcinogenesis (Afaq *et al.*, 2005). In another study (Hora *et al.*, 2003), mice pretreated with pomegranate extract showed 70% less tumour incidence compared to mice that did not receive the extract. The authors of the study indicated pomegranate fruit extract 'possesses anti-skin-tumor promoting effects', and may possess chemopreventive activity 'in a wide range of tumor models'. Additional research demonstrated that pomegranate seed oil was a safe and effective agent against skin cancer and colon cancer tumours, and also inhibited the proliferation of human breast cancer cells up to 90% (Kim *et al.*, 2002). Likewise, pomegranate juice polyphenols have been shown to inhibit cancerous lesion formation by 47% in the mammary gland cells from mice (Kim *et al.*, 2002).

Even more so than its antitumour/anticancer effects, pomegranate is known for its cardiovascular benefits. Human research has shown pomegranate to be effective in reducing several heart risk factors. In one study pomegranate inhibited the oxidation of LDL and slowed the development of atherosclerosis (Aviram *et al.*, 2004). Ten patients supplemented with pomegranate juice for 1 year experienced a 30% reduction in the narrowing of the carotid artery walls. Likewise, diabetic patients with elevated blood lipids who were supplemented with pomegranate juice for 8 weeks experienced significant reductions in their total cholesterol, LDL, LDL: HDL (high-density lipoprotein) ratio, as well as total cholesterol and HDL ratio (Esmaillzadeh *et al.*, 2004).

18.21 *Boerhaavia diffusia* (Punarnava, Family *Nyctaginaceae*)

Boerhaavia diffusa is an herbaceous perennial plant, native of India and Brazil, where it was used for centuries as a medicinal plant by indigenous populations. The root, leaves, aerial parts or the whole plant of *B. diffusa* have been employed for the treatment of various disorders in Ayurvedic herbal medicine (used daily by millions of people in India, Nepal, Sri Lanka and indirectly through it being the major influence on Unani, Chinese and Tibetan medicines).

The first pharmacological studies have demonstrated that the root of Punarnava exhibits a wide range of properties: anti-inflammatory, diuretic, laxative, anti-urethritis, anticonvulsant, antinematodal, antifibrinolytic and antibacterial (Olukoya *et al.*, 1993), antihepatotoxic (Rawat *et al.*, 1997), anthelminthic, febrifuge, antileprotic, anti-asthmatic, antiscabby and antistress activities. An aqueous extract of thinner roots of *B. diffusa* at a dose of 2 mg kg^{-1} exhibited the remarkable protection of various enzymes such as serum glutamic-oxaloacetic transaminase, serum glutamic pyruvic transaminase, and bilirubin in serum against hepatic injury in rats (Rawat *et al.*, 1997).

Currently, various parts of *B. diffusa* are being used for the treatment of numerous disorders in different parts of India. The root of *B. diffusa* is used for the treatment of many diseases, such as liver disorders (jaundice, hepatitis, etc.), gastrointestinal disorders (as laxative), renal disorders (for calculi, cystitis and nephritis), and for the treatment of anaemia and of menstrual syndrome. The drug has recently been used as an adjuvant in an anticancer therapy. *Boerhaavia* leaves are an important source of eicosanoids, stearic and ursolic acids, serine, liirodendrin carbohydrates, proteins and glycoproteins etc. (Aftab *et al.*, 1996). The root of *B. diffusa* contains alkaloids (punarnavine), rotenoids (boeravinones A–F), flavonoids, amino acids, lignans (liriodendrons), β-sitosterols and tetracosanoic, esacosanoic, stearic and ursolic acids.

The most interesting metabolites from the therapeutic point of view are the rotenoids (known as boeravinones A–F) (Lami *et al.*, 1992). The root is mainly used to treat gonorrhea, internal inflammation of all kinds, dyspepsia, oedema, jaundice, menstrual disorders, anaemia, liver, gall bladder and kidney disorders, enlargement of spleen, abdominal pain, abdominal tumours, and cancers, then as a diuretic, digestive aid, laxative and a menstrual promoter. The root powder, when mixed with *mamira* (*Thalictrum foliolosum*), is used to treat eye diseases. It cures corneal ulcers and night blindness, and helps restore virility in men. People in tribal areas use it to hasten childbirth. The juice of *B. diffusa* leaves serves as a lotion in ophthalmia. It is also administered orally as a blood purifier and to relieve muscular pain.

Maximum diuretic and anti-inflammatory activities of Punarnava have been observed in samples collected during the rainy season. Due to the combination of these two activities, Punarnava is regarded therapeutically highly efficacious for the treatment of renal inflammatory diseases and common clinical problems such as nephritic syndrome, oedema, and ascites developing at the early onset of liver cirrhosis and chronic peritonitis. The plant was reported to be efficient for the treatment of abdominal tumours and was proved to be useful as a haematic and as a growth promoter in children fed with milk fortified with the plant drug. In the form of a powder or an aqueous decoction, the plant drug was proved to be beneficial in the treatment of nephritic syndrome and compared well with corticosteroids. It was also demonstrated that the drug decreased the albumin urea, increased the serum protein and lowered serum cholesterol level (Ramabhimaiah *et al.*, 1984).

18.22 *Rosmarinus officinalis* L. (Rosemary, Family *Lamiaceae*)

Rosmarinus officinalis, commonly known as rosemary, is a woody, perennial herb, native to the Mediterranean region. Rosemary is an aromatic evergreen shrub that has leaves similar to hemlock needles. The leaves are used as flavouring agents in foods such as stuffing and roast lamb, pork, chicken and turkey. It is reasonably hardy in cool climates. It can withstand droughts, surviving a severe lack of water for lengthy periods.

The essential oil of rosemary contains several compounds at rather different concentrations. Debersac *et al.* (2001) reported that the major component of dried leaves of *R. officinalis* was monoterpene oxide 1,8 cineole (36.1%). Many monoterpene and sesquiterpene hydrocarbons were present in large amounts (32.2%). All these compounds amounted to 93.4% (w/w) of essential oil. Essential oils derived from rosemary were shown to inhibit osteoclast activity and increase bone density *in vitro* (Putnam *et al.*, 2006). Water-soluble extract from rosemary leaves was rich in rosemarinic acid (1.3%) and flavonoids (3%). The chemical composition of rosemary essential oil can vary between regions and it depends mostly on climate, soil composition, plant organ, age and state of vegetative cycle (Faixova and Faix, 2008).

Bozin *et al.* (2007) tested the antimicrobial and antioxidant activities of the essential oils of *R. officinalis*. Their antimicrobial activity was tested against 13 bacterial strains and six fungi, including *Candica albicans* and five dermatocytes. The highest antibacterial activity of both essential oils was expressed on *Escherichia coli, Salmonella typhi, S. enteritidis* and *Shigella sonnei*. Essential oil of rosemary exhibited a significant rate of antifungal activity. Fu *et al.* (2007) reported that essential oils from clove and rosemary alone and in combination exerted a significant antimicrobial effect against *Staphylococcus epidermidis, E. coli* and *C. albicans*. The cytotoxic effect of rosemary is of great importance in preservation of agricultural or marine products. The antimicrobial efficacy of plant essential oils depends on food composition. Gutierrez *et al.* (2008) found out that essential oils might be more effective against food-borne pathogens and spoilage bacteria when applied to ready to use food containing a high protein level at acidic pH, as well as lower levels of fats or carbohydrates.

18.23 *Senna alexandria* (Senna, Family *Fabaceae*)

Senna alexandria is an ornamental as well as medicinal plant. The bioactive compounds derived from senna are among one of the most commonly used laxative drugs in the Eastern and Western countries for the treatment of constipation (Anonymous, 2004). It is a native of Egypt in particular, in the Nubian region and near Sudan, where it is cultivated commercially. It is also grown in India and Somalia. Historically *S. alexandria* was used in the form of senna pods, or as a tisane made from the leaves, as a laxative (Spiller *et al.*, 2003).

Senna is known for its purgative action. The phyto-constituents principally responsible for its characteristic action are two anthraquinone glycosides, namely sennoside A and sennoside B. Sennoside A and B together are responsible for up to 40–60% activity of crude senna. Sennosides are not restricted to leaflets only; they are present in various vegetative and reproductive structures (underground as well as aerial). Maximum content of sennoside B is found in aerial parts (Evans, 2002). Being β-o-linked glycosides, sennosides remain unabsorbed in the upper gut and unaffected by digestive enzymes. Bacteria present in the large intestine convert these sennosides into the active metabolite, rheinanthrone, that is further oxidized into rhein and sennidins. Orally administered sennosides are mainly excreted as polyquinones in faecal matter. Few metabolites are excreted in urine and bile while some senna metabolites such as rhein accumulate in breast milk (Jafri and Pasricha, 2001). Long-term laxative abuse may result in weakness and orthostatic hypotention in elderly patients. There have been conflicting reports on effects such as intestinal-neuronal damage (Agarwal and Bajpai, 2010).

Senna also contains small quantities of other anthraquinones such as sennosides C and D, rhein-8-glucoside, rhein-8-diglucoside, aloe-emodin, 8-glucoside, anthrone diglucoside and rhein. Additionally senna contains napthalene glycosides (tinnevellin glycoside and 6-hydroxy-musizin-glycoside), flavonoid (kaempferol), phytosterols, myricyl alcohol, salicylic acid, chrysophenic acid, mucilage, resin and calcium oxalate (Kar, 2003; Kokate *et al.*, 2003).

18.24 *Asparagus racemosus* (Shatavari, Family *Liliaceae*)

Asparagus generally occurs in tropical and subtropical regions up to an altitude of 1200 m (Dutta, 2007). By habit, it is a twining parasitic herb. The plant is a climber growing to 1–2 m in length. The genus *Asparagus* has been recently moved from the subfamily *Asparagae* in the family *Liliaceae* to a newly created family *Asparagaceae*. *Asparagus racemosus* is recommended in Ayurvedic texts for prevention and treatment of gastric ulcers as galatogogue and nervine tonic.

The *Asparagus* genus is considered to be of medicinal importance because of the presence of steroidal saponins and sapogenins in various parts of the plant (Goyal *et al.*, 2003). The decoction of root has been used in blood diseases, diarrhoea, dysentery, cough, bronchitis and general debility (Goyal *et al.*, 2003). Reports indicate that the pharmacological activities of root extracts include anti-ulcer (Sairam *et al.*, 2003), antitussive (Mandal *et al.*, 2000a), antioxidant (Kamat *et al.*, 2000) and antibacterial activities (Mandal *et al.*, 2000b). *Asparagus racemosus* has been used extensively as an adaptogen to increase the non-specific resistance of organisms against a variety of stresses (Visavadiya *et al.*, 2009). Besides use in the treatment of diarrhoea and dysentery, the plant also has antioxidant, immune-stimulant, antidyspepsia and antitussive effects (Gautam *et al.*, 2009; Visavadiya *et al.*, 2009; Ojha *et al.*, 2010).

18.25 *Commiphora molmol* (Myrrh, Family *Burseraceae*)

Commiphora molmol is a small tree or large shrub up to 10 feet in height. It peels easily revealing a greenish cortex (inner bark). Myrrh grows along the Red Sea coast and is native to Somalia and Yemen. Over time the tree has been planted farther afield in Yemen, Omar, Somalia, Ethiopia, Egypt and Sudan. It is easily reproduced by cuttings and can be transplanted, but is seldom found in a commercial plantation setting.

There are many resin canals in the periderm, which ooze gum resin when cut. Trees are induced to ooze gum resin by making deep slanted incisions in the thin periderm (outer bark). The newly exuded resin is clear yellow-brown in colour. It quickly hardens and oxidizes to be reddish in colour. True myrrh clumps have a crumbly, dark red interior with a whitish powdery exterior. The best myrrh has little scent and no oily texture.

In a recent study, treatment with *C. molmol*, which was known commercially under the name of Mirazid (MZR) in the Egyptian pharmacies, showed insignificant activity against fascioliasis (Haridy *et al.*, 2003; Botros *et al.*, 2009). Also, it has been reported that myrrh had hypocholesteraemic, antipyretic, antihistaminic and antigastric ulcer properties (Nomicos, 2007). In Egypt, several studies have been conducted to evaluate the antischistosomal (Massoud *et al.*, 1998) and antifascioliasis (Massoud *et al.*, 2001) efficacy of myrrh.

18.26 Conclusions

There is growing interest in correlating the phytochemical constituents of a plant with its pharmacological activity. Scientists have even started correlating the botanical properties of plants with their pharmacological activity. In future, more coordinated multidimensional research aimed at correlating botanical and phytochemical properties to specific pharmacological activities is expected. In addition to the proper utilization of technological advances, a logical interpretation of the codified language of traditional medicine also becomes a necessity in order to further promote research in this field.

References

Afaq, F., Saleem, M., Krueger, C.G., Reed, J.D. and Mukhtar, H. (2005) Anthocyanin-and hydrolyzable tannin-rich pomegranate fruit extract modulates MAPK and NF-kappaB pathways and inhibits skin tumorigenesis in CD-1 mice. *International Journal of Cancer* 113, 423–433.

Aftab, K., Usmani, S.B., Ahmad, S.I. and Usmanghani, K. (1996) Naturally occurring calcium channel blockers-II. *Hamdard Medicus* 39, 44–54.

Agarwal, V. and Bajpai, M. (2010) Pharmacognostical and biological studies on Senna and its products: an overview. *International Journal of Pharma and Biosciences* 1, 1–10.

Aggarwal, R.R. and Dutt, S. (1936) Chemistry, pharmacology and therapeutic actions of *Terminalia arjuna. Proceedings of National Academy of Sciences USA* 6, 305.

Ahmadiani, A., Javan, M., Semnanian, S., Barat, E. and Kamalinejad M. (2001) Anti-inflammatory and antipyretic effects of *Trigonella foenumgraecum* leaves extract in the rat. *Journal of Ethnopharmacology* 75, 283–286.

Akin, M., Aktumsek, A. and Nostro, A. (2010) Antibacterial activity and composition of essential oils of *Eucalyptus camaldulensis* Dehn and *Myrtus communis* L. growing in northern Cyprus. *African Journal of Biotechnology* 9, 531–535.

Alpana, R., Lauria, P., Gupta, R., Kumar, P. and Sharma, V.N. (1997) Hypocholesterolanic effects of *Terminalia arjuna* tree. *Journal of Ethnopharmacology* 55, 165–169.

Ammon, H.P.T. and Muller, A.B. (1985) Forskolin: from an Ayurvedic remedy to a modern agent. *Planta Medica* 51, 473–477.

Anonymous (2003) *Gingiber officinale. Alternative Medicine Review* 8, 331–335.

Anonymous (2004) United states pharmacopoeia 27. Rockville, Maryland. *US pharmacopoeia convention*, 1686.

Atherton, P. (1998) *Aloe vera* revisited. *British Journal of Phytotherapy* 4, 76–83.

Attele, A.S., Wu, J.A. and Yuan, C.S. (1999) Ginseng pharmacology: multiple constituents and multiple actions. *Biochemical Pharmacology* 58, 1685–1693.

Aviram, M., Rosenblat, M., Gaitini, D., Nitecki, S., Hoffman, A., Dornfeld, L., Volkova, N., Presser, D., Attias, J., Liker, H. and Hayek, T. (2004) Pomegranate juice consumption for 3 years by patients with carotid artery stenosis reduces common carotid intima-media thickness, blood pressure and LDL oxidation. *Clinical Nutrition* 23, 423–433.

Badmaev, V., Majeed, M., Conte, A.A. and Parker, J.E. (2002) Diterpene Forscolin (*Coleus forskohlii*, Benth): a possible new compound for reduction of body weight by increasing lean body mass. *NutraCos*, 1, 6–7.

Bagalkotkar, G., Sagineedu, S.R., Saad, M.S. and Stanslas, J. (2006) Phytochemicals from *Phyllanthus niruri* Linn. and their pharmacological properties: a review. *Journal of Pharmacy and Pharmacology* 58, 1559–1570.

Bajaj, Y.P.S. (1995) *Medicinal and Aromatic Plants. Biotechnology in agriculture and forestry*, vol. 8. Springer, Berlin, pp. 194-196.

Bessong, P.O. and Obi, C.L. (2006) Ethnopharmacology of HIV in South Africa - a mini review. *African Journal of Biotechnology* 5, 1693–1699.

Bhattacharya, S.K., Bhattacharya, A., Kumar, A. and Ghosal, S. (2000) Antioxidant activity of *Bacopa monnieri* in rat frontal cortex, striatum and hippocampus. *Phytotherapy Research* 14, 174–179.

Bin-Hafeez, B., Haque, R., Parvez, S., Pandey, S., Sayeed, I. and Raisuddin, S. (2003) Immunomodulatory effects of fenugreek (*Trigonella foeneumgraecum L.*) extract in mice. *International Immunopharmacology* 3, 257–265.

Boim, M.A., Heilberg, I.P. and Schor, N. (2010) *Phyllanthus niruri* as a promising alternative treatment for nephrolithiasis. *International Brazilian Journal of Urology* 36, 657–664.

Bone, K. (1996) *Clinical Applications of Ayurvedic and Chinese Herbs*. Phytotherapy Press, Warwick, Queensland, Australia, pp. 131–133.

Botros, S.S., El-Lakkany, N.M., Badawy, A.A., Mahmoud, S.S., Ebeid, F.A. and Fenwick, A. (2009) Mirazid shows insignificant activity against ovine fascioliasis. *Annals of Tropical Medicine and Parasitology* 103, 605–616.

Bown, D. (1995) *The Royal Horticultural Society Encyclopedia of Herbs and their Uses*. Dorling Kindersley, London, 424 pp.

Bozin, B., Mimica-Dukic, N., Samojlik, I. and Jovin, E. (2007) Antimicrobial and antioxidant properties of rosemary and sage (*Rosmarinus officinalis* L. and *Salvia officinalis* L., *Lamiaceae*) essential oils. *Journal of Agricultural and Food Chemistry* 557, 7879–7885.

Calixto, J.B., Santos, A.R., Cechinel, F.V. and Yunes, R.A. (1998) A review of the plants of the genus *Phyllanthus*: their chemistry, pharmacology, and therapeutic potential. *Medicinal Research Reviews* 18, 225–258.

Chander, R., Khanna, A.K. and Pratap, R. (2002) Antioxidant activity of guggulsterone, the active principal of guggulipid from *Commiphora mukul*. *Journal of Medicinal and Aromatic Plant Sciences* 24, 370–374.

Chander, R., Rizvi, F., Khanna, A.K. and Pratap, R. (2003) Cardioprotective activity of synthetic guggulsterone (E and Z isomers) in isoproterenol induced myocardial ischemia in rats: a comparative study. *Indian Journal of Clinical Biochemistry* 18, 71–79.

Chander, R., Singh, K., Khanna, A.K., Kaul, S.M., Puri, A., Saxena, R., Bhatia, G., Rizvi, F. and Rastogi, A.K. (2004) Antidyslipidemic and antioxidant activities of different fractions of *Terminalia arjuna* stem bark. *Indian Journal of Clinical Biochemistry* 19, 141.

Chaudhary, N.M.A. and Tariq, P. (2006) Antimicrobial activity of *Cinnamomum cassia* against diverse microbial flora with its nutritional and medicinal impacts. *Pakistan Journal of Botany* 38, 169–174.

Chidambara Murthy, K.N., Jayaprakasha, G.K. and Singh, R.P. (2002) Studies on antioxidant activity of pomegranate (*Punica granatum*) peel extract using *in vivo* models. *Journal of Agricultural Food and Chemistry* 50, 4791–4795.

Chopra, R.N., Nayer, S.L. and Chopra, I.C. (1992) *Glossary of Indian Medicinal Plants*, 3rd edn. Council of Scientific and Industrial Research, New Delhi, pp. 7–246.

Chowdhury, A.R., Baberji, R., Mishra, G. and Nigam, S.K. (1983) Chemical composition of *Acacia* seeds. *Journal of the American Oil Chemists' Society* 60, 1893–1894.

Chung, S.H., Choi, C.G. and Park, S.H. (2001) Comparisons between white ginseng radix and rootlet for antidiabetic activity and mechanism in KKAy mice. *Archives of Pharmacological Research* 24, 214–218.

Cohen, J. (2000) Meningococcal disease as a model to evaluate novel antisepsis strategies. *Critical Care Medicine* 28, S64–S67.

Corradin, S.B., Buchmuller-Rouiller, Y.B. and Mauel, J. (1991) Phagocytosis enhances murine macrophages activation by interferon- and tumor necrosis factor-α. *European Journal of Immunology* 21, 2553–2558.

Cralg, W.J. (1999) Health-promoting properties of common herbs. *American Journal of Clinical Nutrition* 70, 4915–4995.

Dateo, J.G.P. and Long, J.L. (1973) Gymnemic acid, the antisaccharine principle of *Gymnema sylvestre* studies on the isolation and heterogeneity of gymnemic acid A1. *Journal of Agricultural Food Chemistry* 21, 899–903.

De Groot, H. and Rauen, U. (1998) Tissue injury by reactive oxygen species and the protective effects of flavonoids. *Fundamental and Clinical Pharmacology* 12, 249–255.

De Souza, N.J. and Shah, V. (1988) Forskolin – an adenylate cyclase activating drug from an Indian herb. *Economic and Medicinal Plant Research* 2, 1–16.

Debersac, P., Heydel, J.M., Amiot, M.J., Goudonnet, H., Artur, Y., Suschetet, M. and Siess, M.H. (2001) Induction of cytochrome P450 and/or detoxication enzymes by various extracts of rosemary: description of specific patterns. *Food Chemistry and Toxicology* 39, 907–918.

Dev, S. (1989) Chemistry of *Commiphora mukul* and development of a hypolipidemic drug. In: Rehman, A. (ed.) *Studies in Natural Product Chemistry*. Elsevier, Amsterdam, pp. 695–719.

Devi, P.U., Sharada, A.C. and Solomon, F.E. (1993) Antitumor and radiosensitizing effects of *Withania somnifera* (Ashwagandha) on a transplantable mouse tumor, Sarcoma-180. *Indian Journal of Experimental Biology* 31, 607–611.

Dutta, I.C. (2007) *Non-Timber Forest Products of Nepal (Identification, Classification, Ethnic Uses and Cultivation)*. Institute of Forestry, Pokhara, Nepal.

El-Tahir, A., Satti, G.M.H. and Khalid, S.A. (1999) Antiplamodial activity of selected sudanese medicinal plants with emphasis on *Acacia nilotica*. *Journal of Phytotherapy* 13, 474–477.

Elliot, W.R. and Jones, D. (1986) *The Encyclopedia of Australian Plants*, Vol. 4. Lothian Publishing Company, Melbourne, Australia.

Eshun, K. and He, Q. (2004) Aloe vera: a valuable ingredient for the food, pharmaceutical and cosmetic industries - a review. *Critical Reviews in Food Science and Nutrition* 44, 91–96.

Esmaillzadeh, A., Tahbaz, F., Gaieni, I., Alavi-Majd, H. and Azadbakht, L. (2004) Concentrated pomegranate juice improves lipid profiles in diabetic patients with hyperlipidemia. *Journal of Medicinal Food* 7, 305–308.

Evans, W.C. (2002) *Trease and Evans Pharmacognosy*, 15th edn. W.B. Saunders, London, 233 pp.

Fabio, A., Corona, A., Forte, E. and Quaglio, P. (2003) Inhibitory activity of spices and essential oils on psychrotrophic bacteria. *Microbiology* 26, 115–120.

Faixova, Z. and Faix, S. (2008) Biological effects of rosemary (*Rosmarinus oficinalis*) essential oil (a review). *Folia Veterinaria* 52, 135–139.

Flier, J.S. (2001) Prevention of obesity reduces the risk of a wide range of health problems. The missing link with obesity? *Nature* 409, 292–293.

Fu, Y., Zu,Y., Chen, L., Shi, X., Wang, Z., Sun, S. and Efferth, T. (2007) Antimicrobial activity of clove and rosemary essential oils alone and in combination. *Phytotherapy Research* 21, 989–994.

Gautam, M., Saha, S., Bani, S., Kaul, A., Mishra, S., Patil, D., Satti, N.K., Suri, K.A., Gairola, S., Suresh, K., Jadhav, S., Qazi, G.N. and Patwardhan, B. (2009) Immunomodulatory activity of *Asparagus racemosus* on systemic Th1/Th2 immunity: implications for immune adjuvant potential. *Journal of Ethnopharmacology* 121, 241–247.

Gende, L., Maggi, M., van Baren, C., di Leo, A., Bandoni, A., Fritz, R. and Eguaras, M. (2010) Antimicrobial and miticide activities of *Eucalyptus globulus* essential oils obtained from different Argentine regions. *Spanish Journal of Agricultural Research* 8, 642–650.

Govindarajan, V.S. (1982) Ginger – chemistry, technology, and quality evaluation: part 1. *Critical Reviews in Food Science and Nutrition* 17, 1–96.

Goyal, R.K., Singh, J. and Lal, H. (2003) *Asparagus racemosus* – an update. *Indian Journal of Medical Sciences* 57, 408–414.

Grandhi, A., Mujumdar, A.M. and Patwardhan, B. (1994) A comparative pharmacological investigation of Ashwagandha and Gingeng. *Journal of Ethnopharmacology* 3, 131–135.

Grant, K.L. and Lutz, R.B. (2000) Ginger. *American Journal of Health System Pharmacy* 57, 945–947.

Gutierrez, J., Barry-Ryan, C. and Bourke, P. (2008) Antimicrobial efficacy of plant essential oil combinations and interactions with food ingredients. *International Journal of Food Microbiology* 124, 91–97.

Halder, S., Bharal, N., Mediratta, P.K., Kaur, I. and Sharma, K.K. (2009) Antiinflammatory, immunomodulatory and antinociceptive activity of *Terminalia arjuna Roxb* bark powder in mice and rats. *Indian Journal of Experimental Biology* 46, 577.

Hamman, J.H. (2008) Composition and applications of *Aloe vera* leaf gel. *Molecules* 13, 1599–1616.

Haridy, F.M., El Garhy, M.F. and Morsy, T.A. (2003) Efficacy of Mirazid (*Commiphora molmol*) against fascioliasis in Egyptian sheep. *Journal of the Egyptian Society of Parasitology* 33, 917–924.

He, Q., Changhong, L., Kojo, E. and Tian, Z. (2005) Quality and safety assurance in the processing of *Aloe vera* gel juice. *Food Control* 16, 95–104.

Hora, J.J., Maydew, E.R., Lansky, E.P. and Dwivedi, C. (2003) Chemopreventive effects of pomegranate seed oil on skin tumor development in CD1 mice. *Journal of Medicinal Food* 6, 157–161.

Hussein, G., Miyashiro, H., Nakamura, N., Hattori, M., Kawahata, T., Otake, T., Kakiuchi, N. and Shimotohno, K. (1999) Inhibitory effect of sudanese medicinal plant extracts on HIV replication and HIV-1 protease protease. *Phytotherapy Research* 13, 31–36.

Hussein, G., Miyashiro, H., Nakamura, N., Hattori, M., Kakiuchi, N. and Shimotohno, K. (2000) Inhibitory effect of sudanese medicinal plant extracts on hepatitis C virus protease. *Phytotherapy Research* 14, 510–516.

Jafri, S. and Pasricha, P.J. (2001) Agents used for diarrhea, constipation and inflammatory bowel disease; agents used for biliary and pancreatic disease. In: Hardman, J.G. and Limbird, L.E. (eds) *Goodman & Gilman's the Pharmacological Basis of Therapeutics*. McGraw-Hill, New York, pp. 1048–1049.

Jayaprakasam, B., Zhang, Y., Seeram, N. and Nair, M. (2003) Growth inhibition of tumor cell lines by withanolides from *Withania somnifera* leaves. *Life Science* 74, 125–132.

Jellin, J.M., Batz, F. and Hitchens, K. (2002) *Natural Medicines Comprehensive Database*. Therapeutic Research Faculty 1530, Stockton, California.

Jia, L. and Zhao, Y. (2009) Current evaluation of the millennium phytomedicine-ginseng (I): etymology, pharmacognosy, phytochemistry, market and regulations. *Current Medicinal Chemistry* 16, 2475–2484.

Jukic, M., Politeo, O. and Milos, M. (2006) Chemical composition and antioxidant effect of free volatile aglycones from nutmeg (*Myristica fragrans* Houtt.) compared to its essential oil. *Chroatica Chemica Acta* 79, 209–214.

Jurenka, J.S. (2008) Therapeutic applications of pomegranate (*Punica granatum* L.): a review. *Alternative Medicine Review* 13, 128–144.

Jyoti, A. and Sharma, D. (2006) Neuroprotective role of *Bacopa monnieri* extract against aluminium-induced oxidative stress in the hippocampus of rat brain. *NeuroToxicology* 27, 451–457.

Kamat, J.P., Boloor, K.K., Devasagayam, T.P. and Venkatachalam, S.R. (2000) Antioxidant properties of *Asparagus racemosus* against damage induced by gamma-radiation in rat liver mitochondria. *Journal of Ethanopharmacology* 71, 425–435.

Kandil, F.E., Elsayeh, N.H., Abou-Douh, A.M., Ishak, M.S. and Mabry, T.J. (1994) Flavonol glycosides and phenolics from *Withania somnifera*. *Phytochemistry* 37, 1215–1216.

Kapoor, L.D. (1990) *Handbook of Ayurvedic Medicinal Plants*. CRC Press, Boca Raton, Florida, pp. 319–320.

Kar, A. (2003) *Pharmacognosy and pharmacobiotechnology*. New Age International, New Delhi, 33, 157–160.

Kaul, S. and Kapoor, N.K. (1989) Reversal of changes of lipid peroxide, xanthine oxidase and superoxide dismutase by cardio-protective drugs in isoproterenol induced myocardial necrosis in rats. *Indian Journal of Experimental Biology* 27, 625–627.

Kaur, K., Arora, S., Kumar, S. and Nagpal, A. (2002) Antimutagenic activity of acetone and methanol fractions of *Terminalia arjuna*. *Journal of Food Chemistry and Toxicology* 40, 1475–1482.

Kennelly, E.J., Flynn, T.J., Mazzola, E.P., Roach, J.A., McCloud, T.G., Danford, D.E. and Betz, J.M. (1999) Detecting potential teratogenic alkaloids from blue cohosh rhizomes using an *in vitro* rat embryo culture. *Journal of Natural Products* 62, 1385–1389.

Khan, S. and Balick, M.J. (2001) Therapeutic plants of Ayurveda: a review of selected clinical and other studies for 166 species. *Journal of Alternative and Complimentary Medicine* 7, 405–515.

Khanna, V.G. and Kannabiran, K. (2007) Larvicidal effect of *Hemidesmus indicus, Gymnema sylvestre*, and *Eclipta prostrata* against *Culex qinquifaciatus* mosquito larvae. *African Journal of Biotechnology* 6, 307–311.

Kim, N.D., Mehta, R., Yu, W., Neeman, I., Livney, T., Amichay, A., Poirier, D., Nicholls, P., Kirby, A., Jiang, W., Mansel, R., Ramachandran, C., Rabi, T., Kaplan, B. and Lansky, E. (2002) Chemopreventive and adjuvant therapeutic potential of pomegranate (*Punica granatum*) for human breast cancer. *Breast Cancer Research and Treatment* 71, 203–217.

Kokate, C.K., Purohit, A.P. and Gokhale, S.B. (2003) *Pharmacognosy*, 25th edn. Nirali prakashan, 77.

Kubmarawa, D., Ajoku, G.A., Enwerem, N.M. and Okorie, D.A. (2007) Preliminary phytochemical and antimicrobial screening of 50 medicinal plants from Nigeria. *African Journal of Biotechnology* 6, 1690–1696.

Kumar, D.S. and Prabhakar, Y.S. (1987) On the ethnomedical significance of the srjun tree, *Terminalia arjuna* (*Roxb.*) Wight & Arnot. *Journal of Enthnopharmacology* 20, 173.

Kumar, G.S., Shetty, A.K., Sambaiah, K. and Salimath, P.V. (2005) Antidiabetic property of fenugreek seed mucilage and spent turmeric in streptozotocin-induced diabetic rats. *Nutrition Research* 25, 1021–1028.

Lami, N., Kadota, S. and Kikuchi, T. (1992) Constituents of the roots of *Boerhaavia diffusa* Linn. IV. Isolation and structure determination of boeravinones D, E and F. *Chemical and Pharmaceutical Bulletin* 39, 1863–1865.

Lansky, E.P. and Newman, R.A. (2007) *Punica granatum* (pomegranate) and its potential for prevention and treatment of inflammation and cancer. *Journal of Ethnopharmacology* 109, 177–206.

Lichtensteiger, C.A., Johnston, N.A. and Beasley, V.R. (1997) *Rhamnus cathartica* (buckthorn) hepatocellular toxicity in mice. *Toxicologic Pathology* 25, 449–452.

Lien, H.C., Sun, W.M., Chen, Y.H., Kim, H., Hasler, W. and Owyang, C. (2003) Effects of ginger on motion sickness and gastric slow wave dysrhythmias induced by circular vection. *American Journal of Physiology Gastrointestinal and Liver Physiology* 284, G481–G489.

Lu, J.M., Yao, Q. and Chen, C. (2009) Ginseng compounds: an update on their molecular mechanisms and medical applications. *Current Vascular Pharmacology* 7, 293–302.

Malik, A., Afaq, A., Sarfaraz, S., Adhami, V.M., Syed, D.N. and Mukhtar, H. (2005) Pomegranate fruit juice for chemoprevention and chemotherapy of prostate cancer. *Proceedings of National Academy of Science* 102, 14813–14818.

Malini, M.M., Lenin, M. and Varalakshmi, P. (2000) Protective effect of triterpenes on calcium oxalate crystal-induced peroxidative changes in experimental urolithiasis. *Pharmacology Research* 41, 413–418.

Mandal, S.C., Kumar, C.K.A., Mohana Lakshmi, S., Sinha, S., Murugesan, T., Saha, B.P. and Pal, M. (2000a) Antitussive effect of *Asparagus racemosus* root against sufur dioxide-induced cough in mice. *Fitoterapia* 71, 686–689.

Mandal, S.C., Nandy, A., Pal, M. and Saha, B.P. (2000b) Evaluation of antibacterial activity of *Asparagus racemosus* wild root. *Phytotherapy Research* 14, 118–119.

Massoud, A., Salama, O. and Bennett, J.L. (1998) Therapeutic efficacy of new schistosomicidal drug, derived from myrrh, in active intestinal schistosomiasis complicated with hepatosplenomegaly. *Proceedings of the 9th International Congress of Parasitology (ICOPA IX)*, Chiba, Japan. Monduzzi Editores, Bologna, pp. 619–623.

Massoud, A.S., Salama, S.E. and Massoud, A. (2001) Preliminary study of therapeutic efficacy of a new fasciolicidal drug derived from *Commiphora molmol* (myrrh). *American Journal of Tropical Medicine Hygiene* 65, 96–99.

McFarlin, B.L., Gibson, M.H., O'Rear, J. and Harman, P. (1999) A national survey of herbal preparation use by nurse-midwives for labor stimulation. Review of the literature and recommendations for practice. *Journal of Nurse-Midwifery* 44, 205–216.

Micali, S., Sighinolfi, M.C., Celia, A., DeStefani, S., Grande, M., Cicero, A.F. and Bianchi, G. (2006) Can *Phyllanthus niruri* affect the efficacy of extracorporeal shock wave lithotripsy for renal stones? A randomized, prospective, long-term study. *Journal of Urology* 176, 1020–1022.

Min, B.S., Cuong, T.D., Hung, T.M., Min, B.K., Shin, B.S. and Woo, M.H. (2011) Inhibitory Effect of Lignans from *Myristica fragrans* on LPS-induced NO Production in RAW264.7 Cells. *Bulletin of the Korean Chemical Society* 32, 40–59.

Mubita, C., Syakalima, M., Chisenga, C., Munyeme, M., Bwalya, M. and Chifumpa, G. (2008) Antibiograms of faecal *Escherichia coli* and *Enterococci* species isolated from pastoralist cattle in the interface areas of the Kafue basin in Zambia. *Veterinarski Archives* 78, 179–185.

Muller-Kobold, A.C., Tulleken, J.E., Zijlstra, J.G., Sluiter, W., Hermans, J., Kallenberg, C.G.M. and Cohen-Tervaert, J.W. (2000) Leukocyte activation in sepsis; correlations with disease state and mortality. *Intensive Care Medicine* 26, 883–892.

Murugaiyah, V. and Chan, K.L. (2006) Antihyperuricemic lignans from the leaves of *Phyllanthus niruri*. *Planta Medica* 72, 1262–1267.

Murugesan, S., Rjeshkannan, C., Sumathi, R., Manivachakam, P. and Babu, D.S. (2012) Bioactivity of root hexane extract of *Coleus forskohlii* Briq. Labiatae: GC/MS/MS Characterization and identification. *European Journal of Experimental Biology* 2, 1469–1473.

Nah, S.Y., Park, H.J. and McCleskey, E.W. (1995) A trace component of ginseng that inhibits Ca2+ channels through a pertussis toxin-sensitive G protein. *Proceedings of the National Academy of Sciences of the United States of America* 92, 8739–8743.

Nishiura, J.L., Campos, A.H., Boim, M.A., Heilberg, I.P. and Schor, N. (2004) *Phyllanthus niruri* normalizes elevated urinary calcium levels in calcium stone forming (CSF) patients. *Urological Research* 32, 362–366.

Nityanand, S. and Kapoor, N.K. (1975) Hypolipidaemic effect of ethyl acetate fraction of *Commiphora mukul* (guggul) in rats. *Indian Journal of Pharmacology* 7, 106.

Nomicos, E.Y. (2007) Myrrh: medical marvel or myth of the Magi? *Holistic Nursing Practice* 21, 308–323.

Ojha, R., Sahu, A.N., Muruganandam, A.V., Singh, G.K. and Krishnamurthy, S. (2010) *Asparagus racemosus* enhances memory and protects against amnesia in rodent models. *Brain & Cognition* 74, 1–9.

Olajide, O.A., Ajayi, F.F., Ekhelar, A.I., Awe, S.O., Makinde, J.M. and Alada, A.R. (1999) Biological effects of *Myristica fragrans* fruits extract in rabbits. *Phytotherapy Research* 13, 344–345.

Olaleye, M.T., Akinmoladun, A.C. and Akindahunsi, A.A. (2006) Antioxidant properties of *Myristica fragrans* (Houtt) and its effect on selected organs of albino rats. *African Journal of Biotechnology* 5, 1274–1278.

Olukoya, D.K., Tdika, N. and Odugbemi, T. (1993) Antibacterial activity of some medicinal plants from Nigeria. *Journal of Ethnopharmacology* 39, 69–72.

Pal, M., Srivastava, M., Soni, D.K., Kumar, A. and Tewari, S.K. (2011) Composition and anti-microbial activity of essential oil of *Myristica fragrans* from Andaman Nicobar Island. *International Journal of Pharmacy and Life Sciences* 2, 1115–1117.

Parthasarathy, S. (1998) Mechanisms by which dietary antioxidants may prevent cardiovascular diseases. *Journal of Medicinal Food* 1, 45–51.

Parthasarathy, V.A., Kandinnan, K. and Srinivasan, V. (2008) *Organic Spices*. New India Publishing Agencies, 694 pp.

Patel, J.R., Tripathi, P., Sharma, V., Chauhan, N.S. and Dixit, V.K. (2011) *Phyllanthus amarus*: Ethnomedicinal uses, phytochemistry and pharmacology: a review. *Journal of Ethnopharmacology* 138, 286–313.

Prakash, A.O., Mather, S. and Mathur, R. (1986) Effect of feeding *Gymnema sylvestre* leaves on blood glucose in beryllium nitrate treated rats. *Journal of Ethnopharmacology* 18, 143–146.

Prakash, J., Gupta, S.K. and Dinda, A.K. (2002) *Withania somnifera* root extract prevents DMBA-induced squamous cell carcinoma of skin in Swiss albino mice. *Nutrition and Cancer* 42, 91–97.

Putnam, S.E., Scutt, A.M., Bicknell, K., Priestley, C.M. and Williamson, E.M. (2006) Natural products as alternative treatments for metabolic bone disorders and for maintenance of bone health (Review article). *Phytotherapy Research* 21, 99–112.

Ramabhimaiah, S., Stalin, D. and Kalaskar, N.J. (1984) Pharmacological investigations of the water soluble fraction of the methanol extracts of *Boerhaavia diffusa* roots. *Indian Drugs* 21, 343–344.

Ramachandran, A., Snehalatha, C., Satvavani, K., Sivasankari, S. and Vijav, V. (2003) Type 2 diabetes in Asian-Indian urban children. *Diabetes Care* 26, 1022–1025.

Ramnath, V., Rekha, P.S. and Sujatha, K.S. (2007) Amelioration of heat stress induced disturbances of antioxidant defense system in chicken by *Brahma rasayana. Evidence-based Complementary and Alternative Medicine* 2, 1–8.

Rawat, A.K.S., Mehrotra, S., Tripathi, S.K. and Shama, U. (1997) Hepatoprotective activity in *punarnava* – a popular Indian ethnomedicine. *Journal of Ethnopharmacology* 56, 61–68.

Roodenrys, S., Booth, D., Bulzomi, S., Phipps, A., Micallef, C. and Smoker, J. (2002) Chronic effects of Brahmi (*Bacopa monnieri*) on human memory. *Neuropsychopharmacology* 27, 279–281.

Rose, J. and Treadway, S. (2000) Herbal support for a healthy cardiovascular system. *Clinical Nutrition Insights* 6, 16.

Rupp, R.H., De Souza, N.J. and Dohadwalla, A.N. (1986) *Proceedings of the International Symposium on Forskolin: Its chemical, biological and medical potential.* Hoechst India Limited, Bombay, pp. 19–30.

Russo, A. and Borrelli, F. (2005) *Bacopa monnieri*, a reputed nootropic plant: an overview. *Phytomedicine* 12, 305–317.

Sairam, K., Priyambada, S., Aryya, N.C. and Goel, R.K. (2003) Gastroduodenal ulcer protective activity of *Asparagus racemosus*: an experimental, biochemical and histological study. *Journal of Ethanopharmacology* 86, 1–10.

Sathya, S., Kokilavani, R. and Gurusamy, K. (2008) Hypoglycemic effect of *Gymnema sylvestre* (retz.,) R.Br leaf in normal and alloxan induced diabetic rats. *Ancient Science of Life* 28, 12–14.

Satyavati, G.V. (1988) Gum guggul (*Commiphora mukul*) – the success story of an ancient insight leading to a modern discovery. *Indian Journal of Medical Research* 87, 327–335.

Satyavati, G.V. (1991) Guggulipid: a promising hypolipidaemic agent from gum guggul (*Commiphora mukul*). In: Wagner, H. and Farnsworth, N.R. (eds) *Economic and Medicinal Plant Research.* New York Academic Press, New York, pp. 47–48.

Seeram, N.P., Aviram, M., Zhang, Y., Henning, S.M., Feng, L., Dreher, M. and Heber, D. (2008) Comparison of antioxidant potency of commonly consumed polyphenol-rich beverages in the United States. *Journal of Agricultural and Food Chemistry* 56, 1415–1422.

Serafini, M., Maiani, G. and Ferro-Luzzi, A. (1998) Alcohol-free red wine enhances plasma antioxidant capacity in humans. *Journal of Nutrition* 128, 1003–1007.

Sharma, S.R., Dwivedi, S.K. and Swarup, D. (1996) Hypoglycaemic and hypolipidaemic effects of *Cinnamomum tamala* Nees leaves. *Indian Journal of Experimental Biology* 34, 372–374.

Shelton, M. (1991) Aloe vera, its chemical and therapeutic properties. *International Journal of Dermatology* 30, 679–683.

Sievenpiper, J.L., Arnason, J.T., Vidgen, E., Leiter, L.A. and Vuksan, V.A. (2004) Systematic quantitative analysis of the literature of the high variability in ginseng (*Panax* spp.): should ginseng be trusted in diabetes? *Diabetes Care* 27, 839–840.

Singh, H.K. and Dhawan, B.N. (1997) Neuropsychopharmacological effects of the Ayurvedic nootropic *Bacopa Monniera* Linn. (Brahmi). *Indian Journal of Pharmacology* 29, 359–365.

Singh, K., Chander, R. and Kapoor, N.K. (1997) Guggulsterone, a potent hypolipidemic, prevents oxidation of low-density lipoprotein. *Phytotherapy Research* 11, 291–294.

Singh, R.B., Niaz, M.A. and Ghosh, S. (1994) Hypolipidemic and antioxidant effects of *Commiphora mukul* as an adjunct to dietary therapy in patients with hypercholesterolemia. *Cardiovascular Drugs and Therapy* 8, 659–664.

Somani, R.S. and Singhai, A.K. (2008) Hypoglycaemic and antidiabetic activities of seeds of *Myristica fragrans* in normoglycaemic and Alloxan-induced diabetic rats. *Asian Journal of Experimental Science* 22, 95–102.

Sotaniemi, E.A., Haapakoski, E. and Rautio, A. (1995) Ginseng therapy in noninsulin-dependent diabetic patients. *Diabetes Care* 18, 1373–1375.

Spiller, H., Winter, M., Weber, J., Krenzelok, E., Anderson, D. and Ryan, M. (2003) Skin breakdown and blisters from senna-containing laxatives in young children. *The Annals of Pharmacotherapy* 37, 636–639.

Steppan, C.M., Bailey, S.T., Bhat, S., Brown, E.J., Banerjee, R.R., Wright, C.M., Patel, H.R., Ahima, R.S. and Lazar, M.A. (2001) The hormone resistin links obesity to diabetes. *Nature* 409, 307–312.

Subhashini, N., Thangathirupathi, A. and Lavanya, N. (2011) Antioxidant activity of *Trigonella foenum graecum* using various *in vitro* and *ex vivo* models. *International Journal of Pharmacy and Pharmaceutical Sciences* 3, 96–102.

Suekawa, M., Ishige, A., Yuasa, K., Sudo, K., Aburada, M. and Hosoya, E. (1984) Pharmacological studies on ginger. I. Pharmacological actions of pungent constituents, (6)-gingerol and (6)-shogaol. *Journal of Pharmacobiodynamics* 7, 836–848.

Tyler, V.E. (1986) Some recent advances in herbal medicine. *Pharmaceutical Development* 7, 203–207.

Vidya, L., Lenin, M. and Varalakshmi, P. (2002) Evaluation of the effect of triterpenes on urinary risk factors of stone formation in pyridoxine deficient hyperoxaluric rats. *Phytotherapy Research* 16, 514–518.

Vinson, J.A., Al Kharrat, H. and Andreoli, L. (2005) Effect of *Aloe vera* preparations on the human bioavailability of vitamins C and E. *Phytomedicine* 12, 760–765.

Visavadiya, N.P., Soni, B., Soni, B. and Madamwar, D. (2009) Suppression of reactive oxygen species and nitric oxide by *Asparagus racemosus* root extract using *in vitro* studies. *Cellular and Molecular Biology* 55, 1083–1095.

Wang, C.Y., Wang, S.Y. and Chen, C. (2008) Increasing antioxidant activity and reducing decay of blueberries by essential oils. *Journal of Agricultural and Food Chemistry* 56, 3587–3592.

Wang, X., Greilberger, J., Ledinski, G., Kager, G., Paigen, B. and Jurgens, G. (2004) The hypolipidemic natural product *Commiphora mukul* and its component guggulsterone inhibit oxidative modification of LDL. *Atherosclerosis* 172, 239–246.

Warren, R.P., Warren, R.M. and Weninger, M.G. (1969) Inhibition of the sweet taste by *Gymnema sylvestre*. *Nature* 223, 94–95.

Woldemariam, T.Z., Betz, J.M. and Houghton, P.J. (1997) Analysis of aporphine and quinolizidine alkaloids from *Caulophyllum thalictroides* by densitometry and HPLC. *Journal of Pharmaceutical and Biomedical Analysis* 15, 839–843.

Xiang, Y.Z., Shang, H.C., Gao, X.M. and Zhang, B.L. (2008) A comparison of the ancient use of ginseng in traditional Chinese medicine with modern pharmacological experiments and clinical trials. *Phytotherapy Research* 22, 851–858.

Zhao, Y.X., Abdelnour, A., Kalland, T. and Tarkowski, A. (1995) Overexpression of the T-cell receptor Vb3 in transgenic mice increase mortality during infection by enterotoxin A-producing *Staphylococcus aureus*. *Infection and Immunity* 63, 4463–4469.

19 Vitamins and Minerals: Roles and Plant Sources

R.L. Singh,* S.P. Vishwakarma and Pankaj Singh
Department of Biochemistry, Dr RML Avadh University, Faizabad, India

19.1 Introduction

Vitamins are organic compounds that are essential nutrients because the body cannot synthesize them. These are used as food or supplementation every day to maintain healthy life. The word vitamin arises from the Latin word 'vita' meaning life. Vitamins play important roles in different metabolic processes throughout the body. For example, vitamins help to release energy from food, keep skin and nerves healthy, and also help to make red blood cells. In fact, hundreds of metabolic activities in the body depend upon enzymes, which are vitamin dependent. Vitamins include biotin, folic acid, niacin, pantothenic acid, riboflavin, thiamine, vitamin A, B_6, B_{12}, C, D, E and K. They are required in very small amounts (Higdon, 2003). Our body fulfils its requirement of vitamins from our diet, with the exception of vitamin D, which can be synthesized with the help of sunlight. Vitamin K is synthesized within the gastrointestinal (GI) tract as can be biotin. Vitamins A, D, E and K are fat soluble and excess quantities can be stored in the body, while vitamins C and those of the B complex, such as thiamine (B_1), riboflavin (B_2), niacin (B_3), pantothenic acid (B_5) vitamin B_6, biotin (B_7), folic acid (B_9) and cyanocobalamin (B_{12}), are water soluble and can not be stored in the body because they may be excreted through urine and transpiration if they are in excess quantity (Ball, 2004). Dietary minerals are inorganic elements that are present in the soil and water. These are also essential for life. Minerals are absorbed by plants or ingested by animals. Dietary minerals are classified as bulk minerals and trace minerals. Bulk minerals are required in relatively larger amounts, and these include calcium, magnesium, phosphorus, potassium, sodium and sulfur. Trace minerals are needed only in very small amounts and include chromium, cobalt, copper, fluorine, iodine, iron, manganese, molybdenum, selenium, zinc etc. (Higdon, 2003; Lieberman and Bruning, 2003). A balanced and varied diet provides the proper amount of vitamins and minerals necessary for smooth functioning of the body.

19.2 Classification of Vitamins

Vitamins are usually classified on the basis of their solubility as either fat-soluble or water-soluble.

* E-mail: drrlsingh@rediffmail.com

19.2.1 Fat-soluble vitamins

Fat-soluble vitamins include vitamins A, D, E and K. These vitamins are absorbed along with ingested dietary fat by the small intestines. They cannot be absorbed unless they are ingested with some fat. These are usually found dissolved in fat present in food and only a small amount of fat is needed to help absorb fat-soluble vitamins. Fat-soluble vitamins may accumulate in the body and lead to hypervitaminosis. Their functions and sources are described in Table 19.1.

19.2.2 Water-soluble vitamins

The water-soluble vitamins are the B complex and vitamin C. The B-complex vitamins include thiamine (B_1), riboflavin (B_2), niacin (B_3), vitamin B_6, folate, vitamin B_{12}, biotin and pantothenic acid. Excess water-soluble vitamins can not be stored in the body. These are absorbed along with water through the GI tract and dissolved in the body fluids such as urine and excreted from the body (Fukuwatari and Shibata, 2008). These are not as readily stored as fat-soluble vitamins, hence consistent daily intake is important for proper metabolic functioning. Water-soluble vitamins, their function and source are described in Table 19.2. Some water-soluble vitamins, such as vitamins B5 and B7, are also synthesized by bacteria (*Pseudomonas* and *Klebsiella* sp.) (Said and Mohammed, 2006).

19.3 Vitamins as Cofactors

In many cases, vitamins act as cofactors (Table 19.3). They are needed in order to allow enzymes to perform their important work of facilitating metabolism in the body. In this case, the vitamins are called *coenzyme*. Most of these cofactors are found in a large variety of species and some are universal to all forms of life. Some examples of coenzyme vitamins include vitamin B_1 in the form of thiamine diphosphate (or cocarboxylase) and vitamin B_6 in the form of pyridoxal 5′-phosphate.

Table 19.1. Functions and sources of fat-soluble vitamins.

Vitamins	Functions	Sources
Vitamin A	Plays an important role in the early embryonic development of all mammals and in proper functioning of the immune system, the rod cells in the retina of the eye and mucous membranes throughout the body	Plant sources: leafy, dark green vegetables; dark orange fruits (apricots, cantaloupe) and vegetables (carrots, winter squash, sweet potatoes, pumpkin) Animal sources: fortified milk, cheese, cream, butter, fortified margarine, eggs, liver
Vitamin D	Increases efficiency of absorption of calcium and phosphorus in the intestinal tract	Egg yolks, liver, fatty fish, fortified milk, fortified margarine; skin can make vitamin D when exposed to sunlight.
Vitamin E	Acts as antioxidant, protects cell walls, modulation of gene expression and inflammatory responses	Plant sources: polyunsaturated plant oils (soybean, maize, cottonseed, safflower), leafy green vegetables, wheatgerm, whole-grain products, nuts and seeds Animal sources: liver, egg yolks
Vitamin K	Helps in blood coagulation and required for the activation of four clotting factors	Plant sources: leafy green vegetables, especially of cabbage family, and also produced in intestinal tract by bacteria Animal sources: milk

Table 19.2. Water-soluble vitamins, their function and sources.

Vitamins	Functions	Sources
Thiamine (Vitamin B_1)	Coenzyme in carbohydrate metabolism, helps to release the energy from food, important in normal functioning of heart, nerves and muscles	All nutritious foods in moderate amounts, such as whole-grain or enriched breads, cereals, legumes, nuts and seeds
Riboflavin (Vitamin B_2)	Coenzyme in protein and energy metabolism, important for normal vision and skin health	Leafy green vegetables, whole-grain, enriched breads and cereals, milk and milk products
Niacin (Vitamin B_3)	Essential for protein and carbohydrate metabolism	Wheatgerm, rice bran, nuts, sunflower seeds, brown rice, green vegetables, Brewer's/ tortula yeasts
Pantothenic acid (Vitamin B_5)	Coenzyme in synthesis of fat, cholesterol, haem and amino acid activation	Legumes, broccoli, kale, sweet potatoes, sweet corn, liver, eggs, milk, beef, cheese also, intestinal bacteria synthesis
Pyridoxine (Vitamin B_6)	Part of an enzyme (pyridoxal phosphatase) needed for protein metabolism; helps make red blood cells	Vegetables, fruits, meat, fish, poultry
Biotin (Vitamin B_7)	Part of an enzyme (carboxylase) needed for energy metabolism	Widespread in foods, also produced in intestinal tract by bacteria
Folic acid (Vitamin B_9)	Part of an enzyme (pyruvate carboxylase) needed for making DNA and new cells, especially red blood cells	Leafy green vegetables, legumes, seeds, orange juice, liver
Cyanocobalamin (Vitamin B_{12})	Promotes growth in children. It aids in the production of erythrocytes (red blood cells). It also enables the body to process carbohydrates and fat	Sunflower seed, comfrey leaves, kelp, banana, groundnut, raw wheatgerm, fortified brewer's yeast
Ascorbic acid (Vitamin C)	Acts as antioxidant, intercellular cement substance, helps in capillary walls and collagen formation and helps in iron absorption and release to tissues for red blood cell formation	Found only in fruits and vegetables, especially citrus fruits, vegetables in the cabbage family, cantaloupe, strawberries, peppers, tomatoes, potatoes, lettuce, papayas, mangoes, kiwifruit

Table 19.3. Vitamins as cofactors.

Vitamin	Coenzyme form	Reaction catalysed
Thiamine	Thiamine pyrophosphate	Aldehyde transfer
Riboflavin	Flavin adenine dinucleotide (FAD)	Oxidation reduction
Niacin	Nicotinamide adenine dinucleotide (NAD+)	Oxidation reduction
Panthothenic acid	Coenzyme A	Acyl group transfer
Pyridoxin	Pyridoxal phosphate	Group transfer from amino acid
Biotin	Biotin-lysine complexes (biocytin)	ATP dependent carboxylation and carboxyl group transfer
Folic acid	Tetrahydrofolate	Transfer of one carbon components, thymine synthesis
Cyanocobalamin	5′-Deoxyadenosyl cobalamin	Transfer of methyl group, intramolecular rearrangements
Ascorbic acid		Antioxidant

19.4 Health Benefits of Vitamins

19.4.1 Vitamin A

Vitamin A (Fig. 19.1) is one of the most important vitamins. It is not a single compound but exists in several forms such as retinol (an alcohol), retinal (an aldehyde), retinoic acid (an acid) and other related compounds (Ross, 2006; Johnson and Russell, 2010). The chemical structure of vitamin A is shown in Fig. 19.1. Vitamin A is needed for healthy gums, teeth, bones and for visual purple production, which is essential for night vision. It helps to prevent diseases such as lung and breast cancer (Mamede *et al.*, 2011) and can also be used therapeutically in the treatment of retinitis pigmentosa and leukaemia (Berson *et al.*, 1993). It also helps to promote growth and longevity, maintains health and vigour and it is essential for normal reproduction, lactation and bearing of children. It promotes appetite and digestion. Several studies showed that people suffering from cystic fibrosis have vitamin A deficiency (Borowitz *et al.*, 2002). Hypervitaminosis or vitamin A toxicity can occur when a high dose of vitamin A is taken (21,600 IU day^{-1} or more) for an extended period of time. The main symptoms of hypervitaminosis include abnormal softening of the skull bone, blurred vision, bone pain, decreased appetite, dizziness, headache, increased intracranial pressure, irritability, liver damage, nausea, skin and hair changes and vomiting. Complications include excessive high levels of calcium (hypercalcaemia), kidney and liver damage.

19.4.2 Vitamin D

Vitamin D (Fig. 19.2) is essential for life in higher animals. It is an oil-soluble vitamin and is known as the sunshine vitamin because it is synthesized when the ultraviolet rays of the sun hit the skin. Cholesterol, a precursor of vitamin D, may change into cholecalciferol (D-3) in the skin, which is similar to naturally occurring vitamin D in fish liver oils. Chemical laboratories have managed to synthesize D-3, as well as a number of stronger members of the vitamin D group including vitamin D-2, D-4, D-5 and D-6. Vitamin D is needed to regulate the absorption of calcium and phosphorus. Thus, it is important for the calcification of bones and teeth (Cranney *et al.*, 2007), promotes growth in children and prevents rickets. Vitamin D is also involved in cell differentiation and growth of keratinocytes and cancer cells. It also plays an important role in the secretion of parathyroid and insulin (Alvarez and Ashraf, 2010). Requirement for vitamin D has never been precisely defined because vitamin D is produced in the skin after exposure to sunlight. Vitamin D is essential for the treatment of osteoporosis, autoimmune diseases, heart diseases, diabetes, hypertension and cancer (Guyton *et al.*, 2003). Hypervitaminosis or vitamin D toxicity is a condition that occurs after taking a high dose of vitamin D (50,000 IU day^{-1} or more) for several months. This level is much higher than the recommended dietary allowance (RDA; 2000 IU day^{-1} for adults). Due to excess of vitamin D, the calcium level increases abnormally in the blood, which is termed hypercalcaemia (Vieth 2007). High levels of calcium cause bone damage and kidney failure. The main symptoms of hypervitaminosis include constipation, decreased appetite (anorexia), dehydration, fatigue, irritability, muscle weakness and vomiting. Complications

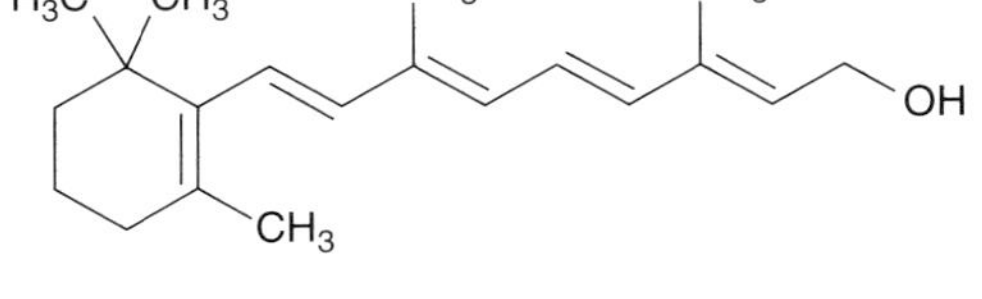

Fig. 19.1. Chemical structure of vitamin A.

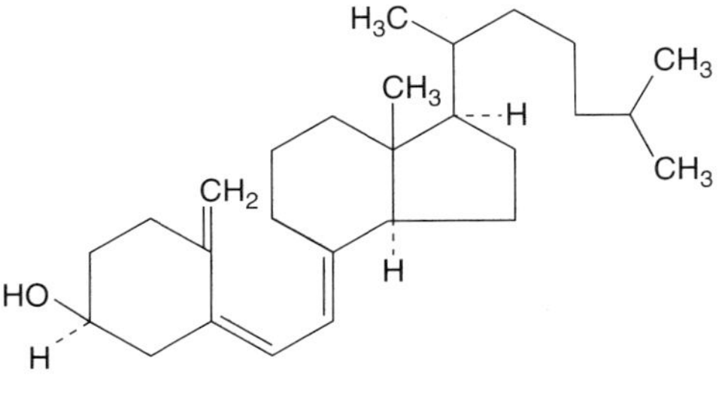

Fig. 19.2. Chemical structure of vitamin D.

include hypercalcaemia, dehydration and kidney damage due to stone formation.

19.4.3 Vitamin E

Vitamin E (tocopherol, Fig. 19.3) is found in cell membranes and fat depots. There are eight stereoisomers of each of the tocopherols, occurring in alpha, beta, gamma and delta forms. The chemical structure of vitamin E is shown in Fig. 19.3. It plays an important role in the protection of polyunsaturated fatty acids (PUFA) from oxidation (Clarke *et al.*, 2008). PUFAs are particularly sensitive to oxidative damage and the protective role of vitamin E is supported by a similar antioxidant protection from vitamin C and selenium. One tocopherol molecule can protect 100 or more PUFA molecules from autoxidative damage (Pryor, 2001). The various forms of vitamin E have different biological activity. Many forms of vitamin E are not present naturally but they may be synthesized. The relative activities of each form are complex. Several studies indicate that naturally occurring vitamin E has more (approximately twice) bioactivity in humans than synthetic vitamin E (Burton *et al.*, 1998). The RDA for vitamin E is 15 mg day^{-1} of naturally occurring α-tocopherol for adults above 19 years of age, during pregnancy 15 mg day^{-1} is recommended and 19 mg day^{-1} for lactation. Vitamin E (tocopherol), besides a role in skin healing, is also involved in prevention of cardiovascular diseases (Keaney *et al.*, 1996) and cancer (Zhang *et al.*, 2002). It also plays an important role in treatment of diabetes cataracts and enhances specific aspects of the immune response (Wang *et al.*, 2004). Hypervitaminosis or vitamin E toxicity is a condition that occurs after taking higher doses of vitamin E (more than 1600 IU day^{-1}). This level is much higher than the RDA (1000 IU day^{-1} for adults).

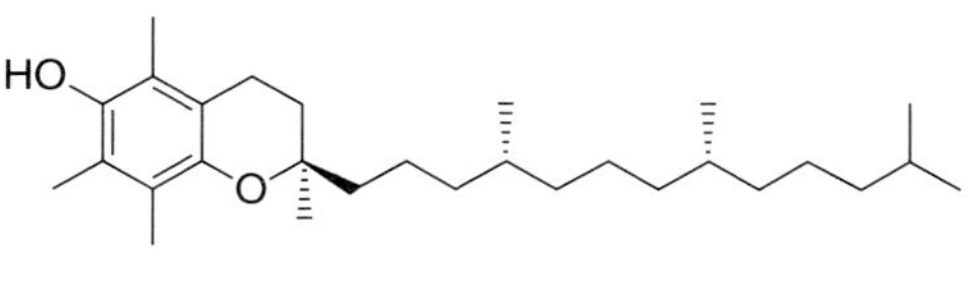

Fig. 19.3. Chemical structure of vitamin E.

The main symptoms of hypervitaminosis include increased bleeding, increased triglyceride, decreased production of thyroid hormones and decreased activity of vitamin K. Complications include haemorrhage, prostate cancer and heart failure.

19.4.4 Vitamin K

Vitamin K (Fig. 19.4) is derived from the first letter of the German word *Koagulation*. The physiological role of vitamin K is in the process of blood coagulation. Its plays an important role in the formation of prothrombin and the bone-forming protein osteocalcin (Giammanco *et al.*, 2012). Vitamin K is also synthesized by bacteria present in human gut but it may not maintain the status of vitamin K because it is biologically inactive. The drug warfarin, widely prescribed as an anticoagulant, functions through inhibition of vitamin K. As a result, alterations in vitamin K intake can influence the efficacy of warfarin. The recommended intake is based on an adequate daily intake of 120 mg day^{-1}. Vitamin K plays a role in the prevention of osteoporosis (Vermeer *et al.*, 1998), vascular calcification (Schurgers *et al.*, 2001) and cardiovascular diseases. There is no adverse effect reported for higher levels of vitamin K intake from food or food supplements. Also, there are no reported toxicity symptoms for vitamin K. One important exception to these toxicity results involves a synthetic form of vitamin K called menadione. The main symptoms of menadione hypervitaminosis include thrombosis, vomiting and jaundice in the newborn. Complications include anaemia, nerve cell damage and haemolysis.

19.4.5 Vitamin B$_1$ (thiamine)

Thiamine (Fig. 19.5) was the first vitamin to be identified. It is involved in the treatment of Alzheimer's disease (Ahmed *et al.*, 2011), congestive heart failure (Wilkinson *et al.*, 2000) and cancer (Comin-Anduix *et al.*, 2001). In modern times, thiamine deficiency is seen most commonly in association with chronic

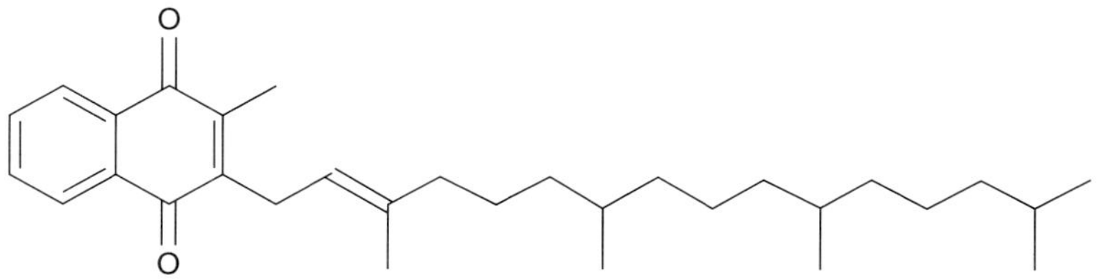

Fig. 19.4. Chemical structure of vitamin K.

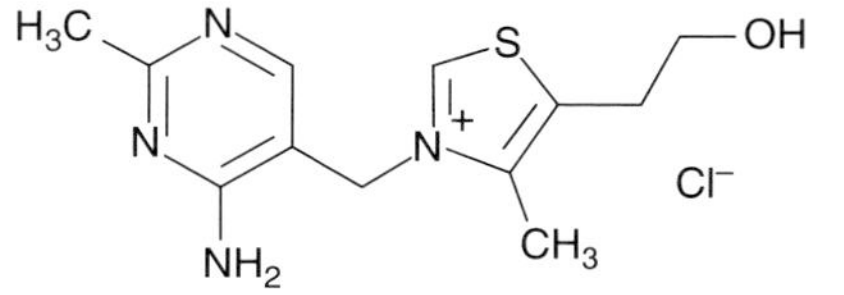

Fig. 19.5. Chemical structure of vitamin B_1.

alcoholism. Only a small percentage of large doses are absorbed, and elevated serum levels result in its active urinary excretion. After an oral dose of the vitamin, peak excretion occurs in about 2 h (Davis *et al.*, 1984). Total body thiamine content in adults is approximately 30 mg with a half-life of 9 to 18 days. The RDA for thiamine in adult women is 1.1 mg day^{-1} and in adult men it is 1.2 mg day^{-1}. The RDA under pregnancy and lactation is 1.4 mg day^{-1}. There are no reports of adverse effects from the consumption of excess thiamine consumed through food or supplements.

19.4.6 Vitamin B_2 (riboflavin)

Riboflavin (Fig. 19.6) deficiency causes sore throat, redness and oedema of the throat and oral mucous membranes, cheilosis (cracking of the skin around the mouth) and glossitis (red tongue) (Rivlin, 2001). Vitamin B_2 deficiency most often occurs in combination with other nutrient deficiencies. The members of the vitamin-B complex are related to each other; for example, for niacin synthesis, riboflavin is required, and riboflavin is necessary for vitamin B_6 for conversion to the active coenzyme (Powers, 2003). Chemical structure of vitamin B_2 is shown in Fig. 19.6. The RDA for riboflavin has been set at 1.3 mg day^{-1} for men and 1.1 mg day^{-1} for women age 70 years and older. For pregnancy, the RDA

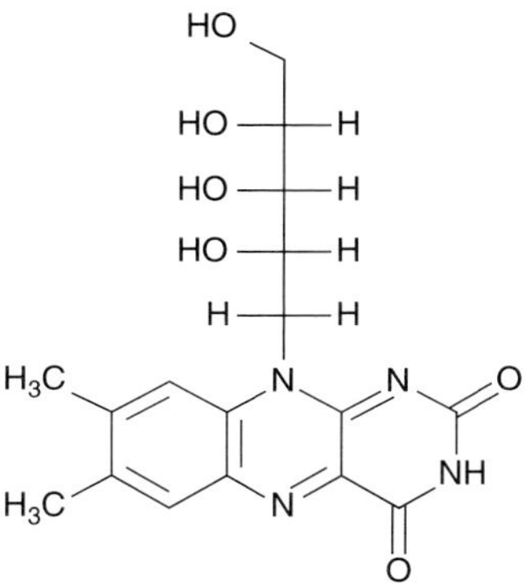

Fig. 19.6. Chemical structure of vitamin B_2.

for riboflavin is set at 1.4 mg day^{-1} and it is 1.6 mg day^{-1} for lactation. The excess riboflavin is not stored in the body and is excreted through urine. No adverse effects associated with riboflavin consumption from food or supplements have been reported. Even a single dose of riboflavin (up to 60 mg oral and 11.6 mg intravenous) has not shown any adverse effect (Zempleni *et al.*, 1996).

19.4.7 Vitamin B_3 (niacin)

Niacin (Fig. 19.7) is the name for both nicotinamide and nicotinic acid, either of which can act as a precursor of nicotinamide coenzymes. The chemical structure of vitamin B_3 is given in Fig. 19.7. Niacin is required for the synthesis of two coenzyme molecules, NAD and NADP, which are involved in several metabolic functions such as fatty acid synthesis etc. The coenzymes, the active form of niacin, are synthesized in all tissues of the body. The concentration of niacin is directly proportional to absorbed nicotinic acid and

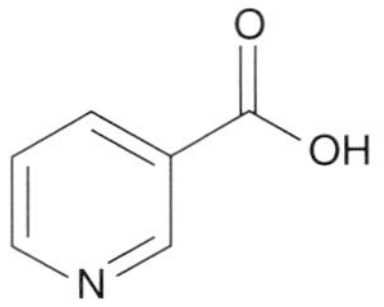

Fig. 19.7. Chemical structure of vitamin B_3.

nicotinamide, as well as conversion of the amino acid tryptophan (60 mg tryptophan = 1 mg niacin). Pellagra is the classical manifestation of niacin deficiency. The RDA for adult males is 16 mg day^{-1} of niacin equivalents, and the RDA for woman aged 19 to over 70 is 14 mg day^{-1}. In pregnant woman the RDA is 18 mg day^{-1} of niacin equivalents and in lactating women it is 17 mg day^{-1}.

19.4.8 Vitamin B_5 (pantothenic acid)

Pantothenic acid (Fig. 19.8) is widespread in our diet. The word pantothenic acid has been derived from the Greek word, meaning 'from everywhere'. Animals and humans are unable to synthesize pantoic acid, a moiety of the vitamin, so it is essential in our diet. Pantothenic acid plays an important role in wound healing (Vaxman *et al.*, 1990). Pantothenic acid is also involved in several metabolic processes, such as oxidative metabolism, cell membrane formation, cholesterol and bile salt production, energy storage and activation of some hormones (Miller *et al.*, 2001; Braun and Cohen, 2005). Many health claims are made regarding the role of pantothenic acid in ameliorating rheumatoid arthritis, lowering cholesterol, enhancing athletic performance and preventing greying of hair (Miller *et al.*, 2001). The Food and Nutrition Board (1998) established an adequate intake level (AI) for pantothenic acid of 5.0 mg day^{-1} for adult men and women, 6.0 mg day^{-1} during pregnancy and 7.0 mg day^{-1} during lactation.

19.4.9 Vitamin B_6 (pyridoxine)

Vitamin B_6 (Fig. 19.9) can lower the concentration of homocysteine and hence lower the risk of cardiovascular disease (Ebbing *et al.*,

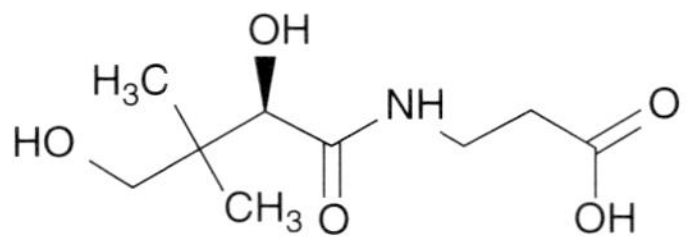

Fig. 19.8. Chemical structure of vitamin B_5.

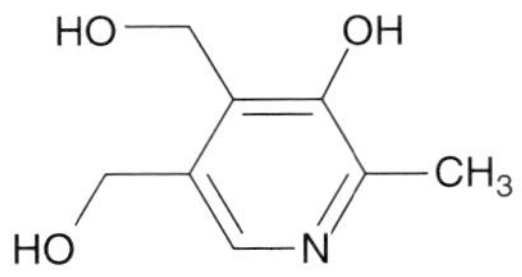

Fig. 19.9. Chemical structure of vitamin B_6.

2010), kidney stones (Mackey *et al.*, 2005) and improve cognitive functions (Balk *et al.*, 2007). It may also be useful in the treatment of pathologies such as premenstrual syndrome (PMS), side effects of oral contraceptives, nausea and vomiting in pregnancy, depression and carpal tunnel syndrome. Vitamin B_6 serves as a coenzyme and plays an important role in the metabolism of amino acids and glycogen. The synergistic effect of vitamin B_6 and folate has been shown to reduce the plasma concentrations of homocysteine and decrease the risk of cardiovascular disease. The RDA for vitamin B_6 is 1.3 mg day^{-1} for adult men and women up to the age 50 years. The RDA for people over 50 years of age is 1.7 mg day^{-1} for men and 1.5 mg day^{-1} for women. The RDA for pregnant women is 1.9 mg day^{-1} and for lactating women, 2.0 mg day^{-1}.

19.4.10 Vitamin B_7 (biotin)

Biotin (Fig. 19.10) plays an important role in fatty acid metabolism and gluconeogenesis because it acts as a coenzyme for this reaction. Besides its role as a carboxylase prosthetic group, it also affects gene expression and has a wide effect on biological functions (Mejia, 2011). Biotin is present in a significant quantity so deficiency is rare but it has been noted in patients on parenteral nutrition without biotin supplementation (Zempleni and Mock, 1999). Lipoic acid and biotin have a similar structure, thus there is competition for intestinal or cellular uptake. The Food and Nutrition

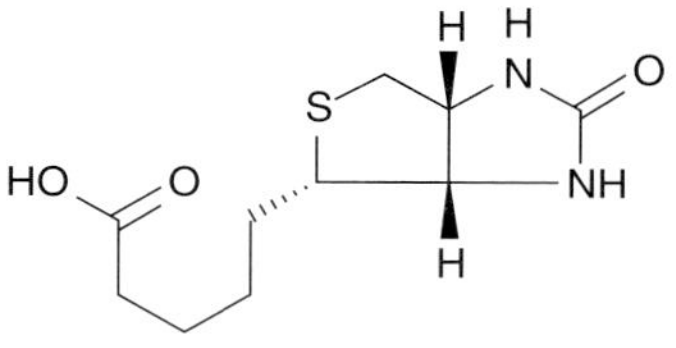

Fig. 19.10. Chemical structure of vitamin B_7.

Board established an AI for biotin due to insufficient data to set an RDA. Adult men and women have an AI of 30 mg day^{-1}. Biotin (vitamin H or B_7) plays an important role in the prevention of some birth defects (Zempleni and Mock, 2000) and in the treatment of diabetes (Larrieta *et al.*, 2012), brittle fingernails (Scheinfeld *et al.*, 2007) and seborrheic dermatitis (Schwartz *et al.*, 2006).

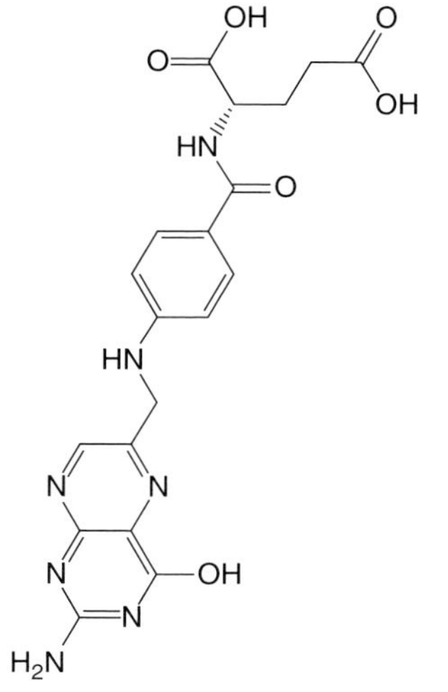

Fig. 19.11. Chemical structure of vitamin B_9.

19.4.11 Vitamin B_9 (folate)

Vitamin B_9 (Fig. 19.11) is present in different chemical forms (Wagner, 1996). The most stable form of folate is folic acid, which rarely occurs in our diet. The main function of folate is in the synthesis of DNA and purine. Folate is also involved in amino acid metabolism. It plays an important role in conversion of homocysteine to methionine. This reaction may lower the risk of cardiovascular diseases because concentration of homocysteine is decreased in plasma (Bazzano, 2009). Pregnant women are at risk of developing folate deficiency because of the heightened demands imposed by increased synthesis of DNA. Low folate status is associated with poor pregnancy outcome, low birth weight and fetal growth retardation (Scholl and Johnson, 2000).

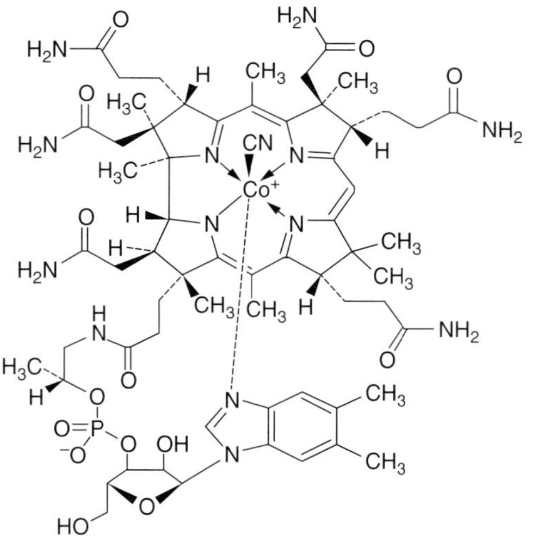

Fig. 19.12. Chemical structure of vitamin B_{12}.

19.4.12 Vitamin B_{12} (cyanocobalamin)

Cyanocobalamin is also known as vitamin B_{12} (Fig. 19.12). It is essential for normal blood formation and neurological function. Vitamin B_{12} is absorbed in our body by different mechanisms, for example in the stomach it is dissociated from proteins in the presence of hydrochloric acid and then it binds with intrinsic factor in the intestine for absorption. If proper absorption of this vitamin does not take place, malabsorption occurs and the resulting condition is called pernicious anaemia. The anaemia caused by deficiency of vitamin B_{12} (completely reversed by addition of B_{12}) is different from anaemia caused by folate deficiency. The RDA of vitamin B_{12} for men and women is 2.4 mg day^{-1} intake from food or supplements. Vitamin B_{12} helps in the prevention of diseases concerning neural tube defects, cardiovascular disease, cancer, depression, Alzheimer's disease and dementia.

19.4.13 Vitamin C (ascorbic acid)

Vitamin C (Fig. 19.13) is a water-soluble white crystalline solid. It has a –OH group and a mono anion which is the most favoured form and hence exists as an ascorbate. Plants and most animals can synthesize ascorbate but humans and other primates do not contain the enzyme gluconolactone oxidase, which is required in the final step of ascorbate synthesis. Hence, humans have to derive ascorbic acid only from the diet. Deficiency of ascorbate in the diet causes scurvy. Ascorbic acid acts as an antioxidant under *in vitro* conditions (Sebastian *et al.*, 2003). Ascorbic acid is a potent antioxidant in animals and plants, and is important in the synthesis of collagen (Sharma *et al.*, 2008). Some evidence indicates that vitamin C plays an important role in inhibiting viral replication (Johnston, 2001). Some epidemiological evidence indicates vitamin C also lower the risk for myocardial infarction (Padayatty and Levine, 2000). However, it has no effect on increasing iron absorption from haem iron (Johnston, 2001). The current requirement of vitamin C is 90 mg day^{-1} for adult men and 75 mg day^{-1} for adult women. During pregnancy the RDA is 85 mg day^{-1} and 120 mg day^{-1} during lactation. The upper limit for vitamin C is 2 g day^{-1}. Vitamin C is involved in prevention of some diseases such as scurvy, lead toxicity, cancer, cataracts and cardiovascular diseases such as coronary heart disease and stroke.

19.5 Dietary Minerals

Dietary minerals are inorganic substances that serve a variety of functions such as cofactors in enzyme-catalysed reactions, in the regulation of acid–base balance (Riond, 2001; Heil, 2010), in nerve conduction and muscle irritability and as structural elements in the body. Some of the more important minerals are calcium, phosphorus, sodium, potassium and iron. Minerals can be divided into two types: macro and micro. Macrominerals are needed in grams per day, whereas microminerals are needed in milligrams or parts per million (PPM). Microminerals are often called trace minerals. Dietary minerals and their sources are summarized in Table 19.4.

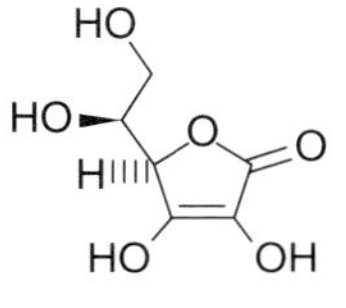

Fig. 19.13. Chemical structure of ascorbic acid.

19.6 Functions of Dietary Minerals

19.6.1 Calcium

Calcium is the most common mineral in the human body. It helps in building bones, teeth, muscle and nerves. It counteracts acidity, aids in vitality and soothes the nerves thus decreasing nervousness. Calcium is needed for the contraction of all muscles. Calcium contracts the heart muscles whereas potassium and sodium relax the heart muscles. Calcium toxicity is rare, but over consumption of calcium supplements may lead to deposits of calcium phosphate in the soft tissues of the body. Higher concentration of phosphate in body can cause hypocalcaemia and deposit of calcium phosphate crystals in various tissues of the body. Deficiency symptoms include osteoporosis (together with vitamin D) (Lanham-New, 2008), rickets (Pettifor, 2008) and osteomalacia (Bhambri *et al.*, 2006).

19.6.2 Magnesium

Magnesium is important in the metabolization of calcium and vitamin C, the maintenance of structure of DNA and RNA, as an activator of enzymes and in the pH balance. Magnesium toxicity causes various GI symptoms, including diarrhoea, abdominal cramps and nausea. As magnesium levels increase, it can result in confusion, loss of appetite, difficulty in breathing and low blood pressure. Deficiency symptoms include hypertension,

Table19.4. Different sources of dietary minerals.

Minerals	Sources
Calcium	Leafy green vegetables, soybeans, fish, meat, milk, egg etc.
Phosphorus	Almonds, wheatgerm, soybeans, black beans, milk, peas, meat, fish, eggs, cottage cheese etc.
Potassium	Spinach, beans, oranges, peas, fruits, nuts, butter, milk, meat etc.
Sodium	Table salt, eggs, meat, milk, cheese, butter, margarine, bacon etc.
Sulfur	Protein, e.g. meat, fish, milk
Manganese	Vegetables and most other foods
Iron	Green vegetables, lentils, potatoes, soybeans, chick peas, black beans, spinach, liver, eggs, meat etc.
Fluorine	Water, milk etc.
Nitrogen	Protein, e.g. meat, fish, milk etc.
Manganese	Vegetables and most other foods; bone development (a growth factor)
Iodine	Seafoods, e.g. fish, shellfish, fish oil
Molybdenum	Brown rice, millet, buckwheat, brewer's yeast, legumes, naturally hard water
Zinc	Whole-grain breads, cereals, liver, eggs, seafood
Copper	Almonds, peas, beans, green leafy vegetables, whole-grain products, prunes, raisins
Cobalt	Liver, red meat

heart diseases (Bo and Pisu, 2008), osteoporosis (Tucker *et al.*, 1999; Stendig-Lindberg *et al.*, 2004), migraine, headaches (Mauskop and Altura, 1998), hypocalcaemia and loss of appetite.

19.6.3 Potassium

Potassium, along with sodium, is needed for relaxation of the heart muscle, as well as for all muscles in the body. It aids in producing alkalinity of the body, stimulates the liver, aids in the heartbeat and is also needed for proper functioning of the nervous system. Potassium toxicity can cause hyperkalaemia and cardiac arrhythmias or even death due to cardiac arrest. However, food consumption may cause mild increases in the concentration of potassium in the bloodstream. Mild higher levels do not causes toxicity due to continuous uptake of potassium by various cells of the body, as well as by the action of the kidneys, which transfer potassium ions from the blood to the urine. Deficiency symptoms include osteoporosis (Zhu *et al.*, 2009), strokes (Bazzano *et al.*, 2001), kidney stones (Melanie *et al.*, 2009), hypertension and skin problems.

19.6.4 Chromium

Chromium is needed to maintain the blood sugar level, glucose metabolization and in the synthesis of fatty acids. Chromium toxicity mostly occurs in industrial workers and it causes irritation in the nose, ulcers or holes may develop in the nasal septum and stomach also. If skin comes in contact with chromium (VI) compounds, it can lead to skin ulcers, allergies, rashes, redness, itching and swelling in the affected area. It can cause respiratory problems such as difficulty in breathing and coughing and may lead to bronchitis or asthma. Deficiency symptoms include cardiovascular diseases (Kobla and Volpe, 2000) and diabetes (Sundaram *et al.*, 2012).

19.6.5 Copper

Copper aids in iron metabolization, which is needed for skin and hair pigmentation, bone formation and in the synthesis of haemoglobin and production of red blood cells. Copper toxicity occurs due to two genetic diseases, i.e. Wilson's and Menkes' disease, both of which are rare but occur 1 in 100,000 births. Both diseases involve mutations resulting in the synthesis of proteins that transport copper through cell membranes. Wilson's disease tends to occur in teenagers and in young adults. In this disease copper accumulates in the liver, kidney and brain resulting in damage to the liver and nervous system. Wilson's disease can be successfully controlled by lifelong treatment with D-penicillamine. Treatment also involves avoiding foods that are high in copper, such as liver, nuts, chocolate and molluscs. Wilson's disease may be treated with zinc supplements to inhibit the absorption of dietary copper. Deficiency symptoms include impairment of immune system functions and osteoporosis (Eaton-Evans *et al.*, 1996).

19.6.6 Fluorine

Fluorine plays an important role in strengthening the bones, teeth and enamel of the teeth and protects against infections. It also increases the metabolization of calcium. Fluoride toxicity occurs due to ingestion (accidental or intentional) of fluoride-containing products. Ingested fluoride initially acts on the intestinal mucosa. It can form hydrofluoric acid in the stomach, which leads to GI irritation or corrosive effect. After absorption, fluoride binds calcium ions and may lead to hypocalcaemia. Fluoride has direct cytotoxic effects and interferes with a number of enzyme systems such as oxidative phosphorylation, glycolysis, coagulation and neurotransmission. Deficiency symptoms include dental caries and osteoporosis (Riggs *et al.*, 1990).

19.6.7 Iodine

Iodine is important in regulating the metabolic function of the thyroid. It plays an important role in prevention of goitre and albuminous toxins in the blood. Iodine toxicity results in impaired thyroid hormone formation, resulting in lower levels of thyroid hormone in the bloodstream. Iodine toxicity produces ulcers on the skin called 'kelp acne' because of its association with eating kelp, an ocean plant that contains high levels of iodine. Iodine toxicity is common in Japan, where large quantities of seaweed are consumed as food source. Deficiency symptoms includes thyroid cancer (Zanzonico and Becker, 2000), goitre and cretinism.

19.6.8 Selenium

Selenium is a trace mineral that is required only in small amounts. Selenium is conjugated with proteins to make selenoproteins, which act as antioxidant. The antioxidant properties of selenoproteins help prevent cellular damage from free radicals. Free radicals are natural by-products of oxygen metabolism that may contribute to the development of chronic diseases such as cancer and heart disease. Other selenoproteins help regulate thyroid function and play an important role in functioning of immune system. Selenium toxicity occurs in some parts of China, where soils contain high levels of selenium that are then found in foods and water. Early signs of selenium toxicity include nausea, weakness and diarrhoea. Deficiency symptoms include general fatigue, hypothyroidism and mental fatigue.

19.6.9 Zinc

Zinc is essential in the synthesis of nucleic acids and aids in the metabolism of vitamins, especially the B-complex factors. It is found in the enzymes that aid digestion and metabolization. Zinc toxicity is rare but can occur in metal workers who are exposed to fumes containing zinc. Excessive dietary supplements of zinc can result in nausea, vomiting, diarrhoea, and copper deficiency because zinc inhibits the absorption of copper. Deficiency symptoms include disturbances of normal physiology, growth and development (Hambidge, 2000), hypozincaemia (abdominal pain, nausea, vomiting, diarrhoea, lethargy, anaemia and dizziness) and anorexia.

19.7 Conclusions

In conclusion, the literature on vitamins and minerals suggests that these are essential nutrients for proper functioning of our body system. Some vitamins act as cofactors or in partnership with other vitamins, minerals, nutrients and other substances in the body such as enzymes. Two of the most essential and powerful vitamin partnerships are the group of eight vitamins that make up the nutritional powerhouse that is collectively known as the vitamin B complex and the group of two vitamins (C and E) that are known as antioxidant vitamins. In addition, there are varieties of other combinations of vitamins that serve essential purposes within the body thus preserving health and enhancing function. Nutritional supplements can ensure that our body has enough nutrients that support these important partnerships.

References

Ahmed, H.H., Shousha, W.G., Hussien, R.M. and Farrag, A.R.H. (2011) Potential role of some nutraceuticals in the regression of Alzheimer's disease in an experimental animal model. *Turkish Journal of Medical Sciences* 41(3), 455–466.

Alvarez, J.A. and Ashraf, A. (2010). Role of vitamin D in insulin secretion and insulin sensitivity for glucose homeostasis., *International Journal of Endocrinology* 2010, 351–385.

Ball, G. (2004) *Vitamins. Their Role in the Human Body*. Blackwell Publishing, New York, pp. 133–187.

Balk, E.M., Raman, G., Tatsioni, A., Chung, M., Lau, J. and Rosenberg, I.H. (2007) Vitamin B6, B12, and folic acid supplementation and cognitive function: a systematic review of randomized trials. *Archives of International Medicine* 167, 21–30.

Bazzano, L.A. (2009) Folic acid supplementation and cardiovascular disease: the state of the Art. *American Journal of the Medical Sciences* 338(1), 48–49.

Bazzano, L.A., He, J., Ogden, L.G., Loria, C., Vupputuri, S., Myers, L. and Whelton, P.K. (2001) Dietary potassium intake and risk of stroke in US men and women. *Stroke* 32, 1473–1480.

Berson, E.L., Rosner, B., Sandberg, M.A., Hayes, K.C., Nicholson, B.W., Weigel-DiFranco, C. and Willett, W. (1993) A randomized trial of vitamin A and vitamin E supplementation for Retinitis pigmentosa. *Archives of Ophthalmology* 111, 761–772.

Bhambri, R., Naik, V., Malhotra, N., Teneja, S., Rastogi, S., Ravishanker, U. and Mithal, A. (2006) Changes in Bone Mineral Density Following Treatment of Osteomalacia. *Journal of Clinical Densitometry* 9(1), 120–127.

Bo, S. and Pisu, E. (2008) Role of dietary magnesium in cardiovascular disease prevention, insulin sensitivity and diabetes. *Current Opinion in Lipidology* 19(1), 50–56.

Borowitz, D., Baker R.D. and Stallings, V. (2002) Consensus report on nutrition for pediatric patients with cystic fibrosis. *Journal of Pediatric Gastroenterology and Nutrition* 35, 246–259.

Braun, L. and Cohen, M. (2005) *Herbs and Natural Supplements: An evidence based guide*. Elsevier, New South Wales, Australia.

Burton, G.W., Traber, M.G., Acuff, R.V., Walters, W., Kayden, H., Hughes, L. and Ingold, K.U. (1998) Human plasma and tissue alpha-tocopherol concentrations in response to supplementation with deuterated natural and synthetic vitamin C. *American Journal of Clinical Nutrition* 67, 669–684.

Clarke, M.W., Burnett, J.R. and Croft, K.D. (2008) Vitamin E in human health and disease. *Critical Reviews in Clinical Laboratory Sciences* 45(5), 417–450.

Comin-Anduix, B., Boren, J. and Martinez, S. (2001) The effect of thiamine supplementation on tumour proliferation. A metabolic control analysis study. *European Journal of Biochemistry* 268(15), 4177–4182.

Cranney, A., Horsley, T., O'Connell, S., Weiler, H., Puil, L., Ooi, D., Atkinson, S., Ward, L., Moher, D., Hanley, D., Fang, M., Yazdi, F., Garritty, C., Sampson, M., Barrowman, N., Tsertsvadze, A. and Mamaladze, V. (2007) Effectiveness and safety of vitamin D in relation to bone health. *Evidence Report/Technology Assessment* (158), 1–235.

Davis, R.E., Icke, G.C., Thom, J. and Riley, W.J. (1984) Intestinal absorption of thiamin in man compared with folate and pyridoxal and its subsequent urinary excretion. *Journal of Nutritional Science and Vitaminology* 30, 475–482.

Eaton-Evans, J., Mellwrath, E.M., Jackson, W.E., McCartney, H. and Strain, J.J. (1996) Copper supplementation and the maintenance of bone mineral density in middle-aged women. *Journal of Trace Element in Experimental Medicine* 9, 87–94.

Ebbing, M., Bonaa, K.H., Arnesen, E., Ueland, P.M., Nordrehaug, J.E. and Rasmussen, K. (2010) Combined analyses and extended follow-up of two randomized controlled homocysteine-lowering B-vitamin trials. *Journal of Internal Medicine* 268, 367–382.

Food and Nutrition Board (1998) *Dietary Reference intakes for Thiamine, Riboflavin, Niacin, Vitamin B_6, Folate, Vitamin B_{12}, Panthothenic acid, Biotin and Choline*. Institute of Medicine, National Academy Press, Washington, DC.

Fukuwatari, T. and Shibata, K. (2008) Urinary water-soluble vitamins and their metabolite contents as nutritional markers for evaluating vitamin intakes in young Japanese women. *Journal of Nutrition Science and Vitaminology* 54, 223–229.

Giammanco, M., Majo, D.D., Leto, G., Flandina, C., Piazza, M.D. and Guardia, M.L. (2012) The Role of Vitamin K in Bone Remodeling and Osteoporosis. *Journal of Food Research* 1(4), 106–123.

Guyton, K.Z., Kensler, T.W. and Posner, G.H. (2003) Vitamin D and vitamin D analogs as cancer chemopreventive agents. *Nutrition Reviews* 61, 227–238.

Hambidge, M. (2000) Human zinc deficiency. *Journal of Nutrition* 130(5S Suppl.), 1344S–1349S.

Heil, D.P. (2010) Acid-base balance and hydration status following consumption of mineral-based alkaline bottled water. *Journal of the International Society of Sport Nutrition* 7, 29.

Higdon, J. (2003) *An Evidence-Based Approach to Dietary Phytochemicals. Health Benefits and Intake Recommendations*. Thieme Medical Publishers, New York.

Johnston, C.S. (2001) Vitamin C: In: Bowman, B.A. and Russell, R.M. (eds) *Present Knowledge of Nutrition*, 8th edn. ILSI Press, Washington, DC, pp. 175–183.

Johnson, E.J. and Russell, R.M. (2010) Beta-Carotene. In: Coates, P.M., Betz, J.M., Blackman, M.R. *et al.* (eds) *Encyclopedia of Dietary Supplements*, 2nd edn. Informa Healthcare, London, pp. 115–120.

Keaney, J.F., Guo, Y., Cunningham, D., Shwaery, G.T., Xu, A. and Vita, J.A. (1996) Vascular Incorporation of Alpha-tocopherol Prevents Endothelial Dysfunction due to Oxidized LDL by Inhibiting Protein Kinase C Stimulation. *Journal of Clinical Investigation* 98, 386–394.

Kobla, H.V. and Volpe, S.L. (2000) Chromium, Exercise, and Body Composition. *Critical Reviews in Food Science and Nutrition* 40, 291–308.

Lanham-New, S.A. (2008) Importance of calcium, vitamin D and vitamin K for osteoporosis prevention and treatment. *Proceedings of the Nutrition Society* 67, 163–176.

Larrieta, E., Vega-Monroy, M.L. and Vital, P. (2012) Effects of biotin deficiency on pancreatic islet morphology, insulin sensitivity and glucose homeostasis. *Journal of Nutritional Biochemistry* 23(4), 392–399.

Lieberman, S. and Bruning, N. (2003) *Real Vitamin and Mineral Book*. Penguin Books, New York.

Mackey, A., Davis, S. and Gregory, J. (2005) Vitamin B6. In: Shils, M., Shike, M., Ross, A., Caballero, B. and Cousins, R. (eds) *Modern Nutrition in Health and Disease*, 10th edn. Lippincott Williams & Wilkins, Baltimore, Maryland.

Mamede, A.C., Tavares, S.D., Abrantes, A.M., Trindade, J., Maia, J.M. and Botelho, M.F. (2011) The role of vitamins in cancer: a review. *Nutrition and Cancer* 63(4), 479–494.

Mauskop, A. and Altura, B.M. (1998) Role of Magnesium in the Pathogenesis and treatment of migraines. *Clinical Neuroscience* 5, 24–27.

Mejia, C.F. (2011) Biological Effects of Pharmacological Concentrations of Biotin. *Journal of Evidence-Based Complementary & Alternative Medicine* 16, 140–148.

Melanie, A., McNally, B.S., Paula, L., Pyzik, B.S., Rubenstein, J.E., Rana, F., Hamdy, E. and Kossoff, H. (2009) Empiric Use of Potassium Citrate Reduces Kidney-Stone Incidence with the Ketogenic Diet. *Pediatrics* 124(2), 300–304.

Miller, J.W., Rogers, L.M. and Rucker, R.B. (2001) Pantothenic Acid. In: Bowman, B.A. and Russell, R.M. (eds) *Present Knowledge of Nutrition*, 8th edn. ILSI Press, Washington, DC, pp. 253–260.

Padayatty, S.J. and Levine, M. (2000) Vitamin C and myocardial infarction: the heart of the matter. *American Journal of Clinical Nutrition* 71, 1027–1028.

Pettifor, J.M. (2008) Vitamin D and/or calcium deficiency rickets in infants and children: a global perspective. *Indian Journal of Medical Research* 127, 245–249.

Powers, H.J. (2003) Riboflavin (vitamin B-2) and health. *American Journal of Clinical Nutrition* 77(6), 1352–1360.

Pryor, W.A. (2001) Vitamin E. In: Bowman, B.A. and Russell, R.M. (eds) *Present Knowledge of Nutrition*, 8th edn. ILSI Press, Washington, DC, pp. 156–163.

Riggs, B.L., Hodgson, S.F., O'Fallon, W.M., Chao, E.Y., Wahner, H.W., Muhs, J.M., Cedel, S.L. and Melton, L.J. (1990) Effect of Fluoride Treatment on the Fracture Rate in Postmenopausal Women with Osteoporosis. *New England Journal of Medicine* 322, 802–809.

Riond, J.L. (2001) Animal nutrition and acid-base balance. *European Journal of Nutrition* 40, 245–254.

Rivlin, R.S. (2001) Riboflavin. In: Bowman, B.A. and Russell, R.M. (eds) *Present Knowledge of Nutrition*, 8th edn. ILSI Press, Washington, DC, pp. 191–198.

Ross, A. (2006) Vitamin A and Carotenoids. In: Shils, M., Shike, M., Ross, A., Caballero, B. and Cousins, R. (eds) *Modern Nutrition in Health and Disease*, 10th edn. Lippincott Williams & Wilkins, Baltimore, Maryland, pp. 351–375.

Said, H.M. and Mohammed, Z.M. (2006) Intestinal absorption of water-soluble vitamins: an update. *Current Opinion in Gastroenterology* 22, 140–146.

Scholl, T.O. and Johnson, W.J. (2000) Folic Acid: Influence on the Outcome of Pregnancy. *American Journal of Clinical Nutrition* 71(Suppl.), 1295–1303.

Scheinfeld, N., Dahdah, M.J. and Scher, R. (2007) Vitamins and minerals: their role in nail health and disease. *The Journal of Drugs in Dermatology* 6(8), 782–787.

Schurgers, L.J., Dissel, P.E., Spronk, H.M., Soute, B.A., Dhore, C.R., Cleutjens, J.P. and Vermeer, C. (2001) Role of Vitamin K and Vitamin K-dependent Proteins in Vascular Calcification. *Zeitschrift fur Kardiologie* 90(Suppl. 3), 57–63.

Schwartz, R.A., Janusz, C.A. and Janniger, C.K. (2006) Seborrheic Dermatitis: An Overview. *American Family Physician* 74(1), 125–132.

Sebastian, J., Katz, P.A., Wang, Y., Eck, P., Kwon, O., Lee, J.H., Chen, S., Corpe, C., Dutta, A., Dutta, S.K. and Levine, M. (2003) Vitamin C as an Antioxidant: Evaluation of Its Role in Disease Prevention. *Journal of the American College of Nutrition* 22(1), 18–35.

Sharma, S.R., Poddar, R., Sen, P. and Andrews, J.T. (2008) Effect of vitamin C on collagen biosynthesis and degree of birefringence in polarization sensitive optical coherence tomography (PS-OCT). *African Journal of Biotechnology* 7(12), 2049–2054.

Stendig-Lindberg, G., Koeller, W., Bauer, A. and Rob, P.M. (2004) Prolonged Magnesium Deficiency Causes Osteoporosis in the Rat. *Journal of the American College of Nutrition* 23(6), 704–711.

Sundaram, B., Singhal, K. and Sandhir, R. (2012) Ameliorating effect of chromium administration on hepatic glucose metabolism in streptozotocin-induced experimental diabetes. *Bio Factors* 38(1), 59–68.

Tucker, K.L., Hannan, M.T., Chen, H., Cupples, L.A., Wilson, P.W. and Kiel, D.P. (1999) Potassium, Magnesium, and Fruit and Vegetable Intakes are Associated with Greater Bone Mineral Density in Elderly Men and Women. *Journal of Clinical Nutrition* 69, 727–736.

Vaxman, F., Chalkiadakis, G., Olender, S., Maldonado, H., Aprahamian, M., Bruch, J.F., Wittmann, T., Volkmar, P. and Grenier, J.F. (1990) Improvement in the Healing of Colonic Anastomoses by Vitamin B5 and C Supplements. Experimental Study in the Rabbit. *Annales de Chirurgie* 44, 512–520.

Vermeer, C., Knapen, M.H. and Schurgers, L.J. (1998) Vitamin K and Metabolic Bone Disease. *Journal of Clinical Pathology* 51, 424–426.

Vieth, R. (2007) Vitamin D toxicity, policy, and science. *Journal of Bone and Mineral Research*, Suppl. 2, V64–8.

Wagner, C. (1996) Symposium on the Subcellular Compartmentation of Folate Metabolism. *Journal of Nutrition* 126, 1228–1234.

Wang, X.L., Rainwater, D.L., Mahaney, M.C. and Stocker, R. (2004) Supplementation with Vitamin E and Coenzyme Q10 Reduces Circulating Markers of Inflammation in Baboons. *Journal of Clinical Nutrition* 80, 649–655.

Wilkinson, T.J., Hanger, H.C., George, P.M. and Sainsbury, R. (2000) Is thiamine deficiency in elderly people related to age or co-morbidity? *Age and Ageing* 29(2), 111–116.

Williams, M.H. (2005) Dietary Supplements and Sports Performance: Minerals. *Journal of the International Society of Sports Nutrition* 2(1), 43–49.

Zanzonico, P.B. and Becker, D.V. (2000) Effects of Time of Administration and Dietary Iodine Levels on Potassium Iodide (KI) Blockade of Thyroid Irradiation by 131I From Radioactive Fallout. *Health Physics* 78, 660–667.

Zempleni, J. and Mock, D.M. (1999) Biotin Biochemistry and Human Requirements. *The Journal of Nutrition Biochemistry* 10, 128–138.

Zempleni, J. and Mock, D.M. (2000) Marginal Biotin Deficiency is Teratogenic. *Proceedings of the Society for Experimental Biology and Medicine* 223, 14–21.

Zempleni, J., Galloway, J.R. and McCormick, D.B. (1996) Pharmacokinetics of Orally and Intravenously Administered Riboflavin in Healthy Humans. *American Journal of Clinical Nutrition* 63, 54–66.

Zhang, Y., Ni, J., Messing, E.M., Chang, E., Yang, C.R. and Yeh, S. (2002) Vitamin E Succinate Inhibits the Function of Androgen Receptor and the Expression of Prostate-specific Antigen in Prostate Cancer Cells. *Proceeding of the National Academy of Sciences USA* 99, 7408–7413.

Zhu, K., Devine, A. and Prince, R.L. (2009) The effects of high potassium consumption on bone mineral density in a prospective cohort study of elderly postmenopausal women. *Osteoporos International* 20, 335–340.

20 Nutrigenomics: Nurturing of Genotype and Role in Human Health

Neeraj Kumar[1]* and Kamal Kishore Maheshwari[2]

[1]*Shri Ram Murti Smarak College of Engineering and Technology (Pharmacy), Bareilly;* [2]*Department of Pharmacy, M.J.P. Rohilkhand University, Bareilly, Uttar Pradesh, India*

20.1 Introduction

20.1.1 Definitions and terms

GENOMICS. The study of the genomes of organisms for determining the entire DNA sequence of organisms and fine-scale genetic mapping (Balammal and Jayachandra Reddy, 2012), while the genome is the set of all genes, regulatory sequences and other information contained within the non-coding regions of DNA of an organism (Roth *et al.*, 1998).

NUTRITIONAL GENOMICS. The science of the relationship between human genome, nutrition and health (Ordovas and Corella, 2004) or the genetic manipulation of plants to create vitamins and minerals that will improve human diet and analysis of an organism's set of genes; hence, it is an area of science that looks at how environmental factors, such as diet, influence the genetic make-up (Ordovas and Mooser, 2004).

NUTRIGENETICS. Nutrigenetics is the interplay between nutrition and genetics of an individual, a branch of science concerned with the effect of heredity on diet and nutrition (Simopoulos, 2010). The term 'nutrigenetics' is used for scientific investigation of impact of changes in inherited traits of nuclear DNA due to any specific metabolic dysfunction that ultimately results in chronic disorder or damage (Simopoulos, 2010; Manzelli, 2012). According to WHO reports diet factors influence occurrence of more than two-thirds of diseases and most of these factors belong to the categories of nutrigenetics. In other words, nutrigenetics concerns individual differences in the reaction to food based on the genetic factors and analyses direct influences of nutrients on gene expression (Svacina, 2007).

PROTEOMICS. The study of structures and functions of protein and makes an analogy with genomics, so proteomics is the study of the genes while proteome is the entire complement of proteins, including the modifications made to a particular set of proteins produced by an organism or system. The word 'proteome' was coined by Marc Wilkins in 1994 by the blend of 'protein' and 'genome' (Wilkins *et al.*, 1996; James, 1997).

* E-mail: neerajsitm@yahoo.com

METABOLOMICS. The systematic study of the unique chemical fingerprints that specific cellular processes leave behind with study of their small-molecule metabolite profiles and increasingly being used in a variety of health applications including pharmacology, pre-clinical drug trials, toxicology, transplant monitoring, newborn screening and clinical chemistry (Nanda *et al.*, 2011), while the metabolome is the collection of all metabolites in a biological cell, tissue, organ or organism, and end products of cellular processes (Daviss, 2005).

GENE EXPRESSION. The process by which information from a gene is used in the synthesis of a functional gene product like proteins, but in non-protein coding genes, such as ribosomal RNA (rRNA), transfer RNA (tRNA) or small nuclear RNA (snRNA) genes, the product is a functional RNA and this process is used by every living thing, including eukaryotes, prokaryotes and viruses, to generate the macromolecules for their body. This process occurs in two major stages: (i) transcription, in which the gene is copied to produce an RNA molecule with essentially the same sequence as the gene; and (ii) protein synthesis, known as translation (Twyman, 2003; Brandenberg *et al.*, 2011).

GENOTYPE. This is the genetic makeup of a cell, an organism, or an individual with reference to a specific character, which is the internally coded, inheritable information, carried by all living organisms and this stored information is used as a blueprint or set of instructions for building and maintaining a living creature (Brandenberg *et al.*, 2011).

PHENOTYPE. The composite of an organism's observable characteristics or traits such as its morphology, development, biochemical or physiological properties, phenology, behaviour and products of behaviour such as physical parts; the sum of the atoms, molecules, macromolecules, cells, structures, metabolism, energy utilization, tissues, organs, reflexes and behaviours of a living organism (Brandenberg *et al.*, 2011).

POLYMORPHISM. Polymorphism in biology occurs when two or more clearly different phenotypes exist in the same population of a species; the occurrence of more than one form or morph (Brandenberg *et al.*, 2011).

ALLELE. An allele is one of two or more forms of a gene or a genetic locus used for an abbreviation of allelomorph and different alleles can result in different observable phenotypic traits, such as different pigmentation (Brandenberg *et al.*, 2011).

EPIGENETIC. A modification of gene expression that is independent of the DNA sequence of the gene (Egger *et al.*, 2004). The current definition of epigenetics is the study of heritable changes in gene expression that occur independent of changes in the primary DNA sequence and these heritable changes are established during differentiation and are stably maintained through multiple cycles of cell division, enabling cells to have distinct identities while containing the same genetic information. This heritability of gene expression patterns is mediated by epigenetic modifications, which include methylation of cytosine bases in DNA, post-translational modifications of histone proteins as well as the positioning of nucleosomes along the DNA (Sharma *et al.*, 2010).

20.1.2 Nutrigenomics

> Let food be thy medicine and medicine thy food
>
> (Hippocrates 400 BC)

Nutrigenomics is the study of how naturally occurring chemicals in foods alter molecular expression of genetic information in each individual. The term nutrigenomics is used for changes in gene expression or its effects due to specific dietary pattern, functional food or supplement on a specific health outcome (Fenech, 2005), so called as the 'next frontier in the post genomic era' (Castle and Ries, 2007). It can be described as the study of the relationship between genes, diet, lifestyle and health, which may regulate gene function like transcription, translation and metabolism, i.e. diet–gene interaction (Ordovas and Mooser, 2004).

Nutrigenomics focuses on the understanding that nutrition influences metabolism and maintenance of the internal equilibrium in the body, and this regulation affects diet-related diseases (Ordovas and Corella, 2004) and offers a powerful and exiting approach to unravel the effect of diet on health. In the past, nutrition research concentrated on nutrient deficiency and impairment of health, but nutrigenomics creates a junction between healthy diet and genomics and it will promotes an increased understanding of how nutrition influences metabolic pathways and homeostasis control.

Biomedical researchers, private sector firms, the public (Caulfield *et al.*, 2008) and the food industry recognize the need for nutrigenomics research as a basis for developing the concept of 'personalized diet' for identifying molecular biomarkers. Over the past few years, there has been rapid increase in the interest in nutrigenomics as a research topic because it is an area that has been viewed as worthy of public funding, both as a topic of basic scientific inquiry and as a field with health care and commercialization possibilities (Ordovas and Corella, 2004). The new scientific understanding of nutrigenomics has led to the increase of commercial development of nutraceutical and functional foods that can modify the negative health effect of individual genetic profiles (Marotta *et al.*, 2012).

The main aim of nutrigenomics is to improve dietary advice, development of health-promoting supplements, preventive strategies and the reduction of healthcare cost (Ordovas and Mooser, 2004). The coming years will likely require patience, realistic expectations and strong advocacy for the needed research funding, and a major focus of nutrition research is on prevention of chronic disease such as cardiovascular disease, metabolic disorders and cancer (Afman and Muller, 2006).

More than simply managing or treating disease or the symptoms associated with disease, nutrigenomics will be used to identify susceptibilities to disease and implement proactive measures to help individuals avoid contracting said disease in the first place, and we can say that nutrigenomics research will lead to development of evidence-based healthy food and lifestyle advice and dietary intervention for contemporary humans (Ordovas and Mooser, 2004). The advent of modern science led to the realization that not only are certain nutrients essential but also that a specific quantity of each is necessary for optimal health, thereby leading to such notions as dietary recommendations, nutritional epidemiology, and the realization that food can directly contribute to disease onset. In this regard the onset of diseases during human development is clearly defined by both environmental influences like diet, smoking education, physical activity etc. and heredity, indicating that both aspects must be considered to optimize health (Ordovas and Corella, 2004).

The excitement about nutrigenomics comes from a growing awareness of the potential for modifications of food or diet to support health and reduce the risk of diet-related diseases, thus by identifying individual genetic predispositions for chronic diseases and the potential for individual's response to dietary intervention, these diseases may be effectively prevented by proper dietary intake. For this, nutrigenomics brings together the science of bioinformatics nutrition, molecular biology genomics, epidemiology and molecular medicine (Neeha and Kinth, 2012).

Nutrigenomics is the application of high-throughput genomics tools to the study of diet–gene interactions in order to identify dietetic components having beneficial or detrimental health effects (Miggiano and De Sanctis, 2006). Traditionally, biomarkers related to onset of disease or organ damage were used to quantify the effects, but now it becomes necessary to quantify phenotype changes that are very close or within the range of health state (Van Ommen *et al.*, 2008) and has primarily focused on nutrient deficiencies and the relation between nutrition and health. The advent of genomics has created unprecedented opportunities for increased understanding of nutrients modulating gene and protein expression and ultimately influence cellular and organizational metabolism (Busstra *et al.*, 2007).

Normally nutrigenomics embodies three normative concepts: (i) food is exclusively interpreted in terms of disease prevention; (ii) striving for health is interpreted as the quantification of risks and prevention of diseases through positive food–gene interactions; and (iii) the normative idea is that disease prevention by the minimization of risks is an individual's task (Korthals, 2011).

Nutritional factors are thought to be the cause of 30–60% of cancers; cases of diabetes, cardiovascular diseases and obesity are increasing rapidly (Zeisel, 2010). The conceptual basis for this new branch of genomic research can best be summarized by the following five tenets of nutrigenomics (DeBusk *et al.*, 2005):

1. Under certain circumstances and in some individuals, diet can be a serious risk factor for a number of diseases.
2. Common dietary chemicals can act on the human genome, either directly or indirectly to alter gene expression or structure.
3. The degree to which diet influences the balance between healthy and disease states may depend on an individual's genetic makeup.
4. Some diet-regulated genes are likely to play a role in the onset, incidence, progression and severity of chronic disease.
5. Dietary intervention based on knowledge of nutritional requirement, nutritional status and genotype can be used to prevent, mitigate or cure chronic diseases.

Nutrigenomics is therefore significant not only as a matter of improving public health but also becomes a tool in nutritional research.

20.1.3 Benefits of nutrigenomics

Scientific studies show that nutrients in food can cause changes in the behaviour of genes and some findings suggest that nutrients can reduce the risk of cancer and other diseases and through it, researchers hope to find ways to use food to prevent, cure and reduce the risk of diet-related disease; benefits include a growth in concern on one's health and the chance to have a personalized nutrition optimized for good health, discovering genetic vulnerabilities, which can be a strong motivating factor to encourage people to make the necessary dietary and lifestyle changes, and the high chances of heeding the advice that they have paid for. Profiling and analysing one's DNA may cost between US$300 and US$3000 and large-scale food corporations are spending fortunes on nutrigenomics, and on development of enhanced or fortified products to deliver personalized diets, and multinational corporations specializing in skin care, anti-ageing and beauty products are using nutrigenomics (Castle and Ries, 2007).

The main aims of nutrigenomics are:

1. Obtaining a personalized dietary regimen may encourage people to become more health conscious.
2. People are more likely to heed advice that they pay for.
3. Discovering genetic susceptibilities can be a strong motivator for making dietary and lifestyle changes.
4. The safe upper and lower limits for essential macronutrients such as proteins, carbohydrates, fats and micronutrients such as vitamins and minerals will be better defined and understood.
5. Diseases may be avoided or ameliorated.
6. Unnecessary vitamins and other dietary supplements can be avoided.
7. People whose health is relatively unaffected by diet can continue to eat foods that they enjoy.
8. Lifespan may be extended.

The following studies are responding to the established nutritional market for seeking new tools to enhance health.

Nutrigenomics: genes can tell us what to eat

The ability of cells to adapt to environmental change by regulation of gene expression is essential for organism survival and organisms vary their gene expression in the absence or presence of nutrients by increasing and decreasing production of cellular proteins necessary for life-sustaining function. A perfect example of this evolutionary process is the development of a gene mutation that alters the ability to tolerate lactose, and adult mammals typically are unable to digest lactose.

Ultimately, the science of nutrigenomics promises to offer health practitioners greater knowledge, enabling them to predict potential genetic responses to nutritional intake and to target and modify associated behaviour (Zeisel *et al.*, 2005).

Nutrigenomics explains omega-3's immune health benefits

Omega-3 fatty acids not only lower low-density lipoprotein (LDL) cholesterol, but also help raise good high-density lipoprotein (HDL), cholesterol which can provide protection against certain cancers, heart diseases, arthritis, degenerative eye disease, and high blood pressure. These fatty acids are found in walnuts, canola oil, and flaxseeds but the best source is cold-water fish. A specific omega-3 fatty acid called eicosapentaenoic acid was shown to reduce expression of inflammatory genes in arthritic canine cells (Balk *et al.*, 2006; Bouwens *et al.*, 2009; Bahadori *et al.*, 2010).

Omega-3 fatty acids are highly concentrated in the brain and appear to be important for brain memory and performance and behavioural function. In fact, infants who do not get enough omega-3 fatty acids from their mothers during pregnancy are at risk for developing vision and nerve problems; symptoms of omega-3 fatty acid deficiency include fatigue, poor memory, dry skin, heart problems, mood swings or depression, and poor circulation. It is important to have the proper ratio of omega-3 and omega-6 in the diet because omega-3 fatty acids help reduce inflammation, and most omega-6 fatty acids tend to promote inflammation (Angerer and Von Schacky, 2000; Aronson *et al.*, 2001; Aben and Danckaerts, 2010).

Nutrigenomics shows blood pressure benefits of cocoa

A new nutrigenomics study shows that the potential of polyphenol compounds in cocoa to reduce blood pressure is related to genotype. Activity of the antiotensin-converting enzyme (ACE), a target for blood pressure medication, was significantly inhibited by dark chocolate containing 72% cocoa, with the degree of inhibition dependent upon the genotype of the human subjects. ACE inhibitors work by inhibiting the conversion of angiotensin-I to the potent vasoconstrictor, angiotensin-II, thereby improving blood flow and blood pressure (Daniells, 2011).

Nutrigenomics shows benefit of magnesium's metabolic actions

Magnesium may up- and down-regulate a number of genes linked to metabolism and shows favourable effects on certain metabolic pathways associated with changes in gene expression (Chacko *et al.*, 2011), and magnesium supplementation was associated with a decrease in levels of C-peptide, a marker of improved insulin sensitivity. The mineral was also linked to down-regulation of certain genes related to metabolic and inflammatory pathways. The report also says that in terms of gene expression, 24 genes were up-regulated and 36 genes were down-regulated in response to magnesium supplementation and some findings also indicated a systemic effect of magnesium supplementation gave measurable physiologic changes in the urinary proteome (Chacko *et al.*, 2011).

Nutrigenomics supports evidence for health benefits of anthocyanins

Anthocyanins, a large subgroup of flavonoids present in many vegetables and fruits, are safe and potent antioxidants. They exhibit diverse potential health benefits including cardioprotection, anti-atherosclerotic activity, anticancer, antidiabetic, and anti-inflammation properties (Konczak and Zhang, 2004). Anthocyanins can cross the blood–brain barrier and distribute in the central nervous system. The studies indicate that anthocyanins represent novel neuroprotective agents and may be beneficial in ameliorating ethanol neurotoxicity (Chen and Luo, 2010). Recently, it was demonstrated that anthocyanins, which are pigments widespread in the plant kingdom, have the potency for anti-obesity in mice and the enhanced adipocytokine secretion and adipocyte gene expression in adipocytes (Tsuda *et al.*, 2005).

Nutrigenomics could provide nutrition-relevant biomarkers

Changes to messenger RNA and the corresponding proteins control the transport of certain nutrients and metabolites in the biochemical pathway. Nutrigenomics could also provide a new set of biomarkers with relevance to nutrition (Van Der Werf *et al.*, 2006).

Benefits of nutrigenomics diet for skin

Many skin problems such as acne, eczema, psoriasis, dry skin and premature ageing of the skin are associated with diet, and inadequate nutrition substantially contributes to the deterioration of such skin conditions and vice versa; with a proper diet, the appearance and health of the skin can be significantly improved. Minimally processed fruits, vegetables, legumes, nuts and seeds, and fermented products from unpasteurized and not homogenized milk contain nutrients necessary for healthy skin, such as vitamins B and E and minerals such as calcium, magnesium, potassium, iron, copper and manganese for certain blood groups and genotypes, while the most advisable foods for all blood groups are flaxseed, almonds and walnuts. In the fruits, a great choice for all blood types is pineapple, blueberries, raspberries and cranberries. Turkey is the only generally available meat that is suitable for all blood types and genotypes. The leading way for beautiful and healthy skin are healthy diet, good lifestyle and products that are tailored to personal nutrigenomics diet profile (Subbiah, 2010).

Health economics of nutrigenomics in weight management

In a theoretical modelling study, Meshkin *et al.* (2008) sought to evaluate the health economics implications of a nutrigenomic product for weight loss for which they constructed a nutrigenomic economic model by linking the published study data related to the efficacy of a product and/or ingredients and validated clinical assessments that have already been tied to health economics data with data involving condition prevalence and overall cost of illness. In this theoretical model, the demonstration is that LG839 (DNA customized nutritional programme) variant positively reduces the cost of illness at the macroeconomic and microeconomic level based upon a cost-effectiveness and cost-benefit analysis, and has forecasted the prognostic health economic implications of a nutrigenomic intervention to demonstrate a theoretical model of nutrigenomic economics. This study is hypothesis-generating and should be used in the definition of protocols to prospectively test the health economic benefits of nutrigenomics.

Nutrigenetic association of the 5-lipoxygenase gene with myocardial infarction

5-Lipoxygenase (5-LO) catalyses the rate-limiting step of the biosynthesis of pro-inflammatory leucotrienes from arachidonic acid and has been associated with atherosclerosis in animal models and humans. Earlier reports stated that variants of a 5-LO promoter repeat polymorphism were associated with carotid atherosclerosis in humans, an effect that was exacerbated by high dietary amino acids but mitigated by high dietary N-3 fatty acids. The 5-LO polymorphism was genotyped by Costa Rican case-control pairs and tested for association with myocardial infarction and, currently, scientists are working with powerful databases to identify variations among genes in individuals and are working to establish correlations for susceptibility to various health conditions, as well as to understand the influence of such genetic variations on responses to dietary components (Allayee *et al.*, 2008).

20.1.4 Persons involved in nutrigenomics

Clinical pharmacologists, biostatisticians and clinicians need to give thoughtful consideration to the type and quantity of evidence to support dosing changes in clinical practice or approved labels intended to improve either the efficacy or safety of a nutrigenomic treatment (DeBusk *et al.*, 2005; German *et al.*, 2005; Afman and Muller, 2006; Trujillo *et al.*, 2006; Lesko, 2007; DeBusk, 2012).

Dieticians

The nutrigenomics practitioner will develop gene-directed nutrition approaches and coach people in how to use food, dietary supplements and lifestyle choices in general in ways that are most appropriate for their genetic makeup. Disease management is expected to become increasingly effective as nutritional genomics is integrated into practice, and even more eagerly anticipated is the opening up of new horizons for healthcare professionals in terms of expertise in health promotion while the ability to identify disease susceptibilities for an individual provides a solid foundation for effective health promotion efforts in ways never before possible (DeBusk, 2012). Two men of the same age eat a diet low in fruits and vegetables and high in sodium and saturated fat; one develops hypertension, hypercholesterolaemia, and eventually atherosclerosis, while the other lives a long life without such chronic disease. In another case, two post-menopausal women consume similar diets low in choline; one develops liver dysfunction due to the choline deficiency, but the other does not. However, because there are several genes involved in the development of these and other polygenic illnesses, dieticians and other healthcare professionals don't fully understand the relationship between diet and disease risk, which stifles our ability to make personalized dietary recommendations as a preventive measure (Baumler, 2012).

Epidemiologists

Epidemiological studies have been helpful in identifying environmental factors associated with incidence or severity of certain diseases. However, these are statistical associations and, as such, do not indicate the exact cause of the disease. Indeed, as the number of environmental variables increases, there is a corresponding need for larger population sizes in order to discriminate between statistically significant and insignificant factors (Malats and Calafell, 2003), so the meta-analysis may be helpful in this regard if studies record similar data elements and use similar environmental survey instruments for their populations. Alternatively, well-designed laboratory animal studies and comparative genomics will be helpful in confirming and extending associations between diet and disease (Baumler, 2012).

Molecular biologists

The diverse tissue- and organ-specific effects of bioactive dietary components include gene expression patterns organization of the chromatin, protein expression patterns including post-translational modifications as well as metabolite profiles (Corthésy-Theulaz *et al.*, 2005) are the pioneer functions of the molecular biologist.

Physicians

The physician with the help of nutrigenomics can see the blueprints and better understand the raw materials required by the body because incomplete or bad food causes toxic by-products that accelerate the ageing and disease processes and free radicals produced by non-specific foods and supplements wreak havoc on our body. In the past two decades, physicians, geneticists and nutritionists have begun to study the effects of genetic variation and gene–nutrient interactions in the management of chronic diseases, such as coronary heart disease, hypertension, cancer, diabetes and obesity; and the role of nutrients in gene expression (Simopoulos, 2002).

Geneticists

Advances in molecular and recombinant DNA technology have led to exquisite studies in the field of genetics and the recognition in a much more specific way, through DNA sequencing and the extent to which genetic variation occurs. The importance of the effects of genetic variation has been extensively studied and applied by pharmacologists in drug development and evaluation of drug metabolism and adverse reactions to drugs (Simopoulos, 2002).

Bioinformatic specialists

The role of bioinformatics in nutrigenomics is multifold, i.e. to create nutrigenomic databases, to set up special ontologies in using available resources, set up and track laboratory samples being tested and their results, pattern recognition, classification and data mining, simulation of complex interactions between genomes, nutrition and health disparities (Schaffer *et al.*, 2006).

Food scientists

Food scientists may use nutrigenomics to provide a balanced and healthy diet for a person and also apply the concept of personalized diet. Steps to increase the nutritional quality of individual foods will assist in personalizing health and in guiding individuals to achieve superior health (German *et al.*, 2011). Modern food scientists are also associated with screening of novel functional bioactives, safety evaluation of food ingredients, control of efficacy and spoilage of food with food processing.

20.1.5 Limitations of nutrigenomics

Nutrigenomics' risks include the knowledge of disease susceptibility may cause high levels of anxiety and stress, genetic testing raises privacy concerns and some companies already sell the results of their genetic profiling to other companies, while those with known genetic susceptibilities may be discriminated against in employment or health insurance. Physicians may not be qualified to interpret nutrigenomic reports and make appropriate decisions based on them, so the demand for nutrigenomic evaluations may eventually overtax the healthcare system. The high cost of the screening and genotype diagnosis of developing novel and functional foods and the poor availability of functional health systems make even the possibility of tailored diets an impossible dream for most populations relying on poorly functioning and poorly resourced health systems (Zeisel *et al.*, 2005). Dietetic practitioners arguably stand to gain the most by developing competency in nutritional genomics. They already have competency in nutrition and professional skills in patient counselling regarding diet and health. As with physicians, financial and other barriers limit comprehensive genetics training in dietetic education, and dietetic practitioners face similar pressures in daily practice that will slow uptake of genetics into their practice (Burton, 2003).

20.2 Technologies Involved in Nutrigenomics

20.2.1 Nutrigenetics

The study of genetic variations on the interaction between diet and health with implications to susceptible subgroups such as people with an enzyme deficiency caused by mutations in the enzyme phenylalanine hydroxylase cannot metabolize foods containing the amino acid phenylalanine and must modify their diets to minimize consumption (Ordovas and Mooser, 2004). This process has several phases that have grown into corresponding new fields within nutrigenetics: transcriptomics, proteinomics and metabolomics. This considers all metabolites in a human cell or organ, and is capable of generating large amounts of data at low cost that detects subtle differences in metabolism that contribute to obesity as well as fluctuations in weight (Mutch *et al.*, 2005).

20.2.2 Transcriptomics

Transcriptomics is the study of the complete set of RNA transcripts produced by the genome at a time, while transcriptome is the set of all RNA molecules, including mRNA, rRNA, tRNA, and other non-coding RNA produced in one or a population of cells (Hocquette *et al.*, 2009). The transcriptome is the total set of transcripts in a given organism, or the specific subset of transcripts present in a particular cell type, which can vary with external environmental conditions. Because it includes all mRNA transcripts in the cell, the transcriptome reflects the genes that are being actively expressed at any given time (Wang *et al.*, 2009). This technique is used for expression profiling, and examines the expression level of mRNAs in a given cell population, often using high-throughput techniques based on

DNA microarray technology. The use of next-generation sequencing technology to study the transcriptome at the nucleotide level is known as RNA-Seq (Gupta *et al.*, 2011).

The transcriptomes can be created by two methods: (i) maps sequence reads on to a reference genome of organism or related species; and (ii) de novo transcriptome assembly, which utilizes algorithms to build assembly software for generation of transcripts from short sequence reads. DNA microarrays can provide a method for comparing on a genome-wide basis the abundance of DNA in the same samples and DNA in spots can only be PCR products that are specific for individual genes. A DNA copy of RNA is made using the enzyme reverse transcriptase and sequencing is now being used instead of gene arrays to quantify DNA levels, at least semi-quantitatively (Katayama *et al.*, 2005).

For understanding of the molecular mechanisms and signalling pathways controlling early embryonic development, the analysis of the transcriptomes of human oocytes and embryos is used for proper embryo selection for *in vitro* fertilization (Subramanian *et al.*, 2005). The analysis of relative mRNA expression levels can be complicated by the fact that relatively small changes in mRNA expression can produce large changes in the total amount of the corresponding protein present in the cell, can be done by Gene Set Enrichment Analysis which identifies co-regulated gene networks rather than individual genes that are up- or down-regulated in different cell populations (Katayama *et al.*, 2005). The number of protein molecules synthesized using a given mRNA molecule as a template is highly dependent on translation-initiation features of the mRNA sequence and the ability of the translation initiation sequence is a key determinant in the recruiting of ribosomes for protein translation (Velculescu *et al.*, 1997).

20.2.3 Metabolomics

The quantitative measurement of the dynamic multi-parametric metabolic response of living systems to pathophysiological stimuli or genetic modification is known as metabolomics. The origin of the word is from the Greek *meta* meaning change and *nomos* meaning a rule set or set of laws (Nicholson, 2006) for the scientific study of chemical processes involving metabolites or systematic study of the unique chemical fingerprints that specific cellular processes leave behind and the study of their small-molecule metabolite profiles (Daviss, 2005). Metabolome represents the collection of all metabolites and end products of cellular processes in a biological cell, tissue, organ or organism (Jordan *et al.*, 2009).

Metabolites are the intermediates or end-products of metabolism and in the context of metabolomics, a metabolite is usually defined as any molecule less than 1 kDa in size (Samuelsson and Larsson, 2008). However, there are exceptions to this depending on the sample and detection method; macromolecules such as lipoproteins and albumin are reliably detected in NMR-based metabolomics studies of blood plasma (Nicholson *et al*, 1995). In human-based metabolomics, it is more common to describe metabolites as being either endogenous or exogenous (Nordstrom *et al.*, 2006). Metabolites of foreign substances such as drugs are termed xenometabolites (Crockford *et al.*, 2008) and metabolome forms a large network of metabolic reactions, where outputs from one enzymatic chemical reaction are inputs to other chemical reactions. Such systems have been described as hypercycles and are used for toxicity assessment/toxicology (Robertson, 2005). Metabolic profiling can be used to detect the physiological changes caused by toxic insult of a chemical and, in many cases, the observed changes can be related to specific syndromes such as a specific lesion in liver or kidney. This is of particular relevance to pharmaceutical companies wanting to test the toxicity of potential drug candidates and if a compound can be eliminated before it reaches clinical trials on the grounds of adverse toxicity, it saves the enormous expense of the trials (Saghatelian *et al.*, 2004; Chiang *et al.*, 2006). It can therefore be an excellent tool for determining the phenotype caused by a genetic manipulation such as gene deletion or insertion (Gibney *et al.*, 2005).

20.2.4 Proteomics

Proteomics is the large scale study of proteins, particularly their structures and functions (Anderson and Anderson, 1998; Blackstock and Weir, 1999), and can give better understanding of an organism: (i) the level of transcription of a gene gives only a rough estimate of its level of expression into a protein (Gygi *et al.*, 1999) and an mRNA produced in abundance may be degraded rapidly or translated inefficiently, resulting in a small amount of protein; (ii) many proteins experience post-translational modifications that profoundly affect their activities, for example some proteins are not active until they become phosphorylated and for study of post-translational modifications, phosphoproteomics and glycoproteomics methods are used; (iii) many transcripts give rise to more than one protein through alternative splicing or alternative post-translational modifications; (iv) many proteins form complexes with other proteins or RNA molecules; and (v) protein degradation rate plays an important role in protein content (Belle *et al.*, 2006).

The practical applications of proteomics are the identification of potential new drugs for the treatment of disease and this relies on genome and proteome information to identify proteins associated with a disease, which computer software can then use as targets for new drugs; e.g. if a certain protein is implicated in a disease and its 3D structure provides the information to design drugs to interfere with the action of the protein (Sreedhar *et al.*, 2011). Another use of proteomics is using specific protein biomarkers to diagnose disease and a number of techniques allow testing for proteins produced during a particular disease, which helps to diagnose the disease quickly by many techniques including western blot, immunehistochemical staining, enzyme-linked immunosorbent assay (ELISA) or mass spectrometry (Klopfleisc *et al.*, 2010).

Secretomics, a branch of proteomics, deals with studies of secretion pathways and secreted proteins using this proteomics approach, and is emerging as an important tool for the discovery of biomarkers of disease (Hathout, 2007). Proteomic technologies such as mass spectrometry are used for improving gene annotations and play an important role in drug discovery, diagnostics and molecular medicine, because it reveals the link between genes, proteins and disease. Advances in proteomics may help scientists eventually create medications that are 'personalized' for different individuals to be more effective and have fewer side effects (Lesko, 2007).

20.3 Nutrients Modulating Genome Expression

Numerous dietary components can alter genetic events in addition to the essential nutrients, such as carbohydrates, amino acids, fatty acids, calcium, zinc, selenium, folate, and vitamins A, C and E. There is a variety of non-essential bioactive components that seem to significantly influence health (Corthésy-Theulaz *et al.*, 2005; Trujillo *et al.*, 2006).

20.3.1 Effect of carbohydrate on gene expression

Glucose, the most abundant monosaccharide in nature, provides a very good example of how organisms have developed regulatory mechanisms to cope with a fluctuating level of nutrient supply (Vaulont *et al.*, 2000). In mammals the response to dietary glucose is complex because it combines effects related to glucose metabolism itself and effects secondary to glucose-dependent hormonal modifications, mainly pancreatic stimulation of insulin secretion and inhibition of glucagon secretion (Vaulont *et al.*, 2000). In the pancreatic cells, glucose is the primary physiological stimulus for the regulation of insulin. In the liver, glucose, in the presence of insulin, induces expression of genes encoding glucose transporters and glycolytic and lipogenic enzymes, e.g. L-type pyruvate kinase, acetyl-CoA carboxylase and fatty acid synthase, and represses genes of the gluconeogenic pathway, such as the phosphoenolpyruvate carboxykinase gene (King, 2012). Although insulin and glucagon were long known as critical in regulating gene expression, it is only recently that glucose also has been shown to play a key role in transcriptional regulation synthesis and secretion (Vaulont *et al.*, 2000).

Feeding high-energy diet to rats leads to early development of obesity and metabolic syndrome, apparently through an inability to cope with the energy density of the diet. Obesity is associated with decrease in mRNA levels for the oxygenic neuropeptides, neuropeptides Y, Agouti-related peptide etc., and the effect of hyperglycaemia on liver angiotensinogen gene expression has found that hyperglycaemia-activated AGT gene expression in liver increased approximately threefold (Gabriely *et al.*, 2001).

20.3.2 Regulation of gene expression by dietary fat

In addition to its role as an energy source and its effects on membrane lipid composition, dietary fat has profound effects on gene expression, leading to changes in metabolism, growth and cell differentiation. The effects of dietary fat on gene expression reflect an adaptive response to changes in the quantity and type of fat ingested (Jump and Clarke, 1999). In mammals, fatty acid regulated transcription factors include peroxisome proliferator activated receptors (PPARα, -β and -γ), HNF-4α, NF-κB and SREBP1c (Nagao and Yanagita, 2008). These factors are regulated by: (i) direct binding of fatty acids, fatty acyl coenzyme A, or oxidized fatty acids; (ii) oxidized fatty acid regulation of G-protein-linked cell surface receptors and activation of signalling cascades targeting the nucleus; or (iii) oxidized fatty acid regulation of intracellular calcium levels, which affect cell signalling cascades targeting the nucleus (Jump and Clarke, 1999).

20.3.3 Role of PUFA on gene expression

Lipogenic enzymes in liver decreased as result of feeding a diet containing 60% linoleic acid. Fatty acids stimulated the expression of adipocyte fatty acid binding protein (ap2) mRNA. In the 3T3-L1 adipocyte cell line, arachidonic acid (n-6) decreased SCD1 mRNA stability in a dose dependent manner (80% maximum repression), as did linoleic and eicosapentanoic acids (Tandon *et al.*, 2012).

20.3.4 Effect of protein on gene expression

Protein is essential for growth, to develop immunity, normal maintenance of body function and structure apart from reproduction and production, and in many developing countries protein insufficiency still remains a major and serious problem (Tandon *et al.*, 2012). The function of protein in the body is not only at a macro level but it also functions at gene level, and a variety or number of genes respond to dietary protein; both protein quantity as well as quality influences gene expression. One study showed that insulin secretion was reduced in rats, which are fed with low protein diet due to reduction in pancreatic beta cell mass, lower response of remaining beta cells to nutrients and lowered protein kinase activity (PKA) (Ferreira *et al.*, 2004), which is involved in potentiating of glucose-induced insulin secretion by gastrointestinal hormones such as glucagon-like peptide-1 and glucose-dependent insulinotropic polypeptide (Jacobo *et al.*, 2009). Feeding a low protein diet to rats altered the many genes expression, which are responsible for proteins related to insulin biosynthesis, secretion and cellular remodelling. Normal insulin secretion is influenced by level of protein kinase C, K^+ channel protein, calcium ion (Ca^{2+}) and PKAa. An increased ATP to ADP ratio achieved through glucose metabolism, closes the K+ ATP channel, which leads to depolarization of β-cells. This results in opening of voltage-dependent Ca^{2+} channels, which results in influx of calcium leading to exocytosis of insulin granules. Feeding a low protein diet and increased expression of PFK in islets results in defective glucose metabolism and it further leads to deceased glucose induced insulin secretion and decreases insulin level; it also acts through decreased movement of intracellular calcium (Tandon *et al.*, 2012).

20.3.5 Influence of amino acids on gene expression

The first step of protein translation is the formation of the 43s pre-initiation complex containing methionyl tRNA, eukaroyotic elongation factor 2 (eIF2), GTP, followed by the association of methionyl tRNA and eIF2–GTP that bind to the 40s ribosomal subunit. Then GTP is hydrolysed late in the initiation process, and eIF2 is released from the ribosome as an inactive eIF2–GTP complex. Formation of eIF2–GTP is mediated by the guanine nucleotide exchange factor eIF2B. The mechanism to regulate eIF2B activity may be at the level of the ribosomal protein S6 and eEF-2, which is phosphorylated in response to many agents, including growth factors and hormones. Amino acids regulate protein translation through modulation of eIF2B activity, 4 E–BP phosphorylation and protein S6 phosphorylation (Tandon *et al.*, 2012).

20.3.6 Effect of minerals on gene expression

Zinc (Zn) is an essential trace element with cofactor functions in a large number of proteins of intermediary metabolism, hormone secretion pathways and immune defence mechanism, and is involved in regulation of small intestinal, thymus and hepatocytes gene expression (Kindermann *et al.*, 2004). MTF-I (metal responsive element factor-I) is a Zn-dependent transcriptional activator that regulates metalothionin I and II through MRE. Zn-dependent KLF4 transcription factor is involved in protein preparation of HT-29 cells. The other proteins that have Zn as a constituent are ATP synathase, cytochrome-c, NADP dehydrogenase I and II. Deficiency of one or more minerals in the diet leads to impaired body functions. Geographical differences in mineral level of soil/plants, such as iron, iodine, selenium deficiency or excess of heavy metal ions, have effects up to gene level such as anaemia (Wu and Wu, 1987; Vallee and Auld, 1990).

20.3.7 Effect of vitamins on gene expression

Vitamins are micronutrients needed in very small quantity and are involved in gene expression. Vitamin A is involved in gene expression of phospho-enol-pyruvate-kinase (PEPCK) and insulin-like growth factor (IGF-9) (Tandon *et al.*, 2012). Vitamin C is involved in hepatic gene expression. PEPCK is involved in conversion of oxaloacetate to phospho-enol-pyruvate, one of the important steps in gluconeogenesis. Vitamin A deficiency condition leads to changes in chromosomal structure of retinoic acid responsive element (RARE), which further leads to change in co-regulator binding and activity. PEPCK-RARE and pre-initiation complex interaction leads to RNA polymerase-II association with PEPCK promoter being reduced due to vitamin A deficiency and finally results in insufficient PEPCK or no PEPCK, which leads to improvement of gluconeogenesis. In vitamin A-sufficient mice PEPCK gene expression is highly induced in the food-deprived state, when blood glucose levels are reduced (Tandon *et al.*, 2012).

Biotin is involved in various essential proteins' or enzymes' synthesis at gene level (Dakshinamurti, 2005). Vitamins B_{12}, B_6 and folic acid converge at the homocysteine metabolic junction where they support the activities of two key enzymes involved in intracellular homocysteine management, methionine synthase (MS) and cystathionine-β-synthase. B_{12} supplementation does not alter mRNA or protein turnover rates but induces translational up-regulation of MS by shifting the mRNA from the ribonucleoprotein to the polysome pool. The B_{12}-responsive element has been localized by deletion analysis using a reporter gene assay to a 70 bp region located at the 3′ end of the 5′-untranslated region of the MS mRNA. The cellular consequence of the B_{12} response is a 2- and 3.5-fold increase in the flux of homocysteine through the MS-dependent transmethylation pathway in HepG2 and 293 cells, respectively (Oltean and Banerjee, 2003).

20.4 Nutrition Gene and Diseases

20.4.1 Metabolic hereditary diseases

Some hereditary disorders of metabolism can be diagnosed in the fetus by using amniocentesis or chorionic villus sampling and blood test or examination of a tissue sample to determine whether a specific enzyme is deficient or missing (Sanders, 2009).

In most inherited metabolic disorders, a single enzyme is either not produced by the body at all, or is produced in a non-working form. Depending on the function of that enzyme, toxic chemicals may build up, or an essential product may not be produced. The code or blueprint to produce an enzyme is usually contained on a pair of genes and most people with inherited metabolic disorders inherit two defective copies of the gene, one from each parent. Both parents are carriers of the bad gene, meaning they carry one defective copy and one normal copy. Inherited metabolic disorders may affect about 1 in 1000 to 2500 newborns. The symptoms of genetic metabolic disorders vary widely depending on the metabolism problem present, including lethargy, poor appetite, abdominal pain, vomiting, weight loss, jaundice, failure to gain weight or grow, developmental delay, seizures, coma, abnormal odour of urine, breath, sweat or saliva. Symptoms may be brought on by foods, medications, dehydration, minor illnesses, or other factors. Hundreds of inherited metabolic disorders have been identified, and new ones continue to be discovered. Some of the more common and important genetic metabolic disorders (Scriver, 2001) include lysosomal storage disorders like Hurler syndrome, Niemann-Pick disease, Tay-Sachs disease, Gaucher disease, Fabry disease, Krabbe disease, galactosaemia, maple syrup urine disease in which deficiency of an enzyme called branched-chain alpha-keto acid dehydrogenase causes build up of amino acids in the body and the urine smells like syrup (Muranjan and Agarwal, 2010), phenylketonuria, where deficiency of the enzyme PAH results in high levels of phenylalanine in the blood (Mitchell and Scriver, 2010), glycogen storage diseases, Friedreich ataxia, i.e. problems related to a protein called frataxin causing nerve damage and often heart problems (Hasan, 2007).

Peroxisomal disorders include Zellweger syndrome (abnormal facial features, enlarged liver and nerve damage in infants) and adrenoleucodystrophy, metal metabolism disorders such as Wilson disease (toxic copper levels accumulate in the liver, brain and other organs) and haemochromatosis (the intestines absorb excessive iron, which builds up in the liver, pancreas, joints and heart, causing damage), organic acidemias, urea cycle disorders including ornithine transcarbamylase deficiency and citrullinemia are a few examples of metabolic hereditary disorders (Hasan, 2007) and inborn errors of metabolism referred as 'silent killers' because they can strike healthy-appearing full-term infants without warning and display hypoglycaemia or poor feeding (Enns, 2005).

Failures of energy production or utilization result from defects in the liver, myocardium, muscle, or brain and disrupt cytoplasmic or mitochondrial energy production. These include the fatty acid oxidation disorders and the congenital lactic acidemias, which present with a variety of findings, but a consistent symptom is hypoglycaemia with clinical features such as lactic acidosis, hypotonia and cardiac involvement (Saudubray *et al.*, 2002). Hypoglycaemia related to fasting can signal a fatty acid oxidation disorder, while hypoglycaemia following eating is characteristic of hereditary fructose intolerance (Garganta and Smith, 2005).

20.4.2 Multifactorial diseases

A multifactorial disease has a combination of distinctive characteristics that can be differentiated from clear-cut Mendelian or sex-limited conditions; the disease can also occur in isolation and to affected children born to unaffected parents. Although familial aggregation is also common, there is no clear Mendelian pattern of inheritance. Environmental influences can increase or decrease the risk of the disease; the disease occurs more frequently in one gender than in the other, but it is not a sex-limited trait. In addition, first-degree relatives of individuals belonging to the more rarely affected gender have a higher risk of bearing the disease. The concordance rate is a measure of the rate at which both twins bear a specific disease. The disease occurs more frequently in a specific ethnic group (i.e. Caucasians, Africans, Asians, Hispanics, etc.) (Lobo, 2008) in monozygotic and dizygotic twins contradicting Mendelian proportions. On a pedigree (genealogical table), polygenic diseases do tend to run in families, but the inheritance does not fit simple patterns as with Mendelian diseases (Burmeister, 1999).

20.4.3 Monogenic and multigenic diseases

Monogenic diseases result from modifications in a single gene occurring in all cells of the body and they affect millions of people worldwide. Scientists currently estimate that over 10,000 human diseases are monogenic (Ikonen, 2006) and according to WHO, the global prevalence of all single gene or monogenic diseases at birth (dominant, recessive and x-linked) is approximately 10/1000 (WHO, 2012). Thalassaemia is a blood-related genetic disorder which involves the absence of or errors in genes responsible for production of haemoglobin, a protein present in the red blood cells. Sickle-cell anaemia is a blood-related disorder that affects the haemoglobin molecule, and causes the entire blood cell to change shape under stressed conditions (Weatherall, 2000). Haemophilia is a hereditary bleeding disorder, in which there is a partial or total lack of an essential blood clotting factor, which results in excessive bleeding and frequent spontaneous bleeding. Haemophilia A is the most common form, referred to as classical haemophilia, and results from the deficiency in clotting factor 8, while haemophilia B is a deficiency in clotting factor 9, a sex-linked recessive disorder (WHO, 2012).

Cystic fibrosis is a genetic disorder that affects the respiratory, digestive and reproductive systems and involves the production of abnormally thick mucous linings in the lungs and can lead to fatal lung infections resulting in various obstructions of the pancreas, hindering digestion (WHO, 2012). Another example is Tay-Sachs disease; a fatal genetic disorder in which harmful quantities of a fatty substance called ganglioside GM2 accumulate in the nerve cells in the brain (WHO, 2012) caused by a decrease in the functioning of the hexosaminidase-A enzyme. Fragile X syndrome is caused by a 'fragile' site at the end of the long arm of the X-chromosome, and is a genetic disorder that manifests itself through a complex range of behavioural and cognitive phenotypes (McMillan, 2006).

20.5 Nutrigenomics and Communication

Nutrient–gene interactions are responsible for maintaining health and preventing or delaying disease. Unbalanced diets for a given genotype lead to chronic diseases such as obesity, diabetes and cardiovascular disease, and are likely to contribute to increased severity and/or early-onset of many age-related diseases. Many nutrition and many genetic studies still fail to properly include both variables in the design, execution and analyses of human, laboratory animal, or cell-culture experiments (Kaput *et al.*, 2006). The complexity of nutrient–gene interactions has led to the realization that strategic international alliances are needed to improve the completeness of nutrigenomic studies, a task beyond the capabilities of a single laboratory team. Eighty-eight researchers from 22 countries recently outlined the issues and challenges for harnessing nutritional genomics for public and personal health. The next step in the process of forming productive international alliances is the development of a virtual centre for organizing collaborations and communications that foster resources sharing, best practices improvement and creation of databases. There is a requirement for a nutrigenomics information portal, a web-based resource for the international nutrigenomics society, and this portal aims at becoming the prime source of information and interaction for nutrigenomics scientists through a collaborative effort (Kaput *et al.*, 2006).

20.6 Nutrigenomics and Bioactive Nutrients

20.6.1 Elk antler velvet

Elk antler velvet (EAV) is the fast-growing, soft cartilaginous tissue that develops out of the frontal bone of *Cervus* species, which rises from skin-covered pedicles before it calicifies and hardens. Antlers are unique in nature and different from horns because they are naturally re-grown and cut off each year. Elk antler velvet, pumped tight with blood and pulsing with hormones, is the most regenerative mammal tissue known, capable of growing over half an inch in 1 day. The active ingredients have been found to include a variety of minerals, proteins, collagens, fatty acids and glycosaminoglycans in varying concentrations with the effects of increased growth (Ko *et al.*, 1986), improved immunity (Suh *et al.*, 1999), antifungal (Park *et al.*, 1998), cardiovascular effects (Clifford *et al.*, 1979), promotion of rapid healing in tissues and bones, relief of symptoms in arthritis and gout, and pain reduction associated with disease or injury to muscles and joints. It is an excellent source of chondronitin sulfate, glucosamine sulfate, type-II collagen and prostaglandins, and has benefits for a free-radical scavenger (Wang *et al.*, 1988), arthritis, antiulcer activity (Wang *et al.*, 1985), anti-infective (Dai *et al.*, 2011), reduces inflammation (Shin *et al.*, 1989), antinarcotic addiction activity (Kim and Lim, 1999) and anti-ageing properties (Chen *et al.*, 1992).

20.6.2 Vegan chyawanprash

India's most famous anti-ageing recipe is chyawanprash, and according to Ayurveda, chyawanprash comes under the category of rasayana (Jose and Kuttan, 2000) used for maintaining youthfulness (Manjunatha *et al.*, 2001), vigour, vitality of the body, keeping away ageing processes, senility and debility, maintains the proper functioning of the cells, rejuvenates the cells and also keeps away diseases (Jose and Kuttan, 2000). This Ayurvedic tonic consisting of about 35 natural herbs including amla (*Embellica officinalis*), the richest natural source of vitamin C, works on the immune system of the body protecting the body against everyday infections such as cough, cold and fever and is hepatoprotective (Jose and Kuttan, 2000) and thus very useful in children, old persons, tubercular patients, bidi smokers (Yadav *et al.*, 2003) and debilitated persons.

20.6.3 Mangosteen

Mangosteen is cultivated in Thailand under the most stringent conditions for this amazing superfood. It contains a class of naturally occurring polyphenolic compounds known as xanthones, which provide beneficial effects on cardiovascular diseases, including ischaemic heart disease, atherosclerosis and hypertension (Lourith and Kanlayavattanukal, 2011), anti-invasive activities (Wang *et al.*, 2012a) and thrombosis (Chin *et al.*, 2011). Xanthones have unique antioxidant properties (Martinez *et al.*, 2011), which help to heal cells damaged by free radicals (Robb-Nicholson, 2012), slow ageing (Ngawhirunpat *et al.*, 2010) and physical (Ryu *et al.*, 2012) and mental deterioration (Robb-Nicholson, 2012). The rind of partially ripe mangosteen fruit yields a polyhydroxy-xanthone derivative termed mangostin and β-mangostin while fully ripe fruits contain the xanthones gartanin, β-disoxygartanin and normangostin, beneficial in various serious ailments such as antifatigue, anti-obesity, antidiabetic (Ryu *et al.*, 2011), anti-anxiety (Shiozaki *et al.*, 2012), antitumour (Kosem *et al.*, 2012), antiseborrheic (Wang *et al.*, 2012b), antiglaucoma, antipyretic, anthelminthic (Keiser *et al.*, 2012), antineuralgia (Reyes-Fermín *et al.*, 2012), antiarthritis, anti-inflammatory (Jang *et al.*, 2012; Liu *et al.*, 2012), anti-ulcer and anticancer (Chang and Yang, 2012; Robb-Nicholson, 2012). Mangosteen also shows inhibitory action against *Mycobacterium tuberculosis* and *Staphylococcus aureus* (Koh *et al.*, 2012) in addition to its antibacterial (Temrangsee *et al.*, 2011), strong antifungal properties and is effective in boosting weak immune systems.

20.6.4 Kaunch

Mucuna prurita Baker (*Fabaceae*), Kaunch (seed) contains alkaloidal contents such as mucuadine, mucuadinine, mucucuadinine, pruriendine, mucunine, mucunadine and nicotine (Saksena and Dixit, 1987) and is used as an Ayurvedic medicine that increases testosterone, libido, reduces spasms, lowers blood sugar, lowers blood pressure, increases urination, relieves pain, reduces inflammation, kills parasites (Meena *et al.*, 2009), calms nerves, reduces fever and lowers cholesterol. It is also used as an aphrodisiac, spermatogenetic (Saksena and Dixit, 1987), androgenic, retentive, L-dopa alternative, menstrual promoter, uterine stimulant, nerve tonic, anti-Parkinson's, (Meena *et al.*, 2009), hypoglycaemic, anabolic etc. (Agrawal *et al.*, 2010) and also produces an antidepressant effect in patients suffering from depressive neurosis. Due to the high concentration of L-dopa in the seeds, it is considered as an alternative to the pharmaceutical medication levodopa in Parkinson's disease; it also has reported anabolic and growth hormone stimulant properties (Agrawal *et al.*, 2010).

20.6.5 Blue lotus flowers

Egyptian medicinal practitioners used this flower to stimulate blood flow, and as an anti-ageing treatment while traditionally it was used to relieve pain, increase memory, increase circulation, promote sexual desire, and create feelings of well-being, euphoria and ecstasy, without the use of narcotics (Emboden, 1981). Approximately 20 phytoconstituents are isolated from Blue lotus (*Nymphaea caerulea*) flowers, for example 2S, 3S, 4S-trihydroxypentanoic acid, and myricetin 3-O-(3′-O-acetyl)-α-L-rhamnoside, along with the known myricetin 3-O-α-L-rhamnoside, myricetin 3-O-β-D-glucoside, quercetin 3-O-(3′-O-acetyl)-α-L-rhamnoside, quercetin 3-O-α-L-rhamnoside, quercetin 3-O-β-D-glucoside, kaempferol 3-O-(3′-O-acetyl)-α-L-rhamnoside, kaempferol 3-O-β-D-glucoside, naringenin, (S)-naringenin 5-O-β-D-glucoside, isosalipurposide, β-sitosterol, β-sitosterol palmitate, 24-methylenecholesterol palmitate, 4 α-methyl-5-α-ergosta-7,24-diene-3β, 4-β-diol, ethyl gallate, gallic acid, p-cumaric acid and 4-methoxybenzoic acid, used as a hypnotic, sedative, euphoric and antispasmodic and also produces an opiate-like intoxication with antioxidant activity (Agnihotri *et al.*, 2008).

20.6.6 Shilajit

Shilajit is a thick rich paste oozing out from the rocks (Agarwal *et al.*, 2007) in the towering cliffs in the Himalayan mountains (Ghosal, 1990), and was used historically for general physical strengthening, anti-ageing (Gaikwad *et al.*, 2012), libido, injury healing, urinary tract rejuvenation, enhanced brain functioning potency, bone healing, kidney rejuvenation, immune system strengthening (Ghosal, 1990), arthritis, hypertension (Gaikwad *et al.*, 2012) and obesity and has unmatched powers of arresting and reversing the ageing process. Shilajit is spermatogenic and ovogenic (Park *et al.*, 2006), and also counteracts diabetes and regulates the blood sugar level and purifies blood (Sharma *et al.*, 2003) and improves functioning of the pancreas, strengthens digestion and promotes the movement of minerals, especially calcium, phosphorus and magnesium, into muscle tissue and bone. It also stimulates the immune system (Ghosal, 1990) and improves restoration after exercise, so counteracting debility and general fatigue (Wilson *et al.*, 2011).

20.6.7 Folate

A gene variant is responsible for increasing homocysteine levels in some people, subsequently leading to a higher risk of cardiovascular diseases and certain cancers. Folate, however, helps to negate this risk. Therefore, people with this identified gene variant are encouraged to consume plenty of folate-rich foods (Pfeiffer *et al.*, 2008; Yang *et al.*, 2008).

20.6.8 Green tea

Green tea is used in Crohn's disease (Alic, 1999), on thermogenesis and energy intake (Belza *et al.*, 2009), human prostate cancer (Bettuzzi *et al.*, 2006), gastrointestinal cancer (Borrelli *et al.*, 2004), skin problems (Katiyar *et al.*, 2000), on weight maintenance after body-weight loss (Kovacs *et al.*, 2004); it reduces body fat and cardiovascular risks (Nagao *et al.*, 2007) and helps prevent breast cancer (Inoue *et al.*, 2001).

20.6.9 Turmeric

Turmeric is used as anti-inflammatory (Arora *et al.*, 1971), for management of neurodegenerative disease (Auddy *et al.*, 2003), on lipid profile (Desphande *et al.*, 1997), cancer chemoprevention (Gescher *et al.*, 2001), specific inhibition of cyclooxygenase-2 (COX-2) expression by dietary curcumin in HT-29 human colon cancer cells (Goel *et al.*, 2001) and inhibition of HIV-1 and HIV-2 proteases (Sui *et al.*, 1993) with antidepressant activity (Yu *et al.*, 2002). Turmeric suppresses a gene that makes inflammatory properties, which is possibly useful in preventing colon cancer and Alzheimer's disease.

20.6.10 Vitamin D

Vitamin D is the sunshine vitamin, synthesized in our skin during sun exposure; most relevant dietary sources of vitamin D are fatty fish and full-fat milk (Boullata, 2012). Most of the vitamin D in blood (80–90%) is bound to α-globulin and transported all over the body to the target organs. Variations in two more genes (*DHCR7* and *CYP2R1*) have been confirmed for Caucasians and both genes encode key enzymes in the vitamin D metabolism pathway (Kulie *et al.*, 2009). Vitamin D mediates expression of human cathelicidin antimicrobial peptide in bronchial epithelial cells (Schrumpf *et al.*, 2012) and modifies the susceptibility to schizophrenia bipolar mood disorder by regulation of dopamine D1 receptor gene expression (Ahmadi *et al.*, 2012). It is also helpful in the management of Alzheimer's disease (Annweiler *et al.*, 2012), multiple sclerosis (Bartosik-Psujek *et al.*, 2010; Holmoy *et al.*, 2012), blood pressure (Caro *et al.*, 2012), common cold (Linder, 2012) and also prevents bone fractures (Paterson, 2012) and osteoporosis (Lakatos, 2011; Curtis and Safford, 2012). Pharmacological studies show that vitamin D prevents progression of peritoneal fibrosis (Hirose *et al.*, 2012), vitiligo (Colucci *et al.*, 2012), asthma (Iqbal and Freishtat, 2011) and polyarthritis (Moghaddami *et al.*, 2012) and enhances the action of the parathyroid gland (Bienaime *et al.*, 2011). It is also useful in treatment of cancer, inflammation (Krishnan and Feldman, 2012) and kidney diseases (Eleftheriadis *et al.*, 2010) and plays a major role in reducing insulin resistance (Teegarden and Donkin, 2009).

20.7 Ethical Considerations in Nutrigenomics

Nutrigenomics lies at the intersection of several fields in which ethical, legal and social issues (ELSIs) arise, such as human genomics and genetics, the molecular nutritional sciences, dietary supplement research and development, medicine and dietetics. As each of these fields progresses, periods of regulatory uncertainty are often encountered and emerging ELSIs must be identified and addressed. With growing investments in human genomics research, much attention has focused on ethical and legal protections for research subjects, including concerns about study design, the collection, use, retention and exchange of biological samples and personal information, the involvement of children or other vulnerable groups, and the obligations of researchers to report results to research subjects and their family members (Castle and Ries, 2007). The 'biohype' around the nutrigenomics foods and nutrigenetics tests is imminent and unavoidable, particularly in the early stages of evolution of a new idea. 'Biohype' is already taking place through the aggressive marketing of nutrigenomics tests to the public, which many consider to be premature, raising concern over ELSIs. Five areas have been identified by international experts in the context of both basic nutrigenomics research and its clinical and commercial uses: (i) health claims benefits arising from nutrigenomics; (ii) managing nutrigenomics information; (iii) delivery methods of nutrigenomics services; (iv) nutrigenomics products; and (v) equitable accessibility to nutrigenomics. Hence it is important to elevate the depth of debate to understand and manage the gap between expectations associated with diet–gene interactions using evidence-based research methodologies (Ghosh, 2009).

Personalized nutrition, i.e. tailoring diet on the basis of genotype, is one possible application of nutrigenomics research. However, until the scientific evidence concerning diet–gene interactions is much more robust, the provision of personalized dietary advice on the basis of specific genotype remains questionable. From the ethical and social perspective, nutrigenomics offers significant opportunities to improve public health by enhancing understanding of the mechanisms through which diet can be used to reduce the risk of common polygenic diseases (Bergmann *et al.*, 2008).

Nutrigenomics is thus surrounded by internal and external uncertainties and it is for end-users to better prepare themselves for a genomics future with fundamental uncertainties than to expect that in the long run the uncertainties will disappear. The issue of uncertainties of genomics has been tackled earlier, e.g. from the perspective of policy making, science management and theory of science. Here we will concentrate on the impact of the uncertainties of nutrigenomics on citizens/consumers by exploring the fruitfulness of an ethical perspective that focuses on how citizens/consumers cope with fundamental uncertainties (Korthals and Komduur, 2010). In terms of global health, the complexity of potential prevention measures based on nutrigenomics knowledge but also ethical issues relating to social justice and to the risks of stigmatization and discrimination are major challenges on which this chapter focuses. While such issues are not unique to nutrigenomics, they appear with a particular strength when it comes to assess the promises of this new field of science (Godard and Hurlimann, 2009).

20.8 Market Implications of Nutrigenomics

Nutrigenomics may be used for people with specific issues such as obesity, diabetes, cardiovascular disease and cancer, and will need medical foods and dietary advice tailored to their genetic profile and require the foods that reduce the risk of allergy and intolerance or new foods to re-set nutrition switches. Personalized functional foods, wellness and performance foods that enhance normal physiological processes, sports performance and mood foods manufacturing is the real implication of nutrigenomics (Ferguson, 2012). Nutrigenomics may be used for *in vitro* screening for new functional food bioactives, quality and authenticity of food processing in microbials, production of food ingredients from microbials and biomarker discovery for humans with genotyping of humans, safety testing in animals and efficacy testing in animals.

References

Aben, A. and Danckaerts, M. (2010) Omega-3 and omega-6 fatty acids in the treatment of children and adolescents with ADHD. *Tijdschrift Voor Psychiatrie* 52(2), 89–97.

Afman, L. and Muller, M. (2006) Nutrigenomics: from molecular nutrition to prevention of disease. *Journal of the American Diabetic Association* 106, 569–576.

Agarwal, S.P., Khanna, R., Karmarkar, R., Anwer, M.K. and Khar, R.K. (2007) Shilajit: a review. *Phytotherapy Research* 21(5), 401–405.

Agnihotri, V.K., Elsohly, H.N., Khan, S.I., Smillie, T.J., Khan, I.A. and Walker, L.A. (2008) Antioxidant constituents of *Nymphaea caerulea* flowers. *Phytochemistry* 69(10), 2061–2066.

Agrawal, A., Agrawal, M. and Rathore, A. (2010) Traditional remedy, Kunch Pak – a review. *International Journal of Pharma and Bio Sciences* 1(3), 1–6.

Ahmadi, S., Mirzaei, K., Hossein-Nezhad, A. and Shariati, G. (2012) Vitamin D receptor FokI genotype may modify the susceptibility to schizophrenia and bipolar mood disorder by regulation of dopamine D1 receptor gene expression. *Minerva Medica* 103(5), 383–391.

Alic, M. (1999) Green tea for remission maintenance in Crohn's disease? *The American Journal of Gastroenterology* 94(6), 1710–1711.

Allayee, H., Baylin, A., Hartiala, J., Wijesuriya, H., Mehrabian, M., Lusis, A.J. and Campos, H. (2008) Nutrigenetic association of the 5-lipoxygenase gene with myocardial infarction. *The American Journal of Clinical Nutrition* 88(4), 934–940.

Anderson, N.L. and Anderson, N.G. (1998) Proteome and proteomics: new technologies, new concepts, and new words. *Electrophoresis* 19(11), 1853–1861.

Angerer, P. and Von Schacky, C. (2000) N-3 polyunsaturated fatty acids and the cardiovascular system. *Current Opinion Lipidology* 11(1), 57–63.

Annweiler, C., Llewellyn, D.J. and Beauchet, O. (2013) Low serum Vitamin D concentrations in Alzheimer's disease: a systematic review and meta-analysis. *Journal of Alzheimer's Disease,* 33(3), 659–674.

Aronson, W.J., Glaspy, J.A., Reddy, S.T., Reese, D., Heber, D. and Bagga, D. (2001) Modulation of omega-3/omega-6 polyunsaturated ratios with dietary fish oils in men with prostate cancer. *Urology* 58(2), 283–288.

Arora, R.B., Basu, N., Kapoor, V. and Jain, A. (1971) Anti-inflammatory studies of *Curcuma longa. The Indian Journal of Medical Research* 59, 1289–1295.

Auddy, B., Ferreira, M., Blasina, F., Lafon, L., Arredondo, F., Dajas, F., Tripathi, P.C., Seal, T. and Murkerjee, B. (2003) Screening of antioxidant activity of three Indian medicinal plants, traditionally used for the management of neurodegenerative diseases. *Journal of Ethnopharmacology* 84, 131–138.

Bahadori, B., Uitz, E., Thonhofer, R., Trummer, M., Pestemer-Lach, I., McCarty, M. and Krejs, G.J. (2010) Omega-3 Fatty acids infusions as adjuvant therapy in rheumatoid arthritis. *Journal of Parenteral and Enteral Nutrition* 34(2), 151–155.

Balammal, G. and Jayachandra Reddy, P. (2012) Proteomics & genomics – a review. *Journal of Science (Biology)* 2(2), 91–95.

Balk, E.M., Lichtenstein, A.H., Chung, M., Kupelnick, B., Chew, P. and Lau, J. (2006) Effects of omega-3 fatty acids on serum markers of cardiovascular disease risk: a systematic review. *Atherosclerosis* 189(1), 19–30.

Bartosik-Psujek, H., Tabarkiewicz, J., Pocinska, K., Stelmasiak, Z. and Rolinski, J. (2010) Immunomodulatory effects of vitamin D on monocyte-derived dendritic cells in multiple sclerosis. *Multiple Sclerosis Journal* 16(12), 1513–1516.

Baumler, M.D. (2012) Nutrigenetics – building a platform for dietitians to offer personalized nutrition. *Today's Dietitian* 14(9), 48–52.

Belle, A., Tanay, A., Bitincka, L., Shamir, R. and O'Shea, E.K. (2006) Quantification of protein half-lives in the budding yeast proteome. *Proceedings of the National Academy of Sciences* 103(35), 13004–13009.

Belza, A., Toubro, S. and Astrup, A. (2009) The effect of caffeine, green tea and tyrosine on thermogenesis and energy intake. *European Journal of Clinical Nutrition* 63(1), 57–64.

Bergmann, M.M., Gorman, U. and Mathers, J.C. (2008) Bioethical considerations for human nutrigenomics. *Annual Review of Nutrition* 28, 447–467.

Bettuzzi, S., Brausi, M., Rizzi, F., Castagnetti, G., Peracchia, G. and Corti, A. (2006) Chemoprevention of human prostate cancer by oral administration of green tea catechins in volunteers with high-grade prostate intraepithelial neoplasia: a preliminary report from a one-year proof-of-principle study. *Cancer Research* 66(2), 1234–1240.

Bienaime, F., Prie, D., Friedlander, G. and Souberbielle, J.C. (2011) Vitamin D metabolism and activity in the parathyroid gland. *Molecular and Cellular Endocrinology* 347(1–2), 30–41.

Blackstock, W.P. and Weir, M.P. (1999) Proteomics: quantitative and physical mapping of cellular proteins. *Trends in Biotechnology* 17(3), 121–127.

Borrelli, F., Capasso, R., Russo, A. and Ernst, E. (2004) Systematic review: green tea and gastrointestinal cancer risk. *Alimentary Pharmacology & Therapeutics* 19(5), 497–510.

Boullata, J.I. (2012) A rational approach to vitamin D supplementation. *Nutrition* 28(11–12), 1204–1205.

Bouwens, M., Van De Rest, O., Dellschaft, N., Grootte Bromhaar, M., De Groot, L.C.P.G.M., Geleijnse, J.M., Muller, M. and Afman, L.A. (2009) Fish-oil supplementation induces antiinflammatory gene expression profiles in human blood mononuclear cells. *The American Journal of Clinical Nutrition* 90, 415–424.

Brandenberg, O., Dhlamini, Z., Sensi, A., Ghosh, K. and Sonnino, A. (2011) *Introduction to Molecular Biology and Genetic Engineering*. Food and Agriculture Organization of the United Nations, Rome.

Burmeister, M. (1999) Basic concepts in the study of diseases with complex genetics. *Biological Psychiatry* 45, 522–532.

Burton, H. (2003) Dietitians education: workshop report. The Welcome Trust and the Cambridge Public Health Genetics Unit, London, pp. 1–9.

Busstra, M.C., Hartog, R., Kersten, S. and Muller, M. (2007) Design guidelines for the development of digital nutrigenomics learning material for heterogeneous target groups. *Advances in Physiology Education* 31(1), 67–75.

Caro, Y., Negron, V. and Palacios, C. (2012) Association between vitamin D levels and blood pressure in a group of Puerto Ricans. *Puerto Rico Health Sciences Journal* 31(3), 123–129.

Castle, D. and Ries, N.M. (2007) Ethical, legal and social issues in nutrigenomics: the challenges of regulating service delivery and building health professional capacity. *Mutation Research* 622, 138–143.

Caulfield, T., Shelly, J., Alfenso, V. and Bubela, T. (2008) Nutrigenomics and the promise of prevention; representation and realities. *Health Law Journal* (Sp. edn) 41, 41–66.

Chacko, S.A., Sul, J., Song, Y., Li, X., LeBlanc, J., You, Y., Butch, A. and Liu, S. (2011) Magnesium supplementation, metabolic and inflammatory markers, and global genomic and proteomic profiling: a randomized, double-blind, controlled, crossover trial in overweight individuals. *The American Journal of Clinical Nutrition* 93(2), 463–473.

Chang, H.F. and Yang, L.L. (2012) Gamma-mangostin, a micronutrient of mangosteen fruit, induces apoptosis in human colon cancer cells. *Molecules* 17(7), 8010–8021.

Chen, G. and Luo, J. (2010) Anthocyanins: are they beneficial in treating ethanol neurotoxicity? *Neurotoxicity Research* 17(1), 91–101.

Chen, X., Jia, Y. and Wang, B.X. (1992) Inhibitory effect of the extract of pilose antler on monoamine oxidase of aged mice. *Journal of Chinese Materia Me* 17(2), 107–110.

Chiang, K.P., Niessen, S., Saghatelian, A. and Cravatt, B.F. (2006) An enzyme that regulates ether lipid signaling pathways in cancer annotated by multidimensional profiling. *Chemistry & Biology* 13(10), 1041–1050.

Chin, Y.W., Shin, E., Hwang, B.Y. and Lee, M.K. (2011) Antifibrotic constituents from *Garcinia mangostana*. *Natural Product Communications* 6(9), 1267–1268.

Clifford, D.H., Lee, M.O., Kim, C.Y. and Lee, D.C. (1979) Can an extract of deer antlers alter cardiovascular dynamics? *American Journal of Chinese Medicine* 7(4), 345–350.

Colucci, R., Lotti, T. and Moretti, S. (2012) Vitiligo: an update on current pharmacotherapy and future directions. *Expert Opinion on Pharmacotherapy* 13(13), 1885–1899.

Corthésy-Theulaz, I., Den Dunnen, J.T., Ferre, P., Geurts, J.M., Muller, M., Van Belzen, N. and Van Ommen, B. (2005) Nutrigenomics: The impact of biomics technology on nutrition research. *Annals of Nutrition and Metabolism* 49, 355–365.

Crockford, D.J., Maher, A.D., Ahmadi, K.R., Barrett, A., Plumb, R.S., Wilson, I.D. and Nicholson, J.K. (2008) 1H NMR and UPLC-MS (E) statistical heterospectroscopy: characterization of drug metabolites (xenometabolome) in epidemiological studies. *Analytical Chemistry* 80(18), 6835–6844.

Curtis, J.R. and Safford, M.M. (2012) Management of osteoporosis among the elderly with other chronic medical conditions. *Drugs & Aging* 29(7), 549–564.

Dai, T.-Y., Wang, C.-H., Chen, K.-N., Huang, I.-N., Hong, W.-S., Wang, S.-Y., Chen, Y.-P., Kuo, C.-Y., and Chen, M.-J. (2011) The anti-infective effects of velvet antler of formosan sambar deer (*Cervus unicolor swinhoei*) on *Staphylococcus aureus*-infected mice. *Evidence-Based Complementary and Alternative Medicine*, 1–9.

Dakshinamurti, K. (2005) Biotin – a regulator of gene expression. *Journal of Nutritional Biochemistry* 16, 419–423.

Daniells, S. (2011) Nutrigenomics shows blood pressure benefits of cocoa. *Journal of Cardiovascular Pharmacology* 57(1), 44–50.

Daviss, B. (2005) Growing pains for metabolomics. *The Scientist* 19(8), 25–28.

DeBusk, R.M. (2012) Nutritional genomics and the nutrigenomics practitioner. Available at: http://www.nugo.org/publicitem.m?key=nip&pgid=28368&trail=/nip/28368 (accessed 2 July 2012).

DeBusk, R.M., Fogarty, C., Ordovas, J.M. and Kornman, K.S. (2005) Nutritional genomics in practice: where do we begin? *Journal of the American Dietetic Association* 105, 589–598.

Desphande, U.R., Joseph, L.J., Manjure, S.S., Samuel, A.L., Pillai, D. and Bhide, S.V. (1997) Effects of tumeric extract on lipid profile in human subjects. *Medical Science Research* 25, 695–698.

Egger, G., Liang, G., Aparicio, A. and Jones, P.A. (2004) Epigenetics in human disease and prospects for epigenetic therapy. *Nature* 429(6990), 457–463.

Eleftheriadis, T., Antoniadi, G., Liakopoulos, V., Antoniadis, N., Stefanidis, I. and Galaktidou, G. (2010) Vitamin D receptor activators and response to injury in kidney diseases. *Journal of Nephrology* 23(5), 514–524.

Emboden, W.A. (1981) Transcultural use of narcotic water lilies in ancient Egyptian and Maya drug ritual. *Journal of Ethnopharmacology* 3(1), 39–83.

Enns, G.M. (2005) Inborn errors of metabolism masquerading as hypoxic-ischemic encephalopathy. *Neo Reviews* 6, 549–557.

Fenech, M. (2005) The genome health clinic and genome health nutrigenomics concepts: diagnosis and nutritional treatment of genome and epigenome damage on an individual basis. *Mutagenesis* 20(4), 255–269.

Ferguson, L. (2012) *Foods and diets tailored to your genes*. Auckland Uni. Services Limited, Auckland, New Zealand.

Ferreira, F., Barbosa, H.C.L., Stoppiglia, L.F., Delghingaro-Augusto, V., Pereira, E.A., Boschero, A.C. and Everardo, M.C. (2004) Decreased insulin secretion in islets from rats fed a low protein diet is associated with a reduced PKAα expression. *Journal of Nutrition* 134(1), 63–67.

Gabriely, I., Yang, X.M., Cases, J.A., Ma, X.H., Rossetti, L. and Barzilai, N. (2001) Hyperglycemia modulates angiotensinogen gene expression. *American Journal of Physiology-Regulatory, Integrative and Comparative Physiology* 281(3), 795–802.

Gaikwad, N.S., Panat, A.V., Deshpande, M.S., Ramya, K., Khalid, P.U. and Augustine, P. (2012) Effect of shilajit on the heart of Daphnia: a preliminary study. *Journal of Ayurveda and Integrative Medicine* 3(1), 3–5.

Garganta, C.L. and Smith, W.E. (2005) Metabolic evaluation of the sick neonate. *Seminars in Perinatology* 29, 164–172.

German, J.B., Watkins, S.M. and Fay, L.B. (2005) Metabolomics in practice: emerging knowledge to guide future dietetic advice toward individualized health. *Journal of the American Dietetic Association* 105, 1425–1432.

German, J.B., Zivkovic, A.M., Dallas, D.C. and Smilowitz, J.T. (2011) Nutrigenomics and personalized diets: what will they mean for food? *Annual Review of Food Science and Technology* 2, 97–123.

Gescher, A.J., Sharma, R.A. and Steward, W.P. (2001) Cancer chemoprevention by dietary constituents: a tale of failure and promise. *The Lancet Oncology* 2, 371–379.

Ghosal, S. (1990) Chemistry of shilajit, an immunomodulatory ayurvedic rasayan. *Pure and Applied Chemistry* 62(7), 1285–1288.

Ghosh, D.K. (2009) Future perspectives of nutrigenomics foods: benefits vs. risks. *Indian Journal of Biochemistry & Biophysics* 46, 31–36.

Gibney, M.J., Walsh, M., Brennan, L., Roche, H.M., German, B. and Van Ommen, B. (2005) Metabolomics in human nutrition: opportunities and challenges. *The American Journal of Clinical Nutrition* 82(3), 497–503.

Godard, B. and Hurlimann, T. (2009) Nutrigenomics for global health: ethical challenges for underserved populations. *Current Pharmacogenomics and Personalized Medicine* 7(3), 205–214.

Goel, A., Boland, C.R. and Chauhan, D.P. (2001) Specific inhibition of cyclooxygenase-2 (COX-2) expression by dietary curcumin in HT-29 human colon cancer cells. *Cancer Letters* 172(2), 111–118.

Gupta, K.S., Kola, J.P.N., Bagchi, D. and Bagchi, M. (2011) Application of toxicogenomics in reproductive and developmental toxicology. In: Ramesh C. Gupta (ed.) *Reproductive and Developmental Toxicology*. Academic Press (Elsevier), Burlington, Massachusetts, pp. 793–794.

Gygi, S.P., Rochon, Y., Franza, R.B. and Aebersold, R. (1999) Correlation between protein and mRNA abundance in yeast. *Molecular and Cellular Biology* 19(3), 1720–1730.

Hasan, O. (2007) Glycogen storage diseases: new perspectives. *World Journal of Gastroenterology* 13(18), 2541–2553.

Hathout, Y. (2007) Approaches to the study of the cell secretome. *Expert Review of Proteomics* 4(2), 239–248.

Hirose, M., Nishino, T., Obata, Y., Nakazawa, M., Nakazawa, Y., Furusu, A., Abe, K., Miyasaki, M., Koji, T. and Kohno, S. (2013) 22-oxacalcitriol prevents progression of peritoneal fibrosis in a mouse model. *Peritoneal Dialysis International* 33(2), 132–142.

Hocquette, J.F., Cassar-Malek, I., Scalbert, A. and Guillou, F. (2009) Contribution of genomics to the understanding of physiological functions. *Journal of Physiology and Pharmacology* 60(3), 5–16.

Holmoy, T., Kampman, M.T. and Smolders, J. (2012) Vitamin D in multiple sclerosis: implications for assessment and treatment. *Expert Review of Neurotherapeutics* 12(9), 1101–1112.

Ikonen, E. (2006) Mechanisms for cellular cholesterol transport: defects and human disease. *Physiological Reviews* 86, 1237–1261.

Inoue, M., Tajima, K., Mizutani, M., Iwata, H., Iwase, T., Miura, S., Hirose, K., Hamajima, N. and Tominaga, S. (2001) Regular consumption of green tea and the risk of breast cancer recurrence: follow-up study from the hospital-based epidemiologic research program at aichi cancer center (HERPACC), Japan. *Cancer Letters* 167(2), 175–182.

Iqbal, S.F. and Freishtat, R.J. (2011) Mechanism of action of vitamin D in the asthmatic lung. *Journal of Investigative Medicine* 59(8), 1200–1202.

Jacobo, S.M.P., Guerra, M.L. and Hockerman, G.H. (2009) Cav 1.2 and Cav 1.3 are differentially coupled to glucagon-like peptide-1 potentiation of glucose-stimulated insulin secretion in the pancreatic-cell line ins-1. *Journal of Pharmacology and Experimental Therapeutics* 331(2), 724–732.

James, P. (1997) Protein identification in the post–genome era: the rapid rise of proteomics. *Quarterly Reviews of Biophysics* 30(4), 279–331.

Jang, H.Y., Kwon, O.K., Oh, S.R., Lee, H.K., Ahn, K.S. and Chin, Y.W. (2012) Mangosteen xanthones mitigate ovalbumin-induced airway inflammation in a mouse model of asthma. *Food and Chemical Toxicology* 50(11), 4042–4050.

Jordan, K.W., Nordenstam, J., Lauwers, G.Y., Rothenberger, D.A., Alavi, K., Garwood, M. and Cheng, L.L. (2009) Metabolomic characterization of human rectal adenocarcinoma with intact tissue magnetic resonance spectroscopy. *Diseases of the Colon & Rectum* 52(3), 520–525.

Jose, J.K. and Kuttan, R. (2000) Hepatoprotective activity of *Emblica officinalis* and chyawanprash. *Journal of Ethnopharmacology* 72, 135–140.

Jump, D.B. and Clarke, S.D. (1999) Regulation of gene expression by dietary fat. *Annual Review of Nutrition* 19, 63–90.

Kaput, J., Astley, S., Renkema, M., Ordovas, J. and Van Ommen, B. (2006) Harnessing Nutrigenomics: Development of web-based communication, databases, resources, and tools. *Genes & Nutrition* 1(1), 5–11.

Katayama, S., Tomaru, Y., Kasukawa, T., Waki, K., Nakanishi, M., Nakamura, M., Nishida, H., Yap, C.C., Suzuki, M., Kawai, J., Suzuki, H., Carninci, P., Hayashizaki, Y., Wells, C., Frith, M., Ravasi, T., Pang, K.C., Hallinan, J., Mattick, J., Hume, D.A., Lipovich, L., Batalov, S., Engström, P.G., Mizuno, Y., Faghihi, M.A., Sandelin, A., Chalk, A.M., Mottagui-Tabar, S., Liang, Z., Lenhard, B. and Wahlestedt, C. (2005) Antisense transcription in the mammalian transcriptome. *Science* 309(5740), 1564–1566.

Katiyar, S.K., Ahmad, N. and Mukhtar, H. (2000) Green tea and skin. *Archives of Dermatology* 136(8), 989–994.

Keiser, J., Vargas, M. and Winter, R. (2012) Anthelminthic properties of mangostin and mangostin diacetate. *Parasitology International* 61(2), 369–371.

Kim, H.S. and Lim, H.K. (1999) Inhibitory effects of velvet antler water extract on morphine-induced conditioned place preference and DA receptor supersensitivity in mice. *Journal of Ethnopharmacology* 66, 25–31.

Kindermann, B., Doring, F., Pfaffl, M. and Daniel, H. (2004) Identification of genes responsive to intracellular zinc depletion in the human colon adenocarcinoma cell line HT-29. *Journal of Nutrition* 134(1), 57–62.

King, M.W. (2012) AMPK: Master metabolic regulator. Available at: http://themedicalbiochemistrypage.org/ampk.php (accessed July 2012).

Klopfleisc, R., Klose, P., Weise, C., Bondzio, A., Multhaup, G., Einspanier, R. and Gruber, A.D. (2010) Proteome of metastatic canine mammary carcinomas: similarities to and differences from human breast cancer. *Journal of Proteome Research* 9(12), 6380–6389.

Ko, K.M., Yip, T.T., Tsao, S.W., Kong, Y.C., Fennessy, P., Belew, M.C. and Porath, J. (1986) Epidermal growth factor from deer (*Cervus elaphus*) submaxillary gland and velvet antler. *General and Comparative Endocrinology* 63, 431–440.

Koh, J.J., Qiu, S., Zou, H., Lakshminarayanan, R., Li, J., Zhou, X., Tang, C., Saraswathi, P., Verma, C., Tan, D.T., Tan, A.L., Liu, S. and Beuerman, R.W. (2013) Rapid bactericidal action of alpha-mangostin against MRSA as an outcome of membrane targeting. *Biochimica et Biophysica Acta*, 1828(2), 834–844.

Konczak, I. and Zhang, W. (2004) Anthocyanins – more than nature's colours. *Journal of Biomedicine and Biotechnology* 5, 239–240.

Korthals, M. (2011) Coevolution of nutrigenomics and society: ethical considerations. *The American Journal of Clinical Nutrition* 94(6), 2025–2029.

Korthals, M. and Komduur, R. (2010) Uncertainties of nutrigenomics and their ethical meaning. *Journal of Agricultural and Environment Ethics* 23(5), 435–454.

Kosem, N., Ichikawa, K., Utsumi, H. and Moongkarndi, P. (2013) *In vivo* toxicity and antitumor activity of mangosteen extract. *Journal of Natural Medicines*, 67(2), 255–263.

Kovacs, E.M., Lejeune, M.P., Nijs, I. and Westerterp-Plantenga, M.S. (2004) Effects of green tea on weight maintenance after body-weight loss. *British Journal of Nutrition* 91(3), 431–437.

Krishnan, A.V. and Feldman, D. (2012) Mechanisms of the anti-cancer and anti-inflammatory actions of vitamin D. *Immunology & Cell Biology* 51, 311–336.

Kulie, T., Groff, A., Redmer, J., Hounshell, J. and Schrager, S. (2009) Vitamin D: an evidence-based review. *The Journal of the American Board of Family Medicine* 22(6), 698–706.

Lakatos, P. (2011) Pharmacologic treatment of osteoporosis 2011. *Orvosi Hetilap* 152(33), 1320–1326.

Lampe, J.W. (2006) For debate: investment in nutrigenomics will advance the role of nutrition in public health. *Cancer Epidemiology, Biomarkers & Prevention* 15(12), 2329–2330.

Lesko, L.J. (2007) Personalized medicine: elusive dream or imminent reality? *Clinical Pharmacology & Therapeutics* 81(6), 807–816.

Linder, J.A. (2012) Vitamin D and the cure for the common cold. *The Journal of the American Medical Association* 308(13), 1375–1376.

Liu, S.H., Lee, L.T., Hu, N.Y., Huang, K.K., Shih, Y.C., Munekazu, I., Li, J.M., Chou, T.Y., Wang, W.H. and Chen, T.S. (2012) Effects of alpha-mangostin on the expression of anti-inflammatory genes in U937 cells. *Chinese Medicine* 7(1), 1–19.

Lobo, I. (2008) Multifactorial inheritance and genetic disease. *Nature Education* 1(1), 1–3.

Lourith, N. and Kanlayavattanakul, M. (2011) Biological activity and stability of mangosteen as a potential natural color. *Bioscience, Biotechnology, and Biochemistry* 75(11), 2257–2259.

Malats, N. and Calafell, F. (2003) Basic glossary on genetic epidemiology. *Journal of Epidemiology and Community Health* 57(7), 480–482.

Manjunatha, S., Jaryal, A.K., Bijlani, R.L., Sachdeva, U. and Gupta, S.K. (2001) Effect of chyawanprash and vitamin C on glucose tolerance and lipoprotein profile. *National Journal of Physiology, Pharmacy and Pharmacology* 45, 71–79.

Manzelli, P. (2012) Nutrigenetics & Nutrigenomics: perspectives for a post-genomic era. Available at: http://www.edscuola.it/archivio/lre/n-DNA_mt-DNA.pdf (accessed 2 July 2012).

Marotta, F., Celep, G.S., Cabeca, A. and Polimeni, A. (2012) Novel concepts on functional foods and nutrigenomics in healthy aging and chronic diseases: a review of fermented papaya preparation research progress. *Functional Foods in Health and Disease* 2(5), 120–136.

Martinez, A., Galano, A. and Vargas, R. (2011) Free radical scavenger properties of α-mangostin: thermodynamics and kinetics of HAT and RAF mechanisms. *The Journal of Physical Chemistry B* 115(43), 12591–12598.

McMillan, J. (2006) *Oski's Pediatrics: Principles and Practice*. Lippincott, Williams & Wilkins, Philadelphia.

Meena, A.J., Bansal, P. and Kumar, S. (2009) Plants – herbal wealth as a potential source of ayurvedic drugs. *Asian Journal of Traditional Medicines* 4(4), 152–170.

Meshkin, B., Chen, T.J.H., Chen, A.L.C., Prihoda, T.J.H., Morrisette, H., Braverman, E.R., Blum, S.H., Cassel, K., Williams, L., Waite, R.L., Downs, B.W., Tung, H., Rhoades, P. and Blum, K. (2008) Health economics of nutrigenomics in weight management. *Gene Therapy and Molecular Biology* 12, 25–30.

Miggiano, G.A. and De Sanctis, R. (2006) Nutritional genomics: toward a personalized diet. *La Clinica Terapeutica* 157(4), 355–361.

Mitchell, J.J. and Scriver, C.R. (2010) Phenylalanine hydroxylase deficiency. In: Pagon, R.A., Bird, T.D., Dolan, C.R., Stephens, K. and Adam, M.P. (eds) *Gene Reviews* [Internet]. University of Washington, Seattle.

Moghaddami, M., Mayrhofer, G., Anderson, P.H., Morris, H.A., Van Der Hoek, M. and Cleland, L.G. (2012) Efficacy and mechanisms of action of vitamin D in experimental polyarthritis. *Immunology & Cell Biology* 90(2), 168–177.

Muranjan, M. and Agarwal, S. (2010) Prenatal diagnosis and newborn screening: relevance in India. *Indian Journal of Practical Pediatrics* 12(2), 131–147.

Mutch, D.M., Wahli, W. and Williamson, G. (2005) Nutrigenomics and nutrigenetics: the emerging faces of nutrition. *The FASEB Journal* 19(12), 1602–1616.

Nagao, K. and Yanagita, T. (2008) Bioactive lipids in metabolic syndrome. *Progress in Lipid Research* 47, 127–146.

Nagao, T., Hase, T. and Tokimitsu, I. (2007) A green tea extract high in catechins reduces body fat and cardiovascular risks in humans. *Obesity (Silver Spring)* 15(6), 1473–1483.

Nanda, T., Das, M., Tripathy, K. and Ravi Teja, Y. (2011) Metabolomics: the future of systems biology. *Journal of Computer Science & Systems Biology* 13, 1–6.

Neeha, V.S. and Kinth, P. (2013) Nutrigenomics research: a review. *Journal of Food Science and Technology* 50(3), 415–428.

Ngawhirunpat, T., Opanasopi, P., Sukma, M., Sittisombut, C., Kat, A., Adachi, I. (2010) Antioxidant, free radical scavenging activity and cytotoxicity of different solvent extracts and their phenolic constituents from the fruit hull of mangosteen (*Garcinia mangostana*). *Pharmaceutical Biology* 48(1), 55–62.

Nicholson, J.K. (2006) Global systems biology, personalized medicine and molecular epidemiology. *Molecular Systems Biology* 2(52), 1–6.

Nicholson, J.K., Foxall, P.J., Spraul, M., Farrant, R.D. and Lindon, J.C. (1995) 750 MHz 1H and 1H-13C NMR spectroscopy of human blood plasma. *Analytical Chemistry* 67(5), 793–811.

Nordstrom, A., Omaille, G., Qin, C. and Siuzdak, G. (2006) Nonlinear data alignment for UPLC-MS and HPLC-MS based metabolomics: quantitative analysis of endogenous and exogenous metabolites in human serum. *Analytical Chemistry* 78(10), 3289–3295.

Oltean, S. and Banerjee, R. (2003) Nutritional modulation of gene expression and homocysteine utilization by vitamin B12. *The Journal of Biological Chemistry* 278(23), 20778–20784.

Ordovas, J.M. and Corella, D. (2004) Nutritional genomics. *Annual Review of Genomics and Human Genetics* 5, 71–118.

Ordovas, J.M. and Mooser, V. (2004) Nutrigenomics and nutrigenetics. *Current Opinion in Lipidology* 15(2), 101–108.

Park, H.S., Jhon, G.J. and Choi, W. (1998) Deer antler extract selectively suppresses hyphal growth in dimorphic fungus *Candida albicans*. *Journal of Microbiology and Biotechnology* 8(3), 291–294.

Park, J.S., Kim, G.Y. and Han, K. (2006) The spermatogenic and ovogenic effects of chronically administered Shilajit to rats. *Journal of Ethnopharmacology* 107(3), 349–353.

Paterson, C.R. (2012) Vitamin D dose requirements for fracture prevention. *The New England Journal of Medicine* 367(14), 1368–1369.

Pfeiffer, C.M., Osterloh, J.D., Kennedy-Stephenson, J., Picciano, M.F., Yetley, E.A., Rader, J.I. and Johnson, C.L. (2008) Trends in circulating concentrations of total homocysteine among US adolescents and adults: findings from the 1991-1994 and 1999-2004 national health and nutrition examination surveys. *Clinical Chemistry* 54(5), 801–13.

Reyes-Fermin, L.M., Gonzalez-Reyes, S., Tarco-Alvarez, N.G., Hernandez-Nava, M., Orozco-Ibarra, M. and Pedraza-Chaverri, J. (2012) Neuroprotective effect of α-mangostin and curcumin against iodoacetate-induced cell death. *Nutritional Neuroscience*, 15(5), 34–41.

Robb-Nicholson, C. (2012) Ask the doctor. I've seen advertisements for mangosteen juice claiming it has lots of antioxidants and health benefits, including anticancer effects. Is there any truth to this? *Harvard Women's Health Watch* 19(9), 1–7.

Robertson, D.G. (2005) Metabonomics in toxicology: a review. *The Journal of Toxicological Sciences* 85(2), 809–822.

Roth, F.P., Hughes, J.D., Estep, P.W. and Churh, G.M. (1998) Finding DNA regulatory motifs within unaligned noncoding sequences clustered by whole genome mRNA quantitation. *Nature Biotechnology* 8, 939–945.

Ryu, H.W., Cho, J.K., Curtis-Long, M.J., Yuk, H.J., Kim, Y.S., Jung, S., Kim, Y.S., Lee, B.W. and Park, K.H. (2011) α-Glucosidase inhibition and antihyperglycemic activity of prenylated xanthones from *Garcinia mangostana*. *Phytochemistry* 17, 2148–2154.

Ryu, H.W., Jeong, S.H., Curtis-Long, M.J., Jung, S., Lee, J.W., Woo, H.S., Cho, J.K. and Park, K.H. (2012) Inhibition Effects of Mangosenone F from *Garcinia mangostana* on Melanin Formation in B16F10 Cells. *Journal of Agricultural and Food Chemistry* 60(34), 8372–8378.

Saghatelian, A., Trauger, S.A., Want, E.J., Hawkins, E.G., Siuzdak, G. and Cravatt, B.F. (2004) Assignment of endogenous substrates to enzymes by global metabolite profiling. *Biochemistry* 43(45), 14332–14339.

Saksena, S. and Dixit, V.K. (1987) Role of total alkaloids of *Mucuna pruriens* Baker in spermatogenesis in albino rats. *Indian Journal of Natural Products* 3, 3–37.

Samuelsson, L.M. and Larsson, D.G. (2008) Contributions from metabolomics to fish research. *Molecular BioSystems* 4(10), 974–979.

Sanders, L.M. (2009) Disorders of amino acid metabolism. Available at: http://www.merckmanuals.com/home/childrens_health_issues/hereditary_metabolic_disorders/disorders_of_amino_acid_metabolism.html (accessed 2 July 2012).

Saudubray, J.M., Nassogne, M.C., De Lonlay, P. and Touati, G. (2002) Clinical approach to inherited metabolic disorders in neonates: an overview. *Seminars in Neonatology* 7, 3–15.

Schaffer, D.J., Dimitrova, N. and Zhang, M. (2006) Bioinformatics: Overview and Research Opportunities. In: Spekowius, G. and Wendler, T. (eds) *Advances in Healthcare Technology*. Springer, the Netherlands, pp. 421–438.

Schrumpf, J.A., Van Sterkenburg, M.A., Verhoosel, R.M., Zuyderduyn, S. and Hiemstra, P.S. (2012) IL-13 exposure enhances vitamin D-mediated expression of the human cathelicidin antimicrobial peptide-hCAP18/LL-37 in bronchial epithelial cells. *Infection and Immunity* 80(12), 4485–4494.

Scriver, C. (2001) *The Metabolic and Molecular Bases of Inherited Disease*, 8th edn. McGraw-Hill Education India, Noida, India.

Sharma, P., Jha, J., Shrinivas, V., Dwivedi, L.K., Suresh, P. and Sinha, M. (2003) Shilajit: evalution of its effects on blood chemistry of normal human subjects. *Ancient Science of Life* 23(2), 114–119.

Sharma, S., Kelly, T.K., and Jones, P.A. (2010) Epigenetics in cancer. *Carcinogenesis* 31(1), 27–36.

Shin, K.H., Lee, E.B., Kim, J.H., Chung, M.S. and Cho, S.Y. (1989) Pharmacological studies on the powdered whole part of unossified antler. *Korean Journal of Pharmacognosy* 20(3), 180–187.

Shiozaki, T., Fukai, M., Hermawati, E., Juliawaty, L.D., Syah, Y.M., Hakim, E.H., Puthongking, P., Suzuki, T., Kinoshita, K., Takahashi, K. and Koyama, K. (2012) Anti-angiogenic effect of α-mangostin. *Journal of Natural Medicines* [Epub ahead of print].

Simopoulos, A.P. (2002) Genetic variation and dietary response: nutrigenetics/nutrigenomics. *Asia Pacific Journal of Clinical Nutrition* 11(6), 117–128.

Simopoulos, A.P. (2010) Nutrigenetics/Nutrigenomics. *Annual Review of Public Health* 31, 53–68.

Sreedhar, A., Prakash, S., Sapna, N. and Kumar, S. (2011) Proteomics – the new era of periodontics. *Journal of Dental Sciences and Research* 2(2), 87–90.

Subbiah, M.T.R. (2010) Application of nutrigenomics in skin health. *The Journal of Clinical and Aesthetic Dermatology* 3(11), 44–46.

Subramanian, A., Tamayo, P., Mootha, V.K., Mukherjee, S., Ebert, B.L., Gillette, M.A., Paulovich, A., Pomeroy, S.L., Golub, T.R., Lander, E.S. and Mesirov, J.P. (2005) Gene set enrichment analysis: a knowledge-based approach for interpreting genome-wide expression profiles. *Proceedings of the National Academy of Sciences* 102(43), 15545–15550.

Suh, J.S., Eun, J.S., So, J.N., Seo, J.I. and Jhon, G.J. (1999) Phagocytic activity of ethyl alcohol fraction of deer antler in murine peritoneal macrophage. *Biological & Pharmaceutical Bulletin* 22(9), 932–935.

Sui, Z., Salto, R., Li, J., Craik, C. and Oritz de Montellano, P.R. (1993) Inhibition of HIV-1 and HIV-2 proteases by curcumin and curcumin boron complexes. *Bioorganic and Medicinal Chemistry* 1, 415–422.

Svacina, S. (2007) Nutrigenetics and nutrigenomics. *Casopis Lekaru Ceskych* 146(11), 837–839.

Tandon, M., Siddique, R.A. and Rai, S.N. (2012) Effect of nutrients on the gene expression: nutria-genomics. Available at: http://www.pitt.edu/~super7/30011-31001/30951.ppt (accessed 2 July 2012).

Teegarden, D. and Donkin, S.S. (2009) Vitamin D: emerging new roles in insulin sensitivity. *Nutrition Research Reviews* 22(1), 82–92.

Temrangsee, P., Kondo, S. and Itharat, A. (2011) Antibacterial activity of extracts from five medicinal plants and their formula against bacteria that cause chronic wound infection. *Journal of The Medical Association of Thailand* 94(7)Suppl., 166–171.

Trujillo, E., Davis, C. and Milner, J. (2006) Nutrigenomics, proteomics, metabolomics, and the practice of dietetics. *Journal of the American Dietetic Association* 106, 403–413.

Tsuda, T., Ueno, Y., Kojo, H., Yoshikawa, T. and Osawa, T. (2005) Gene expression profile of isolated rat adipocytes treated with anthocyanins. *Biochimca et Biophysica Acta* 1733(2–3), 137–147.

Twyman, R. (2003) Gene expression. Available at: http://genome.wellcome.ac.uk/doc_WTD020757.html (accessed 2 July 2012).

Vallee, B.L. and Auld, D.S. (1990) Zinc coordination, function, and structure of zinc enzymes and other proteins. *Biochemistry* 29, 5647–5659.

Van Der Werf, M.J., Schuren, F.H.J., Bijlsma, S., Tas, A.C. and Van Ommen, B. (2006) Nutrigenomics: application of genomics technologies in nutritional sciences and food technology. *Journal of Food Science* 66(6), 772–780.

Van Ommen, B., Keijer, J., Kleemann, R., Elliott, R., Christian, A.D., McArdle, H., Gibney, M. and Muller, M. (2008) The challenges for molecular nutrition research 2: quantification of the nutritional phenotype. *Genes & Nutrition* 3(2), 51–59.

Vaulont, S., Cognet, M.V. and Kahn, A. (2000) Glucose regulation of gene transcription. *The Journal of Biological Chemistry* 275(41), 31555–31558.

Velculescu, V.E., Zhang, L., Zhou, W., Vogelstein, J., Basrai, M.A., Bassett, D.E., Hieter, P., Vogelstein, B. and Kinzler, K.W. (1997) Characterization of the yeast transcriptome. *Cell* 88(2), 243–251.

Wang, B.X., Liu, A.J., Cheng, X.J., Wang, Q.G., Wei, G.R. and Cui, J.C. (1985) Antiulcer activity of the polysaccharides isolated from pilose antler. *Acta Pharmacerutica Sinica* 20(5), 321–325.

Wang, B.X., Zhou, X.H., Qi, S.B., Kaneko, S., Hattori, M., Tsuneo, N. and Nomura, Y. (1988) Stimulating effect of deer antler extract on protein synthesis in senescence accelerated mice in vivo. *Chemical & Pharmaceutical Bulletin* 36(7), 2593–2598.

Wang, J.J., Sanderson, B.J. and Zhang, W. (2012a) Significant anti-invasive activities of α-mangostin from the mangosteen pericarp on two human skin cancer cell lines. *Anticancer Research* 32(9), 3805–3816.

Wang, J.J., Shi, Q.H., Zhang, W. and Sanderson, B.J. (2012b) Anti-skin cancer properties of phenolic-rich extract from the pericarp of mangosteen (*Garcinia mangostana* Linn.). *Food and Chemical Toxicology* 50(9), 3004–3013.

Wang, Z., Gerstein, M. and Snyder, M. (2009) RNA-Seq: a revolutionary tool for transcriptomics. *Nature Reviews Genetics* 10(1), 57–63.

Weatherall, D.J. (2000) Single gene disorders or complex traits: lessons from the thalassaemias and other monogenic diseases. *British Medical Journal* 321, 1117–1120.

WHO (2012) Genes and human disease. Available at: http://www.who.int/genomics/public/geneticdiseases/en/index2.html (accessed 2 October 2012).

Wilkins, M.R., Pasquali, C., Appel, R.D., Ou, K., Golaz, O., Sanchez, J.-C., Yan, J.X., Gooley, A.A., Hughes, G., Humphery-Smith, I., Keith, L.W. and Denis, F.H. (1996) From proteins to proteomes: large scale protein identification by two-dimensional electrophoresis and arnino acid analysis. *Nature Biotechnology* 14(1), 61–65.

Wilson, E., Rajamanickam, G.V., Dubey, G.P., Klose, P., Musial, F., Saha, F.J., Rampp, T., Michalsen, A. and Dobos, G.J. (2011) Review on shilajit used in traditional Indian medicine. *Journal of Ethnopharmacology* 136(1), 1–9.

Wu, F.Y. and Wu, C.W. (1987) Zinc in DNA replication and transcription. *Annual Review of Nutrition* 7, 251–272.

Yadav, J.S., Thakur, S. and Chadha, P. (2003) Chyawanprash awaleha: a genoprotective agent for bidi smokers. *International Journal of Human Genetics* 3(1), 33–38.

Yang, Q.H., Botto, L.D., Gallagher, M., Friedman, J.M., Sanders, C.L., Koontz, D., Nikolova, S., Erickson, J.D. and Steinberg, K. (2008) Prevalence and effects of gene-gene and gene-nutrient interactions on serum folate and serum total homocysteine concentrations in the United States: findings from the third national health and nutrition examination survey DNA bank. *The American Journal of Clinical Nutrition* 88(1), 232–246.

Yu, Z.F., Kong, L.D. and Chen, Y. (2002) Antidepressant activity of aqueous extracts of *Curcuma longa* in mice. *Journal of Ethnopharmacology* 83, 161–165.

Zeisel, S.H. (2010) A grand challenge for nutrigenomics. *Frontiers in Genetics* 1(2), 1–3.

Zeisel, S.H., Freake, H.C., Bauman, D.E., Bier, D.M., Burrin, D.G., German, J.B., Klein, S., Marquis, G.S., Milner, J.A., Pelto, G.H., and Rasmussen, K.M. (2005) The nutritional phenotype in the age of metabolomics. *The Journal of Nutrition* 135(7), 1613–1616.

Index